Canadian Mathematical Society
Société mathématique du Canada

CMS Books in Mathematics
Ouvrages de mathématiques de la SMC

1 HERRMAN/KUČERA/ŠIMŠA Equations and Inequalities

2 ARNOLD Abelian Groups and Representations of Finite Partially Ordered Sets

3 BORWEIN/LEWIS Convex Analysis and Nonlinear Optimization, 2nd Ed.

4 LEVIN/LUBINSKY Orthogonal Polynomials for Exponential Weights

5 KANE Reflection Groups and Invariant Theory

6 PHILLIPS Two Millennia of Mathematics

7 DEUTSCH Best Approximation in Inner Product Spaces

8 FABIAN ET AL. Functional Analysis and Infinite-Dimensional Geometry

9 KŘÍŽEK/LUCA/SOMER 17 Lectures on Fermat Numbers

10 BORWEIN Computational Excursions in Analysis and Number Theory

11 REED/SALES (Editors) Recent Advances in Algorithms and Combinatorics

12 HERMAN/KUČERA/ŠIMŠA Counting and Configurations

13 NAZARETH Differentiable Optimization and Equation Solving

14 PHILLIPS Interpolation and Approximation by Polynomials

15 BEN-ISRAEL/GREVILLE Generalized Inverses, 2nd Ed.

16 ZHAO Dynamical Systems in Population Biology

17 GÖPFERT ET AL. Variational Methods in Partially Ordered Spaces

18 AKIVIS/GOLDBERG Differential Geometry of Varieties with Degenerate Gauss Maps

19 MIKHALEV/SHPILRAIN/YU Combinatorial Methods

20 BORWEIN/ZHU Techniques of Variational Analysis

21 VAN BRUMMELEN/KINYON Mathematics and the Historian's Craft

22 LUCCHETTI Convexity and Well-Posed Problems

23 NICULESCU/PERSSON Convex Functions and Their Applications

24 SINGER Duality for Nonconvex Approximation and Optimization

25 HIGGINSON/PIMM/SINCLAIR Mathematics and the Aesthetic

26 HÁJEK/SANTALUĆIA/VANDERWERFF/ZIZLER Biorthogonal Systems in Banach Spaces

27 BORWEIN/CHOI/ROONEY/WEIRATHMUELLER The Riemann Hypothesis

Peter Borwein, Stephen Choi, Brendan Rooney
and Andrea Weirathmueller (Eds.)

The Riemann Hypothesis

A Resource for the Afficionado and Virtuoso Alike

Springer

Peter Borwein
Department of Mathematics
& Statistics
8888 University Drive
Simon Fraser University
Burnaby, BC, V5A 1S6
CANADA
pborwein@cecm.sfu.ca

Stephen Choi
Department of Mathematics
& Statistics
8888 University Drive
Simon Fraser University
Burnaby, BC, V5A 1S6
CANADA
kkchoi@cecm.sfu.ca

Brendan Rooney
9161 Tenth Avenue
Simon Fraser University
Burnaby, BC, V3N 2T2
CANADA
brooney@irmacs.sfu.ca

Andrea Weirathmueller
727 Windsor Street
University of Western
Ontario
Fredericton, ON E2B 4G3
aweirath@gmail.com

Cover Figure: A "random walk" on the Liouville Function.

ISBN: 978-1-4419-2465-0 e-ISBN: 978-0-387-72126-2

Mathematics Subject Classification (2000): 11-xx 01-xx

Printed on acid-free paper.

9 8 7 6 5 4 3 2 1

springer.com

For Pinot

- P. B.

For my parents, my lovely wife, Shirley, my daughter, Priscilla, and son, Matthew

- S. C.

For my parents Tom and Katrea

- B. R.

For my family

- A. W.

Preface

Since its inclusion as a Millennium Problem, numerous books have been written to introduce the Riemann hypothesis to the general public. In an average local bookstore, it is possible to see titles such as John Derbyshire's *Prime Obsession: Bernhard Riemann and the Greatest Unsolved Problem in Mathematics*, Dan Rockmore's *Stalking the Riemann Hypothesis: The Quest to Find the Hidden Law of Prime Numbers*, and Karl Sabbagh's *The Riemann Hypothesis: The Greatest Unsolved Problem in Mathematics.*

This book is a more advanced introduction to the theory surrounding the Riemann hypothesis. It is a source book, primarily composed of relevant original papers, but also contains a collection of significant results. The text is suitable for a graduate course or seminar, or simply as a reference for anyone interested in this extraordinary conjecture.

The material in Part I (Chapters 1-10) is mostly organized into independent chapters and one can cover the material in many ways. One possibility is to jump to Part II and start with the four expository papers in Chapter 11. The reader who is unfamiliar with the basic theory and algorithms used in the study of the Riemann zeta function may wish to begin with Chapters 2 and 3. The remaining chapters stand on their own and can be covered in any order the reader fancies (obviously with our preference being first to last). We have tried to link the material to the original papers in order to facilitate more in-depth study of the topics presented.

We have divided Part II into two chapters. Chapter 11 consists of four expository papers on the Riemann hypothesis, while Chapter 12 contains the original papers that developed the theory surrounding the Riemann hypothesis.

Presumably the Riemann hypothesis is very difficult, and perhaps none of the current approaches will bear fruit. This makes selecting appropriate papers problematical. There is simply a lack of profound developments and attacks on the full problem. However, there is an intimate connection between the prime number theorem and the Riemann hypothesis. They are connected theoretically and historically, and the Riemann hypothesis may be thought of as a grand generalization of the prime number theorem. There is a large body of theory on the prime number theorem and a progression of solutions. Thus we have chosen various papers that give proofs of the prime number theorem.

While there have been no successful attacks on the Riemann hypothesis, a significant body of evidence has been generated in its support. This evidence is often computational; hence we have included several papers that focus on, or use computation of, the Riemann zeta function. We have also included Weil's proof of the Riemann hypothesis for function fields (Section 12.8) and the deterministic polynomial primality test of Argawal at al. (Section 12.20).

Acknowledgments. We would like to thank the community of authors, publishers, and libraries for their kind permission and assistance in republishing the papers included in Part II. In particular, "On Newman's Quick Way to the Prime Number Theorem" and "Pair Correlation of Zeros and Primes in Short Intervals" are reprinted with kind permission of Springer Science and Business Media, "The Pair Correlation of Zeros of the Zeta Function" is reprinted with kind permission of the American Mathematical Society, and "On the Difference $\pi(x) - \mathrm{Li}(x)$" is reprinted with kind permission of the London Mathematical Society.

We gratefully acknowledge Michael Coons, Karl Dilcher, Ron Ferguson and Alexa van der Waall for many useful comments and corrections that help make this a better book.

Contents

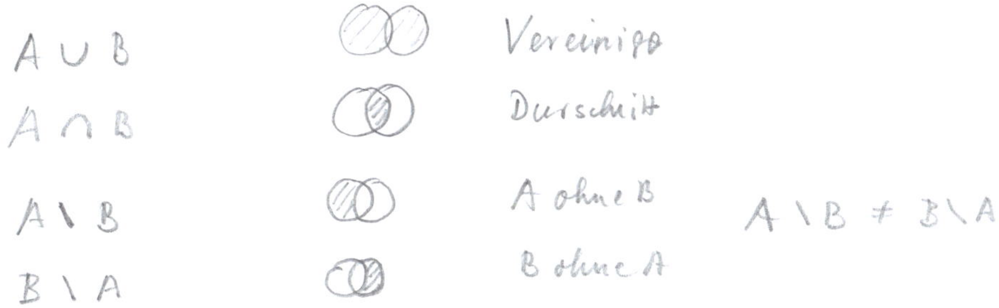

Notation

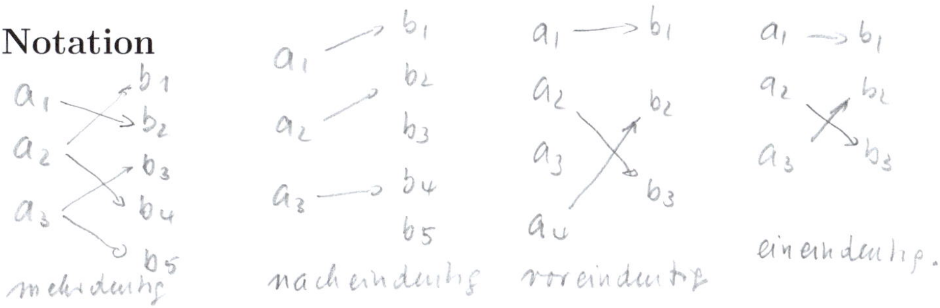

The notation in this book is standard. Specific symbols and functions are defined as needed throughout, and the standard meaning of basic symbols and functions is assumed. The following is a list of symbols that appear frequently in the text, and the meanings we intend them to convey.

$\Rightarrow$	"If ..., then ..." in natural language
$\in$	membership in a set
$\#\{A\}$	the cardinality of the set A
$:=$	defined to be
$x \equiv y \pmod{p}$	x is congruent to y modulo p
$[x]$	the integral part of x
$\{x\}$	the fractional part of x
$\lvert x\rvert$	the absolute value of x
$x!$	for $x \in \mathbb{N}$, $x! = x \cdot (x-1) \cdots 2 \cdot 1$
(n, m)	the greatest common divisor of n and m
$\phi(x)$	Euler's totient function evaluated at x
$\log(x)$	the natural logarithm, $\log_e(x) = \ln(x)$
$\det(A)$	the determinant of matrix A
$\pi(x)$	the number of prime numbers $p \leq x$
$\mathrm{Li}(x)$	the logarithmic integral of x, $\mathrm{Li}(x) := \int_2^x \frac{dt}{\log t}$
$\sum$	summation
$\prod$	product
$\rightarrow$	tends toward
x^+	toward x from the right
x^-	toward x from the left
$f'(x)$	the first derivative of $f(x)$ with respect to x
$\Re(x)$	the real part of x
$\Im(x)$	the imaginary part of x

$\overline{x}$	the complex conjugate of x
$\arg(x)$	the argument of a complex number x
$\Delta_C \arg(f(x))$	the number of changes in the argument of $f(x)$ along the contour C
$\mathbb{N}$	the set of natural numbers $\{1, 2, 3, \ldots\}$
$\mathbb{Z}$	the set of integers
$\mathbb{Z}/p\mathbb{Z}$	the ring of integers modulo p
$\mathbb{R}$	the set of real numbers
$\mathbb{R}^+$	the set of positive real numbers
$\mathbb{C}$	the set of complex numbers
$f(x) = O(g(x))$	$\|f(x)\| \leq A\|g(x)\|$ for some constant A and all values of $x > x_0$ for some x_0
$f(x) = o(g(x))$	$\lim_{x\to\infty} \frac{f(x)}{g(x)} = 0$
$f \ll g$	$\|f(x)\| \leq A\|g(x)\|$ for some constant A and all values of $x > x_0$ for some x_0
$f \ll_\varepsilon g$	$\|f(x)\| \leq A(\varepsilon)\|g(x)\|$ for some given function $A(\varepsilon)$ and all values of $x > x_0$ for some x_0
$f(x) = \Omega(g(x))$	$\|f(x)\| \geq A\|g(x)\|$ for some constant A and all values of $x > x_0$ for some x_0
$f \sim g$	$\lim_{x\to\infty} \frac{f(x)}{g(x)} = 1$

Part I

Introduction to the Riemann Hypothesis

1

Why This Book

One now finds indeed approximately this number of real roots within these limits, and it is very probable that all roots are real. Certainly one would wish for a stricter proof here; I have meanwhile temporarily put aside the search for this after some fleeting futile attempts, as it appears unnecessary for the next objective of my investigation [133].

Bernhard Riemann, 1859

The above comment appears in Riemann's memoir to the Berlin Academy of Sciences (Section 12.2). It seems to be a passing thought, yet it has become, arguably, the most central problem in modern mathematics.

This book presents the Riemann hypothesis, connected problems, and a taste of the related body of theory. The majority of the content is in Part II, while Part I contains a summary and exposition of the main results. It is targeted at the educated nonexpert; most of the material is accessible to an advanced mathematics student, and much is accessible to anyone with some university mathematics.

Part II is a selection of original papers. This collection encompasses several important milestones in the evolution of the theory connected to the Riemann hypothesis. It also includes some authoritative expository papers. These are the "expert witnesses" who offer the most informed commentary on the Riemann hypothesis.

1.1 The Holy Grail

The Riemann hypothesis has been the Holy Grail of mathematics for a century and a half. Bernhard Riemann, one of the extraordinary mathematical talents of the nineteenth century, formulated the problem in 1859. The hypothesis makes a very precise connection between two seemingly unrelated mathematical objects (namely, prime numbers and the zeros of analytic functions). If solved, it would give us profound insight into number theory and, in particular, the nature of prime numbers.

Why is the Riemann hypothesis so important? Why is it the problem that many mathematicians would sell their souls to solve? There are a number of great old unsolved problems in mathematics, but none of them has quite the stature of the Riemann hypothesis. This stature can be attributed to a variety of causes ranging from mathematical to cultural. As with the other old great unsolved problems, the Riemann hypothesis is clearly very difficult. It has resisted solution for 150 years and has been attempted by many of the greatest minds in mathematics.

The problem was highlighted at the 1900 International Congress of Mathematicians, a conference held every four years and the most prestigious international mathematics meeting. David Hilbert, one of the most eminent mathematicians of his generation, raised 23 problems that he thought would shape twentieth century mathematics. This was somewhat self-fulfilling, since solving a Hilbert problem guaranteed instant fame and perhaps local riches. Many of Hilbert's problems have now been solved.

Being one of Hilbert's 23 problems was enough to guarantee the Riemann hypothesis centrality in mathematics for more than a century. Adding to interest in the hypothesis is a million-dollar bounty as a "Millennium Prize Problem" of the Clay Mathematics Institute. That the Riemann hypothesis should be listed as one of seven such mathematical problems (each with a million-dollar prize associated with its solution) indicates not only the contemporary importance of a solution, but also the importance of motivating a new generation of researchers to explore the hypothesis further.

Solving any of the great unsolved problems in mathematics is akin to the first ascent of Everest. It is a formidable achievement, but after the conquest there is sometimes nowhere to go but down. Some of the great problems have proven to be isolated mountain peaks, disconnected from their neighbors. The Riemann hypothesis is quite different in this regard. There is a large body of mathematical speculation that becomes fact if the Riemann hypothesis is solved. We know many statements of the form "if the Riemann hypothesis, then the following interesting mathematical statement", and this is rather different from the solution of problems such as the Fermat problem.

The Riemann hypothesis can be formulated in many diverse and seemingly unrelated ways; this is one of its beauties. The most common formulation is that certain numbers, the zeros of the "Riemann zeta function", all lie on a certain line (precise definitions later). This formulation can, to some extent, be verified numerically.

In one of the largest calculations done to date, it was checked that the first ten trillion of these zeros lie on the correct line. So there are ten trillion pieces of evidence indicating that the Riemann hypothesis is true and not a single piece of evidence indicating that it is false. A physicist might be overwhelmingly pleased with this much evidence in favour of the hypothesis, but to some mathematicians this is hardly evidence at all. However, it is interesting ancillary information.

In order to prove the Riemann hypothesis it is required to show that all of these numbers lie in the right place, not just the first ten trillions. Until such a proof is provided, the Riemann hypothesis cannot be incorporated into the body of mathematical facts and accepted as true by mathematicians (even though it is probably true!). This is not just pedantic fussiness. Certain mathematical phenomena that appear true, and that can be tested in part computationally, are false, but only false past computational range (This is seen in Sections 12.12, 12.9, and 12.14).

Accept for a moment that the Riemann hypothesis is the greatest unsolved problem in mathematics and that the greatest achievement any young graduate student could aspire to is to solve it. Why isn't it better known? Why hasn't it permeated public consciousness in the way black holes and unified field theory have, at least to some extent? Part of the reason for this is that it is hard to state rigorously, or even unambiguously. Some undergraduate mathematics is required in order for one to be familiar enough with the objects involved to even be able to state the hypothesis accurately. Our suspicion is that a large proportion of professional mathematicians could not precisely state the Riemann hypothesis if asked.

If I were to awaken after having slept for a thousand years, my first question would be: Has the Riemann hypothesis been proven?

Attributed to **David Hilbert**

1.2 Riemann's Zeta and Liouville's Lambda

The Riemann zeta function is defined, for $\Re(s) > 1$, by

$$\zeta(s) = \sum_{n=1}^{\infty} \frac{1}{n^s}. \tag{1.2.1}$$

The Riemann hypothesis is usually given as follows: the nontrivial zeros of the Riemann zeta function lie on the line $\Re(s) = \frac{1}{2}$. There is already, of course, the problem that the above series doesn't converge on this line, so one is already talking about an analytic continuation.

Our immediate goal is to give as simple (equivalent) a statement of the Riemann hypothesis as we can. Loosely, the statement is, "the number of integers with an even number of prime factors is the same as the number of integers with an odd number of prime factors." This is made precise in terms of the Liouville function, which gives the parity of the number of prime factors of a positive integer.

Definition 1.1. *The Liouville function is defined by*

$$\lambda(n) = (-1)^{\omega(n)},$$

where $\omega(n)$ is the number of, not necessarily distinct, prime factors of n, counted with multiplicity.

So $\lambda(2) = \lambda(3) = \lambda(5) = \lambda(7) = \lambda(8) = -1$ and $\lambda(1) = \lambda(4) = \lambda(6) = \lambda(9) = \lambda(10) = 1$ and $\lambda(x)$ is completely multiplicative (i.e., $\lambda(xy) = \lambda(x)\lambda(y)$ for any $x, y \in \mathbb{N}$) taking only values ± 1. (Alternatively, one can define λ as the completely multiplicative function with $\lambda(p) = -1$ for any prime p.)

The following connections between the Liouville function and the Riemann hypothesis were explored by Landau in his doctoral thesis of 1899.

Theorem 1.2. *The Riemann hypothesis is equivalent to the statement that for every fixed $\varepsilon > 0$,*

$$\lim_{n\to\infty} \frac{\lambda(1) + \lambda(2) + \cdots + \lambda(n)}{n^{\frac{1}{2}+\varepsilon}} = 0.$$

This translates to the following statement: The Riemann hypothesis is equivalent to the statement that an integer has equal probability of having an odd number or an even number of distinct prime factors (in the precise sense given above). This formulation has inherent intuitive appeal.

We can translate the equivalence once again. The sequence

$$\{\lambda(i)\}_{i=1}^{\infty} = \{1, -1, -1, 1, -1, 1, -1, -1, 1, 1, -1, -1, -1, 1, 1, 1, -1, -1, -1, -1, 1, \ldots\}$$

behaves more or less like a random sequence of 1's and -1's in that the difference between the number of 1's and -1's is not much larger than the square root of the number of terms.

This is an elementary and intuitive statement of the Riemann hypothesis. It is an example of the many diverse reformulations given in Chapter 5. It is relatively easy to formulate, and lends itself to an attractive picture (see Figure 1.1).

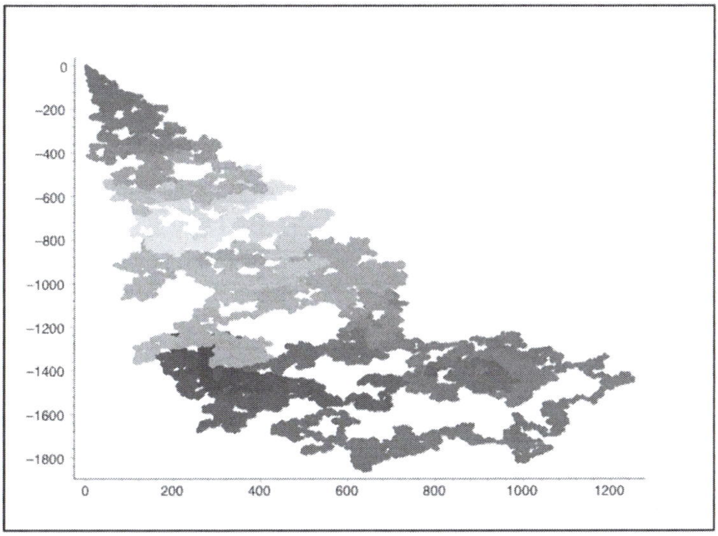

Fig. 1.1. $\{\lambda(i)\}_{i=1}^{\infty}$ plotted as a "random walk," by walking $(\pm 1, \pm 1)$ through pairs of points of the sequence.

1.3 The Prime Number Theorem

The prime number theorem is a jewel of mathematics. It states that the number of primes less than or equal to n is approximately $n/\log n$, and was conjectured by Gauss in 1792, on the basis of substantial computation and insight. One can view the Riemann hypothesis as a precise form of the prime number theorem, in which the rate of convergence is made specific (see Section 5.1).

Theorem 1.3 (The Prime Number Theorem). *Let $\pi(n)$ denote the number of primes less than or equal to n. Then*

$$\lim_{n\to\infty} \frac{\pi(n)}{n/\log(n)} = 1.$$

$\pi(n) \to \frac{n}{\ln n}$

As with the Riemann hypothesis, the prime number theorem can be formulated in terms of the Liouville lambda function, a result also due to Landau in his doctoral thesis of 1899.

Theorem 1.4. *The prime number theorem is equivalent to the statement*

$$\lim_{n\to\infty} \frac{\lambda(1) + \lambda(2) + \cdots + \lambda(n)}{n} = 0.$$

also 1.2 für $\varepsilon = 1/2$

So the prime number theorem is a relatively weak statement of the fact that an integer has equal probability of having an odd number or an even number of distinct prime factors

The prime number theorem was first proved independently by de la Vallée Poussin (see Section 12.4) and Hadamard (see Section 12.3) around 1896, although Chebyshev came close in 1852 (see Section 12.1). The first and easiest proofs are analytic and exploit the rich connections between number theory and complex analysis. It has resisted trivialization, and no really easy proof is known. This is especially true for the so-called elementary proofs, which use little or no complex analysis, just considerable ingenuity and dexterity. The primes arise sporadically and, apparently, relatively randomly, at least in the sense that there is no easy way to find a large prime number with no obvious congruences. So even the amount of structure implied by the prime number theorem is initially surprising.

Included in the original papers in Part II are a variety of proofs of the prime number theorem. Korevaar's expository paper (see Section 12.16) is perhaps the most accessible of these. We focus on the prime number theorem not only for its intrinsic interest, but also because it represents a high point in the quest for resolution of the Riemann hypothesis. Now that we have some notion of what the Riemann hypothesis is, we can move on to the precise analytic formulation.

2

Analytic Preliminaries

The mathematician's patterns, like the painter's or the poet's, must be beautiful; the ideas, like the colours or the words must fit together in a harmonious way. Beauty is the first test: there is no permanent place in this world for ugly mathematics [64].

G. H. Hardy, 1940

In this chapter we develop some of the more important, and beautiful, results in the classical theory of the zeta function. The material is mathematically sophisticated; however, our presentation should be accessible to the reader with a first course in complex analysis. At the very least, the results should be meaningful even if the details are elusive.

We first develop the functional equation for the Riemann zeta function from Riemann's seminal paper, *Ueber die Anzahl der Primzahlen unter einer gegebenen Grösse* (see Section 12.2), as well as some basic properties of the zeta function. Then we present part of de la Vallée Poussin's proof of the prime number theorem (see Section 12.4); in particular, we prove that $\zeta(1+it) \neq 0$ for $t \in \mathbb{R}$. We also develop some of main ideas to numerically verify the Riemann hypothesis; in particular, we prove that $N(T) = \frac{T}{2\pi}\log\frac{T}{2\pi} - \frac{T}{2\pi} + O(\log T)$, where $N(T)$ is the number of zeros of $\zeta(s)$ up to height T. Finally, we give a proof of Hardy's result that there are infinitely many zeros of $\zeta(s)$ on the critical line, that is, the line $\left\{\frac{1}{2}+it, t \in \mathbb{R}\right\}$, from Section 12.5.

2.1 The Riemann Zeta Function

The Riemann hypothesis is a precise statement, and in one sense what it means is clear, but what it's connected with, what it implies, where it comes from, can be very unobvious [138].

M. Huxley

Defining the Riemann zeta function is itself a nontrivial undertaking. The function, while easy enough to define formally, is of sufficient complexity that such a statement would be unenlightening. Instead we "build" the Riemann zeta function in small steps. By and large, we follow the historical development of the zeta function from Euler to Riemann. This development sheds some light on the deep connection between the zeta function and the prime numbers. We begin with the following example of a Dirichlet series.

Let $s = \sigma + it$ ($\sigma, t \in \mathbb{R}$) be a complex number. Consider the Dirichlet series

$$\sum_{n=1}^{\infty} \frac{1}{n^s} = 1 + \frac{1}{2^s} + \frac{1}{3^s} + \cdots . \tag{2.1.1}$$

This series is the first building block of the Riemann zeta function. Notice that if we set $s = 1$, we obtain

$$\sum_{n=1}^{\infty} \frac{1}{n} = 1 + \frac{1}{2} + \frac{1}{3} + \cdots ,$$

the well-known harmonic series, which diverges. Also notice that whenever $\Re(s) \leq 1$, the series will diverge. It is also easy to see that this series converges whenever $\Re(s) > 1$ (this follows from the integral test). So this Dirichlet series defines an analytic function in the region $\Re(s) > 1$. We initially define the Riemann zeta function to be

$$\zeta(s) := \sum_{n=1}^{\infty} \frac{1}{n^s}, \tag{2.1.2}$$

for $\Re(s) > 1$.

Euler was the first to give any substantial analysis of this Dirichlet series. However, Euler confined his analysis to the real line. Euler was also the first to evaluate, to high precision, the values of the series for $s = 2, 3, \ldots, 15, 16$. For example, Euler established the formula,

$$\zeta(2) = 1 + \frac{1}{4} + \frac{1}{9} + \frac{1}{16} + \cdots = \frac{\pi^2}{6}.$$

Riemann was the first to make an intensive study of this series as a function of a complex variable.

Euler's most important contribution to the theory of the zeta function is the Euler product formula. This formula demonstrates explicitly the connection between prime numbers and the zeta function. Euler noticed that every positive integer can be uniquely written as a product of powers of different primes. Thus, for any $n \in \mathbb{N}$, we may write

$$n = \prod_{p_i} p_i^{e_i},$$

where the p_i range over all primes, and the e_i are nonnegative integers. The exponents e_i will vary as n varies, but it is clear that if we consider each $n \in \mathbb{N}$ we will use every possible combination of exponents $e_i \in \mathbb{N} \cup \{0\}$. Thus,

$$\sum_{n=1}^{\infty} \frac{1}{n^s} = \prod_p \left(1 + \frac{1}{p^s} + \frac{1}{p^{2s}} + \cdots \right),$$

where the infinite product is over all the primes. On examining the convergence of both the infinite series and the infinite product, we easily obtain the Euler product formula:

Theorem 2.1 (Euler Product Formula). *For $s = \sigma + it$ and $\sigma > 1$, we have*

$$\zeta(s) = \sum_{n=1}^{\infty} \frac{1}{n^s} = \prod_p \left(1 - \frac{1}{p^s}\right)^{-1}. \tag{2.1.3}$$

The Euler product formula is also called the analytic form of the fundamental theorem of arithmetic. It demonstrates how the Riemann zeta function encodes information on the prime factorization of integers and the distribution of primes.

We can also recast one of our earlier observations in terms of the Euler product formula. Since convergent infinite products never vanish, the Euler product formula yields the following theorem.

Theorem 2.2. *For all $s \in \mathbb{C}$ with $\Re(s) > 1$, we have $\zeta(s) \neq 0$.*

We have seen that the Dirichlet series (2.1.2) diverges for any s with $\Re(s) \leq 1$. In particular, when $s = 1$ the series is the harmonic series. Consequently, the Dirichlet series (2.1.2) does not define the Riemann zeta function outside the region $\Re(s) > 1$. We continue to build the zeta function, as promised. However, first we note that our definition of the zeta function, valid for $\Re(s) > 1$, actually uniquely determines the values of $\zeta(s)$ for all $s \in \mathbb{C}$. This is a consequence of the fact that $\zeta(s)$ is analytic for $\Re(s) > 1$, and continues analytically to the entire plane, with one exceptional point, as explained below.

Recall that analytic continuation allows us to "continue" an analytic function on one domain to an analytic function of a larger domain, uniquely, under certain conditions. (Specifically, given functions f_1, analytic on domain D_1, and f_2, analytic on domain D_2, such that $D_1 \cap D_2 \neq \varnothing$ and $f_1 = f_2$ on $D_1 \cap D_2$, then $f_1 = f_2$ on $D_1 \cup D_2$.) So if we can find a function, analytic on $\mathbb{C}\backslash\{1\}$, that agrees with our Dirichlet series on any domain, D, then we succeed in defining $\zeta(s)$ for all $s \in \mathbb{C}\backslash\{1\}$.

In his 1859 memoir, Riemann proves that the function $\zeta(s)$ can be continued analytically to an analytic function over the whole complex plane, with the exception of $s = 1$, and at $s = 1$, $\zeta(s)$ has a simple pole, with residue 1.

We now define the Riemann zeta function. Following convention, we write $s = \sigma + it$ with $\sigma, t \in \mathbb{R}$ when $s \in \mathbb{C}$.

Definition 2.3. *The Riemann zeta function $\zeta(s)$ is the analytic continuation of the Dirichlet series* (2.1.2) *to the whole complex plane, minus the point* $s = 1$.

Defining the zeta function in this way is concise and correct, but its properties are quite unclear. We continue to build the zeta function by finding the analytic continuation of $\zeta(s)$ explicitly. To start with, when $\Re(s) > 1$, we write

$$\zeta(s) = \sum_{n=1}^{\infty} \frac{1}{n^s} = \sum_{n=1}^{\infty} n \left(\frac{1}{n^s} - \frac{1}{(n+1)^s} \right) = s \sum_{n=1}^{\infty} n \int_n^{n+1} x^{-s-1} dx.$$

Let $x = [x] + \{x\}$, where $[x]$ and $\{x\}$ are the integral and fractional parts of x, respectively.

Since $[x]$ is always the constant n for any x in the interval $[n, n+1)$, we have

$$\zeta(s) = s \sum_{n=1}^{\infty} \int_n^{n+1} [x] x^{-s-1} dx = s \int_1^{\infty} [x] x^{-s-1} dx.$$

By writing $[x] = x - \{x\}$, we obtain

$$\begin{aligned} \zeta(s) &= s \int_1^{\infty} x^{-s} dx - s \int_1^{\infty} \{x\} x^{-s-1} dx \\ &= \frac{s}{s-1} - s \int_1^{\infty} \{x\} x^{-s-1} dx, \qquad \sigma > 1. \end{aligned} \tag{2.1.4}$$

We now observe that since $0 \leq \{x\} < 1$, the improper integral in (2.1.4) converges when $\sigma > 0$ because the integral $\int_1^{\infty} x^{-\sigma-1} dx$ converges. Thus the improper integral in (2.1.4) defines an analytic function of s in the region $\Re(s) > 0$. Therefore, the meromorphic function on the right-hand side of (2.1.4) gives the analytic continuation of $\zeta(s)$ to the region $\Re(s) > 0$, and the term $\frac{s}{s-1}$ gives the simple pole of $\zeta(s)$ at $s = 1$ with residue 1.

Equation (2.1.4) extends the definition of the Riemann zeta function only to the larger region $\Re(s) > 0$. However, Riemann used a similar argument to obtain the analytic continuation to the whole complex plane. He started from the classical definition of the gamma function Γ.

We recall that the gamma function extends the factorial function to the entire complex plane with the exception of the nonpositive integers. The usual definition of the *gamma function*, $\Gamma(s)$, is by means of Euler's integral

$$\Gamma(s) = \int_0^{\infty} e^{-t} t^{s-1} dt$$

but this applies only for $\Re(s) > 0$. Weierstrass' formula

$$\frac{1}{s\Gamma(s)} := e^{\gamma s} \prod_{n=1}^{\infty} \left(1 + \frac{s}{n}\right) e^{-\frac{s}{n}}$$

where γ is Euler's constant, applies in the whole complex plane. The Γ function is analytic on the entire complex plane with the exception of $s = 0, -1, -2, \ldots$, and the residue of $\Gamma(s)$ at $s = -n$ is $\frac{(-1)^n}{n!}$. Note that for $s \in \mathbb{N}$ we have $\Gamma(s) = (s-1)!$.

We have

$$\Gamma\left(\frac{s}{2}\right) = \int_0^\infty e^{-t} t^{\frac{s}{2}-1} dt$$

for $\sigma > 0$. On setting $t = n^2\pi x$, we observe that

$$\pi^{-\frac{s}{2}} \Gamma\left(\frac{s}{2}\right) n^{-s} = \int_0^\infty x^{\frac{s}{2}-1} e^{-n^2\pi x} dx.$$

Hence, with some care on exchanging summation and integration, for $\sigma > 1$,

$$\begin{aligned} \pi^{-\frac{s}{2}} \Gamma\left(\frac{s}{2}\right) \zeta(s) &= \int_0^\infty x^{\frac{s}{2}-1} \left(\sum_{n=1}^{\infty} e^{-n^2\pi x}\right) dx \\ &= \int_0^\infty x^{\frac{s}{2}-1} \left(\frac{\vartheta(x) - 1}{2}\right) dx, \end{aligned}$$

where

$$\vartheta(x) := \sum_{n=-\infty}^{\infty} e^{-n^2\pi x}$$

is the *Jacobi theta function*. The functional equation (also due to Jacobi) for $\vartheta(x)$ is

$$x^{\frac{1}{2}} \vartheta(x) = \vartheta(x^{-1}),$$

and is valid for $x > 0$. This equation is far from obvious; however, the proof lies beyond our focus. The standard proof proceeds using Poisson summation, and can be found in Chapter 2 of [22].

Finally, using the functional equation of $\vartheta(x)$, we obtain

$$\zeta(s) = \frac{\pi^{\frac{s}{2}}}{\Gamma\left(\frac{s}{2}\right)} \left\{ \frac{1}{s(s-1)} + \int_1^\infty \left(x^{\frac{s}{2}-1} + x^{-\frac{s}{2}-\frac{1}{2}}\right) \cdot \left(\frac{\vartheta(x) - 1}{2}\right) dx \right\}. \tag{2.1.5}$$

Due to the exponential decay of $\vartheta(x)$, the improper integral in (2.1.5) converges for *every* $s \in \mathbb{C}$ and hence defines an entire function in $\mathbb{C}$. Therefore, (2.1.5) gives the analytic continuation of $\zeta(s)$ to the whole complex plane, with the exception of $s = 1$.

$$\frac{\vartheta(x)-1}{2} = \sum_{n=1}^{\infty} e^{-n^2\pi x}$$

Theorem 2.4. *The function*

$$\zeta(s) := \frac{\pi^{\frac{s}{2}}}{\Gamma\left(\frac{s}{2}\right)} \left\{ \frac{1}{s(s-1)} + \int_1^{\infty} \left(x^{\frac{s}{2}-1} + x^{-\frac{s}{2}-\frac{1}{2}} \right) \cdot \left(\frac{\vartheta(x)-1}{2} \right) dx \right\}$$

is meromorphic with a simple pole at $s = 1$ with residue 1.

We have now succeeded in our goal of continuing the Dirichlet series (2.1.2) that we started with, to $\zeta(s)$, a meromorphic function on $\mathbb{C}$. We can now consider all complex numbers in our search for the zeros of $\zeta(s)$. We are interested in these zeros because they encode information about the prime numbers. However, not all of the zeros of $\zeta(s)$ are of interest to us. Surprisingly, we can find, with relative ease, an infinite number of zeros, all lying outside of the region $0 \leq \Re(s) \leq 1$. We refer to these zeros as the *trivial zeros* of $\zeta(s)$ and we exclude them from the statement of the Riemann hypothesis.

Before discussing the zeros of $\zeta(s)$ we develop a functional equation for it. Riemann noticed that formula (2.1.5) not only gives the analytic continuation of $\zeta(s)$, but can also be used to derive a functional equation for $\zeta(s)$. He observed that the term $\frac{1}{s(s-1)}$ and the improper integral in (2.1.5) are invariant under the substitution of s by $1-s$. Hence we have the following functional equation:

Theorem 2.5 (The Functional Equation). *For any s in $\mathbb{C}$,*

$$\pi^{-\frac{s}{2}} \Gamma\left(\frac{s}{2}\right) \zeta(s) = \pi^{-\frac{1-s}{2}} \Gamma\left(\frac{1-s}{2}\right) \zeta(1-s).$$

For convenience, and clarity, we will define the function as

$$\xi(s) := \frac{s}{2}(s-1)\pi^{-\frac{s}{2}} \Gamma\left(\frac{s}{2}\right) \zeta(s). \tag{2.1.6}$$

In view of (2.1.5), $\xi(s)$ is an entire function and satisfies the simple functional equation

$$\xi(s) = \xi(1-s). \tag{2.1.7}$$

This shows that $\xi(s)$ is symmetric around the vertical line $\Re(s) = \frac{1}{2}$.

We now have developed the zeta function sufficiently to begin considering its various properties; in particular, the location of its zeros. There are a few assertions we can make based on the elementary theory we have already presented.

We begin our discussion by isolating the trivial zeros of $\zeta(s)$. Recall that the only poles of $\Gamma(s)$ are simple and situated at $s = 0, -1, -2, \ldots$. It follows from (2.1.5) that $\zeta(s)$ has simple zeros at $s = -2, -4, \ldots$ (the pole $s = 0$ of $\Gamma(\frac{s}{2})$ is canceled by the term $\frac{1}{s(s-1)}$). These zeros, arising from the poles of the gamma function, are termed the *trivial zeros*. From the functional equation and Theorem 2.2, all other zeros, the *nontrivial zeros*, lie in the vertical strip $0 \le \Re(s) \le 1$. In view of equation (2.1.6), the nontrivial zeros of $\zeta(s)$ are precisely the zeros of $\xi(s)$, and hence they are symmetric about the vertical line $\Re(s) = \frac{1}{2}$. Also, in view of (2.1.5), they are symmetric about the real axis, $t = 0$. We summarize these results in the following theorem.

Theorem 2.6. *The function $\zeta(s)$ satisfies the following*

1. *$\zeta(s)$ has no zero for $\Re(s) > 1$;*
2. *the only pole of $\zeta(s)$ is at $s = 1$; it has residue 1 and is simple;*
3. *$\zeta(s)$ has trivial zeros at $s = -2, -4, \ldots$;*
4. *the nontrivial zeros lie inside the region $0 \le \Re(s) \le 1$ and are symmetric about both the vertical line $\Re(s) = \frac{1}{2}$ and the real axis $\Im(s) = 0$;*
5. *the zeros of $\xi(s)$ are precisely the nontrivial zeros of $\zeta(s)$.*

The strip $0 \le \Re(s) \le 1$ is called the *critical strip* and the vertical line $\Re(s) = \frac{1}{2}$ is called the *critical line.*

Riemann commented on the zeros of $\zeta(s)$ in his memoir (see the statement at the start of Chapter 1). From his statements the Riemann hypothesis was formulated.

Conjecture 2.7 (The Riemann Hypothesis). *All nontrivial zeros of $\zeta(s)$ lie on the critical line $\Re(s) = \frac{1}{2}$.*

Riemann's eight-page memoir has legendary status in mathematics. It not only proposed the Riemann hypothesis, but also accelerated the development of analytic number theory. Riemann conjectured the asymptotic formula for the number, $N(T)$, of zeros of $\zeta(s)$ in the critical strip with $0 \le \Im(s) < T$ to be

$$N(T) = \frac{T}{2\pi} \log \frac{T}{2\pi} - \frac{T}{2\pi} + O(\log T)$$

(proved by von Mangoldt in 1905). In particular, there are infinitely many nontrivial zeros. Additionally, he conjectured the product representation of $\xi(s)$ to be

$$\xi(s) = e^{A+Bs} \prod_{\rho} \left(1 - \frac{s}{\rho}\right) e^{\frac{s}{\rho}}, \tag{2.1.8}$$

where A, B are constants and ρ runs over all the nontrivial zeros of $\zeta(s)$ (proved by Hadamard in 1893).

2.2 Zero-free Region

One approach to the Riemann hypothesis is to expand the zero-free region as much as possible. However, the proof that the zero-free region includes the vertical line $\Re(s) = 1$ (i.e., $\zeta(1+it) \neq 0$ for all $t \in \mathbb{R}$) is already nontrivial. In fact, this statement is equivalent to the prime number theorem, namely

$$\pi(x) \sim \frac{x}{\log x}, \quad x \to \infty$$

(a problem that required a century of mathematics to solve). Since we wish to focus our attention here on the analysis of $\zeta(s)$, we refer the reader to proofs of this equivalence in Sections 12.3, 12.4, and 12.16.

Theorem 2.8. *For all $t \in \mathbb{R}$, $\zeta(1+it) \neq 0$.*

Proof. In order to prove this result we follow the 1899 approach of de la Vallée Poussin (see Section 12.4). Recall that when $\sigma > 1$, the zeta function is defined by the Dirichlet series (2.1.2) and that the Euler product formula gives us

$$\zeta(s) = \sum_{n=1}^{\infty} \frac{1}{n^s} = \prod_p \left(1 - \frac{1}{p^s}\right)^{-1}, \tag{2.2.1}$$

where $s = \sigma + it$. Taking logarithms of each side of (2.2.1), we obtain

$$\log \zeta(s) = -\sum_p \log \left(1 - \frac{1}{p^s}\right).$$

Using the Taylor expansion of $\log(1-x)$ at $x = 0$, we have

$$\begin{aligned} \log \zeta(s) &= \sum_p \sum_{m=1}^{\infty} m^{-1} p^{-sm} \\ &= \sum_p \sum_{m=1}^{\infty} m^{-1} p^{-\sigma m} p^{-imt} \\ &= \sum_p \sum_{m=1}^{\infty} m^{-1} p^{-\sigma m} e^{-imt \log p}. \end{aligned} \tag{2.2.2}$$

It follows that the real part of $\log \zeta(s)$ is

$$\Re(\log \zeta(s)) = \sum_p \sum_{m=1}^{\infty} m^{-1} p^{-\sigma m} \cos(mt \log p). \tag{2.2.3}$$

Note that by (2.2.3),

$$\begin{aligned}
&3\Re(\log\zeta(\sigma)) + 4\Re(\log\zeta(\sigma+it)) + \Re(\log\zeta(\sigma+2ti)) \\
&= 3\sum_p\sum_{m=1}^{\infty} m^{-1}p^{-\sigma m} + 4\sum_p\sum_{m=1}^{\infty} m^{-1}p^{-\sigma m}\cos(mt\log p) \\
&+\sum_p\sum_{m=1}^{\infty} m^{-1}p^{-\sigma m}\cos(2mt\log p) \\
&= \sum_p\sum_{m=1}^{\infty}\frac{1}{m}\frac{1}{p^{\sigma m}}\Big(3+4\cos(mt\log p)+\cos(2mt\log p)\Big).
\end{aligned}$$

Using $\log W = \log|W| + i\arg(W)$ and the elementary inequality

$$2(1+\cos\theta)^2 = 3 + 4\cos\theta + \cos(2\theta) \geqslant 0, \tag{2.2.4}$$

valid for any $\theta \in \mathbb{R}$, we obtain

$$3\log|\zeta(\sigma)| + 4\log|\zeta(\sigma+it)| + \log|\zeta(\sigma+2ti)| \geqslant 0,$$

or equivalently,

$$|\zeta(\sigma)|^3|\zeta(\sigma+it)|^4|\zeta(\sigma+2it)| \geqslant 1. \tag{2.2.5}$$

Since $\zeta(\sigma)$ has a simple pole at $\sigma = 1$ with residue 1, the Laurent series of $\zeta(\sigma)$ at $\sigma = 1$ is

$$\zeta(\sigma) = \frac{1}{1-\sigma} + a_0 + a_1(\sigma-1) + a_2(\sigma-1)^2 + \cdots = \frac{1}{1-\sigma} + g(\sigma),$$

where $g(\sigma)$ is analytic at $\sigma = 1$. Hence, for $1 < \sigma \leq 2$, we have $|g(\sigma)| \leq A_0$, for some $A_0 > 0$, and

$$|\zeta(\sigma)| = \frac{1}{1-\sigma} + A_0.$$

Now we will show that $\zeta(1+it) \neq 0$ using inequality (2.2.5). To obtain a contradiction, suppose that there is a zero on the line $\sigma = 1$. So $\zeta(1+it) = 0$ for some $t \in \mathbb{R}$, $t \neq 0$. Then by the mean-value theorem,

$$\begin{aligned}
|\zeta(\sigma+it)| &= |\zeta(\sigma+it) - \zeta(1+it)| \\
&= |\sigma-1||\zeta'(\sigma_0+it)|, \qquad 1 < \sigma_0 < \sigma, \\
&\leq A_1(\sigma-1),
\end{aligned}$$

where A_1 is a constant depending only on t. Also, when σ approaches 1 we have $|\zeta(\sigma+2it)| < A_2$, where A_2 again depends only on t. Note that in (2.2.5) the degree of the term $\sigma - 1$, which is 4, is greater than that of the term $\frac{1}{\sigma-1}$, which is 3. So for fixed t, as σ tends to 1^+, we have

$$\begin{aligned}
\lim_{\sigma\to 1^+} &|\zeta(\sigma)|^3|\zeta(\sigma+it)|^4|\zeta(\sigma+2it)| \\
&\le \lim_{\sigma\to 1^+} \left(\frac{1}{\sigma-1}+A_0\right)^3 {A_1}^4(\sigma-1)^4 A_2 \\
&= 0.
\end{aligned}$$

This contradicts (2.2.5). Hence we conclude that $\zeta(1+it)\neq 0$ for any $t\in\mathbb{R}$.

$\square$

This result gives us a critical part of the proof of the prime number theorem, by extending the zero-free region of $\zeta(s)$ to include the line $\Re(s)=1$. It would seem intuitive that by extending the zero-free region even further we could conclude other powerful results on the distribution of the primes. In fact, more explicitly it can be proved that the asymptotic formula,

$$\pi(x)=\int_2^x \frac{dt}{\log t}+O(x^{\Theta}\log x)$$

is equivalent to

$$\zeta(\sigma+it)\neq 0,\ \text{ for } \sigma>\Theta, \tag{2.2.6}$$

where $\frac{1}{2}\le\Theta<1$. In particular, the Riemann hypothesis (that is, $\Theta=\frac{1}{2}$ in (2.2.6)) is equivalent to the statement

$$\pi(x)=\int_2^x \frac{dt}{\log t}+O(x^{\frac{1}{2}}\log x).$$

This formulation gives an immediate method by which to expand the zero-free region. However, we are still unable to improve the zero-free region in the form (2.2.6) for any $\Theta<1$. The best results to date, are of the form proved by Vinogradov and Korobov independently in 1958, is that $\zeta(s)$ has no zeros in the region

$$\sigma\ge 1-\frac{c}{(\log|t|+1)^{\frac{2}{3}}(\log\log(3+|t|))^{\frac{1}{3}}}$$

for some positive constant c [87, 164].

2.3 Counting the Zeros of $\zeta(s)$

Proving that the nontrivial zeros of $\zeta(s)$ have real part $\frac{1}{2}$, and proving that the nontrivial zeros of $\zeta(s)$ lie outside a prescribed region, are both, presumably, extremely difficult. However, we can still glean heuristic evidence for the Riemann hypothesis by counting nontrivial zeros. We can develop tools that allow us to count the number of zeros in the critical strip with imaginary part $|\Im(s)|<T$ for any positive real number T. Once we know how many zeros

should lie in a given region, we can verify the Riemann hypothesis in that region computationally. In this section we develop the theory that will allow us to make this argument.

We begin with the argument principle. The argument principle in complex analysis gives a very useful tool to count the zeros or the poles of a meromorphic function inside a specified region. For a proof of this well-known result in complex analysis see §79 of [32].

The Argument Principle. *Let f be meromorphic in a domain interior to a positively oriented simple closed contour C such that f is analytic and nonzero on C. Then,*

$$\frac{1}{2\pi}\Delta_C \arg(f(s)) = Z - P,$$

where Z is the number of zeros, P is the number of poles of $f(z)$ inside C, counting multiplicities, and $\Delta_C \arg(f(s))$ counts the changes in the argument of $f(s)$ along the contour C.

We apply this principle to count the number of zeros of $\zeta(s)$ within the rectangle $\{\sigma + it \in \mathbb{C} : 0 < \sigma < 1, 0 \le t < T\}$. We denote this number by

$$N(T) := \#\{\sigma + it : 0 < \sigma < 1, 0 \le t < T, \zeta(\sigma + it) = 0\}.$$

As previously mentioned, the following theorem was conjectured by Riemann [134] and proved by von Mangoldt [167]. We follow the exposition of Davenport from [42]. We present the proof in more detail than is typical in this book, since it exemplifies the general form of arguments in this field. Such arguments can be relatively complicated and technical; the reader is forewarned.

Theorem 2.9. *For $N(T)$ defined above, we have*

$$N(T) = \frac{T}{2\pi}\log\frac{T}{2\pi} - \frac{T}{2\pi} + O(\log T).$$

Proof. Instead of working directly with $\zeta(s)$, we will make use of the $\xi(s)$ function, defined in (2.1.6). Since $\xi(s)$ has the same zeros as $\zeta(s)$ in the critical strip, and $\xi(s)$ is an entire function, we can apply the argument principle to $\xi(s)$ instead of $\zeta(s)$. Let R be the positively oriented rectangular contour with vertices $-1, 2, 2 + iT$, and $-1 + iT$. By the argument principle, we have

$$N(T) = \frac{1}{2\pi}\Delta_R \arg(\xi(s)).$$

We now divide R into three subcontours. Let L_1 be the horizontal line segment from -1 to 2. Let L_2 be the contour consisting of the vertical line segment from 2 to $2+iT$ and then the horizontal line segment from $2+iT$ to $\frac{1}{2}+iT$. Finally, let L_3 be the contour consisting of the horizontal line segment from $\frac{1}{2}+iT$ to $-1+iT$ and then the vertical line segment from $-1+iT$ to -1. Now,

$$\Delta_R \arg(\xi(s)) = \Delta_{L_1} \arg(\xi(s)) + \Delta_{L_2} \arg(\xi(s)) + \Delta_{L_3} \arg(\xi(s)). \quad (2.3.1)$$

We wish to trace the argument change of $\xi(s)$ along each contour.

To begin with, there is no argument change along L_1, since the values of $\xi(s)$ here are real, and hence all arguments of $\xi(s)$ are zero. Thus,

$$\Delta_{L_1} \arg(\xi(s)) = 0. \quad (2.3.2)$$

From the functional equation for $\xi(s)$ we have

$$\xi(\sigma+it) = \xi(1-\sigma-it) = \overline{\xi(1-\sigma+it)}.$$

So the argument change of $\xi(s)$ as s moves along L_3 is the same as the argument change of $\xi(s)$ as s moves along L_2. Hence, in conjunction with (2.3.2), equation (2.3.1) becomes

$$N(T) = \frac{1}{2\pi} 2\Delta_{L_2} \arg(\xi(s)) = \frac{1}{\pi}\Delta_{L_2} \arg(\xi(s)). \quad (2.3.3)$$

From the definition (2.1.6) and the basic relation $z\Gamma(z) = \Gamma(z+1)$ we have

$$\xi(s) = (s-1)\pi^{-\frac{s}{2}}\Gamma\left(\frac{s}{2}+1\right)\zeta(s). \quad (2.3.4)$$

Next we work out the argument changes of these four factors of the right-hand side of equation (2.3.4) along L_2 separately.

We begin by considering $\Delta_{L_2} \arg(s-1)$. One has

$$\begin{aligned}\Delta_{L_2} \arg(s-1) &= \arg\left(-\frac{1}{2}+iT\right) - \arg(1) \\ &= \arg\left(-\frac{1}{2}+iT\right) \\ &= \frac{\pi}{2} + \arctan\left(\frac{1}{2T}\right) \\ &= \frac{\pi}{2} + O(T^{-1})\end{aligned}$$

because $\arctan\left(\frac{1}{2T}\right) = O(T^{-1})$.

Next we consider $\Delta_{L_2} \arg(\pi^{-\frac{s}{2}})$:

$$\begin{aligned}\Delta_{L_2} \arg(\pi^{-\frac{s}{2}}) &= \Delta_{L_2} \arg\left(\exp\left(-\frac{s}{2}\log\pi\right)\right)\\ &= \arg\left(\exp\left(-\frac{1}{2}\left(-\frac{1}{2}+iT\right)\log\pi\right)\right)\\ &= \arg\left(\exp\left(\frac{1-2iT}{4}\log\pi\right)\right)\\ &= -\frac{T}{2}\log\pi.\end{aligned}$$

Now we use Stirling's formula to give the asymptotic estimate for $\Gamma(s)$ (see (5) of Chapter 10 in [42]):

$$\log\Gamma(s) = \left(s-\frac{1}{2}\right)\log s - s + \frac{1}{2}\log 2\pi + O(|s|^{-1}), \tag{2.3.5}$$

which is valid when $|s|$ tends to ∞ and the argument satisfies $-\pi+\delta < \arg(s) < \pi-\delta$ for any fixed $\delta > 0$. So we have

$$\begin{aligned}\Delta_{L_2}\arg\left(\Gamma\left(\frac{1}{2}s+1\right)\right) &= \Im\left(\log\Gamma\left(\frac{5}{4}+\frac{iT}{2}\right)\right)\\ &= \Im\left\{\left(\frac{3}{4}+i\frac{T}{4}\right)\log\left(\frac{5}{4}+i\frac{T}{2}\right)-\frac{5}{4}-\frac{iT}{2}+\frac{1}{2}\log 2\pi+O(T^{-1})\right\}\\ &= \frac{1}{2}T\log\frac{T}{2}-\frac{1}{2}T+\frac{3}{8}\pi+O(T^{-1}).\end{aligned}$$

Putting all these together we obtain, from (2.3.3),

$$\begin{aligned}N(T) &= \frac{1}{\pi}\Delta_{L_2}\arg\xi(s)\\ &= \frac{T}{2\pi}\log\frac{T}{2\pi}-\frac{T}{2\pi}+\frac{7}{8}+\mathcal{S}(T)+O(T^{-1}),\end{aligned} \tag{2.3.6}$$

where

$$\mathcal{S}(T) := \frac{1}{\pi}\Delta_{L_2}\arg\zeta(s) = \frac{1}{\pi}\arg\zeta\left(\frac{1}{2}+iT\right). \tag{2.3.7}$$

It now remains to estimate $\mathcal{S}(T)$.

Taking the logarithmic derivative of the Hadamard product representation (2.1.8), we obtain

$$\frac{\xi'(s)}{\xi(s)} = B + \sum_\rho \frac{\frac{-1}{\rho}}{1-\frac{s}{\rho}} + \sum_\rho \frac{1}{\rho} = B + \sum_\rho\left(\frac{1}{s-\rho}+\frac{1}{\rho}\right). \tag{2.3.8}$$

Since $\xi(s)$ is alternatively defined by (2.1.6), we also have

$$\frac{\xi'(s)}{\xi(s)} = \frac{1}{s} + \frac{1}{s-1} + \frac{1}{2}\log\pi + \frac{1}{2}\frac{\Gamma'}{\Gamma}\left(\frac{1}{2}s\right) + \frac{\zeta'(s)}{\zeta(s)}. \tag{2.3.9}$$

Now combining (2.3.8) and (2.3.9), we have

$$-\frac{\zeta'(s)}{\zeta(s)} = \frac{1}{s-1} - B - \frac{1}{2}\log\pi + \frac{1}{2}\frac{\Gamma'}{\Gamma}\left(\frac{1}{2}s+1\right) - \sum_\rho\left(\frac{1}{s-\rho}+\frac{1}{\rho}\right).$$

Given $t \geq 2$ and $1 \leq \sigma \leq 2$, by Stirling's formula for $\Gamma(s)$,

$$\left|\frac{\Gamma'}{\Gamma}\left(\frac{1}{2}s+1\right)\right| \leq A_1 \log t,$$

so we have

$$-\Re\left(\frac{\zeta'}{\zeta}(s)\right) \leq A_2 \log t - \sum_\rho \Re\left(\frac{1}{s-\rho}+\frac{1}{\rho}\right) \tag{2.3.10}$$

for some positive absolute constants A_1 and A_2. If $s = \sigma + it$, $2 \leq t$, and $1 < \sigma \leq 2$, then

$$-\Re\left(\frac{\zeta'}{\zeta}(s)\right) < A_2 \log t - \sum_\rho \Re\left(\frac{1}{s-\rho}+\frac{1}{\rho}\right). \tag{2.3.11}$$

Since $\frac{\zeta'}{\zeta}(s)$ is analytic at $s = 2 + iT$,

$$-\Re\left(\frac{\zeta'}{\zeta}(2+iT)\right) \leq A_3 \tag{2.3.12}$$

for some positive absolute constant A_3. If $\rho = \beta + i\gamma$ and $s = 2 + iT$, then

$$\begin{aligned}\Re\left(\frac{1}{s-\rho}\right) &= \Re\left(\frac{1}{2-\beta+i(T-\gamma)}\right)\\ &= \frac{2-\beta}{(2-\beta)^2+(T-\gamma)^2}\\ &\geq \frac{1}{4+(T-\gamma)^2}\\ &\gg \frac{1}{1+(T-\gamma)^2}\end{aligned}$$

since $0 < \beta < 1$. Also, we have $\Re(\frac{1}{\rho}) = \frac{\beta}{\beta^2+\gamma^2} \geq 0$, so by equations (2.3.11) and (2.3.12),

$$\sum_\rho \frac{1}{1+(T-\gamma)^2} \ll \sum_\rho \Re\left(\frac{1}{s-\rho}\right) \ll \log T.$$

We have proven that for $T \geq 1$,

$$\sum_{\rho} \frac{1}{1+(T-\gamma)^2} = O(\log T). \tag{2.3.13}$$

It immediately follows that

$$\begin{aligned} &\# \{\rho = \beta + i\gamma : 0 < \beta < 1, T \leq \gamma \leq T+1, \zeta(\rho) = 0\} \\ &\leq 2 \sum_{T \leq \gamma \leq T+1} \frac{1}{1+(T-\gamma)^2} \\ &\leq 2 \sum_{\rho} \frac{1}{1+(T-\gamma)^2} \ll \log T. \end{aligned} \tag{2.3.14}$$

For large t and $-1 \leq \sigma \leq 2$,

$$\frac{\zeta'}{\zeta}(s) = O(\log t) + \sum_{\rho} \left(\frac{1}{s-\rho} - \frac{1}{2+it-\rho} \right). \tag{2.3.15}$$

When $|\gamma - t| > 1$, we have

$$\left| \frac{1}{s-\rho} - \frac{1}{2+it-\rho} \right| = \frac{2-\sigma}{|(s-\rho)(2+it-\rho)|} \leq \frac{3}{(\gamma-t)^2},$$

and therefore,

$$\left| \sum_{|\gamma-t|>1} \left(\frac{1}{s-\rho} - \frac{1}{2+it-\rho} \right) \right| \leq \sum_{|\gamma-t|>1} \frac{3}{(\gamma-t)^2} \ll \sum_{\rho} \frac{1}{1+(\gamma-t)^2} \ll \log t.$$

This, combined with (2.3.15), gives

$$\frac{\zeta'}{\zeta}(s) = \sum_{|\gamma-t|\leq 1} \left(\frac{1}{s-\rho} - \frac{1}{2+it-\rho} \right) + O(\log t). \tag{2.3.16}$$

Next, from equation (2.3.7), we have

$$\pi S(T) = \arg \zeta \left(\frac{1}{2} + iT \right) = \int_{\frac{1}{2}+iT}^{2+iT} \frac{\zeta'}{\zeta}(s) ds = \log \zeta(s) \Big|_{\frac{1}{2}+iT}^{2+iT}. \tag{2.3.17}$$

Since $\log \omega = \log |\omega| + i \arg \omega$,

$$\begin{aligned} -\int_{\frac{1}{2}+iT}^{2+iT} \Im \left(\frac{\zeta'}{\zeta}(s) \right) ds &= -\arg(\zeta(s)) \Big|_{\frac{1}{2}+iT}^{2+iT} \\ &= -\arg(\zeta(2+iT)) + \arg\left(\zeta \left(\frac{1}{2} + iT \right) \right). \end{aligned}$$

Therefore we have

$$-\int_{\frac{1}{2}+iT}^{2+iT} \Im\left(\frac{\zeta'}{\zeta}(s)\right) ds = O(1) + \pi \mathcal{S}(T).$$

Thus, with (2.3.16) and (2.3.17),

$$\mathcal{S}(T) \ll \sum_{|\gamma-t|<1} \int_{\frac{1}{2}+iT}^{2+iT} \Im\left(\frac{1}{s-\rho} - \frac{1}{2+it-\rho}\right) ds + \log T,$$

and we have

$$\int_{\frac{1}{2}+iT}^{2+iT} \Im\left(\frac{1}{s-\rho}\right) ds = \Im\left(\log(s-\rho)\right)\Big|_{\frac{1}{2}+iT}^{2+iT} = \arg(s-\rho)\Big|_{\frac{1}{2}+iT}^{2+iT} \ll 1.$$

By (2.3.14) we obtain that

$$\mathcal{S}(T) \ll \sum_{|\gamma-T|<1} 1 + \log T \ll \log T,$$

and finally that

$$N(T) = \frac{T}{2\pi} \log \frac{T}{2\pi} - \frac{T}{2\pi} + O(\log T).$$

This completes the proof.

We now have the theoretical underpinnings for verifying the Riemann hypothesis computationally. We can adapt these results to give us a feasible method of counting the number of zeros of $\zeta(s)$ in the critical strip up to any desired height. Combined with efficient methods of computing values of $\zeta(\frac{1}{2}+it)$ we can gain heuristic evidence in favor of the Riemann hypothesis and, in part, justify over a century of belief in its truth. For a detailed development of these ideas, see Chapter 3.

2.4 Hardy's Theorem

Hardy's theorem is one of the first important results in favor of the Riemann hypothesis. It establishes a fact that is the most basic necessary condition for the Riemann hypothesis to be true. We will now present and prove Hardy's theorem. For Hardy's original paper, see Section 12.5; we follow the exposition of Edwards [51].

Theorem 2.10 (Hardy's Theorem). *There are infinitely many zeros of $\zeta(s)$ on the critical line.*

Proof. The idea of the proof is to apply Mellin inversion to $\xi(s)$, namely if

$$F(s) = \int_{-\infty}^{\infty} f(x)x^{-s}dx,$$

then

$$f(s) = \int_{a-i\infty}^{a+i\infty} F(z)z^{s-1}dz.$$

For an introduction to Mellin inversion see §2.7 of [154].

Recall equations (2.1.5),

$$\zeta(s) = \frac{\pi^{\frac{s}{2}}}{\Gamma\left(\frac{s}{2}\right)}\left\{\frac{1}{s(s-1)} + \int_1^{\infty}\left(x^{\frac{1}{2}s-1} + x^{-\frac{1}{2}s-\frac{1}{2}}\right)\cdot\left(\frac{\vartheta(x)-1}{2}\right)dx\right\},$$

and (2.1.6),

$$\xi(s) = \frac{s}{2}(s-1)\pi^{-\frac{s}{2}}\Gamma\left(\frac{s}{2}\right)\zeta(s).$$

After some substitution and simplification, we realize

$$\frac{2\xi(s)}{s(s-1)} = \int_0^{\infty} u^{-s}\left(\vartheta(u^2) - 1 - \frac{1}{u}\right)du$$

for $0 < \Re(s) < 1$. Application of Mellin inversion formula gives

$$\vartheta(z^2) - 1 - \frac{1}{z} = \frac{1}{2\pi i}\int_{\frac{1}{2}-i\infty}^{\frac{1}{2}+i\infty}\frac{2\xi(s)}{s(s-2)}z^{s-1}ds. \tag{2.4.1}$$

Note that the function $\vartheta(z^2) = \sum_{n=-\infty}^{\infty} e^{-\pi n^2 z^2}$ is defined not only for $z \in \mathbb{R}$, but also for $z \in W$, where

$$W := \left\{z \in \mathbb{C} : -\frac{\pi}{4} < \arg(z) < \frac{\pi}{4}\right\}.$$

We claim that $\vartheta(z^2)$ and all its derivatives approach zero as s approaches $e^{i\pi/4}$ (e.g., along the circle $|z| = 1$ in W). In fact, using the functional equation $z^{\frac{1}{2}}\vartheta(z) = \vartheta(z^{-1})$, we have

$$\begin{aligned}
\vartheta(z^2) &= \sum_{n=-\infty}^{\infty}(-1)^n e^{-\pi n^2(z^2-i)} \\
&= -\vartheta(z^2 - i) + 2\vartheta(4(z^2-i)) \\
&= -\frac{1}{(z^2-i)^{1/2}}\vartheta\left(\frac{1}{z^2-i}\right) + \frac{1}{(z^2-i)^{1/2}}\vartheta\left(\frac{1}{4(z^2-i)}\right) \\
&= \frac{1}{(z^2-i)^{1/2}}\sum_{n\ odd} e^{-\frac{\pi n^2}{4(z^2-i)}}.
\end{aligned}$$

Since $u^k e^{-1/u}$ approaches zero as $u \longrightarrow 0^+$ for any integer k, $\vartheta(z^2)$ and all its derivatives approach zero as z approaches $e^{i\pi/4}$ in the wedge W.

Consider now the integral of the right-hand side of equation (2.4.1). In view of (2.1.4) we have

$$\begin{aligned}\left|\zeta\left(\frac{1}{2}+it\right)\right| &= \left|\frac{1/2+it}{-1/2+it} - (1/2+it)\int_1^\infty \{x\}x^{-\frac{1}{2}-it-1}dx\right| \\ &\leq 1 + |1/2+it|\int_1^\infty x^{-3/2}dx \\ &\ll |t|,\end{aligned}$$

as $t \longrightarrow \pm\infty$. Also, from formula (2.3.5) for the gamma function, we have

$$\log \Gamma\left(\frac{1}{4}+i\frac{t}{2}\right) = \left(-\frac{1}{4}+i\frac{t}{2}\right)\log\left(\frac{1}{4}+i\frac{t}{2}\right) - \frac{1}{2} - i\frac{t}{2} + \frac{1}{2}\log 2\pi + O(|t|^{-1}).$$

It follows that

$$\left|\Gamma\left(\frac{1}{4}+i\frac{t}{2}\right)\right| \ll e^{-\frac{\pi}{4}|t|}$$

as $t \longrightarrow \pm\infty$. Thus we have

$$\left|\xi\left(\frac{1}{2}+it\right)\right| \ll |t|^3 e^{-\frac{\pi}{4}|t|},$$

which shows that the integral in the right-hand side of (2.4.1) converges in the wedge W, and hence is analytic in W.

If the operator $z(d^2/dz^2)z$ is applied to both sides of (2.4.1), then formula (2.4.1) takes a simpler form of

$$H(z) = \frac{1}{2\pi i}\int_{\frac{1}{2}-i\infty}^{\frac{1}{2}+i\infty} 2\xi(s)z^{s-1}ds,$$

where

$$H(z) := z\frac{d^2}{dz^2}z\vartheta(z^2).$$

This shows, similarly to $\vartheta(z^2)$, that $H(z)$ and all of its derivatives approach zero as z approaches $e^{i\pi/4}$.

Using the Taylor series of the exponential function, we obtain

$$\begin{aligned}z^{1/2}H(z) &= \frac{1}{\pi}\int_{-\infty}^{\infty}\xi\left(\frac{1}{2}+it\right)z^{it}dt \\ &= \frac{1}{\pi}\int_{-\infty}^{\infty}\xi\left(\frac{1}{2}+it\right)\sum_{n=0}^{\infty}\frac{(it\log z)^n}{n!}dt \\ &= \sum_{n=0}^{\infty}c_n(i\log z)^n,\end{aligned}$$

where

$$c_n = \frac{1}{\pi n!} \int_{-\infty}^{\infty} \xi\left(\frac{1}{2} + it\right) t^n dt.$$

Since $\xi(s)$ is symmetric about the critical line, $c_{2n+1} = 0$ for all n.

Suppose Hardy's theorem is false. Then $\xi\left(\frac{1}{2} + it\right)$ has a finite number of zeros, and $\xi\left(\frac{1}{2} + it\right)$ does not change sign for all large $t \in \mathbb{R}$. For example, if $\xi\left(\frac{1}{2} + it\right)$ is positive for $t \geq T$, then

$$\begin{aligned}
c_{2n}\pi(2n)! &= 2\int_0^{\infty} \xi\left(\frac{1}{2} + it\right) t^{2n} dt \\
&\geq 2\int_0^{T+2} \xi\left(\frac{1}{2} + it\right) t^{2n} dt \\
&\geq 2\left\{-\int_0^{T} \left|\xi\left(\frac{1}{2} + it\right)\right| T^{2n} dt + \int_{T+1}^{T+2} \xi\left(\frac{1}{2} + it\right) (T+1)^{2n} dt\right\} \\
&\geq A_6(T+1)^{2n} - A_7 T^{2n}
\end{aligned}$$

for some positive constants A_6 and A_7. Hence, $c_{2n} > 0$ for all sufficiently large n. Similarly, if $\xi\left(\frac{1}{2} + it\right)$ is negative for $t \geq T$, then $c_{2n} < 0$ for all sufficiently large n. Thus, if the function $z^{1/2}H(z)$ is differentiated sufficiently many times with respect to $i \log z$, then the right-hand side becomes an even power series in which every term has the same sign. Consequently, as z approaches $e^{i\pi/4}$, the value of this even power series does not approach zero. However, it must approach zero as $d/d(i \log z) = -iz\, d/dz$, and differentiating repeatedly carries $z^{1/2}H(z)$ to a function that approaches zero as z approaches $e^{i\pi/4}$. This contradiction proves Hardy's theorem.

□

As we noted previously, Hardy's theorem establishes a minimal necessary condition for the truth of the Riemann hypothesis. Mathematicians have continued to attack the Riemann hypothesis from this angle, by proving necessary conditions of increasing difficulty. Selberg proved that a positive proportion of the zeros of $\zeta(s)$ lie on the critical line (a result that, in part, earned him the Fields Medal in 1950) [142]. This was improved by Levinson to the result that at least $\frac{1}{3}$ of the zeros lie on the critical line [97]. The best and most recent result in this vein is due to Conrey, who proved that at least $\frac{2}{5}$ of the zeros lie on the critical line [35].

3

Algorithms for Calculating $\zeta(s)$

The Riemann hypothesis has been open since its "proposal" in 1859. The most direct form of attack is to search for a counterexample, and also to search computationally for insight. Mathematicians from the time of Riemann have developed a large body of computational techniques and evidence in support of the Riemann hypothesis. The advent of modern computing empowered mathematicians with new tools, and motivated the development of more efficient algorithms for the computation of $\zeta(s)$.

This chapter presents a basic sketch of the analytic ideas on which these algorithms are based, the three main algorithms used to compute values of $\zeta(s)$, and the analysis required to verify the Riemann hypothesis within large regions of the critical strip. The primarily goal of all of these calculations is to compute zeros on the critical line with the highly unlikely chance of finding a counterexample. For additional details, which are sometimes formidable, the reader is directed to the sources cited within.

3.1 Euler–MacLaurin Summation

The zeta function is difficult to evaluate, especially in the critical strip. For this reason sophisticated methods have been developed and applied in order to perform these evaluations. Euler–MacLaurin summation was one of the first methods used to compute values of $\zeta(s)$. Euler used it in the computation of $\zeta(n)$ for $n = 2, 3, \ldots, 15, 16$. It was also used by Gram, Backlund, and Hutchinson in order to verify the Riemann hypothesis for $t \leq 50$, $t \leq 200$, and $t \leq 300$ respectively ($s = \frac{1}{2} + it$).

Definition 3.1. *[51] The Bernoulli numbers, B_n, are defined by the generating function*

$$\frac{x}{e^x - 1} = \sum_{n=0}^{\infty} \frac{B_n x^n}{n!}.$$

The nth Bernoulli Polynomial, $B_n(x)$, is given by

$$B_n(x) := \sum_{j=0}^{n} \binom{n}{j} B_{n-j} x^j.$$

Euler–MacLaurin Evaluation of $\zeta(s)$. *For $N \geq 1$ one has*

$$\begin{aligned}
\zeta(s) &= \sum_{n=1}^{N-1} n^{-s} + \int_N^\infty x^{-s} dx + \frac{1}{2} N^{-s} - s \int_N^\infty B_1(\{x\}) x^{-s-1} dx \\
&= \sum_{n=1}^{N-1} n^{-s} + \frac{N^{1-s}}{s-1} + \frac{1}{2} N^{-s} + \frac{B_2}{2} s N^{-s-1} + \cdots \\
&\quad \cdots + \frac{B_{2v}}{(2v)!} s(s+1)\cdots(s+2v-2) N^{-s-2v+1} + R_{2v},
\end{aligned}$$

where

$$R_{2v} = -\frac{s(s+1)\cdots(s+2v-1)}{(2v)!} \int_N^\infty B_{2v}(\{x\}) x^{-s-2v} dx.$$

Here $B_i(x)$ denotes the ith Bernoulli polynomial, B_i denotes the ith Bernoulli number, and $\{x\}$ denotes the fractional part of x.

If N is at all large, say N is approximately the same size as $|s|$, then the terms of the series decrease quite rapidly at first, and it is natural to expect that the remainder R_{2v} will be very small. In fact, we have

$$|R_{2v}| \leq \left| \frac{s(s+1)\cdots(s+2v+1) B_{2(v+1)} N^{-\sigma-2v-1}}{2(v+1)!(\sigma+2v+1)} \right|.$$

The application of Euler–MacLaurin summation to $\zeta(s)$ is discussed at length in Chapter 6 of [51].

3.2 Backlund

Around 1912, Backlund developed a method of determining the number of zeros of $\zeta(s)$ in the critical strip $0 < \Re(s) < 1$ up to a given height T [51, p.128]. This is part of proving that the first N zeros of $\zeta(s)$ all lie on the critical line.

Recall from Chapter 2 that $N(T)$ denotes the number of zeros (counting multiplicities) of $\zeta(s)$ in R, the rectangle $\{0 \leq \Re(s) \leq 1,\ 0 \leq \Im(s) \leq T\}$, and ∂R is the boundary of R, oriented positively. In [135] Riemann observed that

$$N(T) = \frac{1}{2\pi i} \int_{\partial R} \frac{\xi'(s)}{\xi(s)} ds$$

and proposed the estimate

$$N(T) \sim \frac{T}{2\pi} \log \frac{T}{2\pi} - \frac{T}{2\pi}.$$

This estimate was first proved by von Mangoldt in 1905 [167]. In [10] Backlund obtained the specific estimate

$$\left| N(T) - \left(\frac{T}{2\pi} \log \frac{T}{2\pi} - \frac{T}{2\pi} + \frac{7}{8} \right) \right| < 0.137 \log T + 0.443 \log\log T + 4.350$$

for all $T \geq 2$.

This estimate can be deduced by using good estimates to R_{2v} as presented in the preceding section. Current computational methods have superseded this estimate.

3.3 Hardy's Function

One way of computing the zeros of any real-valued function is to find small intervals in which that function changes sign. This method cannot be applied naively to arbitrary complex-valued functions. So, in order to calculate zeros of $\zeta(s)$ that lie on the critical line, we would like to find a real-valued function whose zeros are exactly the zeros of $\zeta(s)$ on the critical line. This is achieved by considering the function $\xi(s)$. In particular, we recall that $\xi(s)$ is defined by

$$\xi(s) = \frac{s}{2}(s-1)\pi^{-\frac{s}{2}} \Gamma\left(\frac{s}{2}\right) \zeta(s).$$

Since $\xi(s)$ is a real-valued function on the critical line, we will find the simple zeros of $\xi(s)$ by determining where the function changes sign. We develop Hardy's function following [51, p.119] (also see Section 2.1): for $s = \frac{1}{2} + it$, we have

$$\begin{aligned}
\xi\left(\frac{1}{2} + it\right) &= \left(\frac{1}{4} + \frac{it}{2}\right)\left(-\frac{1}{2} + it\right) \pi^{-\frac{1}{4} - \frac{it}{2}} \Gamma\left(\frac{1}{4} + \frac{it}{2}\right) \zeta\left(\frac{1}{2} + it\right) \\
&= -\frac{1}{2}\left(\frac{1}{4} + t^2\right) \pi^{-\frac{1}{4} - \frac{it}{2}} \Gamma\left(\frac{1}{4} + \frac{it}{2}\right) \zeta\left(\frac{1}{2} + it\right) \\
&= \left[e^{\Re \log \Gamma\left(\frac{1}{4} + \frac{it}{2}\right)} \cdot \pi^{-\frac{1}{4}} \left(\frac{-t^2}{2} - \frac{1}{8} \right) \right] \\
&\quad \times \left[e^{i \Im \log \Gamma\left(\frac{1}{4} + \frac{it}{2}\right)} \cdot \pi^{-\frac{it}{2}} \zeta\left(\frac{1}{2} + it\right) \right].
\end{aligned}$$

Now, since the factor in the first set of brackets is always a negative real number, the sign of $\xi(\frac{1}{2}+it)$ depends entirely on the sign of the second factor. We define Hardy's Z-function to be

$$Z(t) := e^{i\theta(t)}\zeta\left(\frac{1}{2}+it\right),$$

where $\theta(t)$ is given by

$$\theta(t) := \Im\left(\log\Gamma\left(\frac{1}{4}+\frac{it}{2}\right)\right) - \frac{t}{2}\log\pi.$$

Now it is clear that the sign of $\xi(\frac{1}{2}+it)$ is opposite that of $Z(t)$.

At this point it seems that since $\zeta(\frac{1}{2}+it)$ appears in the formula for $Z(t)$ we have not gained any interesting information. However, the Riemann–Siegel formula, which appears in the next section, allows for efficient computation of $Z(t)$ to determine the zeros of $\zeta(\frac{1}{2}+it)$.

3.4 The Riemann–Siegel Formula

The Riemann–Siegel formula was discovered among Riemann's private papers by Siegel in 1932 [144]. It gives an improvement over Euler–MacLaurin summation in approximating values of $\zeta(s)$. It aided computation, and added to the empirical evidence for the Riemann hypothesis. However, it does not give enough information by itself to be used for direct verification of the hypothesis. This is because the formula only gives a fast method of finding zeros of $\zeta(s)$ for which $\Re(s)=\frac{1}{2}$.

We begin with the approximate functional equation for the Riemann zeta function.

Theorem 3.2 (Approximate Functional Equation). *Let $x,y\in\mathbb{R}^+$ with $2\pi xy=|t|$; then for $s=\sigma+it$ with $0\le\sigma\le1$, we have*

$$\zeta(s)=\sum_{n\le x}\frac{1}{n^s}+\chi(s)\sum_{n\le y}\frac{1}{n^{1-s}}+O(x^{-\sigma})+O(|t|^{\frac{1}{2}-\sigma}y^{\sigma-1}),$$

where $\chi(s)$ is given by

$$\chi(s)=2^s\pi^{s-1}\sin\left(\frac{\pi s}{2}\right)\Gamma(1-s).$$

The approximate functional equation can be used to calculate $Z(t)$ in the following way. First we set $x=y=\sqrt{|t|/2\pi}$. This yields

$$\zeta(s)=\sum_{n=1}^{[x]}\frac{1}{n^s}+\chi(s)\sum_{n=1}^{[x]}\frac{1}{n^{1-s}}+E_m(s).$$

Here the error term $E_m(s)$ satisfies $E_m(s) = O(|t|^{-\sigma/2})$. Now we substitute $s = \frac{1}{2} + it$ and multiply by $e^{i\theta(t)}$ to obtain

$$\begin{aligned}Z(t) &= e^{i\theta(t)} \sum_{n=1}^{[x]} \frac{1}{n^{\frac{1}{2}+it}} + e^{-i\theta(t)} \sum_{n=1}^{[x]} \frac{1}{n^{\frac{1}{2}-it}} + O(t^{-\frac{1}{4}}) \\ &= 2\sum_{n=1}^{[x]} \frac{\cos(\theta(t) - t\log n)}{n^{\frac{1}{2}}} + O(t^{-\frac{1}{4}}).\end{aligned}$$

This is the basis of the Riemann–Siegel formula. All that is required to apply this formula to compute $Z(t)$ is a more precise formula for the error term. The formula is derived at length in Chapter 7 of [51].

The Riemann–Siegel Formula. *For all $t \in \mathbb{R}$,*

$$\begin{aligned}Z(t) = 2\sum_{n=1}^{N} &\left(\frac{\cos(\theta(t) - t\log n)}{n^{\frac{1}{2}}}\right) \\ &+ \frac{e^{-i\theta(t)} e^{-\frac{t\pi}{2}}}{(2\pi)^{\frac{1}{2}+it} e^{-\frac{i\pi}{4}} (1 - ie^{-t\pi})} \int_{C_N} \frac{(-x)^{-\frac{1}{2}+it} e^{-Nx} dx}{e^x - 1},\end{aligned}$$

where C_N is a positively oriented closed contour containing all of the points $\pm 2\pi i N$, $\pm 2\pi i(N-1), \ldots,$ $\pm 2\pi i$, and 0.

In practice, algorithms use the truncated series, and a numerical approximation to the integral, in order to realize a useful approximation to $Z(t)$ and thus to $\zeta(\frac{1}{2} + it)$. Note that the term "Riemann–Siegel formula" can refer to the above formula, or to the approximate functional equation given at the start of the section, depending on the source.

3.5 Gram's Law

Gram's Law, which isn't a law, is a heuristics for the location of zeros on the critical line. We have that $Z(t) = e^{i\theta(t)}\zeta\left(\frac{1}{2} + it\right)$, from which it easily follows that

$$\zeta\left(\frac{1}{2} + it\right) = e^{-i\theta(t)} Z(t) = Z(t)\cos\theta(t) - iZ(t)\sin\theta(t).$$

Thus,

$$\Im\left(\zeta\left(\frac{1}{2} + it\right)\right) = -Z(t)\sin\theta(t).$$

The sign changes of $\zeta(\frac{1}{2} + it)$ depend on the sign changes of $Z(t)$ and $\sin\theta(t)$. Gram showed that we can find the zeros of $\sin\theta(t)$ relatively easily; we call these points "Gram points." We define the nth Gram point, g_n, to be the unique real number satisfying $\theta(g_n) = n\pi$. This definition leads us to the formulation of Gram's law.

Gram's Law. *Hardy's function $Z(t)$ satisfies $(-1)^n Z(g_n) > 0$ at the Gram points g_n.*

Although named Gram's law, this statement was originally formulated by Hutchinson [76], and is frequently broken (although for moderate values of t exceptions are surprisingly few).

Gram points that conform to Gram's law are called "good," and those that do not are called "bad." We give the definition of Gram blocks first formulated by Rosser et al. [137]. A Gram block of length k is an interval $B_j = [g_j, g_{j+k})$ such that g_j and g_{j+k} are good Gram points and $g_{j+1}, \dots, g_{j+k-1}$ are bad Gram points. These blocks were introduced to deal with exceptions to Gram's law [26]. This motivates the formulation of Rosser's rule:

Rosser's Rule. *The Gram block B_m satisfies Rosser's rule if $Z(t)$ has at least k zeros in $B_m = [g_m, g_{m+k})$.*

Rosser's rule was proven to fail infinitely often by Lehman in 1970 [93]. However, it still provides a useful tool for counting zeros of $\zeta(s)$ on the critical line. Brent, in [26], gives the following result.

Theorem 3.3. *If K consecutive Gram blocks with union $[g_n, g_p)$ satisfy Rosser's rule, where*

$$K \geq 0.0061 \log^2(g_p) + 0.08 \log(g_p),$$

then

$$N(g_n) \leq n + 1 \quad \text{and} \quad p + 1 \leq N(g_p).$$

These rules lead to computationally feasible methods of verifying the Riemann hypothesis up to a desired height.

3.6 Turing

Alan Turing [157] presented an algorithm that gave higher-precision calculations of the zeta function. However, this development was made redundant with better estimates for the error terms in the Riemann–Siegel formula [126].

Turing also developed a method for calculating $N(T)$ that is much better than Backlund's method outlined in Section 3.2. Turing's method only requires information about the behavior of $\zeta(s)$ on the critical line.

Let g_n denote the nth Gram point, and suppose g_n is a Gram point at which Gram's law fails. So $(-1)^n Z(g_n) > 0$ fails. We want to find a point close to g_n, $g_n + h_n$ such that $(-1)^n Z(g_n + h_n) > 0$. Turing showed that if

$h_m = 0$ and if the values of h_n for n near m are not too large, then $N(g_m)$ must take the value $N(g_m) = m + 1$ [51].

Let $S(N)$ denote the error in the approximation $N(T) \sim \pi^{-1}\theta(T) + 1$, so $S(N) = N(T) - \pi^{-1}\theta(T) - 1$. Then from Turing's results, in order to prove that $S(g_m) = 0$ one need prove only that $-2 < S(g_m) < 2$, since $S(g_m)$ is an even integer [51, p.173].

3.7 The Odlyzko–Schönhage Algorithm

The Odlyzko–Schönhage algorithm is currently the most efficient algorithm for determining values $t \in \mathbb{R}$ for which $\zeta(\frac{1}{2} + it) = 0$. It uses the fact that the most computationally complex part of the evaluation using the Riemann–Siegel formula are computations of the form

$$g(t) = \sum_{k=1}^{M} k^{-it}.$$

The algorithm employs the fast fourier transform (FFT) to convert sums of this type into rational functions. The authors also give a fast algorithm for computing rational function values. The details are found in [126]. The algorithm presented computes the first n zeros of $\zeta(\frac{1}{2} + it)$ in $O(n^{1+\varepsilon})$ (as opposed to $O(n^{\frac{3}{2}})$ using previous methods).

3.8 A Simple Algorithm for the Zeta Function

In the previous section the algorithms are specialized for computations on the critical line. The following algorithm, combined with the reflection formula, allows for computation of $\zeta(s)$ in the entire complex plane. Recall that

$$\eta(s) := \sum_{n=1}^{\infty} \frac{(-1)^{n-1}}{n^s} = (1 - 2^{1-s})\zeta(s).$$

Algorithm 1. *Let*

$$d_k := n \sum_{i=0}^{k} \frac{(n+i-1)!4^i}{(n-i)!(2i)!}.$$

Then

$$\zeta(s) = \frac{-1}{d_n(1 - 2^{1-s})} \sum_{k=0}^{n-1} \frac{(-1)^k(d_k - d_n)}{(k+1)^s} + \gamma_n(s),$$

where for $s = \sigma + it$ *with* $\sigma \geq \frac{1}{2}$,

$$|\gamma_n(s)| \leq \frac{2}{(3+\sqrt{8})^n} \frac{1}{|\Gamma(s)|} \frac{1}{|(1-2^{1-s})|} \leq \frac{3}{(3+\sqrt{8})^n} \frac{(1+2|t|)e^{\frac{|t|\pi}{2}}}{|(1-2^{1-s})|}.$$

The proof of this algorithm relies on the properties of Chebyshev polynomials and can be found in [24].

3.9 Further Reading

A detailed exposition on strategies for evaluating $\zeta(s)$ can be found in [23]. The authors consider the advantages of several algorithms, and several variations of those described above. The material includes time and space complexity benchmarks for specific algorithms and the more general ideas/context from which these algorithms arise.

4

Empirical Evidence

It would be very discouraging if somewhere down the line you could ask a computer if the Riemann Hypothesis is correct and it said, "Yes, it is true, but you won't be able to understand the proof" [75].

Ron Graham

The Riemann hypothesis has endured for more than a century as a widely believed conjecture. There are many reasons why it has endured, and captured the imagination of mathematicians worldwide. In this chapter we will explore the most direct form of evidence for the Riemann hypothesis: empirical evidence. Arguments for the Riemann hypothesis often include its widespread ramifications and appeals to mathematical beauty; however, we also have a large corpus of hard facts. With the advent of powerful computational tools over the last century, mathematicians have increasingly turned to computational evidence to support conjectures, and the Riemann hypothesis is no exception. To date, ten trillion zeros of the Riemann zeta function have been calculated, and all conform to the Riemann hypothesis. It is, of course, impossible to verify the hypothesis through direct computation; however, computational evidence fulfills several objectives. First, and most obviously, it helps to convince mathematicians that the hypothesis is worth attempting to prove. It allows us to prove tangential results that are intricately connected to the Riemann hypothesis (for example, disproving the Mertens conjecture). Finally, numerical data allows us the opportunity to recognize other patterns, and develop new ideas with which to attack the hypothesis (for example, the Hilbert–Pólya conjecture, discussed in Section 4.3).

4.1 Verification in an Interval

In Chapter 2 we presented a thorough definition of the Riemann zeta function using analytic continuation. From that analysis it is clear that calculating zeros of the zeta function is not a simple undertaking. The tools needed to compute zeros are discussed in Chapter 3. However, in order to verify the Riemann hypothesis it is not sufficient merely to compute zeros. A rigorous verification of the hypothesis in a given interval can be effected in the following way [19].

First, we state Cauchy's residue theorem.

Theorem 4.1. *Let C be a positively oriented simple closed contour. If a function f is analytic inside and on C except for a finite number of singular points z_k $(k = 1, 2, \ldots, n)$ inside C, then*

$$\sum_{k=1}^{n} \mathrm{Res}_{z=z_k} f(z) = \frac{1}{2\pi i} \int_C f(z)dz,$$

where $\mathrm{Res}_{z=z_k} f(z)$ is the residue of $f(z)$ at the pole $z = z_k$.

This is a fundamental result in complex analysis and can be found in any text on the subject. For a proof of this theorem, see Section 63 of [32].

Now let $s = \frac{1}{2} + it$ for $t \in \mathbb{R}$, and let $N_1(T)$ be the number of zeros of $\zeta(s)$ in the rectangle R with vertices at $-1 - iT, 2 - iT, 2 + iT, -1 + iT$ (note that $N_1(T) = 2N(T)$, where $N(T)$ is as defined in the first section of Chapter 2). Then by Theorem 4.1 we have

$$N_1(T) - 1 = \frac{1}{2\pi i} \int_{\partial \mathcal{R}} -\frac{\zeta'(s)}{\zeta(s)} ds,$$

as long as T is not the imaginary part of a zero of $\zeta(s)$. The -1 term, which appears on the left-hand side of the equation, corresponds to the pole of $\zeta(s)$ located at $s = 1$.

Both $\zeta(s)$ and $\zeta'(s)$ can be computed to arbitrarily high precision, for example, using the Riemann–Siegel formula or the Euler–Maclaurin summation formula (see Section 3.4). The exact value for $N_1(T) - 1$ is obtained by dividing the integral by $2\pi i$ and rounding off to the nearest integer.

In his 1859 paper [133], Riemann introduced the function

$$\xi(t) = \frac{1}{2} s(s-1)\pi^{-s/2} \Gamma\left(\frac{s}{2}\right) \zeta(s),$$

which is analytic everywhere in $\mathbb{C}$ and has zeros with imaginary part between $-\frac{1}{2}$ and $\frac{1}{2}$ [133]. Riemann "conjectured" that all zeros α of $\xi(t)$ are real, that is, all nontrivial zeros of the Riemann zeta function are of the form $\frac{1}{2} + i\alpha$.

Let $\hat{t}$ be a real variable. Then, in view of the intermediate value theorem in calculus, $\xi(\hat{t})$ is continuous and real, so there will be a zero of odd order between any two points where the sign of $\xi(\hat{t})$ changes. If the number of sign changes of $\xi(\hat{t})$ in $[-T, T]$ equals $N_1(T)$, then all zeros of $\zeta(s)$ in $\mathcal{R}$ are simple and lie on the critical line. It has been verified that the first ten trillion zeros are simple and satisfy the Riemann hypothesis [57].

Conjecture 4.2. *All nontrivial zeros of $\zeta(s)$ are simple.*

All zeros of the Riemann zeta function are believed to be simple [38], though the falsity of this conjecture would not disprove the Riemann hypothesis. However, zeros of higher order would be problematic for many current computational techniques. For further details regarding Riemann hypothesis verification, see Sections 3.2, 3.5, and 3.6.

4.2 A Brief History of Computational Evidence

The first computations of zeros were performed by Riemann himself, prior to the presentation of his paper (see Section 12.2). These computations were never published, but they formed the basis of his famous conjecture. Siegel, in his study of Riemann's notes, uncovered both the fact that Riemann had performed some calculations of the values of $\zeta(s)$, and his brilliant method. The formula Riemann devised was published by Siegel in the 1930s, and it subsequently became known as the Riemann–Siegel formula (see Section 3.4). It has formed the basis of all large-scale computations of the Riemann zeta function [36, pp. 343–344].

The following table outlines the history of computational evidence supporting the Riemann hypothesis.

Year	Number of zeros	Computed by
1859 (approx.)	1	B. Riemann [51, p.159]
1903	15	J. P. Gram [58]
1914	79	R. J. Backlund [10]
1925	138	J. I. Hutchinson [76]
1935	1041	E. C. Titchmarsh [153]
1953	1,104	A. M. Turing [158]
1956	15,000	D. H. Lehmer [95]
1956	25,000	D. H. Lehmer [94]
1958	35,337	N. A. Meller [104]
1966	250,000	R. S. Lehman [92]
1968	3,500,000	J. B. Rosser, et al. [137]
1977	40,000,000	R. P. Brent [168]
1979	81,000,001	R. P. Brent [26]
1982	200,000,001	R. P. Brent, et al [27]
1983	300,000,001	J. van de Lune, H. J. J. te Riele [159]
1986	1,500,000,001	J. van de Lune, et al [160]
2001	10,000,000,000	J. van de Lune (unpublished)
2004	900,000,000,000	S. Wedeniwski [168]
2004	10,000,000,000,000	X. Gourdon [57]

Even a single exception to Riemann's conjecture would have enormously strange consequences for the distribution of prime numbers.... If the Riemann hypothesis turns out to be false, there will be huge oscillations in the distribution of primes. In an orchestra, that would be like one loud instrument that drowns out the others—an aesthetically distasteful situation [20].

Enrico Bombieri

This body of evidence seems impressive; however, there are reasons why one should be wary of computational evidence. The Riemann hypothesis is equivalent to the assertion that for real t, all local maxima of $\xi(t)$ are positive and all local minima of $\xi(t)$ are negative [19]. It has been suggested that a counterexample, if one exists, would be found in the neighborhood of unusually large peaks of $|\zeta(\frac{1}{2}+it)|$. Such peaks do occur, but only at very large heights [19]. This fact, among others, should discourage the reader from being persuaded solely by the numbers in the table above.

4.3 The Riemann Hypothesis and Random Matrices

How could it be that the Riemann zeta function so convincingly mimics a quantum system without being one [33] ?

M. Berry

One of the current ideas regarding the Riemann hypothesis is that the zeros of the zeta function can be interpreted as eigenvalues of certain matrices. This line of thinking is attractive and is potentially a good way to attack the hypothesis, since it gives a possible connection to physical phenomena.

The connection between the Riemann hypothesis and random matrices has its roots in the work of Hilbert and Pólya. Both, independently, enquired whether for zeros $\frac{1}{2}+i\gamma_j$ of $\zeta(s)$, the numbers γ_j belong to a set of eigenvalues of a Hermitian operator.

Not only would the truth of the conjecture indicate that all the numbers γ_j are real, but it would link the zeros to ideas from physics and gives a "natural" reason for all the nontrivial zeros to be on the critical line [17].

Empirical results indicate that the zeros of the Riemann zeta function are indeed distributed like the eigenvalues of certain matrix ensembles, in particular the Gaussian unitary ensemble. This suggests that random matrix theory might provide an avenue for the proof of the Riemann hypothesis.

From Hilbert and Pólya the idea of connecting the zeros of the zeta function to eigenvalues progressed in the work of Montgomery and Dyson.

Recall from Section 2.3 that the number of zeros of $\zeta(s)$ in the critical strip with $0 \leq \Im(s) < T$ is given by

$$N(T) = \frac{T}{2\pi} \log \frac{T}{2\pi} - \frac{T}{2\pi} + O(\log T). \tag{4.3.1}$$

Denote the imaginary parts of the first n zeros above the real axis by

$$\gamma_1 \leq \gamma_2 \leq \gamma_3 \leq \cdots \leq \gamma_n,$$

in order of increasing height (if two zeros are equal in their imaginary parts, then we order them according to their real parts in increasing order). It follows from (4.3.1) that

$$\gamma_n \sim \frac{2\pi n}{\log n}.$$

Under the Riemann hypothesis, Montgomery studied the gaps $\delta_j = \gamma_{j+1} - \gamma_j$ between consecutive zeros, first normalizing them to have mean length 1, asymptotically. The normalized gaps, denoted by $\hat{\delta}_j$, are given by

$$\hat{\delta}_j = \hat{\gamma}_{j+1} - \hat{\gamma}_j \,,$$

where

$$\hat{\gamma}_n = N(\gamma_n)\,.$$

The difference

$$\hat{\gamma}_{j+k} - \hat{\gamma}_j$$

is known as the kth consecutive spacing [147].

Let α and β be positive real numbers, with $\alpha < \beta$. Let $\hat{\gamma}_1 < \cdots < \hat{\gamma}_n$ be the nontrivial zeros of the Riemann zeta function along $\sigma = \frac{1}{2}$, normalized as above. Now let the number of kth consecutive spacings in $[\alpha, \beta]$ be given by

$$\frac{\#\{(j,k)\colon 1 \leq j \leq n, k \geq 0, (\hat{\gamma}_{j+k} - \hat{\gamma}_j) \in [\alpha, \beta]\}}{n}.$$

Montgomery conjectured in the early 1970s that the number of kth consecutive spacings in $[\alpha, \beta]$ behaves asymptotically as follows.

Conjecture 4.3 (Montgomery's Pair Correlation Conjecture).

$$\frac{\#\{(j,k)\colon 1 \leq j \leq n, k \geq 0, (\hat{\gamma}_{j+k} - \hat{\gamma}_j) \in [\alpha, \beta]\}}{n} \sim \int_\alpha^\beta \left(1 - \frac{\sin^2(\pi x)}{(\pi x)^2}\right) dx,$$

as $n \to \infty$ [147, 107].

This conjecture is based on results that are provable assuming the Riemann hypothesis, as well as on conjectures for the distribution of twin primes and other prime pairs [36].

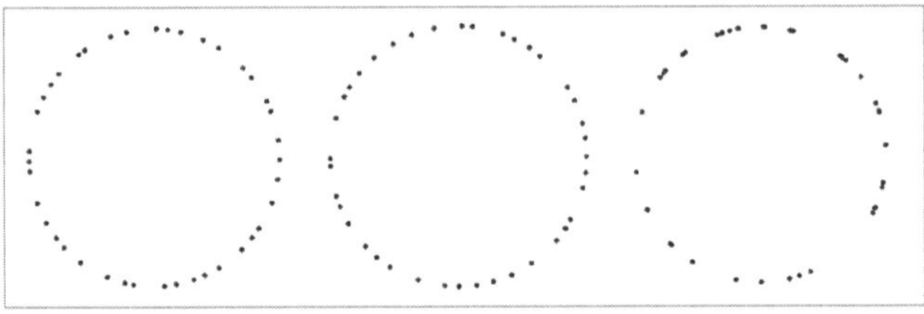

Fig. 4.1. The eigenvalues of a random 40×40 unitary matrix, 40 consecutive zeros of $\zeta(s)$ scaled to wrap once around the circle, and 40 randomly chosen points on the unit circle. Figure from [36] (see Section 11.3).

The function

$$1 - \frac{\sin^2(\pi x)}{(\pi x)^2}$$

is called the *pair correlation function* for zeros of the Riemann zeta function [147].

Part of the folklore surrounding the Riemann hypothesis and random matrices is a meeting between the physicist Freeman Dyson and the mathematician Hugh Montgomery at Princeton [138]. Montgomery showed Dyson the pair correlation function for the zeta function, and the latter recognized that it is also the pair correlation function for suitably normalized eigenvalues in a Gaussian unitary ensemble. For a discussion of the theory of random matrices, including the theory of Gaussian unitary ensembles, see Mehta [103].

Dyson's insight prompted Montgomery to reformulate his conjecture as follows.

Conjecture 4.4 (Montgomery–Odlyzko Law). *The distribution of spacings between nontrivial zeros of the Riemann zeta function is statistically identical to the distribution of eigenvalue spacings in a Gaussian unitary ensemble [84].*

In the 1980s, Odlyzko verified the Riemann hypothesis in large intervals around the 10^{20}th zero of the zeta function [36, 126]. Using the normalization

$$\hat{\delta}_n = (\gamma_{n+1} - \gamma_n)\frac{\log(\gamma_n/2\pi)}{2\pi},$$

Odlyzko calculated the pair correlation for zeros in these intervals (see [122] for Odlyzko's results and other computations).

Figure 4.2 is a plot of the correlation function (solid line) with a superimposed scatterplot of empirical data.

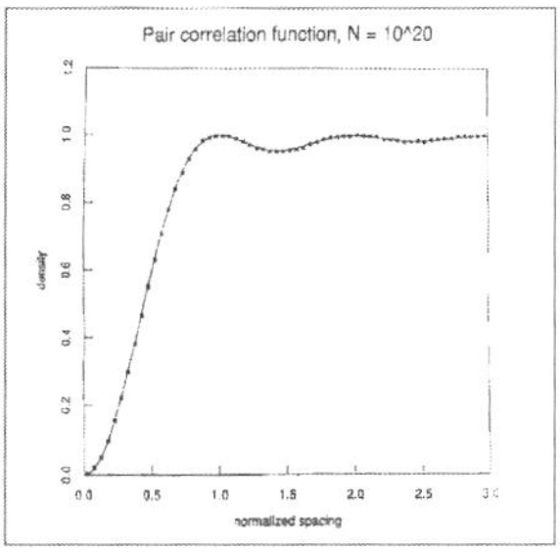

Fig. 4.2. The pair correlation function for GUE (solid) and for 8×10^6 zeros near the 10^{20}th zero of $\zeta(s)$ above the real axis. Figure from [36] (see Section 11.3).

Repulsion between successive zeros is predicted by Montgomery's conjecture. However, some pairs of zeros occur very close together. The zeros at $\frac{1}{2} + (7005 + t)i$ for $t \approx 0.06286617$ and $t \approx 0.1005646$ form one such pair. The appearance of closely spaced nontrivial zeros is known as *Lehmer's phenomenon.*

4.4 The Skewes Number

In Section 4.2, we mentioned that there are reasons to be wary of computational evidence. The Skewes number is a good example of the potentially misleading nature of numerical data, and serves as another warning to those who would be convinced by numbers alone. The Skewes number is the smallest positive integer n such that $\pi(n) < \mathrm{Li}(n)$ fails.

The prime number theorem gives

$$\pi(n) \sim \mathrm{Li}(n),$$

as $n \to \infty$. Preliminary calculations suggested that $\pi(n) < \mathrm{Li}(n)$ for all n, and such was conjectured by many mathematicians including Riemann and Gauss.

This conjecture was refuted in 1912 by Littlewood, who proved that the inequality fails for some n [63]. Two years later, Littlewood proved that in fact, the inequality fails for infinitely many n.

We have seen the strong connection between the Riemann hypothesis and the prime number theorem. As a consequence of this connection, bounds on the Skewes number depend on the validity of the Riemann hypothesis. The following bounds all require the assumption of the Riemann hypothesis. In 1933, Skewes gave the upper bound $n \leq 10^{10^{10^{34}}}$ [145]. In 1966, Lehman showed that there are more than 10^{500} successive integers between 1.53×10^{1165} and

1.65×10^{1165} for which the inequality fails [91]. Using better approximations to the zeros of $\zeta(s)$, te Riele in 1987 showed that there are at least 10^{180} successive integers in $[6.627\times 10^{370}, 6.687\times 10^{370}]$ for which the inequality fails [151]. This was improved to 1.39×10^{316} by Bays and Hudson in 2000 [16]. Skewes also found a bound that does not require the Riemann hypothesis.

First Skewes makes the following hypothesis:

Hypothesis H. *Every complex zero $s = \sigma + it$ of the Riemann zeta function satisfies*

$$\sigma - \frac{1}{2} \leq X^{-3} \log^{-2}(X)$$

provided that $|t| < X^3$.

Note that hypothesis H is weaker than the Riemann hypothesis, so the negation of hypothesis H is stronger than the negation of the Riemann hypothesis. In 1955, assuming the negation of hypothesis H, Skewes computed the upper bound $n \leq 10^{10^{10^{10^3}}}$ [146].

The best current bounds on the Skewes numbers lie well beyond the limits of our computing power. It is entirely possible that a counterexample to the Riemann hypothesis lies beyond our computational limits, and hence that our efforts will inevitably and unfailingly generate positive evidence for a false conjecture.

> *So for all practical purposes, the Riemann zeta function does not show its true colours in the range available by numerical investigations. You should go up to the height 10^{10000} then I would be much more convinced if things were still pointing strongly in the direction of the Riemann hypothesis. So numerical calculations are certainly very impressive, and they are a triumph of computers and numerical analysis, but they are of limited capacity. The Riemann hypothesis is a very delicate mechanism. It works as far as we know for all existing zeros, but we cannot, of course, verify numerically an infinity of zeros, so other theoretical ways of approach must be found, and for the time being they are insufficient to yield any positive conclusion [138].*

Aleksandar Ivić

5

Equivalent Statements

Riemann's insight was that the frequencies of the basic waveforms that approximate the psi function are determined by the places where the zeta function is equal to zero .... To me, that the distribution of prime numbers can be so accurately represented in a harmonic analysis is absolutely amazing and incredibly beautiful. It tells of an arcane music and a secret harmony composed by the prime numbers [20].

Enrico Bombieri

...there is a sense in which we can give a one-line non technical statement of the Riemann Hypothesis: "The primes have music in them" [14].

M. Berry and J. P. Keating

In this chapter we discuss several statements that are equivalent to the Riemann hypothesis. By restating the Riemann hypothesis in different language, and in entirely different disciplines, we gain more possible avenues of attack. We group these equivalences into three categories: equivalences that are entirely number-theoretic, equivalences that are closely related to the analytic properties of the zeta function and other functions, and equivalences that are truly cross-disciplinary. These equivalences range from old to relatively new, from central to arcane, and from deceptively simple to staggeringly complex.

5.1 Number-Theoretic Equivalences

Number-theoretic equivalences of the Riemann hypothesis provide a natural method of explaining the hypothesis to nonmathematicians without appealing to complex analysis. While it is unlikely that any of these equivalences will lead directly to a solution, they provide a sense of how intricately the Riemann zeta function is tied to the primes. We begin by repeating our definition of the Liouville function.

Definition 5.1. *The Liouville function is defined by*

$$\lambda(n) := (-1)^{\omega(n)},$$

where $\omega(n)$ is the number of, not necessarily distinct, prime factors of n, counted with multiplicity.

This leads to an equivalence that can be roughly stated in terms of probability. Namely, the Riemann hypothesis is equivalent to the statement that an integer has an equal probability of having an odd number or an even number of distinct prime factors.

Equivalence 5.2. *The Riemann hypothesis is equivalent to*

$$\lambda(1) + \lambda(2) + \cdots + \lambda(n) \ll n^{1/2+\varepsilon},$$

for every positive ε.

See Section 1.2 for further discussion of this equivalence.

The next equivalence we discuss has a long history. Recall that the prime number theorem states that $\pi(x) \sim \mathrm{Li}(x)$, where $\mathrm{Li}(x)$ is the logarithmic integral, defined as follows.

Definition 5.3. *The logarithmic integral,* Li, *of x is defined as*

$$\mathrm{Li}(x) := \int_2^x \frac{dt}{\log t}.$$

We also define $\pi(x)$, the prime counting function.

Definition 5.4. *The prime counting function, denoted by $\pi(x)$, is the number of primes less than or equal to a real number x.*

The prime number theorem, in the form $\pi(x) \sim \mathrm{Li}(x)$, was conjectured by Gauss in 1849, in a letter to Enke [55]. Gauss made his conjecture on the basis of his calculations (done by hand) of $\pi(x)$ and contemporary tables of values of $\mathrm{Li}(x)$. A short table of his findings appears in his letter to Enke:

Table 5.1. Table of Gauss's calculations of $\pi(x)$ from his letter to Enke [55].

x	Count of primes $< x$	Integral $\int \frac{dn}{\log n}$	Error
500,000	41,556	41,606.4	+50.4
1,000,000	78,501	79,627.5	+126.5
1,500,000	114,112	114,263.1	+151.1
2,000,000	148,883	149,054.8	+171.8
2,500,000	183,016	183,245.0	+229.0
3,000,000	216,745	216,970.6	+225.6

The prime number theorem, proved independently by Hadamard (see Section 12.3) and de la Vallée Poussin (see Section 12.4), requires showing that $\zeta(s) \neq 0$ when $\Re(s) = 1$. However, the truth of the Riemann hypothesis would give us a precise asymptotic estimation to the error in the prime number theorem.

Equivalence 5.5. *The assertion that*

$$\pi(x) = \mathrm{Li}(x) + O(\sqrt{x} \log x)$$

is equivalent to the Riemann hypothesis [19].

The next equivalence involves Mertens' function, for which we will need the Möbius function.

Definition 5.6. *The Möbius function, $\mu(n)$, is defined in the following way:*

$$\mu(n) := \begin{cases} 0 & \text{if } n \text{ has a square factor,} \\ 1 & \text{if } n = 1, \\ (-1)^k & \text{if } n \text{ is a product of } k \text{ distinct primes.} \end{cases}$$

We now define the Mertens function as follows.

Definition 5.7. *The Mertens function, denoted by $M(x)$, is defined by*

$$M(x) := \sum_{n \leq x} \mu(n),$$

where x is real [155, p.370].

In terms of the Mertens function we have the following equivalence.

Equivalence 5.8. *The Riemann hypothesis is equivalent to*

$$M(x) = O(x^{\frac{1}{2}+\varepsilon})$$

for every $\varepsilon > 0$ [155].

Proving that $M(x) = O(x^{\frac{1}{2}+\varepsilon})$ for some $\varepsilon < \frac{1}{2}$ is an open problem and would be an impressive result.

We can also reformulate the Riemann hypothesis in terms of the sum of divisors function. The sum of divisors function is defined as follows.

Definition 5.9. *For $n \in \mathbb{N}$,*

$$\sigma(n) := \sum_{d|n} d.$$

The following equivalence is due to Robin.

Equivalence 5.10. *The Riemann hypothesis is equivalent to the statement that for all* $n > 5040$,

$$\sigma(n) < e^{\gamma} n \log \log n,$$

where γ *is Euler's constant [136].*

Robin also showed, unconditionally, that

$$\sigma(n) < e^{\gamma} n \log \log n + 0.6482 \frac{n}{\log \log n}$$

for all $n \geq 3$.

Building on Robin's work, Lagarias proved another equivalence to the Riemann hypothesis that involves the sum of divisors function.

Equivalence 5.11. *The following statement is equivalent to the Riemann hypothesis:*

$$\sigma(n) \leq H_n + \exp(H_n) \log(H_n),$$

for all $n \geq 1$, *with equality only for* $n = 1$ *[89].*

Here H_n is the nth harmonic number defined as follows.

Definition 5.12. *The* n*th harmonic number is given by*

$$H_n := \sum_{j=1}^{n} \frac{1}{j}.$$

Of particular interest is the fact that Lagarias' equivalence holds for all n, not just asymptotically. Eric Rains verified the inequality for $1 \leq n \leq 5040$.

Using the Mertens function we can recast an earlier equivalence in terms of Farey series. The term Farey series is entirely misleading, since they were not discovered by Farey, nor are they series. Farey series were in fact discovered by Haros, and are defined as follows.

Definition 5.13. *The Farey series* F_n *of order* n *is the set of rationals* $\frac{a}{b}$ *with* $0 \leq a \leq b \leq n$ *and* $(a, b) = 1$, *arranged in increasing order [51].*

For example, the Farey series of order 3 is given by

$$F_3 = \left\{ \frac{0}{1}, \frac{1}{3}, \frac{1}{2}, \frac{2}{3}, \frac{1}{1} \right\}.$$

We denote the jth term of F_n by $F_n(j)$. It is easy to see that the number, m, of terms in the Farey series of order n is $m = 1 + \sum_{j=1}^{n} \phi(j)$, where $\phi(j)$ is the Euler totient function. Now we can formulate the following equivalence.

Equivalence 5.14. *The Riemann hypothesis is equivalent to*

$$\sum_{j=1}^{m} \left| F_n(j) - \frac{j}{m} \right| = O(n^{\frac{1}{2}+\varepsilon}),$$

where $\varepsilon > 0$ *and* $m = \#\{F_n\}$ *[51].*

5.2 Analytic Equivalences

In this section we explore various equivalences that are connected to the Riemann zeta function analytically. These problems vary from direct consequences of the definition of the zeta function to more abstruse reformulations. Generally, these problems are classical, and have arisen in the course of research into the properties of $\zeta(s)$.

We begin by restating the Riemann hypothesis in terms of the Dirichlet eta function, or the alternating zeta function.

Equivalence 5.15. *The Riemann hypothesis is equivalent to the statement that all zeros of the Dirichlet eta function*

$$\eta(s) := \sum_{k=1}^{\infty} \frac{(-1)^{k-1}}{k^s} = (1 - 2^{1-s})\zeta(s)$$

that fall in the critical strip $0 < \Re(s) < 1$ *lie on the critical line* $\Re(s) = \frac{1}{2}$.

We can also consider the convergence of $1/\zeta(s)$ and the values of the derivative, $\zeta'(s)$.

Equivalence 5.16. *The convergence of*

$$\sum_{n=1}^{\infty} \frac{\mu(n)}{n^s}$$

for $\Re(s) > \frac{1}{2}$ *is necessary and sufficient for the Riemann hypothesis [155, pp.369–370]. Note that for* $\Re(s) > 1$,

$$\frac{1}{\zeta(s)} = \sum_{n=1}^{\infty} \frac{\mu(n)}{n^s}.$$

Equivalence 5.17. *The Riemann hypothesis is equivalent to the nonvanishing of* $\zeta'(s)$ *in the region* $0 < \Re(s) < \frac{1}{2}$ *[149].*

Recall from Chapter 2 that in deriving the functional equation for $\zeta(s)$ we defined the ξ function,

$$\xi(s) := \frac{s}{2}(s-1)\pi^{-\frac{s}{2}}\Gamma\left(\frac{s}{2}\right)\zeta(s).$$

Lagarias derived the following criterion on $\xi(s)$ [88].

Equivalence 5.18. *The Riemann hypothesis is equivalent to*

$$\Re\left(\frac{\xi'(s)}{\xi(s)}\right) > 0.$$

Continuing to work with $\xi(s)$, we define λ_n as

$$\lambda_n := \frac{1}{(n-1)!}\frac{d^n}{ds^n}(s^{n-1}\log\xi(s)).$$

Now we can state an equivalence relating the Riemann hypothesis to the value of λ_n.

Equivalence 5.19. *The Riemann hypothesis is equivalent to the nonnegativity of λ_n for all $n \geq 1$ [98].*

Since the Riemann hypothesis is a statement regarding zeros of the zeta function, it is no surprise that we can reformulate the hypothesis into other statements involving the zeros of $\zeta(s)$. The following two equivalences give examples of nontrivial reformulations that use these zeros.

Equivalence 5.20. *The Riemann hypothesis is equivalent to the following statement:*

$$\lambda_n\big|_{s=1} = \sum_{\rho}\left(1-\left(1-\frac{1}{\rho}\right)^n\right) > 0$$

for each $n = 1, 2, 3, \cdots$, where ρ runs over the complex zeros of $\zeta(s)$ [21].

Equivalence 5.21. *The Riemann hypothesis is true if and only if*

$$I := \int_{\frac{1}{2}-i\infty}^{\frac{1}{2}+i\infty} \frac{\log(|\zeta(s)|)}{|s|^2}dt = 0,$$

where $s = \sigma + it$ [14]. Moreover, this integral can be exactly expressed as the sum

$$I = 2\pi \sum_{\Re(\rho)>\frac{1}{2}} \log\left|\frac{\rho}{1-\rho}\right|,$$

where ρ is a zero of $\zeta(s)$ in the region indicated.

Hardy and Littlewood derived another equivalence to the Riemann hypothesis that sums values of $\zeta(s)$, evaluated at odd integers.

Equivalence 5.22. *The Riemann hypothesis holds if and only if*

$$\sum_{k=1}^{\infty} \frac{(-x)^k}{k!\zeta(2k+1)} = O(x^{-\frac{1}{4}}),$$

as $x \to \infty$ [65].

For the next equivalence we will need some definitions. We first define the Xi function, Ξ, as

$$\Xi(iz) := \frac{1}{2}\left(z^2 - \frac{1}{4}\right)\pi^{-z/2-\frac{1}{4}}\Gamma\left(\frac{1}{2}z + \frac{1}{4}\right)\zeta\left(z + \frac{1}{2}\right).$$

We note that $\Xi(\frac{z}{2})/8$ is the Fourier transform of the signal $\Phi(t)$ given by

$$\Phi(t) := \sum_{n=1}^{\infty}(2\pi^2 n^4 e^{9t} - 3\pi n^2 e^{5t})e^{-\pi n^2 e^{4t}},$$

for $t \in \mathbb{R}$ and $t \geq 0$. Now we consider the Fourier transform of $\Phi(t)e^{\lambda t^2}$, which we denote as $H(\lambda, z)$,

$$H(\lambda, z) := \mathcal{F}_t[\Phi(t)e^{\lambda t^2}](z),$$

where $\lambda \in \mathbb{R}$ and $z \in \mathbb{C}$. We can see that $H(0, z) = \Xi(\frac{z}{2})/8$. It was proven in 1950 by de Bruijn that $H(\lambda z)$ has only real zeros for $\lambda \geq \frac{1}{2}$. Furthermore, it is known that there exists a constant Λ such that if $H(\lambda, z) = 0$, then z is real if and only if $\lambda \geq \Lambda$. The value Λ is called the de Bruijn–Newman constant. Now we can state the Riemann hypothesis in terms of the de Bruijn–Newman constant.

Equivalence 5.23. *The Riemann hypothesis is equivalent to the conjecture that $\Lambda \leq 0$ [41, 40].*

There are several lower bounds on Λ, the best at this time is $-2.7 \cdot 10^{-9} < \Lambda$, proven by Odlyzko in 2000 [125]. It had been conjectured by C. M. Newman that Λ satisfies $\Lambda \geq 0$ [112].

Salem, in 1953, gave the following criterion for the truth of the Riemann hypothesis.

Equivalence 5.24. *The Riemann hypothesis holds if and only if the integral equation*

$$\int_{-\infty}^{\infty} \frac{e^{-\sigma y}\varphi(y)}{e^{e^{x-y}} + 1} dy = 0$$

has no bounded solution $\varphi(y)$ other than the trivial solution $\varphi(y) = 0$, for $\frac{1}{2} < \sigma < 1$ [140].

The following elegant equivalence, due to Volchkov, connects the zeros of the Riemann zeta function to Euler's constant γ.

Equivalence 5.25. *The Riemann hypothesis is equivalent to*

$$\int_0^\infty (1-12t^2)(1+4t^2)^{-3} \int_{1/2}^\infty \log|\zeta(\sigma+it)|\, d\sigma\, dt = \pi(3-\gamma)/32,$$

where γ is Euler's constant [165].

Julio Alacántara–Bode, building on the work of Beurling, reformulated the Riemann hypothesis using the Hilbert–Schmidt integral operator.

The Hilbert–Schmidt integral operator A on $L^2(0,1)$ is given by

$$[Af](\theta) := \int_0^1 f(x) \left\{\frac{\theta}{x}\right\} dx,$$

where $\{x\}$ is the fractional part of x. Now the Riemann hypothesis can be stated as follows.

Equivalence 5.26. *The Riemann hypothesis is true if and only if the Hilbert–Schmidt operator A is injective [5].*

We can also restate the Riemann hypothesis in terms of ergodic theory. A theorem of Dani gives that for each $y > 0$, there exist ergodic measures, $m(y)$, on the space $M = \mathrm{PSL}_2(\mathbb{Z})/\mathrm{PSL}_2(\mathbb{R})$, supported on closed orbits of period $1/y$ of the horocyclic flow.

Equivalence 5.27. *The Riemann hypothesis is equivalent to the statement that for any smooth function f on M,*

$$\int_M f\, dm(y) = o(y^{\frac{3}{4}-\varepsilon})$$

for any $\varepsilon > 0$ as $y \to 0$ [162].

5.3 Other Equivalences

At first glance, the remaining equivalences seem to have little to do with the zeros of a complex-valued function.

We begin by defining the Redheffer matrix of order n. The $n \times n$ Redheffer matrix, $R_n := [R_n(i,j)]$, is defined by

$$R_n(i,j) = \begin{cases} 1 & \text{if } j = 1 \text{ or if } i \mid j, \\ 0 & \text{otherwise.} \end{cases}$$

It can easily be shown that $\det R_n = \sum_{k=1}^n \mu(k)$. We can simply use this to state the Riemann hypothesis by employing a previous equivalence.

Equivalence 5.28. *The Riemann hypothesis is true if and only if*

$$\det(R_n) = O(n^{\frac{1}{2}+\varepsilon})$$

for any $\varepsilon > 0$.

This was shown by Redheffer in [132]. Now we can use the Redheffer matrices to translate the Riemann hypothesis into the language of graph theory. We let R_n be the Redheffer matrix as defined above, and set $B_n := R_n - I_n$ (where I_n is the $n \times n$ identity matrix). Now we let G_n be the directed graph whose adjacency matrix is B_n. Finally we let the graph $\overline{G}_n$ be the graph obtained from G_n by adding a loop at node 1 of G_n. Now we can restate the Riemann hypothesis in terms of the cycles of $\overline{G}_n$.

Equivalence 5.29. *The following statement is equivalent to the Riemann hypothesis:*

$$|\#\{\textit{odd cycles in } \overline{G}_n\} - \#\{\textit{even cycles in } \overline{G}_n\}| = O(n^{\frac{1}{2}+\varepsilon})$$

for any $\varepsilon > 0$ *[15].*

The Nyman–Beurling equivalent form translates the Riemann hypothesis into a statement about the span of a set of functions.

Equivalence 5.30. *The closed linear span of* $\{\rho_\alpha(t) : 0 < \alpha < 1\}$ *is* $L^2(0,1)$ *if and only if the Riemann hypothesis is true, where*

$$\rho_\alpha(t) := \left\{\frac{\alpha}{t}\right\} - \alpha\left\{\frac{1}{t}\right\},$$

$\{x\}$ *is the fractional part of* x, *and* $L^2(0,1)$ *is the space of square integrable functions on* $(0,1)$ *[13].*

We can also reformulate the Riemann hypothesis in terms of a problem regarding the order of group elements.

Equivalence 5.31. *The Riemann hypothesis is equivalent to the statement that for sufficiently large* n,

$$\log g(n) < \frac{1}{\sqrt{\mathrm{Li}(n)}},$$

where $g(n)$ *is the maximal order of elements of the symmetric group* S_n *of degree* n *[102].*

Finally, a problem about the rate of convergence of certain discrete measures is equivalent to the Riemann hypothesis. Denote the nonzero reals by $\mathbb{R}^*$, and let $C_c^r(\mathbb{R}^*)$ be the set of all functions $f\colon \mathbb{R}^* \to \mathbb{C}$ that are r times differentiable and have compact support. For $y \in \mathbb{R}^*$ and for $f \in C_c^2(R^*)$, define

$$m_y(f) := \sum_{n \in \mathbb{N}} y\phi(n) f(y^{1/2}n)\,,$$

with ϕ the Euler totient function. Let

$$m_0(f) := \int_0^\infty \left(\frac{6}{\pi^2}\right) u f(u)\, du\,.$$

Equivalence 5.32. *The Riemann hypothesis is true if and only if* $m_y(f) = m_0(f) + o(y^{\frac{3}{4}-\varepsilon})$ *as* $y \to 0$, *for every function* $f \in C_c^2(\mathbb{R}^*)$ *and every* $\varepsilon > 0$ *[162].*

6

Extensions of the Riemann Hypothesis

It is common in mathematics to generalize hard problems to even more difficult ones. Sometimes this is productive and sometimes not. However, in the Riemann hypothesis case this is a valuable enterprise that allows us to see a larger picture. These stronger forms of the Riemann hypothesis allow the proof of many conditional results that appear to be plausible (see Sections 11.1 and 11.2) and add to the heuristic evidence for the Riemann hypothesis.

We begin by once again stating the Riemann hypothesis in its classical form. We then give a variety of extensions to other related problems. There is some variation in the literature on these problems as to names and abbreviations. We attempt to use standard notation as reflected in the source material; however, this is not always possible.

6.1 The Riemann Hypothesis

A usual formulation of the problem is as follows (see [19]):

Conjecture 6.1 (The Riemann Hypothesis). *The nontrivial zeros of $\zeta(s)$ have real part equal to $\frac{1}{2}$.*

The fundamental object under consideration is $\zeta(s)$. As before, the Riemann zeta function, $\zeta(s)$, is the function of the complex variable s, defined in the half-plane $\Re(s) > 1$ by the absolutely convergent series

$$\zeta(s) := \sum_{n=1}^{\infty} \frac{1}{n^s},$$

and in the whole complex plane except $s = 1$ by analytic continuation (see Chapter 2).

This statement of the problem can be simplified by introducing the Dirichlet eta function, also known as the alternating zeta function.

Definition 6.2. *The Dirichlet eta function is defined as*

$$\eta(s) := \sum_{k=1}^{\infty} \frac{(-1)^{k-1}}{k^s} = (1 - 2^{1-s})\zeta(s).$$

Since $\eta(s)$ converges for all $s \in \mathbb{C}$ with $\Re(s) > 0$, one need not consider analytic continuation. The Riemann hypothesis is true if and only if the zeros of $\eta(s)$ in the strip $0 < \Re(s) < 1$ all lie on the line $\Re(s) = \frac{1}{2}$. This equivalence, and other elementary equivalences, are discussed further in Chapter 5.

6.2 The Generalized Riemann Hypothesis

The Riemann zeta function is the object considered by the Riemann hypothesis. However, $\zeta(s)$ is a special case within a larger, more general class of functions. The generalized Riemann hypothesis considers this larger class of functions, the class of Dirichlet L-functions [19]. We will build these functions in the same way we built the Riemann zeta function in Chapter 2. We define a Dirichlet L-series as

$$L(s, \chi_k) := \sum_{n=1}^{\infty} \chi_k(n) n^{-s},$$

where $\chi_k(n)$ is a number-theoretic character modulo k, defined by the properties

$$\begin{aligned} \chi_k(1) &= 1, \\ \chi_k(n) &= \chi_k(n+k), \\ \chi_k(m)\chi_k(n) &= \chi_k(mn), \end{aligned}$$

for all integers m and n, and

$$\chi_k(n) = 0$$

for $(k, n) \neq 1$. A character modulo k is said to be *primitive* if no $\chi_d(n)$ exists such that $\chi_k(n) = \chi_d(n)$ where d divides k and $d \neq k$. The unique $\chi_k(n)$ such that $\chi_k(n) = 1$ for all n with $(k, n) = 1$ is called the principal character modulo k.

A Dirichlet L-function, $L(s, \chi_k)$, is the analytic continuation of the associated Dirichlet L-series. We can see that $\zeta(s)$ belongs in this class of functions, since

$$L(s, \chi_1) = \sum_{n=1}^{\infty} \chi_1(n) n^{-s} = \sum_{n=1}^{\infty} n^{-s} = \zeta(s).$$

We now state the generalized Riemann hypothesis as follows.

Extension 6.3 (The Generalized Riemann Hypothesis). *All nontrivial zeros of $L(s,\chi_k)$ have real part equal to $\frac{1}{2}$ [31, p.3].*

Nontrivial in this case means $L(s,\chi_k)=0$ for $s\in\mathbb{C}$ such that $0<\Re(s)<1$.

We see that the generalized Riemann hypothesis implies the Riemann hypothesis, since $\zeta(s)$ is a member of the class of Dirichlet L-functions. We also note that, unlike $\zeta(s)$, $L(s,\chi_k)$ may have zeros on the line $\Im(s)=0$; however, all of these zeros are known and are considered to be trivial zeros. There is variation in the literature regarding the generalized versus the extended Riemann hypotheses. Also, the theory of global L-functions is rich and extends far beyond the points considered here (see [82] for a more thorough discussion).

6.3 The Extended Riemann Hypothesis

Let p be an odd prime. We define the Legendre symbol, $\left(\frac{n}{p}\right)$, as

$$\left(\frac{n}{p}\right)=\begin{cases}1, & \text{if the congruence } x^2\equiv n \pmod p \text{ has a solution,}\\ 0, & \text{if } p\mid n,\\ -1, & \text{if the congruence } x^2\equiv n \pmod p \text{ has no solution.}\end{cases}$$

Consider the series

$$L_p(s):=\sum_{n=1}^{\infty}\left(\frac{n}{p}\right)\frac{1}{n^s}.$$

The extended Riemann hypothesis considers the extension of $L_p(s)$ to the entire complex plane through analytic continuation.

Extension 6.4 (The Extended Riemann Hypothesis). *The zeros of $L_p(s)$ with $0<\Re(s)<1$ all lie on the line $\Re(s)=\frac{1}{2}$ [31, p.viii].*

Since $\left(\frac{\cdot}{\cdot}\right)$ is a specific example of a number-theoretic character, the extended Riemann hypothesis is an instance of the generalized Riemann hypothesis, presented in Section 6.2.

6.4 An Equivalent Extended Riemann Hypothesis

If we define $\pi(x;k,l)$ to be

$$\pi(x;k,l):=\#\left\{p\ :\ p\le x,\ p \text{ is prime, and } p\equiv l \pmod k\right\},$$

then an equivalent extended Riemann hypothesis is as follows.

Extension 6.5 (An Equivalent Extended Riemann Hypothesis). *For* $(k,l)=1$ *and* $\varepsilon>0$,

$$\pi(x;k,l)=\frac{\mathrm{Li}(x)}{\phi(k)}+O(x^{\frac{1}{2}+\varepsilon}),$$

where $\phi(k)$ *is Euler's totient function, and* $\mathrm{Li}(x):=\int_2^x \frac{dt}{\log t}$.

This statement can be found in [9]. The authors of [9] credit Titchmarsh (circa 1930).

6.5 Another Extended Riemann Hypothesis

Let K be a number field with ring of integers $\mathcal{O}_K$. The Dedekind zeta function of K is given by

$$\zeta_K(s):=\sum_{\mathfrak{a}} N(\mathfrak{a})^{-s}$$

for $\Re(s)>1$, where the sum is over all integral ideals of $\mathcal{O}_K$, and $N(\mathfrak{a})$ is the norm of $\mathfrak{a}$. Again we consider the extension of $\zeta_K(s)$ to the entire complex plane through analytic continuation.

Extension 6.6 (Another Extended Riemann Hypothesis). *All zeros of the Dedekind zeta function of any algebraic number field, with* $0<\Re(s)<1$, *lie on the line* $\Re(s)=\frac{1}{2}$.

This statement includes the Riemann hypothesis, since the Riemann zeta function is the Dedekind zeta function over the field of rational numbers [3].

6.6 The Grand Riemann Hypothesis

In order to state the grand Riemann hypothesis we need to define automorphic L-functions. This requires sophisticated tools, and we refer the interested reader to another source, such as [171], in order to develop the necessary background.

Let $\mathbb{A}$ be the ring of adeles of $\mathbb{Q}$ and let π be an automorphic cuspidal representation of the general linear group $\mathrm{GL}_m(\mathbb{A})$ with central character χ. The representation π is equivalent to $\bigotimes_v \pi_v$ with $v=\infty$ or $v=p$ and π_v an irreducible unitary representation of $\mathrm{GL}_m(\mathbb{Q}_v)$. For each prime and local representation π_p we define

$$L(s,\pi_p):=\prod_{j=1}^{m}\left(1-\alpha_{j,\pi}(p)p^{-s}\right)^{-1},$$

where the m complex parameters $\alpha_{j,\pi}(p)$ are determined by π_p. For $v = \infty$, π_∞ determines parameters $\mu_{j,\pi}(\infty)$ such that

$$L(s, \pi_\infty) := \prod_{j=1}^{m} \Gamma_{\mathbb{R}}(s - \mu_{j,\pi}(\infty)),$$

where

$$\Gamma_{\mathbb{R}}(s) = \pi^{-s/2}\Gamma(s/2).$$

The standard automorphic L-function associated with π is defined by

$$L(s, \pi) := \prod_{p} L(s, \pi_p).$$

Now if we define

$$\Lambda(s, \pi) := L(s, \pi_\infty)L(s, \pi),$$

then $\Lambda(s, \pi)$ is entire and satisfies the functional equation

$$\Lambda(s, \pi) = \epsilon_\pi N_\pi^{\frac{1}{2}-s}\Lambda(1 - s, \tilde{\pi}),$$

where $N_\pi \geq 1$ is an integer, ϵ_π is of modulus 1 and $\tilde{\pi}$ is the contragredient representation $\tilde{\pi}(g) = \pi({}^t g^{-1})$. General conjectures of Langlands assert that these standard L-functions multiplicatively generate all L-functions.

As a simple and explicit example of an automorphic L-function, consider the discriminant

$$\Delta(z) := e^{2\pi i z} \prod_{n=1}^{\infty} \left(1 - e^{2\pi i n z}\right)^{24} := \sum_{n=1}^{\infty} \tau(n) e^{2\pi i n z}.$$

Then $\Delta(z)$ is a meromorphic function in the upper half-plane satisfying the transformation rule

$$\Delta\left(\frac{az + b}{cz + d}\right) = (cz + d)^{12}\Delta(z)$$

for $a, b, c, d \in \mathbb{Z}$ satisfying $ad - bc = 1$. Associated to Δ, we let

$$L(s, \Delta) := \sum_{n=1}^{\infty} \frac{\tau(n)}{n^{11/2}} n^{-s} = \prod_{p} \left(1 - \frac{\tau(p)}{p^{11/2}} p^{-s} + p^{-2s}\right)^{-1}.$$

Now, $L(s, \Delta)$ is entire and satisfies the functional equation

$$\Lambda(s, \Delta) := \Gamma_{\mathbb{R}}\left(s + \frac{11}{2}\right)\Gamma_{\mathbb{R}}\left(s + \frac{13}{2}\right)L(s, \Delta) = \Lambda(1 - s, \Delta).$$

The grand Riemann hypothesis asserts that the zeros of $\Lambda(s, \pi)$ (and in particular, those of $\Lambda(s, \Delta)$) all lie on $\Re(s) = \frac{1}{2}$.

Extension 6.7 (Grand Riemann Hypothesis). *All of the zeros of any automorphic L-function with* $0 < \Re(s) < 1$ *lie on the line* $\Re(s) = \frac{1}{2}$.

Sarnak discusses the grand Riemann hypothesis in more detail in Section 11.2.

The objects considered by extensions of the Riemann hypothesis become more specialized and complicated. We refer the interested reader to more comprehensive resources for further material on these problems [82, 83].

7

Assuming the Riemann Hypothesis and Its Extensions ...

Right now, when we tackle problems without knowing the truth of the Riemann hypothesis, it's as if we have a screwdriver. But when we have it, it'll be more like a bulldozer [86].

Peter Sarnak

The consequences of a proof of the Riemann hypothesis in elementary number theory are apparent, as are the connections to applications such as cryptography. Aside from these considerations, a large body of theory has been built on the Riemann hypothesis. In this chapter we consider some of the more important statements that become true should a proof of the Riemann hypothesis be found.

7.1 Another Proof of The Prime Number Theorem

The prime number theorem was proved independently in 1896 by Hadamard and de la Vallée Poussin, it would be of course, also an immediate consequence of Riemann hypothesis.

Theorem 7.1 (The Prime Number Theorem). $\pi(n) \sim \mathrm{Li}(n)$ *as* $n \to \infty$.

The proof of the prime number theorem relies on showing that $\zeta(s)$ has no zeros of the form $1 + it$ for $t \in \mathbb{R}$. Thus it follows from the truth of the Riemann hypothesis.

It is interesting to note that Erdős and Selberg both found an "elementary" proof of the prime number theorem. Here elementary means that the proofs do not use any advanced complex analysis. These proofs appear in Sections 12.11 and 12.10 respectively. The simplest analytic proof we know of is due to Newman (included in Section 12.15) and is elegantly presented by Korevaar in Section 12.16.

7.2 Goldbach's Conjecture

In 1742, Goldbach conjectured, in a letter to Euler, that every natural number $n \geq 5$ is a sum of three prime numbers (Euler gave the formulation; every even number greater than 3 is a sum of two primes) [82]. This conjecture is referred to as Goldbach's strong conjecture and is one of the oldest unsolved problems in number theory. A related problem is Goldbach's weak conjecture.

Conjecture 7.2 (Goldbach's Weak Conjecture). *Every odd number greater* 7 *can be expressed as a sum of three odd primes.*

Hardy and Littlewood proved that the generalized Riemann hypothesis implies Goldbach's weak conjecture for sufficiently large n [66]. In 1937, Vinogradov gave the following result without assuming any variant of the Riemann hypothesis.

Theorem 7.3. *Every sufficiently large odd number $N \geq N_0$ is the sum of three prime numbers [163].*

The number N_0 is effectively computable. However, the best known N_0 is still far too large to be verified computationally. In 1997 Deshouillers, Effinger, te Riele, and Zinoviev proved the following result.

Theorem 7.4. *Assuming the generalized Riemann hypothesis, every odd number $n > 5$ can be expressed as a sum of three prime numbers [49] .*

7.3 More Goldbach

Hardy and Littlewood [67] proved that if the generalized Riemann hypothesis (see Section 6.2) is true, then almost all even numbers can be expressed as a sum of two primes. Specifically, if $E(N)$ denotes the number of even integers $n < N$ that are not a sum of two primes, then $E(N) = O(N^{\frac{1}{2}+\varepsilon})$ for any $\varepsilon > 0$.

Another important result along these lines was proven by J.J. Chen in 1973.

Theorem 7.5 (Chen's Theorem). *Every sufficiently large even integer is a sum of a prime and number which is a product of at most two primes [30].*

7.4 Primes in a Given Interval

A question of interest in the study of prime numbers concerns the existence of a prime p such that $a < p < b$ for given a and b. Bertrand's postulate of 1845 states that between a and $2a$ there is always a prime number (for $a > 1$). It is not difficult to prove the following under the assumption of the Riemann hypothesis.

Theorem 7.6. *If the Riemann hypothesis is true, then for sufficiently large* x *and for any* $\alpha > \frac{1}{2}$, *there exists a prime,* p, *in* $(x, x + x^{\alpha})$ *[131].*

In 1937, independent of the Riemann hypothesis, Ingham has shown that there is always a prime in the interval $(x, x + x^{\frac{5}{8}})$ for large x.

7.5 The Least Prime in Arithmetic Progressions

The extended Riemann hypothesis can be applied to the problem of computing (or estimating) $\pi(x; k, l)$, where

$$\pi(x; k, l) := \# \{p \ : \ p \text{ is prime}, \ p \leq x, \ \text{and } p \equiv l \pmod{k}\}.$$

Titchmarsh proved the following result given the extended Riemann hypothesis (see Section 6.3).

Theorem 7.7. *If the extended Riemann hypothesis is true, then the least prime* $p \equiv l \pmod{k}$ *is less than* $k^{2+\varepsilon}$, *where* $\varepsilon > 0$ *is arbitrary and* $k > k_0(\varepsilon)$ *[31, p.xiv].*

Unconditionally, the best known result, due to Heath-Brown, is that the least prime p satisfies $p \ll k^{5.5}$ [71].

7.6 Primality Testing

The performance of a variety of algorithms for primality testing relies on the generalized Riemann hypothesis. The probabilistic Miller–Rabin primality test runs in deterministic polynomial time assuming the generalized Riemann hypothesis [106]. Also, the probabilistic Solovay–Strassen algorithm is provably deterministic under the generalized Riemann hypothesis [2]. However, both of these results are superseded by the results of Agrawal, Kayal, and Saxena (presented in Section 12.20). Their paper contains an unconditional deterministic polynomial-time algorithm for primality testing.

7.7 Artin's Primitive Root Conjecture

There is no statement in the literature that is uniquely identified as Artin's conjecture. We present one such conjecture related to primitive roots in $\mathbb{Z}/p\mathbb{Z}$.

Definition 7.8. *Given a prime p an integer a is a primitive root modulo p if $a^{\alpha} \not\equiv 1 \pmod{p}$ for any $0 < \alpha \leq p-2$.*

Artin's conjecture is as follows.

Conjecture 7.9. *Every $a \in \mathbb{Z}$, where a is not square and $a \neq -1$, is a primitive root modulo p for infinitely many primes p.*

The conjecture was proven by Hooley in 1967 assuming the generalized Riemann hypothesis [74]. Artin's conjecture was proven unconditionally for infinitely many a by Ram Murty and Gupta using sieve methods [109, 110]. This result was improved in 1986 by Heath-Brown to the following.

Theorem 7.10. *If q, r, s are three nonzero multiplicatively independent integers such that none of $q, r, s, -3qr, -3qs, -3rs, qrs$ is a square, then the number $n(x)$ of primes $p \leq x$ for which at least one of q, r, s is a primitive root modulo p satisfies $n(x) \gg \frac{x}{\log^2 x}$ [70].*

From this it follows that there are at most two positive primes for which Artin's conjecture fails.

7.8 Bounds on Dirichlet L-Series

Let $L(1, \chi_D)$ be the value of the Dirichlet L-series

$$L(1, \chi_D) = \sum_{n=1}^{\infty} \frac{\chi_D(n)}{n},$$

at 1, where χ_D is a nonprincipal Dirichlet character with modulus D (see Section 6.2).

It is well known [59] that $|L(1, \chi_D)|$ is bounded by

$$D^{-\varepsilon} \ll_{\epsilon} |L(1, \chi_D)| \ll \log D.$$

Littlewood [100], assuming the generalized Riemann hypothesis, gave the following bounds on $|L(1, \chi_D)|$:

$$\frac{1}{\log \log D} \ll |L(1, \chi_D)| \ll \log \log D.$$

7.9 The Lindelöf Hypothesis

The Lindelöf hypothesis states that $\zeta(\frac{1}{2}+it) = O(t^{\varepsilon})$ for any $\varepsilon > 0$, $t \geq 0$. We can make the statement that $\zeta(\sigma + it) = O(|t|^{\max\{1-2\sigma,0\}+\varepsilon})$ for any $\varepsilon > 0$, $0 \leq \sigma \leq 1$. If the Riemann hypothesis holds, then the Lindelöf hypothesis follows; however, it is not known whether the converse is true [155, p.328]. We define the function $N(\sigma, T)$ as follows.

Definition 7.11. *$N(\sigma, T)$ is the number of zeros of $\zeta(\beta+it)$ such that $\beta > \sigma$, for $0 < t \leq T$.*

We can see that if the Riemann hypothesis is true, then $N(\sigma, T) = 0$ if $\sigma \neq \frac{1}{2}$. Also, if we take $N(T)$ to be the number of zeros of $\zeta(\sigma + it)$ in the rectangle $0 \leq \sigma < 1$, $0 \leq t \leq T$, then $N(\sigma, T) \leq N(T)$ [155]. The Lindelöf hypothesis is equivalent to the following statement.

Equivalence 7.12 (An Equivalent Lindelöf Hypothesis). *For every $\sigma > \frac{1}{2}$,*

$$N(\sigma, T+1) - N(\sigma, T) = o(\log T) \quad [11].$$

This statement shows the relevance of the Lindelöf hypothesis in terms of the distribution of the zeros of $\zeta(s)$.

7.10 Titchmarsh's $\mathcal{S}(T)$ Function

In [155], Titchmarsh considers at length some consequences of a proof of the Riemann hypothesis. He introduces the functions $\mathcal{S}(T)$ and $\mathcal{S}_1(T)$, defined as follows.

Definition 7.13. *For a real variable $T \geq 0$, set*

$$\mathcal{S}(T) := \pi^{-1} \arg \zeta \left(\frac{1}{2} + iT \right)$$

and

$$\mathcal{S}_1(T) := \int_0^T \mathcal{S}(t) dt.$$

The connection between $\zeta(s)$ and $\mathcal{S}_1(T)$ is given by the following theorem [155, p.221].

Theorem 7.14. *We have that*

$$\mathcal{S}_1(T) = \frac{1}{\pi} \int_{\frac{1}{2}}^{2} \log |\zeta(\sigma + iT)| d\sigma + O(1).$$

The following results concerning $\mathcal{S}(T)$ and $\mathcal{S}_1(T)$ follow under the assumption of the Riemann hypothesis.

Theorem 7.15. *Under the Riemann hypothesis, for any $\varepsilon > 0$, both of the inequalities*

$$\mathcal{S}(T) > (\log T)^{\frac{1}{2}-\varepsilon}$$

and

$$\mathcal{S}_1(T) < -(\log T)^{\frac{1}{2}-\varepsilon}$$

have solutions for arbitrarily large values of T.

Recall that $f(x) = \Omega(g(x))$ if $|f(x)| \geq A|g(x)|$ for some constant A, and all values of $x > x_0$ for some x_0.

Theorem 7.16. *Under the Riemann hypothesis we have for any $\varepsilon > 0$ that*

$$\mathcal{S}_1(T) = \Omega\left((\log T)^{\frac{1}{2}-\varepsilon}\right).$$

Titchmarsh proves that $\mathcal{S}(T) = O(\log T)$ and $\mathcal{S}_1(T) = O(\log T)$ without the Riemann hypothesis. If we assume the Riemann hypothesis then we can make the following improvements.

Theorem 7.17. *Under the Riemann hypothesis we have*

$$\mathcal{S}(T) = O\left(\frac{\log T}{\log\log T}\right)$$

and

$$\mathcal{S}_1(T) = O\left(\frac{\log T}{(\log\log T)^2}\right).$$

Proofs of these theorems and their connection to the behavior of $\zeta(s)$ are given in detail in [155, §14.10].

7.11 Mean Values of $\zeta(s)$

An ongoing theme in the study of the Riemann zeta function, $\zeta(s)$, has been the estimation of the mean values (or moments) of $\zeta(s)$,

$$I_k(T) := \frac{1}{T}\int_0^T \left|\zeta\left(\frac{1}{2}+it\right)\right|^{2k} dt,$$

for integers $k \geq 1$. Upper bounds for $I_k(T)$ have numerous applications, for example, in zero-density theorems for $\zeta(s)$ and various divisor problems.

In 1918, Hardy and Littlewood [65] proved that

$$I_1(T) \sim \log T$$

as $T \to \infty$. In 1926, Ingham [77] showed that

$$I_2(T) \sim \frac{1}{2\pi^2}(\log T)^4.$$

Although it has been conjectured that

$$I_k(T) \sim c_k(\log T)^{k^2}$$

for some positive constant c_k, no asymptotic estimate for I_k when $k \geq 3$ is known. Conrey and Ghosh (in an unpublished work) conjectured a more precise form of the constant c_k, namely

$$I_k(T) \sim \frac{a(k)g(k)}{\Gamma(k^2+1)}(\log T)^{k^2},$$

where

$$a(k) := \prod_p \left(\left(1 - \frac{1}{p}\right)^{k^2} \sum_{m=0}^{\infty} \left(\frac{\Gamma(m+k)}{m!\Gamma(k)} \right)^2 p^{-m} \right),$$

and $g(k)$ is an integer whenever k is an integer. The results of Hardy–Littlewood and Ingham give $g(1) = 1$ and $g(2) = 2$ respectively. Keating and Snaith [85] suggested that the characteristic polynomial of a large random unitary matrix might be used to model the value distribution of the Riemann zeta function near a large height T. Based on this, Keating and Snaith made an explicit conjecture on the integer $g(k)$ in terms of Barnes's G-function.

Conrey and Ghosh showed that the Riemann hypothesis implies

$$I_k(T) \geq (a(k) + o(1))(\log T)^{k^2}$$

as $T \to \infty$ [37]. Later, Balasubramanian and Ramachandra [12] removed the assumption of the Riemann hypothesis. Their result was subsequently improved by Soundararajan [148] to

$$I_k(T) \geq 2(a_k + o(1))(\log T)^{k^2}$$

for $k \geq 2$.

8

Failed Attempts at Proof

George Pólya once had a young mathematician confide to him that he was working on the great Riemann Hypothesis. "I think about it every day when I wake up in the morning," he said. Pólya sent him a reprint of a faulty proof that had been once been submitted by a mathematician who was convinced he'd solved it, together with a note: "If you want to climb the Matterhorn you might first wish to go to Zermatt where those who have tried are buried" [73].

Lars Hörmander

In this chapter we discuss four famous failed attempts at the Riemann hypothesis. Though flawed, all of these attempts spurred further research into the behavior of the Riemann zeta function.

8.1 Stieltjes and Mertens' Conjecture

Recall that Mertens' function is defined as

$$M(n) := \sum_{k=1}^{n} \mu(k),$$

where $\mu(k)$ is the Möbius function. The statement generally referred to as Mertens' conjecture is that for any $n \geq 1$, $|M(n)| < n^{\frac{1}{2}}$. The truth of Mertens' conjecture implies the truth of the Riemann hypothesis. In 1885, Stieltjes published a note in the *Comptes rendus de l'Acadmie des sciences* in which he claimed to have proved that $M(n) = O(n^{\frac{1}{2}})$. However, he died without publishing his result, and no proof was found in his papers posthumously [48]. Odlyzko and te Riele [127] proved that Mertens' conjecture is false.

They showed that

$$\limsup_{n\to\infty} M(n)n^{-\frac{1}{2}} > 1.06,$$

and

$$\liminf_{n\to\infty} M(n)n^{-\frac{1}{2}} < -1.009.$$

Their proof relies on accurate computation of the first 2000 zeros of $\zeta(s)$. While it is not impossible that $M(n) = O(n^{\frac{1}{2}})$, they feel that it is improbable. The Riemann hypothesis is in fact equivalent to the conjecture that $M(n) = O(n^{\frac{1}{2}+\varepsilon})$, for any $\varepsilon > 0$ (see Chapter 5 for equivalences).

8.2 Hans Rademacher and False Hopes

In 1945, *Time* reported that Hans Rademacher had submitted a flawed proof of the Riemann hypothesis to the journal *Transactions of the American Mathe- matical Society*. The text of the article follows:

> A sure way for any mathematician to achieve immortal fame would be to prove or disprove the Riemann hypothesis. This baffling theory, which deals with prime numbers, is usually stated in Riemann's symbolism as follows: "All the nontrivial zeros of the zeta function of s, a complex variable, lie on the line where sigma is 1/2 (sigma being the real part of s)." The theory was propounded in 1859 by Georg Friedrich Bernhard Riemann (who revolutionized geometry and laid the foundations for Einstein's theory of relativity). No layman has ever been able to understand it and no mathematician has ever proved it.
>
> One day last month electrifying news arrived at the University of Chicago office of Dr. Adrian A. Albert, editor of the *Transactions of the American Mathematical Society*. A wire from the society's secretary, University of Pennsylvania Professor John R. Kline, asked Editor Albert to stop the presses: a paper disproving the Riemann hypothesis was on the way. Its author: Professor Hans Adolf Rademacher, a refugee German mathematician now at Penn.
>
> On the heels of the telegram came a letter from Professor Rademacher himself, reporting that his calculations had been checked and confirmed by famed Mathematician Carl Siegel of Princeton's Institute for Advanced Study. Editor Albert got ready to publish the historic paper in the May issue. U.S. mathematicians, hearing the wildfire rumor, held their breath. Alas for drama, last week the issue went to press without the Rademacher article. At the last moment the professor wired meekly that it was all a mistake; on rechecking. Mathematician Siegel had discovered a flaw (undisclosed) in the Rademacher

reasoning. U.S. mathematicians felt much like the morning after a phony armistice celebration. Sighed Editor Albert: "The whole thing certainly raised a lot of false hopes." [152]

8.3 Turán's Condition

Pál Turán showed that if for all N sufficiently large, the Nth partial sum of $\zeta(s)$ does not vanish for $\sigma > 1$, then the Riemann hypothesis follows [156]. However, Hugh Montgomery proved that this approach will not work, since for any positive $c < \frac{4}{\pi} - 1$, the Nth partial sum of $\zeta(s)$ has zeros in the half-plane $\sigma > 1 + c\frac{\log \log N}{\log N}$ [108].

8.4 Louis de Branges's Approach

Louis de Branges is currently Edward C. Elliott distinguished professor of mathematics at Purdue University. In 1985, de Branges published a proof of another famous open problem, the Bieberbach conjecture. The conjecture considers the class S of all $f(s) = s + a_2 s^2 + a_3 s^3 + \cdots$ that are analytic and univalent.

Theorem 8.1. *For all $n > 1$, and all $f \in S$, $|a_n| \leq n$.*

The proof followed several incomplete attempts by de Branges. Since proving the Bieberbach conjecture, he has claimed several times to have a proof of the Riemann hypothesis. However, none of these are accepted by the mathematical community. In the words of Karl Sabbagh,

> ...another mathematician was quietly working away, ignored or actively shunned by his professional colleagues, on his proof of the Riemann hypothesis, a proof to which he was putting the final touches early in 2002. He was Louis de Branges, a man to be taken seriously, one would think, because he had previously proved another famous theorem–the Bieberbach Conjecture [139, p.115].

An extensive profile of de Branges and one of his earlier proofs can be found in [139]. A press release from Purdue regarding de Branges's latest claim can be found in [25].

The approach used by de Branges has been disputed by Conrey and Li in [39]. The relevant papers by de Branges are [44, 45, 46, 47]. Papers can be found on de Branges's website, [43].

8.5 No Really Good Idea

> *There have probably been very few attempts at proving the Riemann hypothesis, because, simply, no one has ever had any really good idea for how to go about it! [33]*
>
> **Atle Selberg**

Many proofs have been offered by less-illustrious mathematicians. Several popular anecdotes to this effect involve the editor of a mathematical journal staving off enthusiastic amateur mathematicians who are ignorant of the subtleties of complex analysis. Roger Heath-Brown relates his experience with these attempts:

> I receive unsolicited manuscripts quite frequently–one finds a particular person who has an idea, and no matter how many times you point out a mistake they correct that and produce something else that is also a mistake [139, p.112].

Any basic web search will reveal several short and easy "proofs" of the hypothesis. Given the million-dollar award for a proof, we suspect that many more will be offered.

9

Formulas

The following is a reference list of useful formulas involving $\zeta(s)$ and its relatives. Throughout, the variable n is assumed to lie in $\mathbb{N}$.

1. The Euler product formula (p.11):

$$\zeta(s) = \sum_{n=1}^{\infty} \frac{1}{n^s} = \prod_{p} \left(1 - \frac{1}{p^s}\right)^{-1},$$

for $\Re(s) > 1$ and p prime.

2. A globally convergent series representation [69, p.206]:

$$\zeta(s) = \frac{1}{1 - 2^{1-s}} \sum_{n=0}^{\infty} \frac{1}{2^{n+1}} \sum_{k=0}^{n} (-1)^k \binom{n}{k} (k+1)^{-s},$$

for $s \in \mathbb{C} \setminus \{1\}$.

3. A relation with the Möbius function [69, p.209]:

$$\frac{1}{\zeta(s)} = \sum_{n=1}^{\infty} \frac{\mu(n)}{n^s},$$

for $\Re(s) > 1$, where μ is the Möbius function.

4. A quotient relation [68]:

$$\frac{\zeta(2s)}{\zeta(s)} = \sum_{n=1}^{\infty} \frac{\lambda(n)}{n^s},$$

$\Re(s) > 1$, where λ is the Liouville function.

5. Another quotient relation [68, p.255]:

$$\frac{\zeta(s)}{\zeta(2s)} = \sum_{n=1}^{\infty} \frac{\mu^2(n)}{n^s},$$

$\Re(s) > 1$, where μ is the Möbius function.

6. Another quotient relation [68, p.254]:

$$\frac{\zeta^2(s)}{\zeta(2s)} = \sum_{n=1}^{\infty} \frac{2^{\omega(n)}}{n^s},$$

for $\Re(s) > 1$, where $\omega(n)$ is the number of distinct prime factors of n.

7. A formula for $\zeta(2)$ as a binomial sum [60, p.257]:

$$\zeta(2) = 3\sum_{k=1}^{\infty} \frac{1}{k^2 \binom{2k}{k}}.$$

8. A formula for $\zeta(3)$ as a binomial sum [60, p.257]:

$$\zeta(3) = \frac{5}{2}\sum_{k=1}^{\infty} \frac{(-1)^{k-1}}{k^3 \binom{2k}{k}}.$$

9. A formula for $\zeta(4)$ as a binomial sum [60, p.257]:

$$\zeta(4) = \frac{36}{17}\sum_{k=1}^{\infty} \frac{1}{k^4 \binom{2k}{k}}.$$

10. A sum identity:

$$\sum_{n=2}^{\infty} (\zeta(n) - 1) = 1.$$

11. A formula for $\zeta(s)$ at an even integer:

$$\zeta(2n) = \frac{2^{2n-1}|B_{2n}|\pi^{2n}}{(2n)!},$$

where B_n is the nth Bernoulli number.

12. Another formula for $\zeta(s)$ at an even integer:

$$\zeta(2n) = \frac{(-1)^{n+1}2^{2n-3}\pi^{2n}}{(2^{2n}-1)(2n-2)!}\int_0^1 E_{2(n-1)}(x)dx,$$

where E_n is the nth Euler polynomial.

13. A formula for $\zeta(s)$ at an odd integer:

$$\zeta(2n+1) = \frac{(-1)^n 2^{2n-1}\pi^{2n+1}}{(2^{2n+1}-1)(2n)!}\int_0^1 E_{2n}(x)\tan\left(\frac{\pi}{2}x\right)dx,$$

where E_n is the nth Euler polynomial.

14. A relation between $\zeta(s)$ and $\zeta'(s)$:

$$\zeta'(-2n) = \frac{(-1)^n\zeta(2n+1)(2n)!}{2^{2n+1}\pi^{2n}}.$$

15. The analytic continuation for $\zeta(s)$ (see p.13):

$$\zeta(s) = \frac{\pi^{\frac{s}{2}}}{\Gamma\left(\frac{s}{2}\right)}\left(\frac{1}{s(s-1)} + \int_1^\infty \left(x^{\frac{s}{2}-1} + x^{-\frac{s}{2}-\frac{1}{2}}\right)\left(\frac{\vartheta(x)-1}{2}\right)dx\right),$$

for $s \in \mathbb{C}\setminus\{1\}$, and where $\vartheta(x)$ is the Jacobi ϑ function.

dh $\mathbb{C}$ ohne 1

16. A functional equation for $\zeta(s)$ (see p.14):

$$\pi^{-\frac{s}{2}}\Gamma\left(\frac{s}{2}\right)\zeta(s) = \pi^{-\frac{1-s}{2}}\Gamma\left(\frac{1-s}{2}\right)\zeta(1-s),$$

for $s \in \mathbb{C}\setminus\{0,1\}$.

17. The reflection functional equation [63, p.14]:

$$\zeta(1-s) = 2(2\pi)^{-s}\cos\left(\frac{s\pi}{2}\right)\Gamma(s)\zeta(s),$$

for $s \in \mathbb{C} \setminus \{0,1\}$.

18. The definition of $\xi(s)$ (see p.14):

$$\xi(s) = \frac{s}{2}(s-1)\pi^{-\frac{s}{2}}\Gamma\left(\frac{s}{2}\right)\zeta(s),$$

for all $s \in \mathbb{C}$.

19. The functional equation for $\xi(s)$ (see p.14):

$$\xi(s) = \xi(1-s),$$

for all $s \in \mathbb{C}$.

20. The definition of $\eta(s)$ (see p.56):

$$\eta(s) = \sum_{k=1}^{\infty} \frac{(-1)^{k-1}}{k^s} = (1-2^{1-s})\zeta(s),$$

for $\Re(s) > 1$.

21. The definition of $\Xi(iz)$ (see p.51):

$$\Xi(iz) = \frac{1}{2}\left(z^2 - \frac{1}{4}\right)\pi^{-\frac{z}{2}-\frac{1}{4}}\Gamma\left(\frac{z}{2}+\frac{1}{4}\right)\zeta\left(z+\frac{1}{2}\right),$$

for any complex z.

22. The definition of $\mathcal{S}(T)$ (see p.65):

$$\mathcal{S}(T) = \pi^{-1}\arg\zeta\left(\frac{1}{2}+iT\right),$$

for any real T.

23. The definition of $\mathcal{S}_1(T)$ (see p.65):

$$\mathcal{S}_1(T) = \int_0^T \mathcal{S}(t)dt,$$

for $T > 0$.

24. The definition of $Z(t)$ (see p.32):

$$Z(t) = e^{i\left(\Im\left(\log \Gamma\left(\frac{1}{4}+\frac{it}{2}\right)\right)-\frac{t}{2}\log \pi\right)}\zeta\left(\frac{1}{2}+it\right),$$

for any real t.

25. The approximate functional equation (see p.32):

$$\zeta(s) = \sum_{n\le x}\frac{1}{n^s} + \chi(s)\sum_{n\le y}\frac{1}{n^{1-s}} + O(x^{-\sigma}) + O(|t|^{\frac{1}{2}-\sigma}y^{\sigma-1}),$$

where $s = \sigma + it$, $2\pi xy = |t|$ for $x, y \in \mathbb{R}^+$, $0 \le \sigma \le 1$, and

$$\chi(s) = 2^s\pi^{s-1}\sin\left(\frac{\pi s}{2}\right)\Gamma(1-s).$$

26. The Riemann–Siegel formula (p.33):

$$Z(t) = 2\sum_{n=1}^{N}\left(\frac{\cos\left(\theta(t) - t\log n\right)}{n^{\frac{1}{2}}}\right) + O(t^{-\frac{1}{4}}),$$

for $t > 0$, where $N = \left[\sqrt{|t|/2\pi}\right]$.

27. A power series expansion of $\ln \Gamma(1+z)$:

$$\log \Gamma(1+z) = -\log(1+z) - z(\gamma - 1) + \sum_{n=2}^{\infty}(-1)^n(\zeta(n) - 1)\frac{z^n}{n},$$

for $|z| < 1$.

28. An integral representation of $\zeta(s)$:

$$\zeta(s) = \frac{1}{\Gamma(s)} \int_0^\infty \frac{x^{s-1}}{e^x - 1} dx,$$

for $s > 1$.

29. Another integral representation of $\zeta(s)$:

$$\zeta(s) = \frac{1}{(1 - 2^{1-s})\Gamma(s)} \int_0^\infty \frac{x^{s-1}}{e^x + 1} dx,$$

for $s > 1$.

30. A multiple-integral formula for $\zeta(n)$:

$$\zeta(n) = \underbrace{\int_0^1 \cdots \int_0^1}_{n \text{ times}} \frac{dx_1 dx_2 \cdots dx_n}{1 - x_1 x_2 \cdots x_n}.$$

31. The Mellin transform:

$$\int_0^\infty \left\{ \frac{1}{t} \right\} t^{s-1} dt = -\frac{\zeta(s)}{s},$$

for $0 < \Re(s) < 1$, where $\{x\}$ is the fractional part of x.

32. A representation involving a unit square integral:

$$\int_0^1 \int_0^1 \frac{(-\log(xy))^s}{1 - xy} dx\, dy = \Gamma(s+2)\zeta(s+2),$$

for $\Re(s) > 1$.

33. The Laurent series for $\zeta(s)$ at $s = 1$:

$$\zeta(s) = \frac{1}{s-1} + \sum_{n=0}^\infty \frac{(-1)^n}{n!} \gamma_n (s-1)^n,$$

where

$$\gamma_n = \lim_{m \to \infty} \left(\sum_{k=1}^m \frac{(\ln k)^n}{k} - \frac{(\ln m)^{n+1}}{n+1} \right)$$

is the nth Stieltjes constant.

34. A sum identity:

$$\sum_{k=1}^{\infty} \frac{1}{k^2 - x^2} = \sum_{n=0}^{\infty} \zeta(2n+2)x^{2n} = \frac{1 - \pi x \cot(\pi x)}{2x^2},$$

for any complex number x that is not a nonzero integer.

35. A simple pole formula [172, p.271]:

$$\lim_{s \to 1} \left(\zeta(s) - \frac{1}{s-1} \right) = \gamma,$$

where γ is the Euler–Mascheroni constant.

36. A sum identity [69, pp.109, 111–112]:

$$\sum_{k=2}^{\infty} \frac{(-1)^k \zeta(k)}{k} = \gamma,$$

where γ is the Euler–Mascheroni constant.

37. Another sum identity:

$$\sum_{k=2}^{\infty} \zeta(k) x^{k-1} = -\psi_0(1-x) - \gamma,$$

where ψ_0 is the digamma function, and γ is the Euler–Mascheroni constant.

10

Timeline

There is almost three hundred years of history surrounding the Riemann hypothesis and the prime number theorem. While the authoritative history of these ideas has yet to appear, this timeline briefly summarizes the high and low points.

1737 Euler proves the Euler product formula [48]; namely, for real $s > 1$,

$$\sum_{n=1}^{\infty} \frac{1}{n^s} = \prod_{p \text{ prime}} \left(1 - \frac{1}{p^s}\right)^{-1}.$$

1738 Euler invents the "Euler–Maclaurin summation method."

1742 Maclaurin invents the "Euler–Maclaurin summation method."

1742 Goldbach proposes his conjecture to Euler in a letter.

1792 Gauss proposes what would later become the prime number theorem.

1802 Haros discovers and proves results concerning the general properties of Farey series.

1816 Farey rediscovers Farey series and is given credit for their invention. In the somewhat unfair words of Hardy, "*Farey is immortal because he failed to understand a theorem which Haros had proved perfectly fourteen years before* [64]."

1845 Bertrand postulates that for $a > 1$ there is always a prime that lies between a and $2a$.

1850 Chebyshev proves Bertrand's postulate using elementary methods.

1859 Riemann publishes his *Ueber die Anzahl der Primzahlen unter einer gegebenen Grösse,* in which he proposes the Riemann hypothesis. The original statement is as follows:

"*One now finds indeed approximately this number of real roots within these limits, and it is very probable that all roots are real. Certainly one would wish for a stricter proof here; I have meanwhile temporarily put aside the search for this after some fleeting futile attempts, as it appears unnecessary for the next objective of my investigation* [133]."

Riemann's paper contains the functional equation

$$\pi^{-\frac{s}{2}}\Gamma\left(\frac{s}{2}\right)\zeta(s) = \pi^{-\frac{1-s}{2}}\Gamma\left(\frac{1-s}{2}\right)\zeta(1-s).$$

His work contains analysis of the Riemann zeta function. Some of Riemann's analysis is overlooked until it is reinvigorated by Siegel (see 1932).

1885 Stieltjes claims to have a proof of what would later be called Mertens' conjecture. His proof is never published, nor found among his papers posthumously. Mertens' conjecture implies the Riemann hypothesis. This is perhaps the first significant failed attempt at a proof.

1890 Some time after this date Lindelöf proposed the Lindelöf hypothesis. This conjecture concerns the distribution of the zeros of Riemann's zeta function [155, p.328]. It is still unproven.

1896 Hadamard and de la Vallée Poussin independently prove the prime number theorem. The proof relies on showing that $\zeta(s)$ has no zeros of the form $s = 1 + it$ for $t \in \mathbb{R}$.

1897 Mertens publishes his conjecture. (This conjecture is proved incorrect by Odlyzko and te Riele; see 1985.)

1901 Von Koch proves that the Riemann hypothesis is equivalent to the statement

$$\pi(x) = \int_2^x \frac{dt}{\log t} + O(\sqrt{x}\log x).$$

1903 Gram calculates the first 15 zeros of $\zeta(s)$ on the critical line.

1903 In the decade following 1903, Gram, Backlund, and Hutchinson independently apply Euler–Maclaurin summation to calculate $\zeta(s)$ and to verify the Riemann hypothesis for $\Im(s) \leq 300$).

1912 Littlewood proves that Mertens' conjecture implies the Riemann hypothesis.

1912 Littlewood proves that $\pi(n) < \mathrm{Li}(n)$ fails for some n.

1912 Backlund develops a method of determining the number of zeros of $\zeta(s)$ in the critical strip $0 < \Re(s) < 1$ up to a given height [51, p.128]. This method is used through 1932.

1914 Hardy proves that infinitely many zeros of $\zeta(s)$ lie on the critical line.

1914 Backlund calculates the first 79 zeros of $\zeta(s)$ on the critical line.

1914 Littlewood proves that $\pi(n) < \mathrm{Li}(n)$ fails for infinitely many n.

1914 Bohr and Landau prove that if $N(\sigma, T)$ is the number of zeros of $\zeta(s)$ in the rectangle $0 \leq \Im(s) \leq T$, $\sigma < \Re(s) \leq 1$, then $N(\sigma, T) = O(T)$, for any fixed $\sigma \geq \frac{1}{2}$.

1919 Pólya conjectures that the summatory Liouville function,

$$L(x) := \sum_{n=1}^{x} \lambda(n),$$

where $\lambda(n)$ is the Liouville function, satisfies $L(x) \leq 0$ for $x \geq 2$. (This conjecture is proved incorrect by Haselgrove in 1958.)

1920 Carlson proves the density theorem.

Theorem 10.1 (The Density Theorem). *For any* $\varepsilon > 0$ *and* $\frac{1}{2} \leq \sigma \leq 1$*, we have* $N(\sigma, T) = O(T^{4\sigma(1-\sigma)+\varepsilon})$.

1922 Hardy and Littlewood show that the generalized Riemann hypothesis implies Goldbach's weak conjecture.

1923 Hardy and Littlewood [67] prove that if the generalized Riemann hypothesis is true, then almost every even number is the sum of two primes. Specifically, if $E(N)$ denotes the number of even integers less than N that are not a sum of two primes, then $E(N) \ll N^{\frac{1}{2}+\varepsilon}$.

1924 Franel and Landau discover an equivalence to the Riemann hypothesis involving Farey series. The details are not complicated, but are rather lengthy.

1925 Hutchinson calculates the first 138 zeros of $\zeta(s)$ on the critical line.

1928 Littlewood [100] shows that the generalized Riemann hypothesis bounds $L(1, \chi_D)$ as

$$\frac{1}{\log \log D} \ll |L(1, \chi_D)| \ll \log \log D.$$

1932 Siegel analyzes Riemann's private (and public) papers. He finds (among other things) a formula for calculating values of $\zeta(s)$ that is more efficient than Euler–Maclaurin summation. The method is referred to as the Riemann–Siegel formula and is used in some form up to the present.

Siegel is credited with reinvigorating Riemann's most important results regarding $\zeta(s)$. In the words of Edwards, "It is indeed fortunate that Siegel's concept of scholarship derived from the older tradition of respect for the past rather than the contemporary style of novelty [51, p.136]."

1934 Speiser shows that the Riemann hypothesis is equivalent to the nonvanishing of $\zeta'(s)$ in $0 < \sigma < \frac{1}{2}$.

1935 Titchmarsh calculates the first 1041 zeros of $\zeta(s)$ on the critical line.

1937 Vinogradov proves the following result related to Goldbach's conjecture without assuming any variant of the Riemann hypothesis.

Theorem 10.2. *Every sufficiently large odd number is a sum of three prime numbers.*

1940 Ingham shows that $N(\sigma, T) = O\left(T^{3\left(\frac{1-\sigma}{2-\sigma}\right)} \log^5 T\right)$. This is still the best known result for $\frac{1}{2} \leq \sigma \leq \frac{3}{4}$.

1941 Weil proves that the Riemann hypothesis is true for function fields [170].

1942 Ingham publishes a paper building on the conjectures of Mertens and Pólya. He proves that not only do both conjectures imply the truth of the hypothesis, and the simplicity of the zeros, but they also imply a linear dependence between the imaginary parts of the zeros.

1942 Selberg proves that a positive proportion of the zeros of $\zeta(s)$ lie on the critical line.

1943 Alan Turing publishes two important developments. The first is an algorithm for computing $\zeta(s)$ (made obsolete by better estimates to the error terms in the Riemann–Siegel formula). The second is a method for calculating $N(T)$, which gives a tool for verifying the Riemann hypothesis up to a given height.

1945 Time publishes a short article detailing a recent failed attempt at a proof of the Riemann hypothesis. The proof was submitted for review and publication to Transactions of the American Mathematical Society by Hans Rademacher and subsequently withdrawn.

1948 Turán shows that if for all N sufficiently large, the Nth partial sum of $\zeta(s)$ does not vanish for $\sigma > 1$, then the Riemann hypothesis follows [156].

1949 Selberg and Erdős, building on the work of Selberg, both find "elementary" proofs of the prime number theorem.

1951 In [155], Titchmarsh considers at length some consequences of a proof of the Riemann hypothesis. He considers sharper bounds for $\zeta(s)$, as well as the functions $\mathcal{S}(T)$ and $\mathcal{S}_1(T)$.

1951 Sometime before 1951, Titchmarsh found that the extended Riemann hypothesis can be applied in considering the problem of computing (or estimating) $\pi(x; k, l)$. He showed that if the extended Riemann hypothesis is true, then the least $p \equiv l \pmod{k}$ is less than $k^{2+\varepsilon}$, where $\varepsilon > 0$ is arbitrary and $k > k_0(\varepsilon)$. [31, p.xiv]

1953 Turing calculates the first 1104 zeros of $\zeta(s)$ on the critical line.

1955 Skewes bounds the first n such that $\pi(n) < \mathrm{Li}(n)$ fails. This bound is improved in the future, but retains the name "Skewes number."

1955 Beurling finds the Nyman–Beurling equivalent form.

1956 Lehmer calculates the first 15,000 zeros of $\zeta(s)$ on the critical line, and later in the same year the first 25,000 zeros.

1958 Meller calculates the first 35,337 zeros of $\zeta(s)$ on the critical line.

1958 Haselgrove disproves Pólya's conjecture.

1966 Lehman improves Skewes's bound.

1966 Lehman calculates the first 250,000 zeros of $\zeta(s)$ on the critical line.

1967 Hooley proves that Artin's conjecture holds under the assumption of the extended Riemann hypothesis. Artin's conjecture is the following:

Conjecture 10.3. *Every $a \in \mathbb{Z}$, where a is not square and $a \neq -1$, is a primitive root modulo p for infinitely many primes p.*

Hooley's proof appears in [74].

1968 Rosser, Yohe, and Schoenfeld calculate the first 3,500,000 zeros of $\zeta(s)$ on the critical line.

1968 Louis de Branges makes the first of his several attempts to prove the Riemann hypothesis. Other alleged proofs were offered in 1986, 1992, and 1994.

1973 Montgomery conjectures that the correlation for the zeros of the zeta function is

$$1 - \frac{\sin^2(\pi x)}{(\pi x)^2}.$$

1973 Chen proves that every sufficiently large even integer is a sum of a prime and number which is a product of at most two primes.

1974 The probabilistic Solovay–Strassen algorithm for primality testing is published. It can be made deterministic under the generalized Riemann hypothesis [2].

1975 The probabilistic Miller–Rabin algorithm for primality testing is published. It runs in polynomial time under the generalized Riemann hypothesis [2].

1977 Redheffer shows that the Riemann hypothesis is equivalent to the statement that

$$\det(R_n) = O(n^{\frac{1}{2}+\varepsilon})$$

for any $\varepsilon > 0$, where $R_n := [R_n(i,j)]$ is the $n \times n$ matrix with entries

$$R_n(i,j) = \begin{cases} 1 & \text{if } j = 1 \text{ or if } i \mid j \\ 0 & \text{otherwise.} \end{cases}$$

1977 Brent calculates the first 40,000,000 zeros of $\zeta(s)$ on the critical line.

1979 Brent calculates the first 81,000,001 zeros of $\zeta(s)$ on the critical line.

1982 Brent, van de Lune, te Riel, and Winter calculate the first 200,000,001 zeros of $\zeta(s)$ on the critical line.

1983 Van de Lune and te Riele calculate the first 300,000,001 zeros of $\zeta(s)$ on the critical line.

1983 Montgomery proves that the 1948 approach of Turán will not lead to a proof of the Riemann hypothesis. This is because for any positive $c < \frac{4}{\pi} - 1$, the Nth partial sum of $\zeta(s)$ has zeros in the half-plane $\sigma > 1 + c\frac{\log\log N}{\log N}$ [108].

1984 Ram Murty and Gupta prove that Artin's conjecture holds for infinitely many a without assuming any variant of the Riemann hypothesis.

1985 Odlyzko and te Riele prove in [127] that Mertens' conjecture is false. They speculate that, while not impossible, it is improbable that $M(n) = O(n^{\frac{1}{2}})$. The Riemann hypothesis is in fact equivalent to the conjecture $M(n) = O(n^{\frac{1}{2}+\varepsilon})$.

1986 Van de Lune, te Riele, and Winter calculate the first 1,500,000,001 zeros of $\zeta(s)$ on the critical line.

1986 Heath-Brown proves that Artin's conjecture fails for at most two primes.

1987 Te Riele lowers the Skewes number.

1988 Odlyzko and Schönhage publish an algorithm for calculating values of $\zeta(s)$. The Odlyzko–Schönhage algorithm is currently the most efficient algorithm for determining values $t \in \mathbb{R}$ for which $\zeta(\frac{1}{2}+it) = 0$. The algorithm (found in [126]) computes the first n zeros of $\zeta(\frac{1}{2}+it)$ in $O(n^{1+\varepsilon})$ (as opposed to $O(n^{\frac{3}{2}})$ using previous methods).

1988 Barratt, Forcade, and Pollington formulate a graph-theoretic equivalent to the Riemann hypothesis through the use of Redheffer matrices.

1989 Odylzko computes 175 million consecutive zeros around $t = 10^{20}$.

1989 Conrey proves that more than 40% of the nontrivial zeros of $\zeta(s)$ lie on the critical line.

1993 Alcántara-Bode shows that the Riemann hypothesis is true if and only if the Hilbert-Schmidt integral operator on $L^2(0,1)$ is injective.

1994 Verjovsky proves that the Riemann hypothesis is equivalent to a problem about the rate of convergence of certain discrete measures.

1995 Volchkov proves that the statement,

$$\int_0^\infty (1-12t^2)(1+4t^2)^{-3}\int_{\frac{1}{2}}^\infty \log|\zeta(\sigma+it)|d\sigma dt = \frac{\pi(3-\gamma)}{32}$$

is equivalent to the Riemann hypothesis, where γ is Euler's constant.

1995 Amoroso proves that the statement that $\zeta(s)$ does not vanish for $\Re(z) \geq \lambda + \varepsilon$ is equivalent to

$$\tilde{h}(F_N) \ll N^{\lambda+\varepsilon},$$

where

$$\tilde{h}(F_N) = \frac{1}{2\pi}\int_{-\pi}^{\pi} \log^+ |F_N(e^{i\theta})|d\theta$$

and $F_N(z) = \prod_{n\leq N} \Phi_n(z)$, where $\Phi_n(z)$ denotes the nth cyclotomic polynomial, and $\log^+(x) = \max(0, \log x)$.

1997 Hardy and Littlewood's 1922 result concerning Goldbach's conjecture is improved by Deshouillers, Effinger, te Riele, and Zinoviev. They prove the following result.

Theorem 10.4. *Assuming the generalized Riemann hypothesis, every odd number greater than 5 can be expressed as a sum of three prime numbers [49].*

2000 Conrey and Li argue that the approach used by de Branges cannot lead to proof of the Riemann hypothesis [39].

2000 Bays and Hudson lower the Skewes number.

2000 The Riemann hypothesis is named by the Clay Mathematics Institute as one of seven "Millennium Prize Problems." The solution to each problem is worth one million US dollars.

2001 Van de Lune calculates the first 10,000,000,000 zeros of $\zeta(s)$ on the critical line.

2004 Wedeniwski calculates the first 900,000,000,000 zeros of $\zeta(s)$ on the critical line.

2004 Gourdon calculates the first 10,000,000,000,000 zeros of $\zeta(s)$ on the critical line.

Part II

Original Papers

11

Expert Witnesses

Hilbert included the problem of proving the Riemann hypothesis in his list of the most important unsolved problems which confronted mathematics in 1900, and the attempt to solve this problem has occupied the best efforts of many of the best mathematicians of the twentieth century. It is now unquestionably the most celebrated problem in mathematics and it continues to attract the attention of the best mathematicians, not only because it has gone unsolved for so long but also because it appears tantalizingly vulnerable and because its solution would probably bring to light new techniques of far-reaching importance [51].

H. M. Edwards

This chapter contains four expository papers on the Riemann hypothesis. These are our "expert witnesses", and they provide the perspective of specialists in the fields of number theory and complex analysis. The first two papers were commissioned by the Clay Mathematics Institute to serve as official prize descriptions. They give a thorough description of the problem, the surrounding theory, and probable avenues of attack. In the third paper, Conrey gives an account of recent approaches to the Riemann hypothesis, highlighting the connection to random matrix theory. The last paper outlines reasons why mathematicians should remain skeptical of the hypothesis, and possible sources of disproof.

11.1 E. Bombieri (2000–2001)

Problems of the Millennium: The Riemann Hypothesis

> *The failure of the Riemann hypothesis would create havoc in the distribution of prime numbers. This fact alone singles out the Riemann hypothesis as the main open question of prime number theory [69].*
>
> **Enrico Bombieri**

This paper is the official problem description for the Millennium Prize offered by the Clay Mathematics Institute. Bombieri gives the classical formulation of the Riemann hypothesis. He also writes about the history and significance of the Riemann hypothesis to the mathematical community. This paper gives some justification for the prize, by way of heuristic argument for the truth of the hypothesis. It includes an extensive bibliography of standard sources on the Riemann zeta function, as well as sources that consider broader extensions of the Riemann hypothesis.

Problems of the Millennium: the Riemann Hypothesis

E. Bombieri

I. The problem. The Riemann zeta function is the function of the complex variable s, defined in the half-plane[1] $\Re(s) > 1$ by the absolutely convergent series

$$\zeta(s) := \sum_{n=1}^{\infty} \frac{1}{n^s},$$

and in the whole complex plane $\mathbb{C}$ by analytic continuation. As shown by Riemann, $\zeta(s)$ extends to $\mathbb{C}$ as a meromorphic function with only a simple pole at $s = 1$, with residue 1, and satisfies the functional equation

$$\pi^{-s/2}\,\Gamma(\frac{s}{2})\,\zeta(s) = \pi^{-(1-s)/2}\,\Gamma(\frac{1-s}{2})\,\zeta(1-s). \tag{1}$$

In an epoch-making memoir published in 1859, Riemann [Ri] obtained an analytic formula for the number of primes up to a preassigned limit. This formula is expressed in terms of the zeros of the zeta function, namely the solutions $\rho \in \mathbb{C}$ of the equation $\zeta(\rho) = 0$.

In this paper, Riemann introduces the function of the complex variable t defined by

$$\xi(t) = \frac{1}{2}\,s(s-1)\,\pi^{-s/2}\Gamma(\frac{s}{2})\,\zeta(s)$$

with $s = \frac{1}{2} + it$, and shows that $\xi(t)$ is an even entire function of t whose zeros have imaginary part between $-i/2$ and $i/2$. He further states, sketching a proof, that in the range between 0 and T the function $\xi(t)$ has about $(T/2\pi)\log(T/2\pi) - T/2\pi$ zeros. Riemann then continues: "Man findet nun in der That etwa so viel reelle Wurzeln innerhalb dieser Grenzen, und es ist sehr wahrscheinlich, dass alle Wurzeln reell sind.", which can be translated as "Indeed, one finds between those limits about that many real zeros, and it is very likely that all zeros are real."

The statement that all zeros of the function $\xi(t)$ are real is the Riemann hypothesis.

The function $\zeta(s)$ has zeros at the negative even integers $-2, -4, \ldots$ and one refers to them as the *trivial zeros*. The other zeros are the complex numbers $\frac{1}{2} + i\alpha$ where α is a zero of $\xi(t)$. Thus, in terms of the function $\zeta(s)$, we can state

Riemann hypothesis. *The nontrivial zeros of $\zeta(s)$ have real part equal to $\frac{1}{2}$.*

In the opinion of many mathematicians the Riemann hypothesis, and its extension to general classes of L-functions, is probably today the most important open problem in pure mathematics.

II. History and significance of the Riemann hypothesis. For references pertaining to the early history of zeta functions and the theory of prime numbers, we refer to Landau [La] and Edwards [Ed].

[1] We denote by $\Re(s)$ and $\Im(s)$ the real and imaginary part of the complex variable s. The use of the variable s is already in Dirichlet's famous work of 1837 on primes in arithmetic progression.

The connection between prime numbers and the zeta function, by means of the celebrated *Euler product*

$$\zeta(s) = \prod_p (1 - p^{-s})^{-1}$$

valid for $\Re(s) > 1$, appears for the first time in Euler's book *Introductio in Analysin Infinitorum*, published in 1748. Euler also studied the values of $\zeta(s)$ at the even positive and the negative integers, and he divined a functional equation, equivalent to Riemann's functional equation, for the closely related function $\sum(-1)^{n-1}/n^s$ (see the interesting account of Euler's work in Hardy's book [Hard]).

The problem of the distribution of prime numbers received attention for the first time with Gauss and Legendre, at the end of the eighteenth century. Gauss, in a letter to the astronomer Hencke in 1849, stated that he had found in his early years that the number $\pi(x)$ of primes up to x is well approximated by the function[2]

$$\mathrm{Li}(x) = \int_0^x \frac{\mathrm{d}t}{\log t}.$$

In 1837, Dirichlet proved his famous theorem of the existence of infinitely many primes in any arithmetic progression $qn+a$ with q and a positive coprime integers.

On May 24, 1848, Tchebychev read at the Academy of St. Petersburg his first memoir on the distribution of prime numbers, later published in 1850. It contains the first study of the function $\pi(x)$ by analytic methods. Tchebychev begins by taking the logarithm of the Euler product, obtaining[3]

$$-\sum_p \log(1 - \frac{1}{p^s}) + \log(s-1) = \log\big((s-1)\zeta(s)\big), \tag{2}$$

which is his starting point.

Next, he proves the integral formula

$$\zeta(s) - 1 - \frac{1}{s-1} = \frac{1}{\Gamma(s)} \int_0^\infty (\frac{1}{e^x - 1} - \frac{1}{x}) e^{-x} x^{s-1}\, \mathrm{d}x, \tag{3}$$

out of which he deduces that $(s-1)\zeta(s)$ has limit 1, and also has finite derivatives of any order, as s tends to 1 from the right. He then observes that the derivatives of any order of the left-hand side of (2) can be written as a fraction in which the numerator is a polynomial in the derivatives of $(s-1)\zeta(s)$, and the denominator is an integral power of $(s-1)\zeta(s)$, from which it follows that the left-hand side of (2) has finite derivatives of any order, as s tends to 1 from the right. From this, he is able to prove that if there is an asymptotic formula for $\pi(x)$ by means of a finite sum $\sum a_k x/(\log x)^k$, up to an order $O(x/(\log x)^N)$, then $a_k = (k-1)!$ for $k = 1, \ldots, N-1$. This is precisely the asymptotic expansion of the function $\mathrm{Li}(x)$, thus vindicating Gauss's intuition.

A second paper by Tchebychev gave rigorous proofs of explicit upper and lower bounds for $\pi(x)$, of the correct order of magnitude. Here, he introduces the counting functions

$$\vartheta(x) = \sum_{p \le x} \log p, \qquad \psi(x) = \vartheta(x) + \vartheta(\sqrt[2]{x}) + \vartheta(\sqrt[3]{x}) + \ldots$$

---

[2] The integral is a principal value in the sense of Cauchy.

[3] Tchebychev uses $1+\rho$ in place of our s. We write his formulas in modern notation.

and proves the identity[4]

$$\sum_{n \le x} \psi(\frac{x}{n}) = \log [x]! \, .$$

From this identity, he finally obtains numerical upper and lower bounds for $\psi(x)$, $\vartheta(x)$ and $\pi(x)$.

Popular variants of Tchebychev's method, based on the integrality of suitable ratios of factorials, originate much later and cannot be ascribed to Tchebychev.

Riemann's memoir on $\pi(x)$ is really astonishing for the novelty of ideas introduced. He first writes $\zeta(s)$ using the integral formula, valid for $\Re(s) > 1$:

$$\zeta(s) = \frac{1}{\Gamma(s)} \int_0^\infty \frac{e^{-x}}{1 - e^{-x}} \, x^{s-1} \, dx, \tag{4}$$

and then deforms the contour of integration in the complex plane, so as to obtain a representation valid for any s. This gives the analytic continuation and the functional equation of $\zeta(s)$. Then he gives a second proof of the functional equation in the symmetric form (1), introduces the function $\xi(t)$ and states some of its properties as a function of the complex variable t.

Riemann continues by writing the logarithm of the Euler product as an integral transform, valid for $\Re(s) > 1$:

$$\frac{1}{s} \log \zeta(s) = \int_1^\infty \Pi\,(x) x^{-s-1} \, dx \tag{5}$$

where

$$\Pi\,(x) = \pi(x) + \frac{1}{2}\,\pi(\sqrt[2]{x}) + \frac{1}{3}\,\pi(\sqrt[3]{x}) + \dots \, .$$

By Fourier inversion, he is able to express $\Pi\,(x)$ as a complex integral, and compute it using the calculus of residues. The residues occur at the singularities of $\log \zeta(s)$ at $s = 1$ and at the zeros of $\zeta(s)$. Finally an inversion formula expressing $\pi(x)$ in terms of $\Pi\,(x)$ yields Riemann's formula.

This was a remarkable achievement which immediately attracted much attention. Even if Riemann's initial line of attack may have been influenced by Tchebychev (we find several explicit references to Tchebychev in Riemann's unpublished Nachlass[5]) his great contribution was to see how the distribution of prime numbers is determined by the complex zeros of the zeta function.

At first sight, the Riemann hypothesis appears to be only a plausible interesting property of the special function $\zeta(s)$, and Riemann himself seems to take that view. He writes: "Hiervon wäre allerdings ein strenger Beweis zu wünschen; ich habe indess die Aufsuchung desselben nach einigen flüchtigen vergeblichen Versuchen vorläufig bei Seite gelassen, da er für den nächsten Zweck meiner Untersuchung entbehrlich schien.", which can be translated as "Without doubt it would be desirable to have a rigorous proof of this proposition; however I have left this research aside for the time being after some quick unsuccessful attempts, because it appears to be unnecessary for the immediate goal of my study."

[4] Here $[x]$ denotes the integral part of x.

[5] The Nachlass consists of Riemann's unpublished notes and is preserved in the mathematical library of the University of Göttingen. The part regarding the zeta function was analyzed in depth by C.L. Siegel [Sie].

On the other hand, one should not draw from this comment the conclusion that the Riemann hypothesis was for Riemann only a casual remark of minor interest. The validity of the Riemann hypothesis is equivalent to saying that the deviation of the number of primes from the mean $\mathrm{Li}(x)$ is

$$\pi(x) = \mathrm{Li}(x) + O\big(\sqrt{x}\log x\big);$$

the error term cannot be improved by much, since it is known to oscillate in both directions to order at least $\mathrm{Li}(\sqrt{x})\log\log\log x$ (Littlewood). In view of Riemann's comments at the end of his memoir about the approximation of $\pi(x)$ by $\mathrm{Li}(x)$, it is quite likely that he saw how his hypothesis was central to the question of how good an approximation to $\pi(x)$ one may get from his formula.

The failure of the Riemann hypothesis would create havoc in the distribution of prime numbers. This fact alone singles out the Riemann hypothesis as the main open question of prime number theory.

The Riemann hypothesis has become a central problem of pure mathematics, and not just because of its fundamental consequences for the law of distribution of prime numbers. One reason is that the Riemann zeta function is not an isolated object, rather is the prototype of a general class of functions, called *L-functions*, associated with algebraic (automorphic representations) or arithmetical objects (arithmetic varieties); we shall refer to them as *global L-functions*. They are Dirichlet series with a suitable Euler product, and are expected to satisfy an appropriate functional equation and a Riemann hypothesis. The factors of the Euler product may also be considered as some kind of zeta functions of a local nature, which also should satisfy an appropriate Riemann hypothesis (the so-called Ramanujan property). The most important properties of the algebraic or arithmetical objects underlying an L-function can or should be described in terms of the location of its zeros and poles, and values at special points.

The consequences of a Riemann hypothesis for global L-functions are important and varied. We mention here, to indicate the variety of situations to which it can be applied, an extremely strong effective form of Tchebotarev's density theorem for number fields, the non-trivial representability of 0 by a non-singular cubic form in 5 or more variables (provided it satisfies the appropriate necessary congruence conditions for solubility, Hooley), and Miller's deterministic polynomial time primality test. On the other hand, many deep results in number theory which are consequences of a general Riemann hypothesis can be shown to hold independently of it, thus adding considerable weight to the validity of the conjecture.

It is outside the scope of this article even to outline the definition of global L-functions, referring instead to Iwaniec and Sarnak [IS] for a survey of the expected properties satisfied by them; it suffices here to say that the study of the analytic properties of these functions presents extraordinary difficulties.

Already the analytic continuation of L-functions as meromorphic or entire functions is known only in special cases. For example, the functional equation for the L-function of an elliptic curve over $\mathbb{Q}$ and for its twists by Dirichlet characters is an easy consequence of, and is equivalent to, the existence of a parametrization of the curve by means of modular functions for a Hecke group $\Gamma_0(N)$; the real difficulty lies in establishing this modularity. No one knows how to prove this functional equation by analytic methods. However the modularity of elliptic curves over $\mathbb{Q}$ has been established directly, first in the semistable case in the spectacular work of Wiles [Wi] and Taylor and Wiles [TW] leading to the solution of Fermat's Last Theorem, and then in the general case in a recent preprint by Breuil, Conrad, Diamond and Taylor.

Not all L-functions are directly associated to arithmetic or geometric objects. The simplest example of L-functions not of arithmetic/geometric nature are those arising from Maass waveforms for a Riemann surface X uniformized by an arithmetic subgroup Γ of $\mathrm{PGL}(2,\mathbb{R})$. They are pull-backs $f(z)$, to the universal covering space $\Im(z) > 0$ of X, of simultaneous eigenfunctions for the action of the hyperbolic Laplacian and of the Hecke operators on X.

The most important case is again the group $\Gamma_0(N)$. In this case one can introduce a notion of *primitive* waveform, analogous to the notion of primitive Dirichlet character, meaning that the waveform is not induced from another waveform for a $\Gamma_0(N')$ with N' a proper divisor of N. For a primitive waveform, the action of the Hecke operators T_n is defined for every n and the L-function can be defined as $\sum \lambda_f(n)n^{-s}$ where $\lambda_f(n)$ is the eigenvalue of T_n acting on the waveform $f(z)$. Such an L-function has an Euler product and satisfies a functional equation analogous to that for $\zeta(s)$. It is also expected that it satisfies a Riemann hypothesis.

Not a single example of validity or failure of a Riemann hypothesis for an L-function is known up to this date. The Riemann hypothesis for $\zeta(s)$ does not seem to be any easier than for Dirichlet L-functions (except possibly for non-trivial real zeros), leading to the view that its solution may require attacking much more general problems, by means of entirely new ideas.

III. Evidence for the Riemann hypothesis. Notwithstanding some skepticism voiced in the past, based perhaps more on the number of failed attempts to a proof rather than on solid heuristics, it is fair to say that today there is quite a bit of evidence in its favor. We have already emphasized that the general Riemann hypothesis is consistent with our present knowledge of number theory. There is also specific evidence of a more direct nature, which we shall now examine.

First, strong numerical evidence.

Interestingly enough, the first numerical computation of the first few zeros of the zeta function already appears in Riemann's Nachlass. A rigorous verification of the Riemann hypothesis in a given range can be done numerically as follows. The number $N(T)$ of zeros of $\zeta(s)$ in the rectangle $\mathcal{R}$ with vertices at $-1-iT$, $2-iT$, $2+iT$, $-1+iT$ is given by Cauchy's integral

$$N(T)-1=\frac{1}{2\pi i}\int_{\partial\mathcal{R}}-\frac{\zeta'}{\zeta}(s)\,\mathrm{d}s,$$

provided T is not the imaginary part of a zero (the -1 in the left-hand side of this formula is due to the simple pole of $\zeta(s)$ at $s=1$). The zeta function and its derivative can be computed to arbitrary high precision using the MacLaurin summation formula or the Riemann-Siegel formula [Sie]; the quantity $N(T)-1$, which is an integer, is then computed exactly by dividing by $2\pi i$ the numerical evaluation of the integral, and rounding off its real part to the nearest integer (this is only of theoretical interest and much better methods are available in practice for computing $N(T)$ exactly). On the other hand, since $\xi(t)$ is continuous and real for real t, there will be a zero of odd order between any two points at which $\xi(t)$ changes sign. By judiciously choosing sample points, one can detect sign changes of $\xi(t)$ in the interval $[-T, T]$. If the number of sign changes equals $N(T)$, one concludes that all zeros of $\zeta(s)$ in $\mathcal{R}$ are simple and satisfy the Riemann hypothesis. In this way, it has been shown by van de Lune, te Riele and Winter [LRW] that the first 1.5 billion zeros of $\zeta(s)$, arranged by increasing positive imaginary part, are simple and satisfy the Riemann hypothesis.

The Riemann hypothesis is equivalent to the statement that all local maxima of $\xi(t)$ are positive and all local minima are negative, and it has been suggested that if a counterexample exists then it should be in the neighborhood of unusually large peaks of $|\zeta(\frac{1}{2}+it)|$. The above range for T is $T \cong 5 \times 10^8$ and is not large enough for $|\zeta(\frac{1}{2}+it)|$ to exhibit these peaks which are known to occur eventually. However, further calculations done by Odlyzko [Od] in selected intervals show that the Riemann hypothesis holds for over 3×10^8 zeros at heights up to[6] 2×10^{20}. These calculations also strongly support independent conjectures by Dyson and Montgomery [Mo] concerning the distribution of spacings between zeros.

Computing zeros of L-functions is more difficult, but this has been done in several cases, which include examples of Dirichlet L-functions, L-functions of elliptic curves, Maass L-functions and nonabelian Artin L-functions arising from number fields of small degree. No exception to a generalized Riemann hypothesis has been found in this way.

Second, it is known that hypothetical exceptions to the Riemann hypothesis must be rare if we move away from the line $\Re(s) = \frac{1}{2}$.

Let $N(\alpha, T)$ be the number of zeros of $\zeta(s)$ in the rectangle $\alpha \leq \Re(s) \leq 2$, $0 \leq \Im(s) \leq T$. The prototype result goes back to Bohr and Landau in 1914, namely $N(\alpha, T) = O(T)$ for any fixed α with $\frac{1}{2} < \alpha < 1$. A significant improvement of the result of Bohr and Landau was obtained by Carlson in 1920, obtaining the *density theorem* $N(\alpha, T) = O(T^{4\alpha(1-\alpha)+\varepsilon})$ for any fixed $\varepsilon > 0$. The fact that the exponent here is strictly less than 1 is important for arithmetic applications, for example in the study of primes in short intervals. The exponent in Carlson's theorem has gone through several successive refinements for various ranges of α, in particular in the range $\frac{3}{4} < \alpha < 1$. Curiously enough, the best exponent known up to date in the range $\frac{1}{2} < \alpha \leq \frac{3}{4}$ remains Ingham's exponent $3(1-\alpha)/(2-\alpha)$, obtained in 1940. For references to these results, the reader may consult the recent revision by Heath-Brown of the classical monograph of Titchmarsh [Ti], and the book by Ivic [Iv].

Third, it is known that more than 40% of nontrivial zeros of $\zeta(s)$ are simple and satisfy the Riemann hypothesis (Selberg [Sel], Levinson [Le], Conrey [Conr]). Most of these results have been extended to other L-functions, including all Dirichlet L-functions and L-functions associated to modular forms or Maass waveforms.

IV. Further evidence: varieties over finite fields. It may be said that the best evidence in favor of the Riemann hypothesis derives from the corresponding theory which has been developed in the context of algebraic varieties over finite fields. The simplest situation is as follows.

Let C be a nonsingular projective curve over a finite field $\mathbb{F}_q$ with $q = p^a$ elements, of characteristic p. Let $\mathrm{Div}(C)$ be the additive group of divisors on C

[6] The most recent calculations by Odlyzko, which are approaching completion, will explore completely the interval $[10^{22}, 10^{22} + 10^{10}]$.

defined over $\mathbb{F}_q$, in other words formal finite sums $\mathfrak{a} = \sum a_i P_i$ with $a_i \in \mathbb{Z}$ and P_i points of C defined over a finite extension of $\mathbb{F}_q$, such that $\phi(\mathfrak{a}) = \mathfrak{a}$ where ϕ is the Frobenius endomorphism on C raising coordinates to the q-th power. The quantity $\deg(\mathfrak{a}) = \sum a_i$ is the degree of the divisor $\mathfrak{a}$. The divisor $\mathfrak{a}$ is called effective if every a_i is a positive integer; in this case, we write $\mathfrak{a} > 0$. Finally, a prime divisor $\mathfrak{p}$ is a positive divisor which cannot be expressed as the sum of two positive divisors. By definition, the norm of a divisor $\mathfrak{a}$ is $N\mathfrak{a} = q^{\deg(\mathfrak{a})}$.

The zeta function of the curve C, as defined by E. Artin, H. Hasse and F.K. Schmidt, is

$$\zeta(s, C) = \sum_{\mathfrak{a}>0} \frac{1}{N\mathfrak{a}^s}.$$

This function has an Euler product

$$\zeta(s, C) = \prod_{\mathfrak{p}} (1 - N\mathfrak{p}^{-s})^{-1}$$

and a functional equation

$$q^{(g-1)s}\zeta(s, C) = q^{(g-1)(1-s)}\zeta(1 - s, C)$$

where g is the genus of the curve C; it is a consequence of the Riemann-Roch theorem. The function $\zeta(s, C)$ is a rational function of the variable $t = q^{-s}$, hence is periodic[7] with period $2\pi i/\log q$, and has simple poles at the points $s = 2\pi i m/\log q$ and $s = 1 + 2\pi i m/\log q$ for $m \in \mathbb{Z}$. Expressed in terms of the variable t, the zeta function becomes a rational function $Z(t, C)$ of t, with simple poles at $t = 1$ and $t = q^{-1}$. The use of the variable t, rather than q^{-s}, is more natural in the geometric case and we refer to Zeta functions, with a capital Z, to indicate the corresponding objects.

The Riemann hypothesis for $\zeta(s, C)$ is the statement that all its zeros have real part equal to $\frac{1}{2}$; in terms of the Zeta function $Z(t, C)$, which has a numerator of degree $2g$, has zeros of absolute value $q^{-\frac{1}{2}}$.

This is easy to verify if $g = 0$, because the numerator is 1. If $g = 1$, a proof was obtained by Hasse in 1934. The general case of arbitrary genus g was finally settled by Weil in the early 1940s (see his letter to E. Artin of July 10, 1942 where he gives a complete sketch of the theory of correspondences on a curve [We1]); his results were eventually published in book form in 1948 [We2].

Through his researches, Weil was led to the formulation of sweeping conjectures about Zeta functions of general algebraic varieties over finite fields, relating their properties to the topological structure of the underlying algebraic variety. Here the Riemann hypothesis, in a simplified form, is the statement that the reciprocals of the zeros and poles of the Zeta function (the so-called *characteristic roots*) have absolute value $q^{d/2}$ with d a positive integer or 0, and are interpreted as eigenvalues of the Frobenius automorphism acting on the cohomology of the variety. After M. Artin, A. Grothendieck and J.-L. Verdier developed the fundamental tool of étale cohomology, the proof of the corresponding Riemann hypothesis for Zeta functions of arbitrary varieties over finite fields was finally obtained by Deligne [Del1], [Del2]. Deligne's theorem surely ranks as one of the crowning achievements

[7] Similarly, $\zeta(s)$ is almost periodic in any half-plane $\Re(s) \geq 1 + \delta$, $\delta > 0$.

of twentieth century mathematics. Its numerous applications to the solution of long-standing problems in number theory, algebraic geometry and discrete mathematics are witness to the significance of these general Riemann hypotheses.

In our opinion, these results in the geometric setting cannot be ignored as not relevant to the understanding of the classical Riemann hypothesis; the analogies are too compelling to be dismissed outright.

V. Further evidence: the explicit formula. A conceptually important generalization of Riemann's explicit formula for $\pi(x)$, obtained by Weil [We3] in 1952, offers a clue to what may lie still undiscovered behind the problem.

Consider the class $\mathcal{W}$ of complex-valued functions $f(x)$ on the positive half-line $\mathbb{R}_+$, continuous and continuously differentiable except for finitely many points at which both $f(x)$ and $f'(x)$ have at most a discontinuity of the first kind, and at which the value of $f(x)$ and $f'(x)$ is defined as the average of the right and left limits there. Suppose also that there is $\delta > 0$ such that $f(x) = O(x^\delta)$ as $x \to 0+$ and $f(x) = O(x^{-1-\delta})$ as $x \to +\infty$.

Let $\widetilde{f}(s)$ be the Mellin transform

$$\widetilde{f}(s) = \int_0^\infty f(x)\, x^s \, \frac{dx}{x},$$

which is an analytic function of s for $-\delta < \Re(s) < 1 + \delta$.

For the Riemann zeta function, Weil's formula can be stated as follows. Let $\Lambda(n) = \log p$ if $n = p^a$ is a power of a prime p, and 0 otherwise. We have

Explicit Formula. *For $f \in \mathcal{W}$ we have*

$$\widetilde{f}(0) - \sum_\rho \widetilde{f}(\rho) + \widetilde{f}(1) = \sum_{n=1}^\infty \Lambda(n)\{f(n) + \frac{1}{n} f(\frac{1}{n})\} + (\log 4\pi + \gamma) f(1)$$
$$+ \int_1^\infty \{f(x) + \frac{1}{x} f(\frac{1}{x}) - \frac{2}{x} f(1)\} \frac{dx}{x - x^{-1}}.$$

Here the first sum ranges over all nontrivial zeros of $\zeta(s)$ and is understood as

$$\lim_{T \to +\infty} \sum_{|\Im(\rho)| < T} \widetilde{f}(\rho).$$

In his paper, Weil showed that there is a corresponding formula for zeta and L-functions of number fields as well as for Zeta functions of curves over finite fields. The terms in the right-hand side of the equation can be written as a sum of terms of local nature, associated to the absolute values of the underlying number field, or function field in the case of curves over a field of positive characteristic. Moreover, in the latter case the explicit formula can be deduced from the Lefschetz fixed point formula, applied to the Frobenius endomorphism on the curve C. The three terms in the left-hand side, namely $\widetilde{f}(0)$, $\sum \widetilde{f}(\rho)$, $\widetilde{f}(1)$, now correspond to the trace of the Frobenius automorphism on the l-adic cohomology of C (the interesting term $\sum \widetilde{f}(\rho)$ corresponds to the trace on H^1), while the right-hand side corresponds to the number of fixed points of the Frobenius endomorphism, namely the prime divisors of degree 1 on C.

Weil also proved that the Riemann hypothesis is equivalent to the negativity of the right-hand side for all functions $f(x)$ of type

$$f(x) = \int_0^\infty g(xy)\,\overline{g(y)}\,\mathrm{d}y,$$

whenever $g \in \mathcal{W}$ satisfies the additional conditions

$$\int_0^\infty g(x)\,\frac{\mathrm{d}x}{x} = \int_0^\infty g(x)\,\mathrm{d}x = 0.$$

In the geometric case of curves over a finite field, this negativity is a rather easy consequence of the *algebraic index theorem* for surfaces, namely:

Algebraic Index Theorem. *Let X be a projective nonsingular surface defined over an algebraically closed field. Then the self-intersection quadratic form $(D \cdot D)$, restricted to the group of divisors D on X of degree 0 in the projective embedding of X, is negative semidefinite.*

The algebraic index theorem for surfaces is essentially due to Severi[8] in 1906 [Sev,§2,Teo.I]. The proof uses the Riemann-Roch theorem on X and the finiteness of families of curves on X of a given degree; no other proof by algebraic methods is known up to now, although much later several authors independently rediscovered Severi's argument.

The algebraic index theorem for nonsingular projective varieties of even dimension over the complex numbers was first formulated and proved by Hodge, as a consequence of his theory of harmonic forms. No algebraic proof of Hodge's theorem is known, and it remains a fundamental open problem to extend it to the case of varieties over fields of positive characteristic.

The work of Montgomery [Mo], Odlyzko [Od] and Rudnick and Sarnak [RS] on correlations for spacings of zeros of $\xi(t)$ suggests that L-functions can be grouped into a few families, in each of which the spacing correlation is universal; the conjectured spacing correlation is the same as for the limiting distribution of eigenvalues of random orthogonal, unitary or symplectic matrices in suitable universal families, as the dimension goes to ∞. All this is compatible with the view expressed by Hilbert and Pólya that the zeros of $\xi(t)$ could be the eigenvalues of a self-adjoint linear operator on an appropriate Hilbert space. It should also be noted that a corresponding unconditional theory for the spacing correlations of characteristic roots of Zeta functions of families of algebraic varieties over a finite field, has been developed by Katz and Sarnak [KS], using methods introduced by Deligne in his proof of the Riemann hypothesis for varieties over finite fields. Thus the problem of spacing correlations for zeros of L-functions appears to lie very deep.

All this leads to several basic questions.

Is there a theory in the global case, playing the same role as cohomology does for Zeta functions of varieties over a field of positive characteristic? Is there an analogue of a Frobenius automorphism in the classical case? Is there a general index theorem by which one can prove the classical Riemann hypothesis? We are

[8] Severi showed that a divisor D on X is algebraically equivalent to 0 up to torsion, if it has degree 0 and $(D \cdot D) = 0$. His proof holds, without modifications, under the weaker assumption $(D \cdot D) \geq 0$, which yields the index theorem.

here in the realm of conjectures and speculation. In the adelic setting propounded by Tate and Weil, the papers [Conn], [Den], [Hara] offer glimpses of a possible setup for these basic problems.

On the other hand, there are L-functions, such as those attached to Maass waveforms, which do not seem to originate from geometry and for which we still expect a Riemann hypothesis to be valid. For them, we do not have algebraic and geometric models to guide our thinking, and entirely new ideas may be needed to study these intriguing objects.

INSTITUTE FOR ADVANCED STUDY, PRINCETON, NJ 08540

REFERENCES

[Conn] A. CONNES, Trace formula in noncommutative geometry and the zeros of the Riemann zeta function, *Selecta Math. (NS)* **5** (1999), 29–106.

[Conr] J.B. CONREY, More than two fifths of the zeros of the Riemann zeta function are on the critical line, *J. reine angew. Math.* **399** (1989), 1–26.

[Del1] P. DELIGNE, La Conjecture de Weil I, *Publications Math. IHES* **43** (1974), 273–308.

[Del2] P. DELIGNE, La Conjecture de Weil II, *Publications Math. IHES* **52** (1980), 137–252.

[Den] C. DENINGER, Some analogies between number theory and dynamical systems on foliated spaces, *Proc. Int. Congress Math. Berlin 1998*, Vol. I, 163–186.

[Ed] H.M. EDWARDS, *Riemann's Zeta Function*, Academic Press, New York - London 1974.

[Hara] S. HARAN, Index theory, potential theory, and the Riemann hypothesis, *L-functions and Arithmetic, Durham 1990*, LMS Lecture Notes **153** (1991), 257–270.

[Hard] G.H. HARDY, *Divergent Series*, Oxford Univ. Press 1949, Ch. II, 23–26.

[IS] H. IWANIEC AND P. SARNAK, Perspectives on the Analytic Theory of L-Functions, to appear in proceedings of the conference *Visions 2000*, Tel Aviv.

[Iv] A. IVIČ, *The Riemann Zeta-Function - The Theory of the Riemann Zeta-Function with Applications*, John Wiley & Sons Inc., New York - Chichester - Brisbane - Toronto - Singapore 1985.

[KS] N.M. KATZ AND P. SARNAK, Random matrices, Frobenius eigenvalues and monodromy, *Amer. Math. Soc. Coll. Publ.* **49**, Amer. Math. Soc., Providence RI 1999.

[La] E. LANDAU, *Primzahlen*, Zwei Bd., IInd ed., with an Appendix by Dr. Paul T. Bateman, Chelsea, New York 1953.

[Le] N. LEVINSON, More than one-third of the zeros of the Riemann zeta-function are on $\sigma = 1/2$, *Adv. Math.* **13** (1974), 383–436.

[LRW] J. VAN DE LUNE, J.J. TE RIELE AND D.T. WINTER, On the zeros of the Riemann zeta function in the critical strip, IV, *Math. of Comp.* **46** (1986), 667–681.

[Mo] H.L. MONTGOMERY, Distribution of the Zeros of the Riemann Zeta Function, *Proceedings Int. Cong. Math. Vancouver 1974*, Vol. I, 379–381.

[Od] A.M. ODLYZKO, Supercomputers and the Riemann zeta function, *Supercomputing 89: Supercomputing Structures & Computations, Proc. 4-th Intern. Conf. on Supercomputing*, L.P. Kartashev and S.I. Kartashev (eds.), International Supercomputing Institute 1989, 348–352.

[Ri] B. RIEMANN, Ueber die Anzahl der Primzahlen unter einer gegebenen Grösse, *Monat. der Königl. Preuss. Akad. der Wissen. zu Berlin aus der Jahre 1859* (1860), 671–680; also, *Gesammelte math. Werke und wissensch. Nachlass*, 2. Aufl. 1892, 145–155.

[RS] Z. RUDNICK AND P. SARNAK, Zeros of principal L-functions and random matrix theory, *Duke Math. J.* **82** (1996), 269–322.

[Sel] A. SELBERG, On the zeros of the zeta-function of Riemann, *Der Kong. Norske Vidensk. Selsk. Forhand.* **15** (1942), 59–62; also, *Collected Papers*, Springer-Verlag, Berlin - Heidelberg - New York 1989, Vol. I, 156–159.

[Sev] F. SEVERI, Sulla totalità delle curve algebriche tracciate sopra una superficie algebrica, *Math. Annalen* **62** (1906), 194–225.

[Sie] C.L. SIEGEL, Über Riemanns Nachlaß zur analytischen Zahlentheorie, *Quellen und Studien zur Geschichte der Mathematik, Astronomie und Physik* **2** (1932), 45–80; also *Gesammelte Abhandlungen*, Springer-Verlag, Berlin - Heidelberg - New York 1966, Bd. I, 275–310.

[Ti] E.C. TITCHMARSH, *The Theory of the Riemann Zeta Function*, 2nd ed. revised by R.D. Heath-Brown, Oxford Univ. Press 1986.

[TW] R. TAYLOR AND A. WILES, Ring theoretic properties of certain Hecke algebras, *Annals of Math.* **141** (1995), 553–572.

[We1] A. WEIL, *Œuvres Scientifiques–Collected Papers*, corrected 2nd printing, Springer-Verlag, New York - Berlin 1980, Vol. I, 280-298.

[We2] A. WEIL, *Sur les Courbes Algébriques et les Variétés qui s'en déduisent*, Hermann & Cie, Paris 1948.

[We3] A. WEIL, Sur les "formules explicites" de la théorie des nombres premiers, *Meddelanden Från Lunds Univ. Mat. Sem.* (dedié à M. Riesz), (1952), 252-265; also, *Œuvres Scientifiques–Collected Papers*, corrected 2nd printing, Springer-Verlag, New York - Berlin 1980, Vol. II, 48–61.

[Wi] A. WILES, Modular elliptic curves and Fermat's Last Theorem, *Annals of Math.* **141** (1995), 443–551.

11.2 P. Sarnak (2004)

Problems of the Millennium: The Riemann Hypothesis

> *The Riemann hypothesis is the central problem and it implies many, many things. One thing that makes it rather unusual in mathematics today is that there must be over five hundred papers — somebody should go and count — which start "Assume the Riemann hypothesis", and the conclusion is fantastic. And those [conclusions] would then become theorems ... With this one solution you would have proven five hundred theorems or more at once [138].*
>
> **Peter Sarnak**

This paper was written for the 2004 annual report of the Clay Mathematics Institute. In this paper Peter Sarnak elaborates on Bombieri's official problem description. Sarnak focuses his attention on the grand Riemann hypothesis and presents some connected problems. He also states his skepticism toward the random matrix approach.

Problems of the Millennium: The Riemann Hypothesis (2004)

by

Peter Sarnak
Princeton University & Courant Institute of Math. Sciences.

The Riemann Hypothesis (RH) is one of the seven millennium prize problems put forth by the Clay Mathematical Institute in 2000. Bombieri's statement [Bo1] written for that occasion is excellent. My plan here is to expand on some of his comments as well as to discuss some recent developments. RH has attracted further attention of late in view of the fact that three popular books were written about it in 2003 and a fourth one is forthcoming in 2005. Unlike the case of Fermat's Last Theorem, where the work of Frey, Serre and Ribet redefined the outlook on the problem by connecting it to a mainstream conjecture in the theory of modular forms (the "Shimura-Taniyama-Weil Conjecture"), in the case of RH we cannot point to any such dramatic advance. However, that doesn't mean that things have stood still.

For definiteness we recall the statement of RH. For $\Re(s) > 1$ the zeta function is defined by

$$\zeta(s) = \sum_{n=1}^{\infty} n^{-s} = \prod_{p} (1 - p^{-s})^{-1}, \tag{1}$$

the product being over the prime numbers. Riemann showed how to continue zeta analytically in s and he established the Functional Equation:

$$\Lambda(s) := \pi^{-s/2}\,\Gamma\left(\frac{s}{2}\right)\,\zeta(s) = \Lambda(1-s)\,, \tag{2}$$

Γ being the Gamma function. RH is the assertion that all the zeros of $\Lambda(s)$ are on the line of symmetry for the functional equation, that is on $\Re(s) = \frac{1}{2}$. Elegant, crisp, falsifiable and far-reaching this conjecture is the epitome of what a good conjecture should be. Moreover, its generalizations to other zeta functions (see below) have many striking consequences making the conjecture even more important. Add to this the fact that the problem has resisted the efforts of many of the finest mathematicians and that it now carries a financial reward, it is not surprising that it has attracted so much attention.

We begin by describing the generalizations of RH which lead to the Grand Riemann Hypothesis. After that we discuss various consequences and developments. The first extension is to the zeta functions of Dirichlet, which he introduced in order to study primes in arithmetic progressions and which go by the name L-functions. Let χ be a primitive Dirichlet character of modulus q (i.e. $\chi(mn) = \chi(m)\,\chi(n)$, $\chi(1) = 1$, $\chi(m + bq) = \chi(m)$) then L is defined by

$$L(s,\chi) = \sum_{n=1}^{\infty} \chi(n) n^{-s} = \prod_{p} (1 - \chi(p)p^{-s})^{-1}\,. \tag{3}$$

As with zeta, $L(s,\chi)$ extends to an entire function and satisfies the functional equation:

$$\Lambda(s,\chi) := \pi^{-(s+a_\chi)/2}\,\Gamma\left(\frac{s+a_\chi}{2}\right) L(s,\chi) \;=\; \epsilon_\chi q^{\frac{1}{2}-s}\,\Lambda(1-s,\bar{\chi})\,, \tag{4}$$

where $a_\chi = (1+\chi(-1))/2$ and ϵ_χ is the sign of a Gauss sum (see [Da]). These $L(s,\chi)$'s give all the degree one L-functions (i.e., in the product over primes each local factor at p is the inverse of a polynomial of degree one in p^{-s}). The general L-function of degree m comes from an automorphic form on the general linear group of m by m invertible matrices, GL_m. As an explicit example of a degree two L-function, consider the discriminant

$$\triangle(z) \;=\; e^{2\pi i z}\prod_{n=1}^{\infty}\left(1-e^{2\pi i n z}\right)^{24} := \sum_{n=1}^{\infty}\tau(n)\,e^{2\pi i n z}\,, \tag{5}$$

for z in the upper half plane. $\triangle(z)$ is a holomorphic cusp form of weight 12 for the modular group. That is

$$\triangle\left(\frac{az+b}{cz+d}\right) \;=\; (cz+d)^{12}\,\triangle(z)$$

for $a,b,c,d\in\mathbb{Z}$, $ad-bc=1$. Associated to $\triangle$ is its degree two L-function,

$$L(s,\triangle) \;=\; \sum_{n=1}^{\infty}\frac{\tau(n)}{n^{11/2}}\,n^{-s} \;=\; \prod_{p}\left(1-\frac{\tau(p)}{p^{11/2}}\,p^{-s}\;+\;p^{-2s}\right)^{-1}. \tag{6}$$

$L(s,\triangle)$ is entire and satisfies the functional equation

$$\Lambda(s,\triangle) := \Gamma_{\mathbb{R}}\left(s+\frac{11}{2}\right)\Gamma_{\mathbb{R}}\left(s+\frac{13}{2}\right) L(s,\triangle) \;=\; \Lambda(1-s,\triangle)\,,$$

where

$$\Gamma_{\mathbb{R}}(s) \;=\; \pi^{-s/2}\,\Gamma(s/2)\,. \tag{7}$$

The definition in the general case requires more sophisticated tools, which can be found in [We1] for example. Let $\mathbb{A}$ be the ring of adeles of $\mathbb{Q}$ and let π be an automorphic cuspidal representation of $GL_m(\mathbb{A})$ with central character χ (see [J]). The representation π is equivalent to $\underset{v}{\otimes}\,\pi_v$ with $v=\infty(\mathbb{Q}_\infty=\mathbb{R})$ or $v=p$ and π_v an irreducible unitary representation of $GL_m(\mathbb{Q}_v)$. Corresponding to each prime p and local representation π_p one forms the local factor $L(s,\pi_p)$ which takes the form:

$$L(s,\pi_p) \;=\; \Pi_{j=1}^{m}\left(1-\alpha_{j,\pi}(p)p^{-s}\right)^{-1} \tag{8}$$

for m complex parameters $\alpha_{j,\pi}(p)$ determined by π_p [T]. Similarly for $v=\infty$, π_∞ determines parameters $\mu_{j,\pi}(\infty)$ such that

$$L(s,\pi_\infty) \;=\; \Pi_{j=1}^{m}\,\Gamma_{\mathbb{R}}(s-\mu_{j,\pi}(\infty))\,. \tag{9}$$

The standard L-function associated to π is defined using this data by

$$L(s,\pi) = \prod_p L(s,\pi_p)\,. \tag{10}$$

The completed L-function is given as before by

$$\Lambda(s,\pi) = L(s,\pi_\infty)\,L(s,\pi)\,. \tag{11}$$

From the automorphy of π one can show that $\Lambda(s,\pi)$ is entire and satisfies a Functional Equation (see [J])

$$\Lambda(s,\pi) = \epsilon_\pi\, N_\pi^{\frac{1}{2}-s}\,\Lambda(1-s,\tilde{\pi}), \tag{12}$$

where $N_\pi \geq 1$ is an integer and is called the conductor of π, ϵ_π is of modulus 1 and is computable in terms of Gauss sums and $\tilde{\pi}$ is the contragredient representation $\tilde{\pi}(g) = \pi({}^tg^{-1})$.

General conjectures of Langlands assert these standard L-functions multiplicatively generate all L-functions (in particular Dedekind Zeta Functions, Artin L-functions, Hasse-Weil Zeta Functions, ...). So at least conjecturally one is reduced to the study of these. We can now state the general hypothesis (GRH).

GRAND RIEMANN HYPOTHESIS

Let π be as above then the zeros of $\Lambda(s,\pi)$ all lie on $\Re(s) = \frac{1}{2}$.

$L(s,\pi)$ is a generating function made out of the data π_p for each prime p and GRH naturally gives very sharp information about the variation of π_p with p. Many of its applications make direct use of this. However, there are also many applications where it is the size of $L(s,\pi)$ on $\Re(s) = \frac{1}{2}$ that is the critical issue (this coming from the ever growing number of formulae which relate values of L-functions at special points such as $s = \frac{1}{2}$ to arithmetic information and to periods, see [Wa] and [Wat] for example). GRH implies uniform and sharp bounds. Define the analytic conductor C_π of π to be $N_\pi \Pi_{j=1}^m (1 + |\mu_{j,\pi}(\infty)|)$. It measures the complexity of π. In what follows we fix m. For $X \geq 1$ the number of π with $C_\pi \leq X$ is finite. It is known [Mo] that for each $\epsilon > 0$ there is a constant B_ϵ such that $|L(\sigma + it, \pi)| \leq B_\epsilon C_\pi^\epsilon$ for $\sigma \geq 1 + \epsilon$ and $t \in \mathbb{R}$. The Grand Lindelöf Hypothesis (GLH) is the assertion that such bounds continue to hold for $\sigma \geq \frac{1}{2}$. Precisely, that for $\epsilon > 0$ there is B'_ϵ such that for $t \in \mathbb{R}$

$$|L\left(\frac{1}{2} + it, \pi\right)| \leq B'_\epsilon((1+|t|)C_\pi)^\epsilon\,. \tag{13}$$

GRH implies GLH with an effective B'_ϵ for $\epsilon > 0$.

It would be interesting to compile the long list of known consequences (many of which are quite indirect) of GRH and GLH. At the top of the list would be the prime number theorem with a sharp remainder term, the connection going back to Riemann's paper. He gives a formula for the number of primes less than X from which the prime number theorem would follow if one knew some deeper information about the zeros of zeta. In fact, the eventual proof of the prime number theorem went via this route with Hadamard and de la Vallée Poussin showing that $\zeta(s) \neq 0$ for $\Re(s) \geq 1$ (this is also known for all the $L(s, \pi)$'s). Other consequences of GRH for the family of Dirichlet L-functions followed, including the surprising result of Goldbach type due to Hardy and Littlewood, who showed that every sufficiently large odd integer is a sum of three primes. Later I. Vinogradov developed some fundamental novel methods which allowed him to prove this result unconditionally. A more recent application, again of GRH for Dirichlet L-functions, which was noted in Bombieri [Bo1] is the algorithm of Miller which determines whether a large integer n is prime in $O((\log n)^4)$ steps. Two years ago, Agrawal-Kayal and Saxena [A-K-S] put forth a primality testing algorithm (like Miller's it is based on Fermat's little Theorem) which they show without recourse to any unproven hypothesis, runs in $O((\log n)^{15/2})$ steps (the important point being that it is polynomial in $\log n$). Thus in practice GRH is used as a very reliable working hypothesis, which in many cases has been removed.

We give four examples of results in the theory of primes, Diophantine equations and mathematical physics which so far have been established only under GRH.

(A) (Serre), [Se, p. 632 and 715]: Given two non-isogeneous elliptic curves E and E' over $\mathbb{Q}$, there is a prime p which is $O((\log N_E N_{E'})^2)$ for which E and E' have good reduction at p and their number of points mod p are different. Here N_E is the conductor of E.

(B) (Artin's primitive root conjecture): If $b \neq \pm 1$ or a perfect square then b is a primitive root for infinitely many primes p [Ho].

(C) Ramanujan points out that congruence tests prevent $x^2 + y^2 + 10z^2$ representing any positive integer of the form $4^\lambda(16\mu + 6)$. He asks which numbers not of this form are not represented and lists 16 such. There are in fact exactly 18 such exceptions [O-S], 3, 7, 21, 31, 33, 43, 67, 79, 87, 133, 217, 219, 223, 253, 307, 391, 679, 2719. (The story for a general ternary quadratic form is similar but a little more complicated)

(D) The problem of the rate of equidistribution in the semi-classical limit of quantum eigenstates for a Hamiltonian which is classically chaotic, is a central one in "quantum chaos." The only Hamiltonian for which progress has been achieved is that of geodesic motion on an arithmetic hyperbolic surface where the equidistribution and its exact rate are known assuming GLH [Wat].

Towards problems (B), (C) and (D) slightly weaker but impressive unconditional results are known. In [HB] following [G-M] it is shown that there are at most three squarefree positive b's for which (B) is not true. Following the advance in [I], a general result about ternary quadratic forms is proven in [D-SP] from which it follows that the list of exceptions in (C) is finite. The method does not give an effective bound for the largest possible exception. As for (D) for arithmetic hyperbolic surfaces, the equidistribution problem is all but settled in [Li] using ergodic theoretic methods.

The passage to such results is often based on approximations to GRH and GLH that have been proven. The basic approximations to GRH that are known are zero free regions (the best ones for zeta being based on I. Vinogradov's method (see [I-K] for example)), and zero density theorems which give upper bounds for the total number of zeros that a family of L-functions can have in a box $\sigma_0 \leq \sigma < 1$ and $|t| \leq T$, where $\sigma_0 \geq \frac{1}{2}$ and $T > 0$. The best known approximation of GRH for the L-functions $L(s, \chi)$ is the Bombieri-A.Vinogradov Theorem [Bo2], [V]. It gives sharp bounds for the average equidistribution of primes in progressions and it has often served as a complete substitute for GRH in applications. The approximations to GLH which are decisive in various applications (including (C) above) are the so-called subconvex estimates (see [I-S]) which are known in some generality. Neither the zero density theorems nor the subconvex estimates are known for general families of automorphic L-functions and establishing them would be an important achievement.

As pointed out in [Bo1] there is an analogue of RH for curves and more generally varieties defined over finite fields. These analogues in their full generality are known as the Weil Conjectures. They have had a major impact and they have been proven! There are at least two fundamental steps in the proof. The first is the linearization of the problem [A-G-V], that is the realization of the zeros of the corresponding zeta functions in terms of eigenvalues of a linear transformation (specifically the eigenvalues of Frobenius acting on an associated cohomology group). This is then used together with a second step [De] which involves deforming the given variety in a family. The family has a symmetry group (the monodromy group) which is used together with its high tensor power representations and a positivity argument to prove the Weil Conjectures for each member of the family at once (see [K] for an overview). Like GRH, the Weil Conjectures have found far-reaching and often unexpected applications.

The success of this analogue is for many a guide as to what to expect in the case of GRH itself. Apparently both Hilbert and Polya (see the letter from Polya to Odlyzko [Od1]) suggested that there might be a spectral interpretation of the zeros of the zeta function in which the corresponding operator (after a change of variable) is self-adjoint, hence rendering the zeros on the line $\sigma = \frac{1}{2}$. While the evidence for the spectral nature of the zeros of any $\Lambda(s, \pi)$ has grown dramatically in recent years, I don't believe that the self-adjointness idea is very likely. It is not the source of the proof of the Weil Conjectures. In fact, it is the deformation in families idea that lies behind much of the progress in the known approximations to GRH and GLH and which seems more promising.

The spectral nature of the zeros emerges clearly when studying the local statistical fluctuations of the high zeros of a given $\Lambda(s, \pi)$ or the low-lying zeros of a given family of L-functions. These statistical distributions are apparently dictated by random matrix ensembles [K-S]. For a given $\Lambda(s, \pi)$ the local fluctuations of high zeros are universal and follow the laws of fluctuations of the eigenvalue distribution for matrices in the Gaussian Unitary Ensemble (GUE). We call this the "Montgomery-Odlyzko Law." For low-lying zeros in a family there is a symmetry type that one can associate with the family from which the densities and fluctuations can be predicted. The analogues of these fluctuation questions are understood in the setting of varieties over finite fields [K-S], and this in fact motivated the developments for the $\Lambda(s, \pi)$'s. Numerical experimentations starting with [Od2] have given further striking confirmation of this statistical fluctuation phenomenon. Based on this and a more sophisticated analysis, Keating and Snaith have put forth very elegant conjectures for the asymptotics of the moments of $L(\frac{1}{2} + it, \pi)$ for $|t| \leq T$ a $T \to \infty$, as well as for $L(\frac{1}{2}, \pi)$ as π varies over a family (see [C-F-K-S-R] and also [D-G-H], the latter being an approach based on multiple Dirichlet series). The numerical and theoretical confirmation of these conjectures is again very striking.

The above gives ample evidence for there being a spectral interpretation of the zeros of an L-function and indeed such interpretations have been given. One which has proven to be important is via Eisenstein series [La]. These are automorphic forms associated with boundaries of arithmetic quotients of semi-simple groups. These series have poles at the zeros of L-functions (hence the zeros can in this way be thought of as resonances for a spectral problem). Combining this with a positivity argument using the inner product formula for Eisenstein series ("Maass Selberg Formula") yields effective zero free regions in $\Re(s) \leq 1$ for all L-functions whose analytic continuation and functional equations are known (see [G-L-S]). As mentioned earlier, general conjectures of Langlands assert that the L-functions in question should be products of standard L-functions but this is far from being established. This spectral method for non-vanishing applies to L-functions for which the generalizations of the method of de la Vallée Poussin do not work and so at least for now, this method should be considered as the most powerful one towards GRH.

Connes [Co] defines an action of the idele class group $\mathbb{A}^*/\mathbb{Q}^*$ on a singular space of adeles (the action is basically multiplication over addition). In order to make sense of the space he uses various Sobolev spaces and shows that the decomposition of this abelian group action corresponds to the zeros of all the $\Lambda(s,\chi)$'s which lie on $\Re(s) = \frac{1}{2}$. This artificial use of function spaces in order to pick up only the eigenvalues of the action that lie on a line, makes it difficult to give a spectral interpretation. In a recent paper Meyer [Me] fixes this problem by using a much larger space of functions. He shows that the decomposition of the action on his space has eigenvalues corresponding to all the zeros of all the $\Lambda(s,\chi)$'s whether they lie on $\Re(s) = \frac{1}{2}$ or not. In this way he gives a spectral interpretation of the zeros of Dirichlet's L-functions. Taking traces of the action he derives (following Connes) the explicit formula of Weil-Guinand-Riemann [We2]. This derivation is quite different from the usual complex analytic one which is done by taking logarithmic derivatives, shifting contours and using the functional equation. Whether this spectral interpretation can be used to prove anything new about L-functions remains to be seen. The explicit formula itself has been very useful in the study of the distribution of the zeros of L-functions. In [Bo3] an in-depth investigation of the related Weil quadratic functional is undertaken.

There have been many equivalences of RH that have been found over the years. Without their having led to any new information about the zeta function and without being able to look into the future, it is difficult to know if such equivalences constitute any real progress. In an evolving paper, de Branges constructs a Hilbert space of entire functions from an L-function. He relates positivity properties of associated kernel functions to the zeros of the L-function. Whether this approach can be used to give any information about the zeros of the L-function is unclear. In any case, it should be noted that the positivity condition that he would like to verify in his recent attempts, is false since it implies statements about the zeros of the zeta function which are demonstrably false [C-L].

We end with some philosophical comments. We have highlighted GRH as the central problem. One should perhaps entertain the possibility that RH is true but GRH is false or even that RH is false. In fact Odlyzko, who has computed the 10^{23}-rd zero of zeta and billions of its neighbors, notes that the fact that the first 10^{13} zeros are on the line $\Re(s) = \frac{1}{2}$ (Gourdon 2004) should not by itself be taken as very convincing evidence for RH. One reason to be cautious is the following. The function $S(t)$ which measures the deviation of the number of zeros of height at most t from the expected number of such zeros, satisfies $|S(t)| < 1$ for $t < 280$ (and hence immediately all the zeros in this range are on the half line), while the largest observed value of $S(t)$ in the range that it has been computed, is about 3.2. On the other hand it is known that the mean-square of $S(t)$ is asymptotically $\frac{1}{2\pi^2} \log\log t$ ([Sel]), and hence $S(t)$ gets large but does so very slowly. One can argue that unless $S(t)$ is reasonably large (say 100), one has not as yet seen the true state of affairs as far as the behavior of the zeros of zeta. Needless to say, the height t for which $S(t)$ is of this size is far beyond the computational capabilities of present day machines. Returning to the expected scenario that RH is true, it is quite possible (even likely) that the first proof of RH that is given will not generalize to the $\Lambda(s,\pi)$ cases. Most people believe that any such proof for zeta would at least apply to the family of Dirichlet L-functions.* In fact it would be quite disappointing if it didn't since the juicy applications really start with this family. The millennium prize correctly focuses on the original basic case of zeta, however a closer reading of the fine-print of the rules shows them to be more guarded in the case that a disproof is given. As far as the issue of what might or might not be true, some may feel that GRH is true for the $\Lambda(s,\pi)$'s for which π is arithmetic in nature (in particular that the coefficients in the local factor $L(s,\pi_p)$ should be algebraic) but that for the more transcendental π's such as general Maass forms, that it may fail. I am an optimist and don't contemplate such a world.

To conclude, we point the reader to the interesting recent article by Conrey on the Riemann Hypothesis [Conr] as well as the comprehensive treatment by Iwaniec and Kowalski [I-K] of the modern tools, which are used in the study of many of the topics mentioned above.

Bibliography

[A-K-S] M. Agrawal, N. Kayal and N. Saxena, "Primes is in P" *Annals of Math.*, **160**, (2004), 781-793.

[A-G-V] M. Artin, A. Grothendieck and J-L. Verdier, **SG4** "Théorie des topos et cohomologie étale des schémas," **SLN**, **269, 270, 305**, (1972).

[Bo1] E. Bombieri, "Problems of the millennium: The Riemann Hypothesis," *CLAY*, (2000).

[Bo2] E. Bombieri, "On the large sieve," *Mathematika*, **12**, (1965), 201-225.

[Bo3] E. Bombieri, "A variational approach to the explicit formula," *CPAM*, **56**, (2003), 1151-1164.

[Co] A. Connes, "Trace formula in noncommutative geometry and the zeros of the Riemann zeta function," *Selecta Math*, **5**, (1999), 29-106.

[C-F-K-S-R] J. Conrey, D. Farmer, J. Keating, N. Snaith and M. Rubinstein, "Integral moments of ζ and L-functions," (2004).

*Hardy (Collected Papers, Vol 1, page 560) assures us that latter will be proven within a week of a proof of the former.

[Conr] B. Conrey,"The Riemann Hypothesis," *Notices of AMS*, March, 2003, 341-353.

[C-L] B. Conrey and X. Li, "A note on some positivity conditions related to the zeta and L-functions," *IMRN*, (2000), 929-940.

[Da] H. Davenport, "Multiplicative Number Theory," Springer, (1974).

[D-G-H] A. Diaconu, D. Goldfeld and J. Hoffstein, "Multiple Dirichlet series and moments of zeta and L-functions," *Compositio Math.*, **139**, (2003), 297-360.

[D-SP] W. Duke and R. Schulze-Pillot, "Representation of integers by positive ternary quadratic forms and equidistribution of lattice points on spheres," *Invent. Math.*, **99**, (1990), 49-57.

[De] P. Deligne, Publ. *I.H.E.S.*, **43**, "La conjecture de Weil I," (1974), 273-307.

[G-L-S] S. Gelbart, E. Lapid and P. Sarnak, "A new method for lower bounds for L-functions," *C.R.A.S.*, **339**, (2004), 91-94.

[G-M] R. Gupta and R. Murty, "A remark on Artin's conjecture," *Invent. Math*, **78**, (1984), 127-130.

[Ho] C. Hooley, "On Artin's conjecture," *J. reine angew Math*, **225**, (1967), 30-87.

[HB] R. Heath-Brown, "Artin's conjecture for primitive roots," *Quart. J. Math., Oxford*, **37**, (1986), 27-38.

[I] H. Iwaniec, "Fourier coefficients of modular forms of half integral weight," *Invent. Math.*, **87**, (1987), 385-401.

[I-K] H. Iwaniec and E. Kowalski, "Analytic Number Theory," *AMS Colloquium Publications*, **53**, (2004).

[I-S] H. Iwaniec and P. Sarnak, "Perspectives on the analytic theory of L-functions," *GAFA*, (2000), 705-741.

[J] H. Jacquet, "Principal L-functions of the linear group," *Proc. Symp. Pure Math.*, Vol. 33, (1979), 63-86.

[K] N. Katz, "An overview of Deligne's proof of the Riemann hypothesis for varieties over finite fields," *Proc. Symp. Pure Math.*, Vol. XXVIII, (1976), 275-305.

[K-S] N. Katz and P. Sarnak, "Zeroes of zeta functions and symmetry," *BAMS*, **36**, (1999), 1-26.

[La] R. Langlands, "On the functional equations satisfied by Eisenstein series," Springer Lecture Notes Math., **544**, (1976).

[Li] E. Lindenstrauss, "Invariant measures and arithmetic quantum unique ergodicity," *Ann. of Math.* to appear.

[Me] R. Meyer, "On a representation of the idele class group related to primes and zeros of L-functions," (2004).

[Mo] G. Molteni, "Upper and lower bounds at $s = 1$ of certain Dirichlet series with Euler product," *Duke Math. Jnl.*, **111**, (2002), 133-158.

[Od1] A. Odlyzko, www.dtc.umu.edu/~odlyzko/polya.

[Od2] A. Odlyzko, "The 10^{20}-th zero of the Riemann Zeta Function and 70 Million of its Neighbors," www.dtc.umn.edu/~odlyzko/unpublished.

[O-S] K. Ono and K. Soundararajan, "Ramanujan's ternary quadratic form," *Invent. Math.*, **130**, (1997), 415-454.

[Sel] A. Selberg, "On the remainder in the formulas for $N(T)$, the number of zeros of $\zeta(s)$ in the strip $0 < t < T$," Collected Works, Vol. I (Springer 1989), 179-203.

[Se] J-P. Serre, "Quelques applications du théoreme de densite' de Chebotarev,", (563-641), Collected Papers, Vol. III, (1972-1984), Springer Verlag, 1986.

[T] J. Tate, "Number theoretic backround," *Proc. Sym. Pure Math.*, Vol. 23, part 2, (1979), 3-27.

[V] A.I. Vinogradov, "The density hypothesis for Dirichlet L-series," *Izv. Akad. Nauk*, SSSR, Ser. Mat. **29**, 1965, 903-934.

[Wa] L. Waldspurger, "Sur les coefficients de Fourier des formes modulaires de poids demi-entier," *Jnl. de Mathematiques Pure et Applique's*, **60**, (1981), 375-484.

[Wat] T. Watson, "Rankin triple products and quantum chaos," Princeton Thesis, (2001).

[We1] A. Weil, "Basic number theory," Springer, (1974).

[We2] A. Weil, "Sur les "formules explicites" de la théorie des nombers premiers," (48-61), Collected Papers, Vol. II, (1951-1964), Springer Verlag, (1980).

April 18, 2005

11.3 J. B. Conrey (2003)

The Riemann Hypothesis

> *It's a whole beautiful subject and the Riemann zeta function is just the first one of these, but it's just the tip of the iceberg. They are just the most amazing objects, these L-functions — the fact that they exist, and have these incredible properties and are tied up with all these arithmetical things — and it's just a beautiful subject. Discovering these things is like discovering a gemstone or something. You're amazed that this thing exists, has these properties and can do this [138].*
>
> **J. Brian Conrey**

J. Brian Conrey presents a variety of material on the Riemann hypothesis motivated by three workshops sponsored by the American Institute of Mathematics. This paper gives a detailed account of the various approaches that mathematicians have undertaken to attack the Riemann hypothesis. Conrey reports on the most active areas of current research into the hypothesis, and focuses on the areas that currently enjoy the most interest. Particularly, he highlights recent work on the connections between random matrix theory, the Riemann zeta function, and L-functions. He also discusses the Landau–Siegel zero and presents the recent innovative approach of Iwaniec toward its elimination.

The Riemann Hypothesis

J. Brian Conrey

Hilbert, in his 1900 address to the Paris International Congress of Mathematicians, listed the Riemann Hypothesis as one of his 23 problems for mathematicians of the twentieth century to work on. Now we find it is up to twenty-first century mathematicians! The Riemann Hypothesis (RH) has been around for more than 140 years, and yet now is arguably the most exciting time in its history to be working on RH. Recent years have seen an explosion of research stemming from the confluence of several areas of mathematics and physics.

In the past six years the American Institute of Mathematics (AIM) has sponsored three workshops whose focus has been RH. The first (RHI) was in Seattle in August 1996 at the University of Washington. The second (RHII) was in Vienna in October 1998 at the Erwin Schrödinger Institute, and the third (RHIII) was in New York in May 2002 at the Courant Institute of Mathematical Sciences. The intent of these workshops was to stimulate thinking and discussion about one of the most challenging problems of mathematics and to consider many different approaches. Are we any closer to solving the Riemann Hypothesis after these efforts? Possibly. Have we learned anything about the zeta-function as a result of these workshops? Definitely. Several of the participants from the workshops are collaborating on the website (`http://www.aimath.org/WWN/rh/`) which provides an overview of the subject.

J. Brian Conrey is director of the American Institute of Mathematics. His email address is `conrey@aimath.org`.

Here I hope to outline some of the approaches to RH and to convey some of the excitement of working in this area at the present moment. To begin, let us examine the Riemann Hypothesis itself. In 1859 in the seminal paper "Ueber die Anzahl der Primzahlen unter eine gegebener Grösse", G. B. F. Riemann outlined the basic analytic properties of the zeta-function

$$\zeta(s) := 1 + \frac{1}{2^s} + \frac{1}{3^s} + \cdots = \sum_{n=1}^{\infty} \frac{1}{n^s}.$$

The series converges in the half-plane where the real part of s is larger than 1. Riemann proved that $\zeta(s)$ has an analytic continuation to the whole plane apart from a simple pole at $s = 1$. Moreover, he proved that $\zeta(s)$ satisfies an amazing *functional equation*, which in its symmetric form is given by

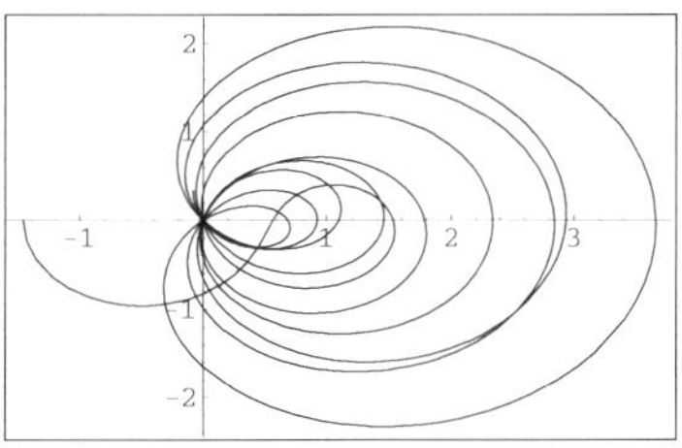

Figure 1. $\zeta(\frac{1}{2} + it)$ **for** $0 < t < 50$.

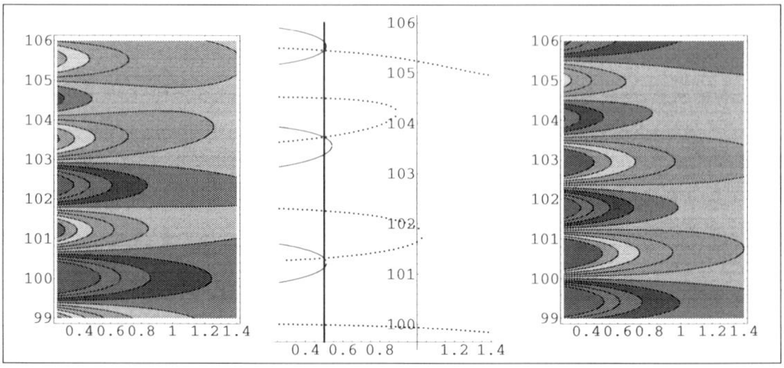

Figure 2. Contour plot of $\Re\zeta(s)$, the curves $\Re\zeta(s) = 0$ (solid) and $\Im\zeta(s) = 0$ (dotted), contour plot of $\Im\zeta(s)$.

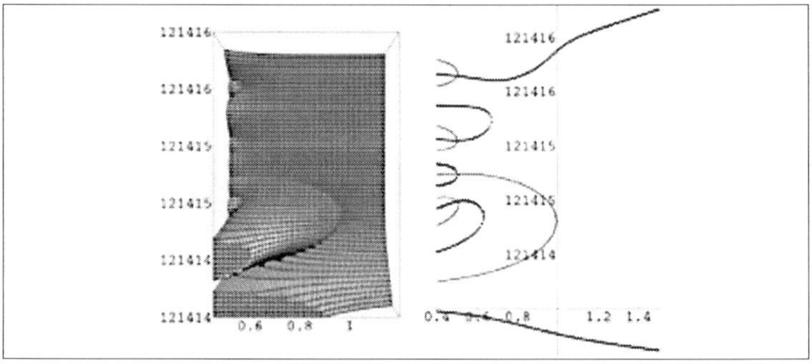

Figure 3. 3-D plot of $|\Re\zeta(s)|$, and the curves $\Re\zeta(s) = 0$ (solid) and $\Im\zeta(s) = 0$ (dotted). This may be the first place in the critical strip where the curves $\Re\zeta(s) = 0$ loop around each other.

$$\xi(s) := \tfrac{1}{2}s(s-1)\pi^{-\frac{s}{2}}\Gamma\left(\frac{s}{2}\right)\zeta(s) = \xi(1-s),$$

where $\Gamma(s)$ is the usual Gamma-function.

The zeta-function had been studied previously by Euler and others, but only as a function of a real variable. In particular, Euler noticed that

$$\begin{aligned}\zeta(s) &= \left(1 + \frac{1}{2^s} + \frac{1}{4^s} + \frac{1}{8^s} + \dots\right)\\ &\quad\times\left(1 + \frac{1}{3^s} + \frac{1}{9^s} + \dots\right)\left(1 + \frac{1}{5^s} + \dots\right)\dots\\ &= \prod_p \left(1 - \frac{1}{p^s}\right)^{-1},\end{aligned}$$

where the infinite product (called the *Euler product*) is over all the prime numbers. The product converges when the real part of s is greater than 1. It is an analytic version of the fundamental theorem of arithmetic, which states that every integer can be factored into primes in a unique way. Euler used this product to prove that the sum of the reciprocals of the primes diverges. The Euler product suggests Riemann's interest in the zeta-function: he was trying to prove a conjecture made by Legendre and, in a more precise form, by Gauss:

$$\pi(x) := \#\{\text{primes less than } x\} \sim \int_2^x \frac{dt}{\log t}.$$

Riemann made great progress toward proving Gauss's conjecture. He realized that the distribution of the prime numbers depends on the distribution of the complex zeros of the zeta-function. The Euler product implies that there are no zeros of $\zeta(s)$ with real part greater than 1; the functional equation implies that there are no zeros with real part less than 0, apart from the *trivial zeros* at

$s = -2, -4, -6, \ldots$. Thus, all of the complex zeros are in the *critical strip* $0 \le \Re s \le 1$. Riemann gave an explicit formula for $\pi(x)$ in terms of the complex zeros $\rho = \beta + i\gamma$ of $\zeta(s)$. A simpler variant of his formula is

$$\psi(x) := \sum_{n<x} \Lambda(n) = x - \sum_{\rho} \frac{x^\rho}{\rho} - \log 2\pi - \tfrac{1}{2}\log(1 - x^{-2}),$$

valid for x not a prime power, where the von Mangoldt function $\Lambda(n) = \log p$ if $n = p^k$ for some k and $\Lambda(n) = 0$ otherwise. Note that the sum is not absolutely convergent; if it were, then $\sum_{n<x} \Lambda(n)$ would have to be a continuous function of x, which it clearly is not. Consequently, there must be infinitely many zeros ρ. The sum over ρ is with multiplicity and is to be interpreted as $\lim_{T\to\infty} \sum_{|\rho|<T}$. Note also that $|x^\rho| = x^\beta$; thus it was necessary to show that $\beta < 1$ in order to conclude that $\sum_{n\le x} \Lambda(n) \sim x$, which is a restatement of Gauss's conjecture.

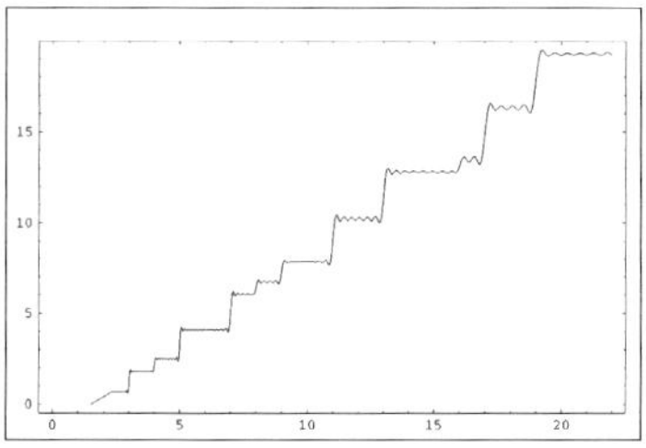

Figure 4. Explicit formula for $\psi(x)$ using the first 100 pairs of zeros.

The functional equation shows that the complex zeros are symmetric with respect to the line $\Re s = \frac{1}{2}$. Riemann calculated the first few complex zeros $\frac{1}{2} + i14.134\ldots, \frac{1}{2} + i21.022\ldots$ and proved that the number $N(T)$ of zeros with imaginary parts between 0 and T is

$$N(T) = \frac{T}{2\pi}\log\frac{T}{2\pi e} + \frac{7}{8} + S(T) + O(1/T),$$

where $S(T) = \frac{1}{\pi}\arg\zeta(1/2 + iT)$ is computed by continuous variation starting from $\arg\zeta(2) = 0$ and proceeding along straight lines, first up to $2 + iT$ and then to $1/2 + iT$. Riemann also proved that $S(T) = O(\log T)$. Note for future reference that at a height T the average gap between zero heights is $\sim 2\pi/\log T$. Riemann suggested that the number $N_0(T)$ of zeros of $\zeta(1/2 + it)$ with $0 < t \le T$ seemed to be about

$$\frac{T}{2\pi}\log\frac{T}{2\pi e}$$

and then made his conjecture that all of the zeros of $\zeta(s)$ in fact lie on the 1/2-line; this is the Riemann Hypothesis.

Riemann's effort came close to proving Gauss's conjecture. The final step was left to Hadamard and de la Vallée Poussin, who proved independently in 1896 that $\zeta(s)$ does not vanish when the real part of s is equal to 1 and from that fact deduced Gauss's conjecture, now called the Prime Number Theorem.

Initial Ideas

It is not difficult to show that RH is equivalent to the assertion that for every $\epsilon > 0$,

$$\pi(x) = \int_2^x \frac{dt}{\log t} + O(x^{1/2+\epsilon}).$$

However, it is difficult to see another way to approach $\pi(x)$ and so get information about the zeros.

Another easy equivalent to RH is the assertion that $M(x) = O(x^{1/2+\epsilon})$ for every $\epsilon > 0$, where

$$M(x) = \sum_{n<x} \mu(n)$$

and $\mu(n)$ is the Möbius function whose definition can be inferred from its generating Dirichlet series $1/\zeta$:

$$\frac{1}{\zeta(s)} = \sum_{n=1}^{\infty} \frac{\mu(n)}{n^s} = \prod_p \left(1 - \frac{1}{p^s}\right).$$

Thus, if $p_1, \ldots, p_k$ are distinct primes, then $\mu(p_1 \ldots p_k) = (-1)^k$; also $\mu(n) = 0$ if $p^2 \mid n$ for some prime p. This series converges absolutely when $\Re s > 1$. If the estimate $M(x) = O(x^{1/2+\epsilon})$ holds for every $\epsilon > 0$, then it follows by partial summation that the series converges for every s with real part greater than 1/2; in particular, there can be no zeros of $\zeta(s)$ in this open half-plane, because zeros of $\zeta(s)$ are poles of $1/\zeta(s)$. The converse, that RH implies this estimate for $M(x)$, is also not difficult to show.

Instead of analyzing $\pi(x)$ directly, it might seem easier to work with $M(x)$ and prove the above estimate, perhaps by some kind of combinatorial reasoning. In fact, Stieltjes let it be known that he had such a proof. Hadamard, in his famous 1896 proof of the Prime Number Theorem, refers to Stieltjes's claim and somewhat apologetically offers his much weaker theorem that $\zeta(s)$ does not vanish on the 1-line in the hope that the simplicity of his proof will be useful. Stieltjes never published his proof.

Mertens made the stronger conjecture that

$$|M(x)| \le \sqrt{x};$$

clearly this implies RH. However, Mertens's conjecture was disproved by Odlyzko and te Riele in 1985. The estimate $M(x) = O(\sqrt{x})$ is also likely to

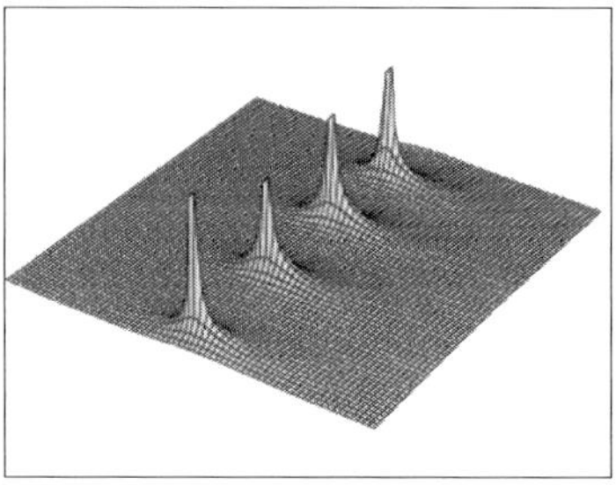

Figure 5. $1/|\zeta(x+iy)|$ **for** $0 < x < 1$ **and** $16502.4 < y < 16505$.

be false, but a proof of its falsity has not yet been found.

Subsequent Efforts

In England in the early 1900s the difficulty of the question was not yet appreciated. Barnes assigned RH to Littlewood as a thesis problem. Littlewood independently discovered some of the developments that had already occurred on the continent. Hardy, Littlewood, Ingham, and other British mathematicians were responsible for many of the results on the zeta-function in the first quarter of the century. Hardy and Littlewood gave the first proof that infinitely many of the zeros are on the 1/2-line. They found what they called the *approximate functional equation* for $\zeta(s)$. Later, Siegel uncovered a very precise version of this formula while studying Riemann's notes in the Göttingen library; the formula is now called the Riemann-Siegel formula and gives the starting point for all large-scale calculations of $\zeta(s)$. Hardy and Littlewood gave an asymptotic evaluation of the second moment of $\zeta(\frac{1}{2}+it)$; Ingham proved the asymptotics for the fourth moment.

Much effort has also been expended on the unproved Lindelöf hypothesis, which is a consequence of RH. The Lindelöf hypothesis asserts that for every $\epsilon > 0$,

$$\zeta(1/2+it) = O(t^{\epsilon}) \qquad \text{as } t \to \infty.$$

Hardy and Littlewood proved that $\zeta(1/2+it) = O(t^{1/4+\epsilon})$. This bound is now called the "convexity bound", since it follows from the functional equation together with general principles of complex analysis (the maximum modulus principle in the form of the Phragmén-Lindelöf theorem). Weyl improved the bound to $t^{1/6+\epsilon}$ with his new ideas for estimating special trigonometrical sums, now called Weyl sums.

Hardy grew to love the problem. He and Littlewood wrote at least ten papers on the zeta-function. Hardy once included proving RH on a list of New Year's goals he set for himself. Hardy even used RH as a defense: he once sent a postcard to his colleague Harald Bohr prior to crossing the English Channel one stormy night, claiming that he had solved RH. Even though Hardy was an atheist, he was relatively certain that God, if he did exist, would not allow the ferry to sink under circumstances so favorable to Hardy!

Hilbert seems to have had somewhat contradictory views about the difficulty of RH. On one occasion he compared three unsolved problems: the transcendence of $2^{\sqrt{2}}$, Fermat's Last Theorem, and the Riemann Hypothesis. In his view, RH would likely be solved in a few years, Fermat's Last Theorem possibly in his lifetime, and the transcendence question possibly never. Amazingly, the transcendence question was resolved a few years later by Gelfond and Schneider, and, of course, Andrew Wiles recently proved Fermat's Last Theorem. Another time Hilbert remarked that if he were to awake after a sleep of five hundred years, the first question he would ask was whether RH was solved.

Near the end of his career, Hans Rademacher, best known for his exact formula for the number of partitions of an integer, thought he had disproved RH. Siegel had checked the work, which was based on the deduction that a certain function would absurdly have an analytic continuation if RH were true. The mathematics community tried to get *Time* magazine interested in the story. It transpired that *Time* became interested and published an article only after it was discovered that Rademacher's proof was incorrect.

Evidence for RH

Here are some reasons to believe RH.

- Billions of zeros cannot be wrong. Recent work by van de Lune has shown that the first 10 billion zeros are on the line. Also, there is a distributed computing project organized by Sebastian Wedeniwski—a screen-saver type of program—that many people subscribe to, which claims to have verified that the first 100 billion zeros are on the line. Andrew Odlyzko has calculated millions of zeros near zeros number 10^{20}, 10^{21}, and 10^{22} (available on his website).
- Almost all of the zeros are very near the 1/2-line. In fact, it has been proved that more than 99 percent of zeros $\rho = \beta + i\gamma$ satisfy $|\beta - \frac{1}{2}| \le 8/\log|\gamma|$.
- Many zeros can be proved to be on the line. Selberg got a positive proportion, and N. Levinson showed at least 1/3; that proportion has been improved to 40 percent. Also, RH implies that all zeros of all derivatives of $\xi(s)$ are on the 1/2-line. It has been shown that more than 99 percent of the zeros of the third derivative $\xi'''(s)$ are on the 1/2-line. Near the end of his life, Levinson thought he had a method that allowed for a converse to Rolle's theorem in

this situation, implying that if $\xi'(s)$ has at least a certain proportion of zeros on the line, then so does ξ and similarly for ξ'' to ξ' and so on. However, no one has been able to make this argument work.
- Probabilistic arguments. For almost all random sequences of -1's and $+1$'s, the associated summatory function up to x is bounded by $x^{1/2+\epsilon}$. The Möbius sequence appears to be fairly random.
- Symmetry of the primes. RH tells us that the primes are distributed in as nice a way as possible. If RH were false, there would be some strange irregularities in the distribution of primes; the first zero off the line would be a very important mathematical constant. It seems unlikely that nature is that perverse!

Various Approaches

There is an often-told story that Hilbert and Pólya independently suggested that the way to prove RH was to interpret the zeros spectrally, that is, to find a naturally occurring Hermitian operator whose eigenvalues are the nontrivial zeros of $\zeta(1/2+it)$. Then RH would follow, since Hermitian operators have real eigenvalues. This idea has been one of the main approaches that has been tried repeatedly.

We describe an assortment of other interesting approaches to RH.

Pólya's Analysis

Pólya investigated a chain of ideas that began with Riemann: namely, studying the Fourier transform of $\Xi(t) := \xi(\frac{1}{2}+it)$, which as a consequence of the functional equation is real for real t and an even function of t. RH is the assertion that all zeros of Ξ are real. The Fourier transform can be computed explicitly:

$$\begin{aligned}\Phi(t) &:= \int_{-\infty}^{\infty} \Xi(u)e^{itu}\,du \\ &= \sum_{n=1}^{\infty}(2n^4\pi^2\exp(9t/2) - 3n^2\pi\exp(5t/2)) \\ &\qquad \times \exp\left(-\pi n^2 e^{2t}\right).\end{aligned}$$

It can be shown that Φ and Φ' are positive for positive t. One idea is to systematically study classes of reasonable functions whose Fourier transforms have all real zeros and then try to prove that $\Xi(t)$ is in the class. A sample theorem in this direction is due to de Bruijn:

Let $f(t)$ be an even nonconstant entire function of t such that $f(t) \geq 0$ for real t and $f'(t) = \exp(\gamma t^2)g(t)$, where $\gamma \geq 0$ and $g(t)$ is an entire function of genus ≤ 1 with purely imaginary zeros only. Then $\Psi(z) = \int_{-\infty}^{\infty} \exp\{-f(t)\}e^{izt}dt$ has real zeros only.

In particular, all the zeros of the Fourier transform of a first approximation (see Titchmarsh for details)

$$\begin{aligned}\phi(t) &= (2\pi\cosh\frac{9t}{2} - 3\cosh\frac{5t}{2}) \\ &\quad\times\exp(-2\pi\cosh 2t)\end{aligned}$$

to $\Phi(t)$ are real. These ideas have been further explored by de Bruijn, Newman, D. Hejhal, and others. Hejhal (1990) has shown that almost all of the zeros of the Fourier transform of any partial sum of $\Phi(t)$ are real.

Probabilistic Models

Researchers working in probability are intrigued by the fact that the ξ-function arises as an expectation in a moment of a Brownian bridge:

$$2\xi(s) = E(Y^s)$$

where

$$Y := \sqrt{\frac{2}{\pi}}\left(\max_{t\in[0,1]} b_t - \min_{t\in[0,1]} b_t\right)$$

with $b_t = \beta_t - t\beta_1$ where β_t is standard Brownian motion. See a paper of Biane, Pitman, and Yor (*Bull. Amer. Math. Soc.* (N.S.) **38** (2001), 435–65).

Functional Analysis: The Nyman-Beurling Approach

This approach begins with the following theorem of Nyman, a student of Beurling.

RH holds if and only if

$$\mathrm{span}_{L^2(0,1)}\{\eta_\alpha, 0<\alpha<1\} = L^2(0,1)$$

where

$$\eta_\alpha(t) = \{\alpha/t\} - \alpha\{1/t\}$$

and $\{x\} = x - [x]$ is the fractional part of x.

This has been extended by Baez-Duarte, who showed that one may restrict attention to integral values of $1/\alpha$. Balazard and Saias have rephrased this in a nice way:

RH holds if and only if

$$\inf_A \int_{-\infty}^{\infty}\left|1 - A(\tfrac{1}{2}+it)\zeta(\tfrac{1}{2}+it)\right|^2 \frac{dt}{\frac{1}{4}+t^2} = 0,$$

where the infimum is over all Dirichlet polynomials A.

Let d_N be the infimum over all Dirichlet polynomials

$$A(s) = \sum_{n=1}^{N} a_n n^{-s}$$

of length N. They conjecture that $d_N \sim C/\log N$, where $C = \sum_\rho 1/|\rho|^2$. Burnol has proved that

$$d_n \geq \frac{1}{\log N}\sum_{\rho \text{ on the line}} \frac{m_\rho}{|\rho|^2},$$

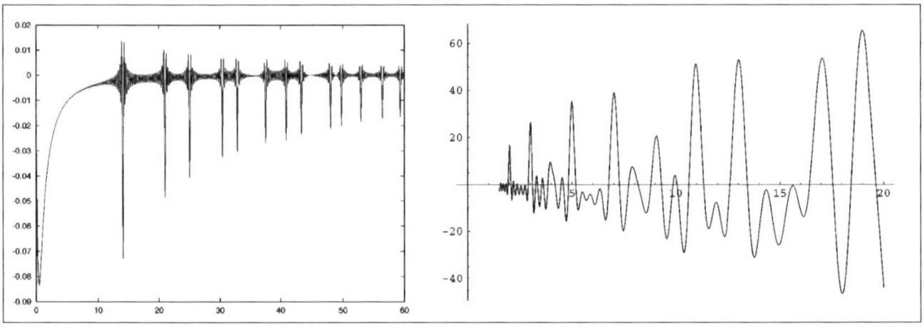

Figure 6. Duality: The Fourier transform of the error term in the Prime Number Theorem (note the spikes at ordinates of zeros) and the sum over zeros $-\sum x^{\rho}$ with $|\rho| < 100$ (note the peaks at primes and prime powers).

where m_ρ is the multiplicity of the zero ρ. If RH holds and all the zeros are simple, then clearly these two bounds are the same.

Weil's Explicit Formula and Positivity Criterion

André Weil proved the following formula, which is a generalization of Riemann's formula mentioned above and which specifically illustrates the dependence between primes and zeros. Suppose h is an even function that is holomorphic in the strip $|\Im t| \le 1/2 + \delta$ and that satisfies $h(t) = O((1 + |t|)^{-2-\delta})$ for some $\delta > 0$, and let

$$g(u) = \frac{1}{2\pi}\int_{-\infty}^{\infty} h(r)e^{-iur}\,dr.$$

Then we have the following duality between primes and zeros of ζ:

$$\sum_{\gamma} h(\gamma) = 2h(\tfrac{i}{2}) - g(0)\log \pi + \frac{1}{2\pi}\int_{-\infty}^{\infty} h(r)\frac{\Gamma'}{\Gamma}(\tfrac{1}{4} + \tfrac{1}{2}ir)\,dr - 2\sum_{n=1}^{\infty}\frac{\Lambda(n)}{\sqrt{n}}g(\log n).$$

In this formula, a zero is written as $\rho = 1/2 + i\gamma$ where $\gamma \in \mathbb{C}$; of course RH is the assertion that every γ is real. Using this duality Weil gave a criterion for RH:

RH holds if and only if

$$\sum_{\gamma} h(\gamma) > 0$$

for every (admissible) function h of the form $h(r) = h_0(r)\overline{h_0(\bar{r})}$.

Xian-Jin Li has given a very nice criterion which, in effect, says that one may restrict attention to a specific sequence h_n:

The Riemann Hypothesis is true if and only if $\lambda_n \ge 0$ *for each* $n = 1, 2, \ldots$ *where*

$$\lambda_n = \sum_{\rho}(1 - (1 - 1/\rho)^n).$$

As usual, the sum over zeros is $\lim_{T\to\infty}\sum_{|\rho|<T}$. Another expression for λ_n is

$$\lambda_n = \frac{1}{(n-1)!}\frac{d^n}{ds^n}(s^{n-1}\log \xi(s))\Big|_{s=1}.$$

It would be interesting to find an interpretation (geometric?) for these λ_n, or perhaps those associated with a different L-function, to make their positivity transparent.

Selberg's Trace Formula

Selberg, perhaps looking for a spectral interpretation of the zeros of $\zeta(s)$, proved a trace formula for the Laplace operator acting on the space of real-analytic functions defined on the upper half-plane $\mathcal{H} = \{x + iy : y > 0\}$ and invariant under the group $SL(2,\mathbb{Z})$ of linear fractional transformations with integer entries and determinant one, which acts discontinuously on $\mathcal{H}$. This invariance is expressed as

$$f\left(\frac{az+b}{cz+d}\right) = f(z);$$

the Laplace operator in this case is

$$\Delta = -y^2\left(\frac{\partial^2}{\partial x^2} + \frac{\partial^2}{\partial x^2}\right).$$

The spectrum of Δ splits into a continuous part and a discrete part. The eigenvalues λ are all positive and, by convention, are usually expressed as $\lambda = s(1 - s)$. The continuous part consists of all $s = 1/2 + it$, $t \ge 0$, and we write the discrete part as $s_j = \frac{1}{2} + ir_j$. Then

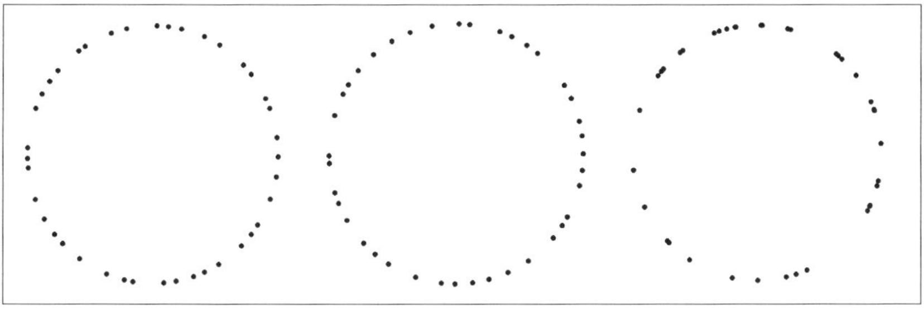

Figure 7. The eigenvalues of a random 40 x 40 unitary matrix, 40 consecutive zeros of $\zeta(s)$ scaled to wrap once around the circle, and 40 randomly chosen points on the unit circle.

$$\sum_{j=1}^{\infty} h(r_j) = -h(0) - g(0)\log\tfrac{\pi}{2} - \frac{1}{2\pi}\int_{-\infty}^{\infty} h(r)G(r)\,dr$$
$$+ 2\sum_{n=1}^{\infty}\frac{\Lambda(n)}{n}g(2\log n)$$
$$+ \sum_{P}\sum_{\ell=1}^{\infty}\frac{g(\ell\log P)\log P}{P^{\ell/2}-P^{-\ell/2}}$$

where g, h, and Λ are as in Weil's formula and

$$G(r) = \frac{\Gamma'}{\Gamma}(\tfrac{1}{2}+ir) + \frac{\Gamma'}{\Gamma}(1+ir) - \frac{\pi}{6}r\tanh\pi r$$
$$+ \frac{\pi}{\cosh\pi r}(\tfrac{1}{8} + \tfrac{\sqrt{3}}{9}\cosh\tfrac{\pi r}{3}).$$

The final sum is over the norms P of prime geodesics of $SL(2,\mathbb{Z})\backslash\mathcal{H}$. The values taken on by P are of the form $(n+\sqrt{n^2-4})^2/4$, $n \geq 3$, with certain multiplicities (the class number $h(n^2-4)$). H. Haas was one of the first people to compute the eigenvalues $r_1 = 9.533\ldots$, $r_2 = 12.173\ldots$, $r_3 = 13.779\ldots$ of $SL(2,\mathbb{Z})$ in 1977 in his University of Heidelberg Diplomarbeit. Soon after, Hejhal was visiting San Diego, and Audrey Terras pointed out to him that Haas's list contained the numbers $14.134\ldots, 21.022\ldots$: the ordinates of the first few zeros of $\zeta(s)$ were lurking amongst the eigenvalues! Hejhal discovered the ordinates of the zeros of $L(s,\chi_3)$ (see section 7) on the list too. He unraveled this perplexing mystery about six months later. It turned out that the spurious eigenvalues were associated to "pseudo cusp forms" and appeared because of the method of computation used. If the zeros had appeared legitimately, RH would have followed because $\lambda = \rho(1-\rho)$ is positive. (The 1979 IHÉS preprint by P. Cartier and Hejhal contains additional details of the story.)

The trace formula resembles the explicit formula in certain ways. Many researchers have attempted to interpret Weil's explicit formula in terms of Selberg's trace formula.

Some Other Equivalences of Interest

Here are a few other easy-to-state equivalences of RH:

- Hardy and Littlewood (1918): *RH holds if and only if*

$$\sum_{k=1}^{\infty}\frac{(-x)^k}{k!\,\zeta(2k+1)} = O(x^{-1/4}) \quad \text{as } x \to \infty.$$

- Redheffer (1977): *RH holds if and only if for every $\epsilon > 0$ there is a $C(\epsilon) > 0$ such that $|\det(A(n))| < C(\epsilon)n^{1/2+\epsilon}$, where $A(n)$ is the $n \times n$ matrix of 0's and 1's defined by $A(i,j) = 1$ if $j = 1$ or if i divides j, and $A(i,j) = 0$ otherwise.* It is known that $A(n)$ has $n - [n\log 2] - 1$ eigenvalues equal to 1. Also, A has a real eigenvalue (the spectral radius) which is approximately $\sqrt{n}$, a negative eigenvalue which is approximately $-\sqrt{n}$, and the remaining eigenvalues are small.
- Lagarias (2002): Let $\sigma(n)$ denote the sum of the positive divisors of n. *RH holds if and only if*

$$\sigma(n) \leq H_n + \exp(H_n)\log H_n$$

for every n, where $H_n = 1 + \frac{1}{2} + \frac{1}{3} + \cdots + \frac{1}{n}$.

Other Zeta- and L-Functions

Over the years striking analogies have been observed between the Riemann zeta-function and other zeta- or L-functions. While these functions are seemingly independent of each other, there is growing evidence that they are all somehow connected in a way that we do not fully understand. In any event, trying to understand, or at least classify, all of the objects which we believe satisfy RH is a reasonable thing to do. The rest of the article will give a glimpse in this direction and perhaps a clue to the future.

First, some examples of other functions that we believe satisfy RH. The simplest after ζ is the Dirichlet L-function for the nontrivial character of conductor 3:

$$L(s,\chi_3) = 1 - \frac{1}{2^s} + \frac{1}{4^s} - \frac{1}{5^s} + \frac{1}{7^s} - \frac{1}{8^s} + \ldots.$$

This can be written as an Euler product

$$\prod_{p=1 \bmod 3} (1-p^{-s})^{-1} \prod_{p=2 \bmod 3} (1+p^{-s})^{-1},$$

it satisfies the functional equation

$$\xi(s,\chi_3) := \left(\frac{\pi}{3}\right)^{-\frac{s}{2}} \Gamma\left(\tfrac{s+1}{2}\right) L(s,\chi_3) = \xi(1-s,\chi_3),$$

and it is expected to have all of its nontrivial zeros on the 1/2-line. A similar construction works for any primitive Dirichlet character.

Dedekind, Hecke, Artin, and others developed the theory of zeta-functions associated with number fields and their characters. These have functional equations and Euler products, and are expected to satisfy a Riemann Hypothesis. Ramanujan's tau-function defined implicitly by

$$x\prod_{n=1}^{\infty}(1-x^n)^{24} = \sum_{n=1}^{\infty}\tau(n)x^n$$

also yields an L-function. The associated Fourier series $\Delta(z) := \sum_{n=1}^{\infty} \tau(n)\exp(2\pi i n z)$ satisfies

$$\Delta\left(\frac{az+b}{cz+d}\right) = (cz+d)^{12}\Delta(z)$$

for all integers a, b, c, d with $ad - bc = 1$. A function satisfying these equations is called a *modular form* of weight 12. The associated L-function

$$\begin{aligned} L_\Delta(s) &:= \sum_{n=1}^{\infty} \frac{\tau(n)/n^{11/2}}{n^s} \\ &= \prod_p \left(1 - \frac{\tau(p)/p^{11/2}}{p^s} + \frac{1}{p^{2s}}\right)^{-1} \end{aligned}$$

satisfies the functional equation

$$\xi_\Delta := (2\pi)^{-s}\Gamma(s+11/2)L_\Delta(s) = \xi_\Delta(1-s),$$

and all of its complex zeros are expected to be on the 1/2-line.

Another example is the L-function associated to an elliptic curve $E: y^2 = x^3 + Ax + B$, where A and B are integers. The associated L-function, called the Hasse-Weil L-function, is

$$\begin{aligned} L_E(s) &= \sum_{n=1}^{\infty} \frac{a(n)/n^{1/2}}{n^s} \\ &= \prod_{p\nmid N} \left(1 - \frac{a(p)/p^{1/2}}{p^s} + \frac{1}{p^{2s}}\right)^{-1} \\ &\quad \times \prod_{p\mid N}\left(1 - \frac{a(p)/p^{1/2}}{p^s}\right)^{-1}, \end{aligned}$$

where N is the conductor of the curve. The coefficients a_n are constructed easily from a_p for prime p; in turn the a_p are given by $a_p = p - N_p$, where N_p is the number of solutions of E when considered modulo p. The work of Wiles and others proved that these L-functions are associated to modular forms of weight 2. This modularity implies the functional equation

$$\xi_E(s) := (2\pi/\sqrt{N})^{-s}\Gamma(s+1/2)L_E(s) = \xi_E(1-s).$$

It is believed that all of the complex zeros of $L_E(s)$ are on the 1/2-line. A similar construction ought to work for other sets of polynomial equations, but so far this has not been proved.

What is the most general situation in which we expect the Riemann Hypothesis to hold? The Langlands program is an attempt to understand all L-functions and to relate them to automorphic forms. At the very least a Dirichlet series that is a candidate for RH must have an Euler product and a functional equation of the right shape. Selberg has given a set of four precise axioms which are believed to characterize the L-functions for which RH holds. Examples have been given that show the necessity of most of the conditions in his axioms.

L-Functions and Random Matrix Theory

An area of investigation which has stimulated much recent work is the connection between the Riemann zeta-function and Random Matrix Theory (RMT). This work does not seem to be leading in the direction of a proof of RH, but it is convincing evidence that the spectral interpretation of the zeros sought by Hilbert and Pólya is an idea with merit. Moreover, the connection between zeta theory and RMT has resulted in a very detailed model of $\zeta(s)$ and its value distribution.

Montgomery's Pair Correlation Conjecture

In 1972 Hugh Montgomery was investigating the spacings between zeros of the zeta-function in an attempt to solve the class number problem. He formulated his Pair Correlation Conjecture based in part on what he could prove assuming RH and in part on RH plus conjectures for the distribution of twin primes and other prime pairs. This conjecture asserts that

$$\sum_{\frac{2\pi\alpha}{\log T} < \gamma - \gamma' < \frac{2\pi\beta}{\log T}} 1 \sim N(T)\int_\alpha^\beta \left(1 - \left(\frac{\sin \pi u}{\pi u}\right)^2\right) du.$$

The sum on the left counts the number of pairs $0 < \gamma, \gamma' < T$ of ordinates of zeros with normalized spacing between positive numbers $\alpha < \beta$. Montgomery had stopped in Princeton on his way from St. Louis, where he had presented this result at an AMS symposium, to Cambridge University, where he was a graduate student. Chowla persuaded him to show this result to Freeman Dyson at afternoon tea at the Institute for Advanced Study. Dyson immediately identified the integrand $1 - \left(\frac{\sin \pi u}{\pi u}\right)^2$ as the pair correlation function for eigenvalues of large random Hermitian matrices measured with a Gaussian measure—the Gaussian Unitary Ensemble that physicists had long been studying in connection with the distribution of energy levels in large systems of particles. With this insight, Montgomery went on to conjecture that perhaps all the statistics, not just the pair correlation statistic, would match up for zeta-zeros and eigenvalues of Hermitian matrices. This conjecture is called the GUE conjecture. It has the flavor of a spectral interpretation of the zeros, though it gives no indication of what the particular operator is.

Odlyzko's Calculations

In the 1980s Odlyzko began an intensive numerical study of the statistics of the zeros of $\zeta(s)$. Based on a new algorithm developed by Odlyzko and Schönhage that allowed them to compute a value of $\zeta(1/2+it)$ in an average time of t^{ϵ} steps, he computed millions of zeros at heights around 10^{20} and spectacularly confirmed the GUE conjecture.

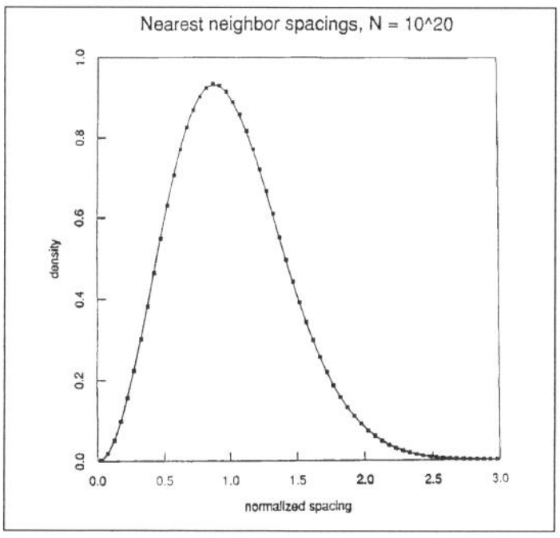

Figure 8a. The nearest neighbor spacing for GUE (solid) and for 7.8×10^{7} zeros of $\zeta(s)$ near the 10^{20} zero (scatterplot). Graphic by A. Odlyzko.

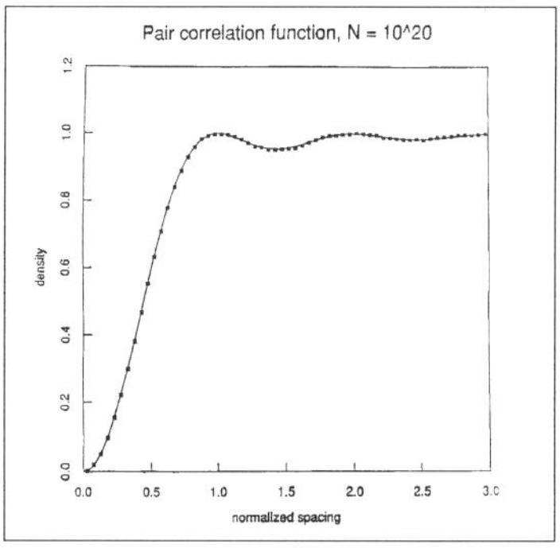

Figure 8b.The pair-correlation function for GUE (solid) and for 8×10^{6} zeros of $\zeta(s)$ near the 10^{20} zero (scatterplot). Graphic by A. Odlyzko.

Moments of Zeta

More recently, RMT has led to a conjecture for moments of ζ on the critical line. Let

$$I_k(T)=\frac{1}{T}\int_0^T|\zeta(1/2+it)|^{2k}\,dt.$$

Asymptotic formulas for I_1 and I_2 were found by Hardy and Littlewood and Ingham by 1926. In 1995 Ghosh and I formulated a conjecture for I_3 and set up a notation to clarify the part missing from our understanding of I_k. After scaling out the arithmetic parts, we identified a factor g_k which we could not predict. The factor is $g_1=1$ and $g_2=2$ for the second and fourth moments and conjecturally $g_3=42$ for the sixth moment. At RHI in Seattle, Sarnak proposed to Keating that he find a random matrix explanation for these numbers. By 1998 Gonek and I had found a number-theoretic way to conjecture the answer for the eighth moment, namely $g_4=24024$. At RHII in Vienna, Keating announced that he and Snaith had a conjecture for all of the moments which agreed with g_1, g_2, and g_3. Keating, Snaith, and I—moments before Keating's lecture—checked (amid great excitement!) that the Keating and Snaith conjecture also produced $g_4=24024$.

The idea of Keating and Snaith was that if the eigenvalues of unitary matrices model zeta zeros, then perhaps the characteristic polynomials of unitary matrices model zeta values. They were able to compute—exactly—the moments of the characteristic polynomials of unitary matrices averaged with respect to Haar measure by using Selberg's integral, which is a formula found in the 1940s by Selberg that vastly generalizes the integral for the beta-function. Keating and Snaith proposed that

$$g_k=k^2!\prod_{j=0}^{k-1}\frac{j!}{(j+k)!}.$$

Farmer and I (2000) proved that g_k is always an integer and found that it has an interesting prime factorization.

Families

At RHI in Seattle, Sarnak gave a lecture on families of L-functions based on work that he and Katz were doing. They discovered a way to identify a symmetry type (unitary, orthogonal, or symplectic) with various families of L-functions. Their work was based on studying families of zeta-functions over finite fields (for which RH was already proved by Weil for curves and by Deligne for general varieties). For these zeta-functions, Katz and Sarnak proved that the zeros of the family were distributed exactly

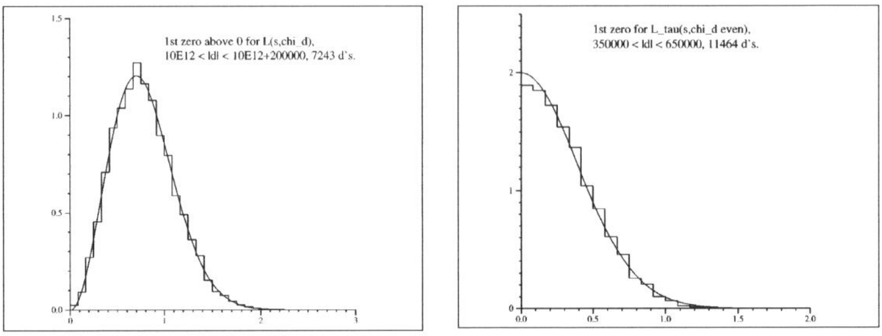

Figure 9. A comparison of the distribution of the lowest lying zero for two families of L-functions. In each case one first needs to suitably normalize the zeros. The first figure compares the distribution of the lowest zero of $L(s, \chi_d)$, Dirichlet L-functions, for several thousand d's of size 10^{12}, against the distribution of the zero closest to 1 for large unitary symplectic matrices. In the second picture we show the same statistic, but for several thousand even quadratic twists d of size 500,000, of the Ramanujan τ cusp form L-function. This is compared to the distribution of the zero closest to 1 for large orthogonal matrices with even characteristic polynomial (in the latter family, one needs to distinguish between even and odd twists). Graphics by M. Rubenstein.

as the RMT distributions of the monodromy group associated with the family.

Katz and Sarnak stress that the proofs of Weil and Deligne use families of zeta-functions over finite fields to prove RH for an individual zeta-function. The modelling of families of L-functions by ensembles of random matrix theory gives evidence for a spectral interpretation of the zeros, which may prove important if families are ultimately used to prove RH. At this point, however, we do not know what plays the role of the monodromy groups in this situation.

RMT and Families

Keating and Snaith extended their conjectures to moments of families of L-functions by computing moments of characteristic polynomials of symplectic and orthogonal matrices, each with their own Haar measure. (It should be mentioned that the orthogonal and symplectic circular ensembles used by the physicists do not use Haar measure and so have different answers. Katz and Sarnak figured out that Haar measure must be used to model L-functions.)

Further works by Farmer, Keating, Rubinstein, Snaith, and this author have led to precise conjectures for all of the main terms in moments for many families of L-functions. These results are so precise that they lead to further conjectures about the distribution of values of the L-functions. We can even predict how frequently we find double zeros at the center of the critical strip of L-functions within certain families.

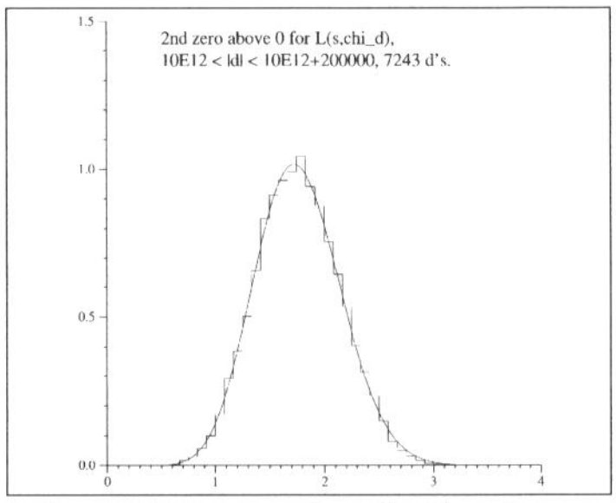

Figure 10. The second zero for $L(s, \chi_d)$ as compared to the RMT prediction. Graphic by M. Rubenstein.

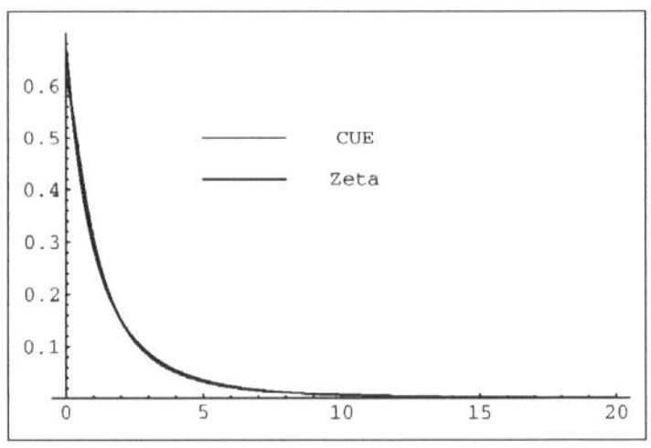

Figure 11. The distribution of values of $|\zeta(1/2 + it)|$ near $t = 10^6$ compared with the distribution of values of characteristic polynomials of 12×12 unitary matrices. Graphic by N. Snaith.

The Conspiracy of L-Functions

There is a growing body of evidence that there is a conspiracy among L-functions—a conspiracy which is preventing us from solving RH!

The first clue that zeta- and L-functions even know about each other appears perhaps in works of Deuring and Heilbronn in their study of one of the most intriguing problems in all of mathematics: Gauss's class number problem. Gauss asked whether the number of equivalence classes of binary quadratic forms of discriminant $d < 0$ goes to ∞ as d goes to $-\infty$.

The equivalence class of a quadratic form $Q(m,n) = am^2 + bmn + cn^2$ of discriminant $d = b^2 - 4ac$ consists of all of the quadratic forms obtained by a linear substitution $m \to \alpha m + \beta n$, $n \to \gamma m + \delta n$, where $\alpha, \beta, \gamma, \delta$ are integers with $\alpha\delta - \beta\gamma = 1$. The number $h(d)$ of these equivalence classes is called the class number and is known to be finite. Equivalently, $h(d)$ is the number of ideal classes of the imaginary quadratic field $Q(\sqrt{d})$. The history of Gauss's problem is extremely interesting; it has many twists and turns and is not yet finished—we seem to be players in the middle of a mystery novel.

Deuring and Heilbronn were trying to solve Gauss's problem. The main tool they were using was the beautiful class number formula of Dirichlet, $h(d) = \sqrt{|d|}\, L(1, \chi_d)/\pi$ ($|d| > 4$), which gives the class number in terms of the value of the L-function at 1, which is at the edge of the critical strip. So the question boils down to giving a lower bound for $L(1, \chi_d)$; this question, in turn, can be resolved by proving that there is no real zero of $L(s, \chi_d)$ very near to 1.

Hecke had shown that the truth of RH for $L(s, \chi_d)$ implies that $h(d) \to \infty$. Then Deuring proved that the falsity of RH for $\zeta(s)$ implies that $h(d) > 1$ for large $|d|$. Finally, Heilbronn showed that the falsity of RH for $L(s, \chi)$ for any χ implied that $h(d) \to \infty$. These results together proved Gauss's conjecture and gave a first indication of a connection between the zeros of $\zeta(s)$ and those of $L(s, \chi_d)$!

Later Landau showed that a hypothetical zero of $L(s, \chi_{d_1})$ very near to 1 implies that no other $L(s, \chi_d)$, $d \neq d_1$, could have such a zero, further illustrating that zeros of $L(s, \chi_d)$ know about each other. Siegel strengthened this approach to show that for every $\epsilon > 0$ there is a $c(\epsilon) > 0$ such that no zero β of $L(s, \chi_d)$ satisfies $\beta > 1 - c(\epsilon)|d|^{-\epsilon}$. The problem with the arguments of Landau and Siegel is that the constant $c(\epsilon)$ cannot be effectively computed, and so the bound cannot be used to actually calculate the list of discriminants d with a given class number, which presumably is what Gauss wanted. The ineffectivity comes about from the assumption that some L-function actually has a real zero near 1. Such a hypothetical zero of some L-function, which no one believes exists, is called a Landau-Siegel zero.

In fact, one can show that if there is some d_1 such that $L(s, \chi_{d_1})$ has a zero at $\beta < 1$, then it follows that $h(d) > c|d|^{\beta - 1/2}/\log|d|$ for all other d, where $c > 0$ can be effectively computed. Thus, the closer to 1 the hypothetical zero is, the stronger the result. But note also that any zero bigger than 1/2 would give a result. The basic idea behind this approach is that if there is an $L(s, \chi_d)$ with a zero near 1, then $\chi_d(p) = -1$ for many small primes. In other words, χ_d mimics the Möbius function $\mu(n)$ for small n. This is consistent with the fact that

$$\sum_{n=1}^{\infty} \frac{\mu(n)}{n^s}$$

has a zero at $s = 1$ (since $\zeta(s)$ has a pole at $s = 1$).

The Landau-Siegel Zero

Much effort has gone toward trying to eliminate the Landau-Siegel zero described above and so find an effective solution to Gauss's problem. However, the L-function conspiracy blocks every attempt exactly at the point where success appears to be in sight. We begin to suspect that the battle for RH will not be won without getting to the bottom of this conspiracy. Here are some tangible examples which give a glimpse of this tangled web.

The Brun-Titchmarsh theorem. Let $\pi(x;q,a)$ denote the number of primes less than or equal to x that lie in the arithmetic progression $a \bmod q$. Sieve methods can show that for any $1 \leq q < x$ the inequality

$$\pi(x;q,a) \leq 2\frac{x}{\phi(q)\log(x/q)}$$

holds, where ϕ is Euler's phi-function. It is believed that the same theorem should be true with 2 replaced by any number larger than 1 and sufficiently large x. Any lowering of the constant 2 would eliminate the Landau-Siegel zero. In particular, Motohashi [1979] proved that if $1 - \delta$ is a real zero of $L(s, \chi_q)$, then if for $x \geq q^c$ the Brun-Titchmarsh theorem is valid in the form $\pi(x;q,a) \leq (2 - \alpha)x/(\phi(q)\log(x/q))$, where $\alpha > 0$ is an absolute constant, then $\delta \geq c'\xi/\log q$, where c and c' are certain numerical constants.

The Alternative Hypothesis. This is an alternative to the GUE model for the distribution of zeros. It proposes the existence of a function $f(T)$ that goes to 0 as $T \to \infty$ such that if any two consecutive ordinates γ and γ' of zeros of ζ larger than some T_0 are given, then the normalized gap $2\pi(\gamma \log \gamma - \gamma' \log \gamma')$ between γ and γ' is within $f(T_0)$ of half of an integer. This hypothesis is clearly absurd! However, ruling this out would eliminate the Landau-Siegel zero (Conrey-Iwaniec (2002)), and so for all we know it could be true.

If one could prove, for example, that there is a $\delta > 0$ such that for all sufficiently large T there is a pair of consecutive zeros with ordinates between T and $2T$ whose distance apart is less than $1/2 - \delta$ times the average spacing, then the alternative hypothesis would be violated. Random matrix theory predicts the exact distribution of these neighbor spacings and shows that we should expect that about 11 percent of the time the neighbor gaps are smaller than $1/2$ of the average. These ideas were what led Montgomery to consider the pair-correlation of the zeros of $\zeta(s)$ mentioned above. He showed that there are arbitrarily large pairs of zeros that are as close together as 0.68 of the average spacing. Later works have gotten this bound down to 0.5152. There are indications that using work of Rudnick and Sarnak on higher correlations of the zeros of ζ, one might be able to reach 0.5, but 0.5 is definitely a limit (more like a brick wall!) of all of the known methods.

Vanishing of modular L-functions. The most spectacular example is the work of Iwaniec and Sarnak. They showed that if one could prove that there is a $\delta > 0$ such that more than $1/2 + \delta$ of the modular L-functions of a fixed weight, large level, and even functional equation do not vanish, then the Landau-Siegel zero could be eliminated. It is predicted that all but an infinitesimal proportion of these values are nonzero; they just needed one-half plus δ of them to be nonzero. They can prove that 50 percent do not vanish, but despite their best efforts they cannot get that extra little tiny bit needed to eliminate the Landau-Siegel zero.

A Clue and a Partial Victory

The only approach that has made an impact on the Landau-Siegel zero problem is an idea of Goldfeld. In 1974 Goldfeld, anticipated somewhat by Friedlander, realized that while a zero at $1/2$ would barely fail to produce a lower bound for the class number tending to infinity, a multiple zero at $1/2$ would produce a lower bound which, while not a positive power of $|d|$, still goes to ∞. Moreover, it was believed—by virtue of the Birch and Swinnerton-Dyer conjecture—that zeros of high multiplicity do exist and the place to look for them is among L-functions associated to elliptic curves with large rank. However, it was not until 1985 that Gross and Zagier demonstrated conclusively that there exist L-functions with triple zeros at $1/2$. This led to the lower bound that for any $\epsilon > 0$ there is an effectively computable $c_1(\epsilon) > 0$ such that $h(d) > c_1(\epsilon)(\log|d|)^{1-\epsilon}$. This is a long way from the expected $h(d) > c\sqrt{|d|}/\log|d|$, but it did solve Gauss's problem. The clue that it gave us was to study exotic L-functions, or extremal L-functions, which have zeros of high multiplicity at the center. At present, our best hope for finding these L-functions is to look at elliptic curves with many rational points.

Iwaniec's Approach

Iwaniec, in his lecture at RHIII, proposed a way to take advantage of the above ideas. In a nutshell, his idea is to take a family of L-functions having a multiple zero at $1/2$ and use this family to obtain useful approximations for the Möbius function $\mu(n)$ as a linear combination of the coefficients of the L-functions from the family. In this way, the Möbius function is tamed. One example of a family considered by Iwaniec is the family of L-functions associated to the elliptic curves

$$E_{A,B^2} : y^2 = x^3 + Ax + B^2,$$

which have a rational point $(B, 0)$ and so have rank at least one. Considering A and B in certain arithmetic progressions shows that the associated L-function must have a double zero at the center.

Iwaniec presented three conjectures which together would eliminate the Landau-Siegel zero. The main two theorems needed to complete his program are a bound for the second moment

$$\sum_{A \approx X^{1/3}, B \approx X^{1/4}} L_{A,B^2}(1/2)^2 = O\left(X^{7/12}(\log X)^C\right)$$

of this family together with a good estimate (square-root cancellation uniform in M, N, and q) for the incomplete exponential sum

$$\sum_{M<m<2M, N<n<2N} \chi_q(mn) \exp\left(2\pi i \frac{m^3 n^{-4}}{q}\right),$$

the kind of estimate that for a completed exponential sum follows from the RH for varieties proved by Deligne. Iwaniec has similar, but more complicated, constructions that would lead to a quasi-Riemann hypothesis, producing a concrete $\beta < 1$ such that there are no zeros to the right of the line through β.

Iwaniec's approach will likely reduce the question of RH, which is ostensibly about zeros or poles, into several subsidiary questions that have a much different flavor, such as finding upper bound estimates for moments and values of L-functions. This approach offers hope of attack by methods from analytic number theory.

Conclusion

A major difficulty in trying to construct a proof of RH through analysis is that the zeros of L-functions behave so much differently from zeros of many of the special functions we are used to seeing in mathematics and mathematical physics. For example, it is known that the zeta-function does not satisfy any differential equation. The functions which do arise as solutions of some of the classical differential equations, such as Bessel functions, hypergeometric functions, etc., have zeros which are fairly regularly spaced. A similar remark holds for the zeros of solutions of classical differential equations regarded as a function of a parameter in the differential equation. For instance, in the Pólya theorem above comparing $\phi(t)$ with $\Phi(t)$, the zeros are actually zeros of a Bessel function of fixed argument regarded as a function of the index. Again the zeros are regularly spaced.

On the other hand, the zeros of L-functions are much more irregularly spaced. For example, the RMT models predict that for any $\epsilon > 0$ there are infinitely many pairs of zeros ρ and ρ' such that $|\rho - \rho'| < |\rho|^{-1/3+\epsilon}$. Generally it is believed that all zeros of all L-functions are linearly independent (in particular, simple), except that certain L-functions can have a zero at $s = 1/2$ of high multiplicity. The conjecture of Birch and Swinnerton-Dyer asserts that the multiplicity of the zero of the L-function associated with a given elliptic curve is equal to the rank of the group of rational points on the elliptic curve. It is known that the latter can be as large as 26, and it is generally believed to get arbitrarily large. None of the methods from analysis seem capable of dealing with such exotic phenomena.

It is my belief that RH is a genuinely arithmetic question that likely will not succumb to methods of analysis. There is a growing body of evidence indicating that one needs to consider families of L-functions in order to make progress on this difficult question. If so, then number theorists are on the right track to an eventual proof of RH, but we are still lacking many of the tools. The ingredients for a proof of RH may well be moment theorems for a new family of L-functions not yet explored; modularity of Hasse-Weil L-functions for many varieties, like that proved by Wiles and others for elliptic curves; and new estimates for exponential sums, which could come out of arithmetic geometry. The study of L-functions is still in its beginning stages. We only recently learned the modularity of the L-functions associated to elliptic curves; it would be very helpful to understand the L-functions for more complicated curves and generally for varieties. It would be useful to systematically compute many new examples of L-functions to get a glimpse of what is out there waiting to be discovered. The exotic behavior of the multiple zeros of L-functions associated to elliptic curves with many rational points could be just the beginning of the story.

Acknowledgements

I would like to thank David Farmer, K. Soundararajan, Roger Heath-Brown, Brianna Conrey, and Harold Boas for their valuable comments during the preparation of this article. Also, I thank Andrew Odlyzko, Michael Rubenstein, Nina Snaith, and Sandra Frost for their assistance with the graphics in the article.

—*J.B.C.*

References

- E. Bombieri, Problems of the Millennium: The Riemann Hypothesis, http://claymath.org/prizeproblems/riemann.htm.
- Harold Davenport, *Multiplicative Number Theory*, third edition, revised and with a preface by Hugh L. Montgomery, Grad. Texts in Math., vol. 74, Springer-Verlag, New York, 2000.
- P. A. Deift, *Orthogonal Polynomials and Random Matrices: A Riemann-Hilbert Approach*, Courant Lecture Notes in Math., vol. 3, New York University, Courant Institute of Mathematical Sciences, New York; Amer. Math. Soc., Providence, RI, 1999.
- Henryk Iwaniec, *Introduction to the Spectral Theory of Automorphic Forms*, Bibl. Rev. Mat. Iberoamericana [Library of the Revista Matemática Iberoamericana], Rev. Mat. Iberoamericana, Madrid, 1995.
- ———, *Topics in Classical Automorphic Forms*, Grad. Stud. Math., vol. 17, Amer. Math. Soc., Providence, RI, 1997.
- Nicholas M. Katz and Peter Sarnak, *Random Matrices, Frobenius Eigenvalues, and Monodromy*, Amer. Math. Soc. Colloq. Publ., vol. 45, Amer. Math. Soc., Providence, RI, 1999.
- Madan Lal Mehta, *Random Matrices*, second edition, Academic Press, Inc., Boston, MA, 1991.
- A. M. Odlyzko, http://www.dtc.umn.edu/~odlyzko/.
- Atle Selberg, *Old and New Conjectures and Results about a Class of Dirichlet Series, Collected Papers*, Vol. II, with a foreword by K. Chandrasekharan, Springer-Verlag, Berlin, 1991.
- E. C. Titchmarsh, *The Theory of the Riemann Zeta-Function*, second edition, edited and with a preface by D. R. Heath-Brown, The Clarendon Press, Oxford University Press, New York, 1986.

11.4 A. Ivić (2003)

On Some Reasons for Doubting the Riemann Hypothesis

> *... I don't believe or disbelieve the Riemann hypothesis. I have a certain amount of data and a certain amount of facts. These facts tell me definitely that the thing has not been settled. Until it's been settled it's a hypothesis, that's all. I would like the Riemann hypothesis to be true, like any decent mathematician, because it's a thing of beauty, a thing of elegance, a thing that would simplify many proofs and so forth, but that's all [138].*
>
> **Aleksandar Ivić**

In this article Ivić discusses several arguments against the truth of the Riemann hypothesis. Ivić points to heuristics on the moments of zeta functions as evidence that the hypothesis is false. He discusses this and other conditional disproofs of the Riemann hypothesis. This paper questions the widely held belief in the Riemann hypothesis and cautions mathematicians against accepting the hypothesis as fact, even in light of the empirical data.

ON SOME REASONS FOR DOUBTING THE RIEMANN HYPOTHESIS

Aleksandar Ivić

Abstract. *Several arguments against the truth of the Riemann hypothesis are extensively discussed. These include the Lehmer phenomenon, the Davenport–Heilbronn zeta-function, large and mean values of* $|\zeta(\frac{1}{2}+it)|$ *on the critical line, and zeros of a class of convolution functions closely related to* $\zeta(\frac{1}{2}+it)$. *The first two topics are classical, and the remaining ones are connected with the author's recent research. By means of convolution functions a conditional disproof of the Riemann hypothesis is given.*

0. Foreword ("Audiatur et altera pars")

This is the unabridged version of the work that was presented at the Bordeaux Conference in honour of the Prime Number Theorem Centenary, Bordeaux, January 26, 1996 and later during the 39th Taniguchi International Symposium on Mathematics "*Analytic Number Theory*", May 13-17, 1996 in Kyoto and its forum, May 20-24, 1996. The abridged printed version, with a somewhat different title, is [62]. The multiplicities of zeros are treated in [64]. A plausible conjecture for the coefficients of the main term in the asymptotic formula for the $2k$-th moment of $|\zeta(\frac{1}{2}+it)|$ (see (4.1)–(4.2)) is given in [67].

In the years that have passed after the writing of the first version of this paper, it appears that the subject of the Riemann Hypothesis has only gained in interest and importance. This seem particularly true in view of the Clay Mathematical Institute prize of one million dollars for the proof of the Riemann Hypothesis, which is called as one of the mathematical "Problems of the Millenium". A comprehensive account is to be found in E. Bombieri's paper [65]. It is the author's belief that the present work can still be of interest, especially since the Riemann Hypothesis may be still very far from being settled. Inasmuch the Riemann Hypothesis is commonly believed to be true, and for several valid reasons, I feel that the arguments that disfavour it should also be pointed out.

One of the reasons that the original work had to be shortened and revised before being published is the remark that "The Riemann hypothesis is in the process of being proved" by powerful methods from Random matrix theory (see e.g., B. Conrey's survey article [66]). Random matrix theory has undisputably found its place in the theory of $\zeta(s)$ and allied functions (op. cit. [66], [67]). However, almost ten years have passed since its advent, but the Riemann hypothesis seems as distant now as it was then.

1. Introduction

A central place in Analytic number theory is occupied by the Riemann zeta-function $\zeta(s)$, defined for $\Re e\, s > 1$ by

$$\zeta(s) = \sum_{n=1}^{\infty} n^{-s} = \prod_{p\,\text{prime}} (1-p^{-s})^{-1}, \tag{1.1}$$

and otherwise by analytic continuation. It admits meromorphic continuation to the whole complex plane, its only singularity being the simple pole $s=1$ with residue 1. For general information on $\zeta(s)$ the reader is referred to the monographs [7], [16], and [61]. From the functional equation

$$\zeta(s) = \chi(s)\zeta(1-s),\ \chi(s) = 2^s\pi^{s-1}\sin\left(\frac{\pi s}{2}\right)\Gamma(1-s), \tag{1.2}$$

which is valid for any complex s, it follows that $\zeta(s)$ has zeros at $s = -2, -4, \ldots$. These zeros are traditionally called the "trivial" zeros of $\zeta(s)$, to distinguish them from the complex zeros of $\zeta(s)$, of which the smallest ones (in absolute value) are $\frac{1}{2} \pm 14.134725\ldots i$. It is well-known that all complex zeros of $\zeta(s)$ lie in the so-called "critical strip" $0 < \sigma = \Re e\, s < 1$, and if $N(T)$ denotes the number of zeros $\rho = \beta + i\gamma$ (β, γ real) of $\zeta(s)$ for which $0 < \gamma \le T$, then

$$N(T) = \frac{T}{2\pi}\log\left(\frac{T}{2\pi}\right) - \frac{T}{2\pi} + \frac{7}{8} + S(T) + O\left(\frac{1}{T}\right) \tag{1.3}$$

with

$$S(T) = \frac{1}{\pi}\arg\zeta(\tfrac{1}{2} + iT) = O(\log T). \tag{1.4}$$

This is the so-called Riemann–von Mangoldt formula. *The Riemann hypothesis* (henceforth RH for short) is the conjecture, stated by B. Riemann in his epoch-making memoir [52], that *very likely all complex zeros of $\zeta(s)$ have real parts equal to* $1/2$. For this reason the line $\sigma = 1/2$ is called the "critical line" in the theory of $\zeta(s)$. Notice that Riemann was rather cautious in formulating the RH, and that he used the wording "very likely" ("sehr wahrscheinlich" in the German original) in connection with it. Riemann goes on to say in his paper: "One would of course like to have a rigorous proof of this, but I have put aside the search for such a proof after some fleeting vain attempts because it is not necessary for the immediate objective of my investigation". The RH is undoubtedly one of the most celebrated and difficult open problems in whole Mathematics. Its proof (or disproof) would have very important consequences in multiplicative number theory, especially in problems involving the distribution of primes. It would also very likely lead to generalizations to many other zeta-functions (Dirichlet series) having similar properties as $\zeta(s)$.

The RH can be put into many equivalent forms. One of the classical is

$$\pi(x) = \mathrm{li}\, x + O(\sqrt{x}\log x), \tag{1.5}$$

where $\pi(x)$ is the number of primes not exceeding x (≥ 2) and

$$\mathrm{li}\, x = \int_0^x \frac{\mathrm{d}t}{\log t} = \lim_{\varepsilon\to 0+}\left(\int_0^{1-\varepsilon} + \int_{1+\varepsilon}^x\right)\frac{\mathrm{d}t}{\log t} = \sum_{n=1}^N \frac{(n-1)!x}{\log^n x} + O\left(\frac{x}{\log^{N+1} x}\right) \tag{1.6}$$

for any fixed integer $N \ge 1$. One can give a purely arithmetic equivalent of the RH without mentioning primes. Namely we can define recursively the Möbius function $\mu(n)$ as

$$\mu(1) = 1,\ \mu(n) = -\sum_{d|n, d<n}\mu(d) \qquad (n > 1).$$

Then the RH is equivalent to the following assertion: For any given integer $k \ge 1$ there exists an integer $N_0 = N_0(k)$ such that, for integers $N \ge N_0$, one has

$$\left(\sum_{n=1}^N \mu(n)\right)^{2k} \le N^{k+1}. \tag{1.7}$$

The above definition of $\mu(n)$ is elementary and avoids primes. A non-elementary definition of $\mu(n)$ is through the series representation

$$\sum_{n=1}^\infty \mu(n)n^{-s} = \frac{1}{\zeta(s)} \qquad (\Re e\, s > 1), \tag{1.8}$$

and an equivalent form of the RH is that (1.8) holds for $\sigma > 1/2$. The inequality (1.7) is in fact the bound

$$\sum_{n\le x}\mu(n) \ll_\varepsilon x^{\frac{1}{2}+\varepsilon} \tag{1.9}$$

in disguise, where ε corresponds to $1/(2k)$, x to N, and the $2k$–th power avoids absolute values. The bound (1.9) (see [16] and [61]) is one of the classical equivalents of the RH. The sharper bound

$$\Big|\sum_{n\le x}\mu(n)\Big| < \sqrt{x} \qquad (x>1)$$

was proposed in 1897 by Mertens on the basis of numerical evidence, and later became known in the literature as *the Mertens conjecture*. It was disproved in 1985 by A.M. Odlyzko and H.J.J. te Riele [47].

Instead of working with the complex zeros of $\zeta(s)$ on the critical line it is convenient to introduce the function

$$Z(t) = \chi^{-1/2}(\tfrac12+it)\zeta(\tfrac12+it), \tag{1.10}$$

where $\chi(s)$ is given by (1.2). Since $\chi(s)\chi(1-s)=1$ and $\overline{\Gamma(s)}=\Gamma(\bar s)$, it follows that $|Z(t)|=|\zeta(\frac12+it)|$, $Z(t)$ is even and

$$\overline{Z(t)} = \chi^{-1/2}(\tfrac12-it)\zeta(\tfrac12-it) = \chi^{1/2}(\tfrac12-it)\zeta(\tfrac12+it) = Z(t).$$

Hence $Z(t)$ is real if t is real, and the zeros of $Z(t)$ correspond to the zeros of $\zeta(s)$ on the critical line. Let us denote by $0<\gamma_1\le\gamma_2\le\dots$ the positive zeros of $Z(t)$ with multiplicities counted (all known zeros are simple). If the RH is true, then it is known (see [61]) that

$$S(T) = O\Big(\frac{\log T}{\log\log T}\Big), \tag{1.11}$$

and this seemingly small improvement over (1.4) is significant. If (1.11) holds, then from (1.3) one infers that $N(T+H)-N(T)>0$ for $H=C/\log\log T$, suitable $C>0$ and $T\ge T_0$. Consequently we have the bound, on the RH,

$$\gamma_{n+1}-\gamma_n \ll \frac{1}{\log\log\gamma_n} \tag{1.12}$$

for the gap between consecutive zeros on the critical line. For some unconditional results on $\gamma_{n+1}-\gamma_n$, see [17], [18] and [25].

We do not know exactly what motivated Riemann to conjecture the RH. Some mathematicians, like Felix Klein, thought that he was inspired by a sense of general beauty and symmetry in Mathematics. Although doubtlessly the truth of the RH would provide such harmonious symmetry, we also know now that Riemann undertook rather extensive numerical calculations concerning $\zeta(s)$ and its zeros. C.L. Siegel [57] studied Riemann's unpublished notes, kept in the Göttingen library. It turned out that Riemann had computed several zeros of the zeta-function and had a deep understanding of its analytic behaviour. Siegel provided rigorous proof of a formula that had its genesis in Riemann's work. It came to be known later as *the Riemann–Siegel formula* (see [16], [57] and [61]) and, in a weakened form, it says that

$$Z(t) = 2\sum_{n\le(t/2\pi)^{1/2}} n^{-1/2}\cos\Big(t\log\frac{\sqrt{t/2\pi}}{n}-\frac t2-\frac\pi8\Big)+O(t^{-1/4}), \tag{1.13}$$

where the O-term in (1.13) is actually best possible, namely it is $\Omega_\pm(t^{-1/4})$. As usual $f(x)=\Omega_\pm(g(x))$ (for $g(x)>0$ when $x\ge x_0$) means that we have $f(x)=\Omega_+(x)$ and $f(x)=\Omega_-(x)$, namely that both

$$\limsup_{x\to\infty}\frac{f(x)}{g(x)}>0,\quad \liminf_{x\to\infty}\frac{f(x)}{g(x)}<0$$

are true. The Riemann–Siegel formula is an indispensable tool in the theory of $\zeta(s)$, both for theoretical investigations and for the numerical calculations of the zeros.

Perhaps the most important concrete reason for believing the RH is the impressive numerical evidence in its favour. There exists a large and rich literature on numerical calculations involving $\zeta(s)$ and its zeros (see [38], [44], [45], [46], [51], which contain references to further work). This literature reflects the development of Mathematics in general, and of Numerical analysis and Analytic number theory in particular. Suffice to say that it is known that the first 1.5 billion complex zeros of $\zeta(s)$ in the upper half-plane are simple and do have real parts equal to 1/2, as predicted by the RH. Moreover, many large blocks of zeros of much greater height have been thoroughly investigated, and all known zeros satisfy the RH. However, one should be very careful in relying on numerical evidence in Analytic number theory. A classical example for this is the inequality $\pi(x) < \operatorname{li} x$ (see (1.5) and (1.6)), noticed already by Gauss, which is known to be true for all x for which the functions in question have been actually computed. But the inequality $\pi(x) < \operatorname{li} x$ is false; not only does $\pi(x) - \operatorname{li} x$ assume positive values for some arbitrarily large values of x, but J.E. Littlewood [37] proved that

$$\pi(x) = \operatorname{li} x + \Omega_{\pm}\left(\sqrt{x}\,\frac{\log\log\log x}{\log x}\right).$$

By extending the methods of R. Sherman Lehman [56], H.J.J. te Riele [50] showed that $\pi(x) < \operatorname{li} x$ fails for some (unspecified) $x < 6.69 \times 10^{370}$. For values of t which are this large we may hope that $Z(t)$ will also show its true asymptotic behaviour. Nevertheless, we cannot compute by today's methods the values of $Z(t)$ for t this large, actually even $t = 10^{100}$ seems out of reach at present. To assess why the values of t where $Z(t)$ will "really" exhibit its true behaviour must be "very large", it suffices to compare (1.4) and (1.11) and note that the corresponding bounds differ by a factor of $\log\log T$, which is a very slowly varying function.

Just as there are deep reasons for believing the RH, there are also serious grounds for doubting its truth, although the author certainly makes no claims to possess a disproof of the RH. It is in the folklore that several famous mathematicians, which include P. Turán and J.E. Littlewood, believed that the RH is not true. The aim of this paper is to state and analyze some of the arguments which cast doubt on the truth of the RH. In subsequent sections we shall deal with the Lehmer phenomenon, the Davenport-Heilbronn zeta-function, mean value formulas on the critical line, large values on the critical line and the distribution of zeros of a class of convolution functions. These independent topics appear to me to be among the most salient ones which point against the truth of the RH. The first two of them, the Lehmer phenomenon and the Davenport-Heilbronn zeta-function, are classical and fairly well known. The remaining ones are rather new and are connected with the author's research, and for these reasons the emphasis will be on them. A sharp asymptotic formula for the convolution function $M_{Z,f}(t)$, related to $Z(t)$, is given in Section 8. Finally a conditional disproof of the RH, based on the use of the functions $M_{Z,f}(t)$, is given at the end of the paper in Section 9. Of course, nothing short of *rigorous* proof or disproof will settle the truth of the RH.

Acknowledgement. I want to thank Professors M. Jutila, K. Matsumoto, Y. Motohashi and A.M. Odlyzko for valuable remarks.

2. Lehmer's phenomenon

The function $Z(t)$, defined by (1.10), has a negative local maximum $-0.52625\ldots$ at $t = 2.47575\ldots$. This is the only known occurrence of a negative local maximum, while no positive local minimum is known. *Lehmer's phenomenon* (named after D.H. Lehmer, who in his works [35], [36] made significant contributions to the subject) is the fact (see [46] for a thorough discussion) that the graph of $Z(t)$ sometimes barely crosses the t-axis. This means that the absolute value of the maximum or minimum of $Z(t)$ between its two consecutive zeros is small. For instance, A.M. Odlyzko found (in the version of [46] available to the author, but Odlyzko kindly informed me that many more examples occur in the computations that are going on now) 1976 values of n such that $|Z(\frac{1}{2}\gamma_n + \frac{1}{2}\gamma_{n+1})| < 0.0005$ in the block that he investigated. Several extreme examples are also given by van de Lune et al. in [38]. The Lehmer phenomenon shows the delicacy of the RH, and the possibility that a counterexample to the RH may be found numerically. For should it happen that, for $t \ge t_0$, $Z(t)$ attains a negative local maximum or a positive local minimum, then the RH would be disproved. This assertion follows (see [7]) from the following

Proposition 1. If the RH is true, then the graph of $Z'(t)/Z(t)$ is monotonically decreasing between the zeros of $Z(t)$ for $t \ge t_0$.

Namely suppose that $Z(t)$ has a negative local maximum or a positive local minimum between its two consecutive zeros γ_n and γ_{n+1}. Then $Z'(t)$ would have at least two distinct zeros x_1 and x_2 ($x_1 < x_2$) in (γ_n, γ_{n+1}), and hence so would $Z'(t)/Z(t)$. But we have

$$\frac{Z'(x_1)}{Z(x_1)} < \frac{Z'(x_2)}{Z(x_2)},$$

which is a contradiction, since $Z'(x_1) = Z'(x_2) = 0$.

To prove Proposition 1 consider the function

$$\xi(s) := \frac{1}{2}s(s-1)\pi^{-s/2}\Gamma(\frac{s}{2})\zeta(s),$$

so that $\xi(s)$ is an entire function of order one (see Ch. 1 of [16]), and one has unconditionally

$$\frac{\xi'(s)}{\xi(s)} = B + \sum_\rho \left(\frac{1}{s-\rho} + \frac{1}{\rho}\right) \tag{2.1}$$

with

$$B = \log 2 + \frac{1}{2}\log\pi - 1 - \frac{1}{2}C_0,$$

where ρ denotes complex zeros of $\zeta(s)$ and $C_0 == \Gamma'(1)$ is Euler's constant. By (1.2) it follows that

$$Z(t) = \chi^{-1/2}(\tfrac{1}{2}+it)\zeta(\tfrac{1}{2}+it) = \frac{\pi^{-it/2}\Gamma(\frac{1}{4}+\frac{1}{2}it)\zeta(\frac{1}{2}+it)}{|\Gamma(\frac{1}{4}+\frac{1}{2}it)|},$$

so that we may write

$$\xi(\tfrac{1}{2}+it) = -f(t)Z(t), \quad f(t) := \tfrac{1}{2}\pi^{-1/4}(t^2+\tfrac{1}{4})|\Gamma(\tfrac{1}{4}+\tfrac{1}{2}it)|.$$

Consequently logarithmic differentiation gives

$$\frac{Z'(t)}{Z(t)} = -\frac{f'(t)}{f(t)} + i\frac{\xi'(\frac{1}{2}+it)}{\xi(\frac{1}{2}+it)}. \tag{2.2}$$

Assume now that the RH is true. Then by using (2.1) with $\rho = \frac{1}{2}+i\gamma, s = \frac{1}{2}+it$ we obtain, if $t \ne \gamma$,

$$\left(\frac{i\xi'(\frac{1}{2}+it)}{\xi(\frac{1}{2}+it)}\right)' = -\sum_\gamma \frac{1}{(t-\gamma)^2} < -C(\log\log t)^2 \quad (C>0)$$

for $t \ge t_0$, since (1.12) holds. On the other hand, by using Stirling's formula for the gamma-function and $\log|z| = \Re\mathrm{e}\,\log z$, it is readily found that

$$\frac{\mathrm{d}}{\mathrm{d}t}\left(\frac{f'(t)}{f(t)}\right) \ll \frac{1}{t},$$

so that from (2.2) it follows that $(Z'(t)/Z(t))' < 0$ if $t \ge t_0$, which implies Proposition 1. Actually the value of t_0 may be easily effectively determined and seen not to exceed 1000. Since $Z(t)$ has no positive local minimum or negative local maximum for $3 \le t \le 1000$, it follows that the RH is false if we find (numerically) the occurrence of a single negative local maximum (besides the one at $t = 2.47575\ldots$) or a positive local minimum of $Z(t)$. It seems appropriate to quote in concluding Edwards [7], who says that Lehmer's phenomenon "must give pause to even the most convinced believer of the Riemann hypothesis".

3. The Davenport-Heilbronn zeta-function

This is a zeta-function (Dirichlet series) which satisfies a functional equation similar to the classical functional equation (1.2) for $\zeta(s)$. It has other analogies with $\zeta(s)$, like having infinitely many zeros on the critical line $\sigma = 1/2$, but for this zeta-function the analogue of the RH does not hold. This function was introduced by H. Davenport and H. Heilbronn [6] as

$$f(s) = 5^{-s}\big(\zeta(s,\frac{1}{5}) + \tan\theta\,\zeta(s,\frac{2}{5}) - \tan\theta\,\zeta(s,\frac{3}{5}) - \zeta(s,\frac{4}{5})\big), \tag{3.1}$$

where $\theta = \arctan\big(\sqrt{10-2\sqrt{5}}-2\big)/(\sqrt{5}-1)$ and, for $\Re e\, s > 1$,

$$\zeta(s,a) = \sum_{n=0}^{\infty}(n+a)^{-s} \qquad (0 < a \le 1)$$

is the familiar *Hurwitz zeta-function*, defined for $\Re e\ s \le 1$ by analytic continuation. With the above choice of θ (see [6], [32] or [61]) it can be shown that $f(s)$ satisfies the functional equation

$$f(s) = X(s)f(1-s), \quad X(s) = \frac{2\Gamma(1-s)\cos(\frac{\pi s}{2})}{5^{s-\frac{1}{2}}(2\pi)^{1-s}}, \tag{3.2}$$

whose analogy with the functional equation (1.2) for $\zeta(s)$ is evident. Let $1/2 < \sigma_1 < \sigma_2 < 1$. Then it can be shown (see Ch. 6 of [32]) that $f(s)$ has infinitely many zeros in the strip $\sigma_1 < \sigma = \Re e\, s < \sigma_2$, and it also has (see Ch. 10 of [61]) an infinity of zeros in the half-plane $\sigma > 1$, while from the product representation in (1.1) it follows that $\zeta(s) \ne 0$ for $\sigma > 1$, so that in the half-plane $\sigma > 1$ the behaviour of zeros of $\zeta(s)$ and $f(s)$ is different. Actually the number of zeros of $f(s)$ for which $\sigma > 1$ and $0 < t = \Im m\, s \le T$ is $\gg T$, and similarly each rectangle $0 < t \le T, 1/2 < \sigma_1 < \sigma \le \sigma_2 \le 1$ contains at least $c(\sigma_1,\sigma_2)T$ zeros of $f(s)$. R. Spira [58] found that $0.808517 + 85.699348i$ (the values are approximate) is a zero of $f(s)$ lying in the critical strip $0 < \sigma < 1$, but not on the critical line $\sigma = 1/2$. On the other hand, A.A. Karatsuba [31] proved that the number of zeros $\frac{1}{2} + i\gamma$ of $f(s)$ for which $0 < \gamma \le T$ is at least $T(\log T)^{1/2-\varepsilon}$ for any given $\varepsilon > 0$ and $T \ge T_0(\varepsilon)$. This bound is weaker than A. Selberg's classical result [53] that there are $\gg T \log T$ zeros $\frac{1}{2} + i\gamma$ of $\zeta(s)$ for which $0 < \gamma \le T$. From the Riemann–von Mangoldt formula (1.3) it follows that, up to the value of the $\ll$–constant, Selberg's result on $\zeta(s)$ is best possible. There are certainly $\ll T \log T$ zeros $\frac{1}{2} + i\gamma$ of $f(s)$ for which $0 < \gamma \le T$ and it may be that almost all of them lie on the critical line $\sigma = 1/2$, although this has not been proved yet. The Davenport-Heilbronn zeta-function is not the only example of a zeta-function that exhibits the phenomena described above, and many so-called *Epstein zeta-functions* also have complex zeros off their respective critical lines (see the paper of E. Bombieri and D. Hejhal [5] for some interesting results).

What is the most important difference between $\zeta(s)$ and $f(s)$ which is accountable for the difference of distribution of zeros of the two functions, which occurs at least in the region $\sigma > 1$? It is most likely that the answer is the lack of the Euler product for $f(s)$, similar to the one in (1.1) for $\zeta(s)$. But $f(s)$ can be written as a linear combination of two L-functions which have Euler products (with a common factor) and this fact plays the crucial rôle in Karatsuba's proof of the lower bound result for the number of zeros of $f(s)$. In any case one can argue that it may likely happen that the influence of the Euler product for $\zeta(s)$ will not extend all the way to the line $\sigma = 1/2$. In other words, the existence of zeta-functions such as $f(s)$, which share many common properties with $\zeta(s)$, but which have infinitely many zeros off the critical line, certainly disfavours the RH.

Perhaps one should at this point mention the *Selberg zeta-function* $\mathcal{Z}(s)$ (see [55]). This is an entire function which enjoys several common properties with $\zeta(s)$, like the functional equation and the Euler product. For $\mathcal{Z}(s)$ the corresponding analogue of the RH is true, but it should be stressed that $\mathcal{Z}(s)$ is *not* a classical Dirichlet series. Its Euler product

$$\mathcal{Z}(s) = \prod_{\{P_0\}} \prod_{k=0}^{\infty} (1 - N(P_0)^{-s-k}) \qquad (\Re e\, s > 1)$$

is not a product over the rational primes, but over norms of certain conjugacy classes of groups. Also $\mathcal{Z}(s)$ is an entire function of order 2, while $(s-1)\zeta(s)$ is an entire function of order 1. For these reasons $\mathcal{Z}(s)$ cannot be compared too closely to $\zeta(s)$.

4. Mean value formulas on the critical line

For $k \geq 1$ a fixed integer, let us write the $2k$-th moment of $|\zeta(\frac{1}{2}+it)|$ as

$$\int_0^T |\zeta(\tfrac{1}{2}+it)|^{2k}\,dt = T\,P_{k^2}(\log T) + E_k(T), \tag{4.1}$$

where for some suitable coefficients $a_{j,k}$ one has

$$P_{k^2}(y) = \sum_{j=0}^{k^2} a_{j,k}y^j. \tag{4.2}$$

An extensive literature exists on $E_k(T)$, especially on $E_1(T) \equiv E(T)$ (see F.V. Atkinson's classical paper [2]), and the reader is referred to [20] for a comprehensive account. It is known that

$$P_1(y) = y + 2C_0 - 1 - \log(2\pi),$$

and $P_4(y)$ is a quartic polynomial whose leading coefficient equals $1/(2\pi^2)$ (see [22] for an explicit evaluation of its coefficients). One hopes that

$$E_k(T) = o(T) \qquad (T \to \infty) \tag{4.3}$$

will hold for each fixed integer $k \geq 1$, but so far this is known to be true only in the cases $k=1$ and $k=2$, when $E_k(T)$ is a true error term in the asymptotic formula (4.1). In fact heretofore it has not been clear how to define properly (even on heuristic grounds) the values of $a_{j,k}$ in (4.2) for $k \geq 3$ (see [24] for an extensive discussion concerning the case $k=3$). The connection between $E_k(T)$ and the RH is indirect, namely there is a connection with the *Lindelöf hypothesis* (LH for short). The LH is also a famous unsettled problem, and it states that

$$\zeta(\tfrac{1}{2}+it) \ll_\varepsilon t^\varepsilon \tag{4.4}$$

for any given $\varepsilon > 0$ and $t \geq t_0 > 0$ (since $\overline{\zeta(\frac{1}{2}+it)} = \zeta(\frac{1}{2}-it)$, t may be assumed to be positive). It is well-known (see [61] for a proof) that the RH implies

$$\zeta(\tfrac{1}{2}+it) \ll \exp\Big(\frac{A\log t}{\log\log t}\Big) \qquad (A>0,\ t \geq t_0), \tag{4.5}$$

so that obviouly the RH implies the LH. In the other direction it is unknown whether the LH (or (4.5)) implies the RH. However, it is known that the LH has considerable influence on the distribution of zeros of $\zeta(s)$. If $N(\sigma,T)$ denotes the number of zeros $\rho=\beta+i\gamma$ of $\zeta(s)$ for which $\sigma\le\beta$ and $|\gamma|\le T$, then it is known (see Ch. 11 of [16]), that the LH implies that $N(\sigma,T)\ll T^{2-2\sigma+\varepsilon}$ for $1/2\le\sigma\le 1$ (this is a form of *the density hypothesis*) and $N(\frac34+\delta,T)\ll T^\varepsilon$, where $\varepsilon=\varepsilon(\delta)$ may be arbitrarily small for any $0<\delta<\frac14$.

The best unconditional bound for the order of $\zeta(s)$ on the critical line, known at the time of the writing of this text is

$$\zeta(\tfrac12+it)\ll_\varepsilon t^{c+\varepsilon} \tag{4.6}$$

with $c=89/570=0.15614\ldots$. This is due to M.N. Huxley [13], and represents the last in a long series of improvements over the past 80 years. The result is obtained by intricate estimates of exponential sums of the type $\sum_{N<n\le 2N} n^{it}\quad (N\ll\sqrt{t})$, and the value $c=0.15$ appears to be the limit of the method.

Estimates for $E_k(T)$ in (4.1) (both pointwise and in the mean sense) have many applications. From the knowledge about the order of $E_k(T)$ one can deduce a bound for $\zeta(\frac12+iT)$ via the estimate

$$\zeta(\tfrac12+iT)\ll(\log T)^{(k^2+1)/(2k)}+\big(\log T\max_{t\in[T-1,T+1]}|E_k(t)|\big)^{1/(2k)}, \tag{4.7}$$

which is Lemma 4.2 of [20]. Thus the best known upper bound

$$E(T)\equiv E_1(T)\ll T^{72/227}(\log T)^{679/227} \tag{4.8}$$

of M.N. Huxley [14] yields (4.6) with $c=36/227=0.15859\ldots$. Similarly the sharpest known bound

$$E_2(T)\ll T^{2/3}\log^C T\qquad (C>0) \tag{4.9}$$

of Y. Motohashi and the author (see [20], [26], [28]) yields (4.6) with the classical value $c=1/6$ of Hardy and Littlewood. Since the difficulties in evaluating the left-hand side of (4.1) greatly increase as k increases, it is reasonable to expect that the best estimate for $\zeta(\frac12+iT)$ that one can get from (4.7) will be when $k=1$.

The LH is equivalent to the bound

$$\int_0^T|\zeta(\tfrac12+it)|^{2k}\,\mathrm{d}t\ll_{k,\varepsilon}T^{1+\varepsilon} \tag{4.10}$$

for any $k\ge1$ and any $\varepsilon>0$, which in turn is the same as

$$E_k(T)\ll_{k,\varepsilon}T^{1+\varepsilon}. \tag{4.11}$$

The enormous difficulty in settling the truth of the LH, and so *a fortiori* of the RH, is best reflected in the relatively modest upper bounds for the integrals in (4.10) (see Ch. 8 of [16] for sharpest known results). On the other hand, we have Ω-results in the case $k=1,2$, which show that $E_1(T)$ and $E_2(T)$ cannot be always small. Thus J.L. Hafner and the author [11], [12] proved that

$$E_1(T)=\Omega_+\Big((T\log T)^{\frac14}(\log\log T)^{\frac{3+\log 4}{4}}\mathrm{e}^{-C\sqrt{\log\log\log T}}\Big) \tag{4.12}$$

and

$$E_1(T)=\Omega_-\Big(T^{\frac14}\exp\Big(\frac{D(\log\log T)^{\frac14}}{(\log\log\log T)^{\frac34}}\Big)\Big) \tag{4.13}$$

for some absolute constants $C, D > 0$. Moreover the author [19] proved that there exist constants $A, B > 0$ such that, for $T \geq T_0$, every interval $[T, T + B\sqrt{T}]$ contains points t_1, t_2 for which

$$E_1(t_1) > At_1^{1/4}, \quad E_1(t_2) < -At_2^{1/4}.$$

Numerical investigations concerning $E_1(T)$ were carried out by H.J.J. te Riele and the author [29].

The Ω–result

$$E_2(T) = \Omega(\sqrt{T}) \tag{4.14}$$

(meaning $\lim_{T\to\infty} E_2(T)T^{-1/2} \neq 0$) was proved by Y. Motohashi and the author (see [26], [28] and Ch. 5 of [20]). The method of proof involved differences of values of the functions $E_2(T)$, so that (4.14) was the limit of the method. The basis of this, as well of other recent investigations involving $E_2(T)$, is Y. Motohashi's fundamental explicit formula for

$$(\Delta\sqrt{\pi})^{-1}\int_{-\infty}^{\infty} |\zeta(\tfrac{1}{2} + it + iT)|^4 \,\mathrm{e}^{-(t/\Delta)^2}\,\mathrm{d}t \qquad (\Delta > 0), \tag{4.15}$$

obtained by deep methods involving spectral theory of the non-Euclidean Laplacian (see [40], [41], [43], [63] and Ch. 5 of [16]). On p. 310 of [20] it was pointed out that a stronger result than (4.14), namely

$$\limsup_{T\to\infty} |E_2(T)|T^{-1/2} = +\infty$$

follows if certain quantities connected with the discrete spectrum of the non-Euclidean Laplacian are linearly independent over the integers. Y. Motohashi [42] recently unconditionally improved (4.14) by showing that

$$E_2(T) = \Omega_{\pm}(\sqrt{T}) \tag{4.16}$$

holds. Namely he proved that the function

$$Z_2(\xi) := \int_1^{\infty} |\zeta(\tfrac{1}{2} + it)|^4\, t^{-\xi}\,\mathrm{d}t,$$

defined initially as a function of the complex variable ξ for $\Re\mathrm{e}\,\xi > 1$, is meromorphic over the whole complex plane. In the half-plane $\Re\mathrm{e}\,\xi > 0$ it has a pole of order five at $\xi = 1$, infinitely many simple poles of the form $\frac{1}{2} \pm \kappa i$, while the remaining poles for $\Re\mathrm{e}\,\xi > 0$ are of the form $\rho/2, \zeta(\rho) = 0$. Here $\kappa^2 + \frac{1}{4}$ is in the discrete spectrum of the non-Euclidean Laplacian with respect to the full modular group. By using (4.1) and integration by parts it follows that

$$Z_2(\xi) = C + \xi\int_1^{\infty} P_4(\log t)t^{-\xi}\,\mathrm{d}t + \xi\int_1^{\infty} E_2(t)t^{-\xi-1}\,\mathrm{d}t \tag{4.17}$$

with a suitable constant C, where the integrals are certainly absolutely convergent for $\Re\mathrm{e}\,\xi > 1$ (actually the second for $\Re\mathrm{e}\,\xi > 1/2$ in view of (4.20)). Now (4.16) is an immediate consequence of (4.17) and the following version of a classical result of E. Landau (see [1] for a proof).

Proposition 2. Let $g(x)$ be a continuous function such that

$$G(\xi) := \int_1^{\infty} g(x)x^{-\xi-1}\,\mathrm{d}x$$

converges absolutely for some ξ. Let us suppose that $G(\xi)$ admits analytic continuation to a domain including the half-line $[\sigma, \infty)$, while it has a simple pole at $\xi = \sigma + i\delta$ $(\delta \neq 0)$, with residue γ. Then

$$\limsup_{x\to\infty} g(x)x^{-\sigma} \geq |\gamma|, \quad \liminf_{x\to\infty} g(x)x^{-\sigma} \leq -|\gamma|.$$

It should be pointed out that (4.14) shows that the well-known analogy between $E_1(T)$ and $\Delta_2(x)$ ($=\Delta(x)$, the error term in the formula for $\sum_{n\le x} d(n)$), which is discussed e.g., in Ch. 15 of [16], cannot be extended to general $E_k(T)$ and $\Delta_k(x)$. The latter function denotes the error term in the asymptotic formula for $\sum_{n\le x} d_k(n)$, where $d_k(n)$ is the general divisor function generated by $\zeta^k(s)$. The LH is equivalent to either $\alpha_k \le 1/2$ $(k \ge 2)$ or $\beta_k = (k-1)/(2k)$ $(k \ge 2)$, where α_k and β_k are the infima of the numbers a_k and b_k for which

$$\Delta_k(x) \ll x^{a_k}, \quad \int_1^x \Delta_k^2(y)\,dy \ll x^{b_k}$$

hold, respectively. We know that $\beta_k = (k-1)/(2k)$ for $k = 2, 3, 4$, and it is generally conjectured that $\alpha_k = \beta_k = (k-1)/(2k)$ for any k. At first I thought that, analogously to the conjecture for α_k and β_k, the upper bound for general $E_k(T)$ should be of such a form as to yield the LH when $k \to \infty$, but in view of (4.14) I am certain that this cannot be the case.

It may be asked then how do the Ω-results for $E_1(T)$ and $E_2(T)$ affect the LH, and thus indirectly the RH? A reasonable conjecture is that these Ω-results lie fairly close to the truth, in other words that

$$E_k(T) = O_{k,\varepsilon}(T^{\frac{k}{4}+\varepsilon}) \tag{4.18}$$

holds for $k = 1, 2$. This view is suggested by estimates in the mean for the functions in question. Namely the author [15] proved that

$$\int_1^T |E_1(t)|^A\,dt \ll_\varepsilon T^{1+\frac{A}{4}+\varepsilon} \qquad (0 \le A \le \frac{35}{4}), \tag{4.19}$$

and the range for A for which (4.19) holds can be slightly increased by using the best known estimate (4.6) in the course of the proof. Also Y. Motohashi and the author [27], [28] proved that

$$\int_0^T E_2(t)\,dt \ll T^{3/2}, \quad \int_0^T E_2^2(t)\,dt \ll T^2 \log^C T \quad (C > 0). \tag{4.20}$$

The bounds (4.19) and (4.20) show indeed that, in the mean sense, the bound (4.18) does hold when $k = 1, 2$. Curiously enough, it does not seem possible to show that the RH implies (4.18) for $k \le 3$. If (4.18) holds for any k, then in view of (4.7) we would obtain (4.6) with the hitherto sharpest bound $c \le 1/8$, or equivalently $\mu(1/2) \le 1/8$, where for any real σ one defines

$$\mu(\sigma) = \limsup_{t\to\infty} \frac{\log|\zeta(\sigma+it)|}{\log t},$$

and it will be clear from the context that no confusion can arise with the Möbius function. What can one expect about the order of magnitude of $E_k(T)$ for $k \ge 3$? It was already mentioned that the structure of $E_k(T)$ becomes increasingly complex as k increases. Thus we should not expect a smaller exponent than $k/4$ in (4.18) for $k \ge 3$, as it would by (4.7) yield a result of the type $\mu(1/2) < 1/8$, which in view of the Ω-results is not obtainable from (4.18) when $k = 1, 2$. Hence by analogy with the cases $k = 1, 2$ one would be led to conjecture that

$$E_k(T) = \Omega(T^{k/4}) \tag{4.21}$$

holds for any fixed $k \ge 1$. But already for $k = 5$ (4.21) yields, in view of (4.1),

$$\int_0^T |\zeta(\tfrac{1}{2}+it)|^{10}\,dt = \Omega_+(T^{5/4}), \tag{4.22}$$

which contradicts (4.10), thereby disproving *both the LH and the RH.* It would be of great interest to obtain more detailed information on $E_k(T)$ in the cases when $k = 3$ and especially when $k = 4$, as the latter probably represents a turning point in the asymptotic behaviour of mean values of $|\zeta(\frac{1}{2} + it)|$. Namely the above phenomenon strongly suggests that either the LH fails, or the shape of the asymptotic formula for the left-hand side of (4.1) changes (in a yet completely unknown way) when $k = 4$. In [24] the author proved that $E_3(T) \ll_\varepsilon T^{1+\varepsilon}$ conditionally, that is, provided that a certain conjecture involving the ternary additive divisor problem holds. Y. Motohashi ([40] p. 339, and [42]) proposes, on heuristic grounds based on analogy with explicit formulas known in the cases $k = 1, 2$, a formula for the analogue of (4.15) for the sixth moment, and also conjectures (4.21) for $k = 3$. Concerning the eighth moment, it should be mentioned that N.V. Kuznetsov [33] had an interesting approach based on applications of spectral theory, but unfortunately his proof of

$$\int_0^T |\zeta(\tfrac{1}{2} + it)|^8 \, dt \ll T \log^C T \tag{4.23}$$

had several gaps (see the author's review in Zbl. 745.11040 and the Addendum of Y. Motohashi [40]), so that (4.23) is still a conjecture. If (4.23) is true, then one must have $C \geq 16$ in (4.23), since by a result of K. Ramachandra (see [16] and [48]) one has, for any rational number $k \geq 0$,

$$\int_0^T |\zeta(\tfrac{1}{2} + it)|^{2k} \, dt \gg_k T(\log T)^{k^2}.$$

The LH (see [61]) is equivalent to the statement that $\mu(\sigma) = 1/2 - \sigma$ for $\sigma < 1/2$, and $\mu(\sigma) = 0$ for $\sigma \geq 1/2$. If the LH is not true, what would then the graph of $\mu(\sigma)$ look like? If the LH fails, it is most likely that $\mu(1/2) = 1/8$ is true. Since $\mu(\sigma)$ is (unconditionally) a non-increasing, convex function of σ,

$$\mu(\sigma) = \frac{1}{2} - \sigma \quad (\sigma \leq 0), \quad \mu(\sigma) = 0 \quad (\sigma \geq 1),$$

and by the functional equation one has $\mu(\sigma) = \frac{1}{2} - \sigma + \mu(1 - \sigma)$ $(0 < \sigma < 1)$, perhaps one would have

$$\mu(\sigma) = \begin{cases} \frac{1}{2} - \sigma & \sigma \leq \frac{1}{4}, \\ \frac{3}{8} - \frac{\sigma}{2} & \frac{1}{4} \leq \sigma < \frac{3}{4}, \\ 0 & \sigma \geq \frac{3}{4}, \end{cases} \tag{4.24}$$

or the slightly weaker

$$\mu(\sigma) = \begin{cases} \frac{1}{2} - \sigma & \sigma \leq 0, \\ \frac{2-3\sigma}{4} & 0 < \sigma < \frac{1}{2}, \\ \frac{(1-\sigma)}{4} & \frac{1}{2} \leq \sigma \leq 1, \\ 0 & \sigma > 1. \end{cases} \tag{4.25}$$

A third candidate is

$$\mu(\sigma) = \begin{cases} \frac{1}{2} - \sigma & \sigma \leq 0, \\ \frac{1}{2}(1 - \sigma)^2 & 0 < \sigma < 1, \\ 0 & \sigma \geq 1, \end{cases} \tag{4.26}$$

which is a quadratic function of σ in the critical strip. Note that (4.26) sharpens (4.25) for $0 < \sigma < 1$, except when $\sigma = 1/2$, when (4.24)–(4.26) all yield $\mu(\sigma) = 1/8$. So far no exact value of $\mu(\sigma)$ is known when σ lies in the critical strip $0 < \sigma < 1$.

5. Large values on the critical line

One thing that has constantly made the author skeptical about the truth of the RH is: How to draw the graph of $Z(t)$ when t is large? By this the following is meant. R. Balasubramanian and K. Ramachandra (see [3], [4], [48], [49]) proved unconditionally that

$$\max_{T\le t\le T+H}|\zeta(\tfrac{1}{2}+it)| > \exp\Bigl(\frac{3}{4}\Bigl(\frac{\log H}{\log\log H}\Bigr)^{1/2}\Bigr) \tag{5.1}$$

for $T \ge T_0$ and $\log\log T \ll H \le T$, and probably on the RH this can be further improved (but no results seem to exist yet). Anyway (5.1) shows that $|Z(t)|$ assumes large values relatively often. On the other hand, on the RH one expects that the bound in (1.11) can be also further reduced, very likely (see [46]) to

$$S(T) \ll_\varepsilon (\log T)^{\frac{1}{2}+\varepsilon}. \tag{5.2}$$

Namely, on the RH, H.L. Montgomery [39] proved that

$$S(T) = \Omega_\pm\Bigl(\Bigl(\frac{\log T}{\log\log T}\Bigr)^{1/2}\Bigr),$$

which is in accord with (5.2). K.-M. Tsang [59], improving a classical result of A. Selberg [54], has shown that one has unconditionally

$$S(T) = \Omega_\pm\Bigl(\Bigl(\frac{\log T}{\log\log T}\Bigr)^{1/3}\Bigr).$$

Also K.-M. Tsang [60] proved that (unconditionally; $\pm$ means that the result holds both with the + and the − sign)

$$\Bigl(\sup_{T\le t\le 2T}\log|\zeta(\tfrac{1}{2}+it)|\Bigr)\Bigl(\sup_{T\le t\le 2T}\pm S(t)\Bigr) \gg \frac{\log T}{\log\log T},$$

which shows that either $|\zeta(\frac{1}{2}+it)|$ or $|S(t)|$ must assume large values in $[T, 2T]$. It may be pointed out that the calculations relating to the values of $S(T)$ (see e.g., [45], [46]) show that all known values of $S(T)$ are relatively small. In other words they are not anywhere near the values predicted by the above Ω–results, which is one more reason that supports the view that the values for which $\zeta(s)$ will exhibit its true asymptotic behaviour must be *really very large*.

If on the RH (5.2) is true, then clearly (1.12) can be improved to

$$\gamma_{n+1} - \gamma_n \ll_\varepsilon (\log \gamma_n)^{\varepsilon-1/2}. \tag{5.3}$$

This means that, as $n \to \infty$, the gap between the consecutive zeros of $Z(t)$ tends to zero not so slowly. Now take $H = T$ in (5.1), and let t_0 be the point in $[T, 2T]$ where the maximum in (5.1) is attained. This point falls into an interval of length $\ll (\log T)^{\varepsilon-1/2}$ between two consecutive zeros, so that in the vicinity of t_0 the function $Z(t)$ must have very large oscillations, which will be carried over to $Z'(t), Z''(t), \dots$ etc. For example, for $T = 10^{5000}$ we shall have

$$|Z(t_0)| > 2.68 \times 10^{11}, \tag{5.4}$$

while $(\log T)^{-1/2} = 0.00932\dots$, which shows how large the oscillations of $Z(t)$ near t_0 will be. Moreover, M. Jutila [30] unconditionally proved the following

Proposition 3. There exist positive constants a_1, a_2 and a_3 such that, for $T \ge 10$, we have

$$\exp(a_1(\log\log T)^{1/2}) \le |Z(t)| \le \exp(a_2(\log\log T)^{1/2})$$

in a subset of measure at least $a_3 T$ of the interval $[0, T]$.

For $T = 10^{5000}$ one has $e^{(\log\log T)^{1/2}} = 21.28446\ldots$, and Proposition 3 shows that relatively large values of $|Z(t)|$ are plentiful, and in the vicinity of the respective t's again $Z(t)$ (and its derivatives) must oscillate a lot. The RH and (5.3) imply that, as $t \to \infty$, the graph of $Z(t)$ will consist of tightly packed spikes, which will be more and more condensed as t increases, with larger and large oscillations. This I find hardly conceivable. Of course, it could happen that the RH is true and that (5.3) is not.

6. A class of convolution functions

It does not appear easy to put the discussion of Section 5 into a quantitative form. We shall follow now the method developed by the author in [21] and [23] and try to make a self-contained presentation, resulting in the proof of Theorem 1 (Sec. 8) and Theorem 2 (Sec. 9). The basic idea is to connect the order of $Z(t)$ with the distribution of its zeros and the order of its derivatives (see (7.5)). However it turned out that if one works directly with $Z(t)$, then one encounters several difficulties. One is that we do not know yet whether the zeros of $Z(t)$ are all distinct (simple), even on the RH (which implies by (1.11) only the fairly weak bound that the multiplicities of zeros up to height T are $\ll \log T/\log\log T$). This difficulty is technical, and we may bypass it by using a suitable form of divided differences from Numerical analysis, as will be shown a little later in Section 7. A.A. Lavrik [34] proved the useful result that, uniformly for $0 \le k \le \frac{1}{2}\log t$, one has

(6.1)

$$Z^{(k)}(t) = 2\sum_{n\le(t/2\pi)^{1/2}} n^{-1/2}\Big(\log\frac{(t/2\pi)^{1/2}}{n}\Big)^k \cos\Big(t\log\frac{(t/2\pi)^{1/2}}{n} - \frac{t}{2} - \frac{\pi}{8} + \frac{\pi k}{2}\Big) + O\Big(t^{-1/4}\big(\tfrac{3}{2}\log t\big)^{k+1}\Big).$$

The range for which (6.1) holds is large, but it is difficult to obtain good uniform bounds for $Z^{(k)}(t)$ from (6.1). To overcome this obstacle the author introduced in [21] the class of convolution functions

$$M_{Z,f}(t) := \int_{-\infty}^{\infty} Z(t-x) f_G(x)\,\mathrm{d}x = \int_{-\infty}^{\infty} Z(t+x) f\Big(\frac{x}{G}\Big)\,\mathrm{d}x, \tag{6.2}$$

where $G > 0$, $f_G(x) = f(x/G)$, and $f(x)$ (≥ 0) is an even function belonging to the class of smooth (C^∞) functions $f(x)$ called S_α^β by Gel'fand and Shilov [9]. The functions $f(x)$ satisfy for any real x the inequalities

$$|x^k f^{(q)}(x)| \le CA^k B^q k^{k\alpha} q^{q\beta} \qquad (k, q = 0, 1, 2, \ldots) \tag{6.3}$$

with suitable constants $A, B, C > 0$ depending on f alone. For $\alpha = 0$ it follows that $f(x)$ is of bounded support, namely it vanishes for $|x| \ge A$. For $\alpha > 0$ the condition (6.3) is equivalent (see [9]) to the condition

$$|f^{(q)}(x)| \le CB^q q^{q\beta}\exp(-a|x|^{1/\alpha}) \qquad (a = \alpha/(eA^{1/\alpha})) \tag{6.4}$$

for all x and $q \ge 0$. We shall denote by E_α^β the subclass of S_α^β with $\alpha > 0$ consisting of even functions $f(x)$ such that $f(x)$ is not the zero-function. It is shown in [9] that S_α^β is non-empty if $\beta \ge 0$ and $\alpha + \beta \ge 1$. If these conditions hold then E_α^β is also non-empty, since $f(-x) \in S_\alpha^\beta$ if $f(x) \in S_\alpha^\beta$, and $f(x) + f(-x)$ is always even.

One of the main properties of the convolution function $M_{Z,f}(t)$, which follows by k-fold integration by parts from (6.2), is that for any integer $k \ge 0$

$$M_{Z,f}^{(k)}(t) = M_{Z^{(k)},f}(t) = \int_{-\infty}^{\infty} Z^{(k)}(t+x) f\Big(\frac{x}{G}\Big)\,\mathrm{d}x = \Big(\frac{-1}{G}\Big)^k \int_{-\infty}^{\infty} Z(t+x) f^{(k)}\Big(\frac{x}{G}\Big)\,\mathrm{d}x. \tag{6.5}$$

This relation shows that the order of $M^{(k)}$ depends only on the orders of Z and $f^{(k)}$, and the latter is by (6.4) of exponential decay, which is very useful in dealing with convergence problems etc. The salient point of our approach is that the difficulties inherent in the distribution of zeros of $Z(t)$ are transposed to the distribution of zeros of $M_{Z,f}(t)$, and for the latter function (6.5) provides good uniform control of its derivatives.

Several analogies between $Z(t)$ and $M_{Z,f}(t)$ are established in [21], especially in connection with mean values and the distribution of their respective zeros. We shall retain here the notation introduced in [21], so that $N_M(T)$ denotes the number of zeros of $M_{Z,f}(t)$ in $(0,T]$, with multiplicities counted. If $f(x) \in E^\beta_\alpha$, $f(x) \ge 0$ and $G = \delta/\log(T/(2\pi))$ with suitable $\delta > 0$, then Theorem 4 of [21] says that

$$N_M(T+V) - N_M(T-V) \gg \frac{V}{\log T}, \quad V = T^{c+\varepsilon}, \quad c = 0.329021\ldots, \tag{6.6}$$

for any given $\varepsilon > 0$. The nonnegativity of $f(x)$ was needed in the proof of this result. For the function $Z(t)$ the analogous result is that

$$N_0(T+V) - N_0(T-V) \gg V\log T, \quad V = T^{c+\varepsilon}, \quad c = 0.329021\ldots, \tag{6.7}$$

where as usual $N_0(T)$ denotes the number of zeros of $Z(t)$ (or of $\zeta(\frac12+it)$) in $(0,T]$, with multiplicities counted. Thus the fundamental problem in the theory of $\zeta(s)$ is to estimate $N(T) - N_0(T)$, and the RH may be reformulated as $N(T) = N_0(T)$ for $T > 0$. The bound (6.7) was proved by A.A. Karatsuba (see [32] for a detailed account). As explained in [21], the bound (6.6) probably falls short (by a factor of $\log^2 T$) from the expected (true) order of magnitude for the number of zeros of $M_{Z,f}(t)$ in $[T-V, T+V]$. This is due to the method of proof of (6.6), which is not as strong as the classical method of A. Selberg [54] (see also Ch. 10 of [61]). The function $N_M(T)$ seems much more difficult to handle than $N_0(T)$ or $N(T)$. The latter can be conveniently expressed (see [16] or [61]) by means of a complex integral from which one infers then (1.3) with the bound (1.4). I was unable to find an analogue of the integral representation for $N_M(T)$. Note that the bound on the right-hand side of (6.7) is actually of the best possible order of magnitude.

In the sequel we shall need the following technical result, which we state as

Lemma 1. *If* $L = (\log T)^{\frac12+\varepsilon}, P = \sqrt{\frac{T}{2\pi}}, 0 < G < 1, L \ll V \le T^{\frac13}, f(x) \in E^\beta_\alpha$, *then*

$$\int\limits_{T-VL}^{T+VL} |M_{Z,f}(t)| e^{-(T-t)^2V^{-2}}\,dt \ge GV\{|\widehat{f}(\frac{G}{2\pi}\log P)| + O(T^{-1/4} + V^2T^{-3/4}L^2)\}. \tag{6.8}$$

Proof. In (6.8) $\widehat{f}(x)$ denotes the Fourier transform of $f(x)$, namely

$$\widehat{f}(x) = \int_{-\infty}^{\infty} f(u)e^{2\pi ixu}\,du = \int_{-\infty}^{\infty} f(u)\cos(2\pi xu)\,du + i\int_{-\infty}^{\infty} f(u)\sin(2\pi xu)\,du = \int_{-\infty}^{\infty} f(u)\cos(2\pi xu)\,du$$

since $f(x)$ is even. From the Riemann–Siegel formula (1.13) we have, if $|T-t| \le VL$ and $|x| \le \log^C T$ $(C > 0)$,

$$Z(t+x) = 2\sum_{n\le P} n^{-1/2}\cos\Big((t+x)\log\frac{((t+x)/(2\pi))^{1/2}}{n} - \frac{t+x}{2} - \frac{\pi}{8}\Big) + O(T^{-1/4}).$$

Simplifying the argument of the cosine by Taylor's formula it follows that

$$Z(t+x) = 2\sum_{n\le P} n^{-1/2}\cos\left((t+x)\log\frac{P}{n} - \frac{T}{2} - \frac{\pi}{8}\right) + O(T^{-1/4} + V^2T^{-3/4}L^2). \tag{6.9}$$

Hence from (6.2) and (6.9) we have, since

$$\cos(\alpha+\beta) = \cos\alpha\cos\beta - \sin\alpha\sin\beta$$

and $f(x)$ is even,

$$M_{Z,f}(t) = 2G\sum_{n\le P} n^{-1/2}\cos(t\log\frac{P}{n} - \frac{T}{2} - \frac{\pi}{8})\int_{-\infty}^{\infty} f(x)\cos(Gx\log\frac{P}{n})\,\mathrm{d}x + O(GT^{-1/4} + GV^2T^{-3/4}L^2)$$

$$(6.10)\qquad = 2G\sum_{n\le P} n^{-1/2}\cos(t\log\frac{P}{n} - \frac{T}{2} - \frac{\pi}{8})\hat{f}(\frac{G}{2\pi}\log\frac{P}{n}) + O(GT^{-1/4} + GV^2T^{-3/4}L^2).$$

Therefore we obtain from (6.10)

$$\int_{T-VL}^{T+VL} |M_{Z,f}(t)|\mathrm{e}^{-(T-t)^2V^{-2}}\,\mathrm{d}t \ \ge\ GI + O(GVT^{-1/4} + GV^3T^{-3/4}L^2),$$

say, where

$$I := \int_{T-VL}^{T+VL} \Big|\sum_{n\le P} n^{-1/2}\hat{f}(\frac{G}{2\pi}\log\frac{P}{n})\Big(\exp(it\log\frac{P}{n} - \frac{iT}{2} - \frac{i\pi}{8}) + \exp(-it\log\frac{P}{n} + \frac{iT}{2} + \frac{i\pi}{8})\Big)\Big|\mathrm{e}^{-(T-t)^2V^{-2}}\,\mathrm{d}t.$$

By using the fact that $|\exp(it\log P - \frac{iT}{2} - \frac{i\pi}{8})| = 1$ and the classical integral

$$\int_{-\infty}^{\infty}\exp(Ax - Bx^2)\,\mathrm{d}x = \sqrt{\frac{\pi}{B}}\exp(\frac{A^2}{4B}) \qquad (\Re e\, B > 0)$$

we shall obtain

$$I \ \ge\ |I_1 + I_2|,$$

where

$$\begin{aligned} I_1 &= \int_{T-VL}^{T+VL}\sum_{n\le P} n^{-1/2}\hat{f}(\frac{G}{2\pi}\log\frac{P}{n})\exp(-it\log n - (T-t)^2V^{-2})\,\mathrm{d}t \\ &= \sum_{n\le P} n^{-1/2}\hat{f}(\frac{G}{2\pi}\log\frac{P}{n})\exp(-iT\log n)\int_{-VL}^{VL}\exp(-ix\log n - x^2V^{-2})\,\mathrm{d}x \\ &= \sum_{n\le P} n^{-1/2}\hat{f}(\frac{G}{2\pi}\log\frac{P}{n})\Big\{\sqrt{\pi}V\exp(-iT\log n - \frac{1}{4}V^2\log^2 n) + O(\exp(-\log^{1+2\varepsilon}T))\Big\} \\ &= \sqrt{\pi}V\hat{f}(\frac{G}{2\pi}\log P) + O(T^{-C}) \end{aligned}$$

for any fixed $C > 0$. Similarly we find that

$$I_2 = \int_{T-VL}^{T+VL}\sum_{n\le P} n^{-1/2}\hat{f}(\frac{G}{2\pi}\log\frac{P}{n})\exp\big(-it\log(\frac{T}{2\pi n}) + iT + \frac{\pi i}{4} - (T-t)^2V^{-2}\big)\,\mathrm{d}t =$$

$$\sum_{n\le P} n^{-1/2}\widehat{f}(\frac{G}{2\pi}\log\frac{P}{n})\exp\bigl(-it\log(\frac{T}{2\pi n})+iT+\frac{\pi i}{4}\bigr)\Bigl(\sqrt{\pi}V\exp\{-\frac{1}{4}(V\log(\frac{T}{2\pi n}))^2\}+O\{\exp(-\log^{1+2\varepsilon}T)\}\Bigr)$$

$$= O(T^{-C})$$

again for any fixed $C>0$, since $\log(\frac{T}{2\pi n})\ge\log\frac{T}{2\pi P}=\frac{1}{2}\log(\frac{T}{2\pi})$. ¿From the above estimates (6.8) follows.

7. Technical preparation

In this section we shall lay the groundwork for the investigation of the distribution of zeros of $Z(t)$ via the convolution functions $M_{Z,f}(t)$. To do this we shall first briefly outline a method based on a generalized form of the mean value theorem from the differential calculus. This can be conveniently obtained from the expression for the n-th divided difference associated to the function $F(x)$, namely

$$[x,x_1,x_2,\cdots,x_n]:=$$

$$=\frac{F(x)}{(x-x_1)(x-x_2)\cdots(x-x_n)}+\frac{F(x_1)}{(x_1-x)(x_1-x_2)\cdots(x_1-x_n)}+\cdots+\frac{F(x_n)}{(x_n-x)(x_n-x_1)\cdots(x_n-x_{n-1})}$$

where $x_i\ne x_j$ if $i\ne j$, and $F(t)$ is a real-valued function of the real variable t. We have the representation

$$[x,x_1,x_2,\cdots,x_n]= \tag{7.1}$$

$$=\int_0^1\int_0^{t_1}\cdots\int_0^{t_{n-1}}F^{(n)}\bigl(x_1+(x_2-x_1)t_1+\cdots+(x_n-x_{n-1})t_{n-1}+(x-x_n)t_n\bigr)\,\mathrm{d}t_n\cdots\mathrm{d}t_1=\frac{F^{(n)}(\xi)}{n!}$$

if $F(t)\in C^n[I],\xi=\xi(x,x_1,\cdots,x_n)$ and I is the smallest interval containing all the points $x,x_1,\cdots,x_n$. If we suppose additionally that $F(x_j)=0$ for $j=1,\cdots,n$, then on comparing the two expressions for $[x,x_1,x_2,\cdots,x_n]$, it follows that

$$F(x)=(x-x_1)(x-x_2)\cdots(x-x_n)\frac{F^{(n)}(\xi)}{n!}, \tag{7.2}$$

where $\xi=\xi(x)$ if we consider $x_1,\cdots,x_n$ as fixed and x as a variable. The underlying idea is that, if the (distinct) zeros x_j of $F(x)$ are sufficiently close to one another, then (7.2) may lead to a contradiction if $F(x)$ is assumed to be large and one has good bounds for its derivatives.

To obtain the analogue of (7.2) when the points x_j are not necessarily distinct, note that if $F(z)$ is a regular function of the complex variable z in a region which contains the distinct points $x,x_1,\cdots,x_n$, then for a suitable closed contour $\mathcal{C}$ containing these points one obtains by the residue theorem

$$[x,x_1,x_2,\cdots,x_n]=\frac{1}{2\pi i}\int_{\mathcal{C}}\frac{F(z)}{(z-x)(z-x_1)\cdots(z-x_n)}\,\mathrm{d}z.$$

A comparison with (7.1) yields then

$$\frac{1}{2\pi i}\int_{\mathcal{C}}\frac{F(z)}{(z-x)(z-x_1)\cdots(z-x_n)}\,\mathrm{d}z \tag{7.3}$$

$$=\int_0^1\int_0^{t_1}\cdots\int_0^{t_{n-1}}F^{(n)}\bigl(x_1+(x_2-x_1)t_1+\cdots+(x_n-x_{n-1})t_{n-1}+(x-x_n)t_n\bigr)\,\mathrm{d}t_n\cdots\mathrm{d}t_1.$$

Now (7.3) was derived on the assumption that the points $x, x_1, \cdots, x_n$ are distinct. But as both sides of (7.3) are regular functions of $x, x_1, \cdots, x_n$ in some region, this assumption may be dropped by analytic continuation. Thus let the points $x, x_1, \cdots, x_n$ coincide with the (distinct) points z_k, where the multiplicity of z_k is denoted by $p_k; k = 0, 1, \cdots, \nu; \sum_{k=0}^{\nu} p_k = n+1$. If we set

$$z_0 = x,\ p_0 = 1,\ Q(z) = \prod_{k=1}^{\nu} (z - z_k)^{p_k},$$

then the complex integral in (7.3) may be evaluated by the residue theorem (see Ch.1 of A.O. Gel'fond [10]). It equals

$$\frac{F(x)}{Q(x)} - \sum_{k=1}^{\nu} \sum_{m=0}^{p_k-1} \frac{F^{(p_k-m-1)}(z_k)}{(p_k-m-1)!} \sum_{s=0}^{m} \frac{1}{(m-s)!} \frac{d^{m-s}}{dz^{m-s}} \Big(\frac{(z-z_k)^{p_k}}{Q(z)}\Big)\Big|_{z=z_k} \cdot (x - x_k)^{-s-1}. \tag{7.4}$$

If $x_1, x_2, \cdots, x_n$ are the zeros of $F(z)$, then $F^{(p_k-m-1)}(z_k) = 0$ for $k = 1, \cdots, \nu$ and $m = 0, \cdots, p_{k-1}$, since p_k is the multiplicity of (the zero) z_k. Hence if $F(x) \neq 0$, then on comparing (7.1), (7.3) and (7.4) one obtains

$$|F(x)| \le \prod_{k=1}^{n} |x - x_k| \frac{|F^{(n)}(\xi)|}{n!} \qquad (\xi = \xi(x)), \tag{7.5}$$

and of course (7.5) is trivial if $F(x) = 0$.

Now we shall apply (7.5) to $F(t) = M_{Z,f}(t), f(x) \in E_\alpha^\beta$, with n replaced by k, to obtain

$$|M_{Z,f}(t)| \le \prod_{t-H\le\gamma\le t+H} |\gamma - t| \frac{|M_{Z,f}^{(k)}(\tau)|}{k!}, \tag{7.6}$$

where γ denotes the zeros of $M_{Z,f}(t)$ in $[t-H, t+H], \tau = \tau(t, H) \in [t-H, t+H], |t - T| \le T^{1/2+\varepsilon}$ and $k = k(t, H)$ is the number of zeros of $M_{Z,f}(t)$ in $[t-H, t+H]$. We shall choose

$$H = \frac{A \log_3 T}{\log_2 T} \qquad (\log_r T = \log(\log_{r-1} T),\ \log_1 T \equiv \log T) \tag{7.7}$$

for a sufficiently large $A > 0$. One intuitively feels that, with a suitable choice (see (8.1) and (8.2)) of G and f, the functions $N(T)$ and $N_M(T)$ will not differ by much. Thus we shall suppose that the analogues of (1.3) and (1.11) hold for $N_M(T)$, namely that

$$N_M(T) = \frac{T}{2\pi} \log(\frac{T}{2\pi}) - \frac{T}{2\pi} + S_M(T) + O(1) \tag{7.8}$$

with a continuous function $S_M(T)$ satisfying

$$S_M(T) = O\Big(\frac{\log T}{\log\log T}\Big), \tag{7.9}$$

although it is hard to imagine what should be the appropriate analogue for $S_M(T)$ of the defining relation $S(T) = \frac{1}{\pi} \arg\zeta(\frac{1}{2} + iT)$ in (1.4). We also suppose that

$$\int_T^{T+U} (S_M(t+H) - S_M(t-H))^{2m}\, dt \ll U(\log(2 + H\log T))^m \tag{7.10}$$

holds for any fixed integer $m \ge 1, T^a < U \le T, 1/2 < a \le 1, 0 < H < 1$. Such a result holds unconditionally (even in the form of an asymptotic formula) if $S_M(T)$ is replaced by $S(T)$, as shown in the works of A. Fujii

[8] and K.-M. Tsang [59]. Thus it seems plausible that (7.10) will also hold. It was already mentioned that it is reasonable to expect that $S(T)$ and $S_M(T)$ will be close to one another. One feels that this "closeness" should hold also in the mean sense, and that instead of (7.10) one could impose a condition which links directly $S_M(T)$ and $S(T)$, such as that for any fixed integer $m \ge 1$ one has

$$\int_T^{T+U} (S_M(t) - S(t))^{2m}\,\mathrm{d}t \ll U(\log\log T)^m \qquad (T^a < U \le T,\ \tfrac12 < a \le 1). \tag{7.11}$$

If (7.8) holds, then

$$k = N_M(t+H) - N_M(t-H) + O(1) \tag{7.12}$$

$$= \frac{(t+H)}{2\pi}\log(\frac{t+H}{2\pi}) - \frac{t+H}{2\pi} - \frac{t-H}{2\pi}\log(\frac{t-H}{2\pi}) + \frac{t-H}{2\pi} + S_M(t+H) - S_M(t-H) + O(1)$$

$$= \frac{H}{\pi}\log(\frac{T}{2\pi}) + S_M(t+H) - S_M(t-H) + O(1).$$

To bound from above the product in (7.6) we proceed as follows. First we have trivially

$$\prod_{|\gamma-t|\le 1/\log_2 T} |\gamma - t| \le 1.$$

The remaining portions of the product with $t - H \le \gamma < t - 1/\log_2 T$ and $t + 1/\log_2 T < \gamma \le t + H$ are treated analogously, so we shall consider in detail only the latter. We have

$$\log\Big(\prod_{t+1/\log_2 T<\gamma\le t+H} |\gamma - t|\Big)$$

$$= \sum_{t+1/\log_2 T<\gamma\le t+H} \log(\gamma - t) = \int_{t+1/\log_2 T+0}^{t+H} \log(u-t)\,\mathrm{d}N_M(u)$$

$$= \frac{1}{2\pi}\int_{t+1/\log_2 T+0}^{t+H} \log(u-t)\log(\frac{u}{2\pi})\,du + \int_{t+1/\log_2 T+0}^{t+H} \log(u-t)\,\mathrm{d}(S_M(u)+O(1)).$$

By using integration by parts and (7.9) it follows that

$$\int_{t+1/\log_2 T+0}^{t+H} \log(u-t)\,\mathrm{d}(S_M(u)+O(1)) = O\Big(\frac{\log T\log_3 T}{\log_2 T}\Big) - \int_{t+1/\log_2 T}^{t+H} \frac{S_M(u)+O(1)}{u-t}\,\mathrm{d}u \ll \frac{\log T\log_3 T}{\log_2 T},$$

and we have

$$\frac{1}{2\pi}\int_{t+1/\log_2 T+0}^{t+H} \log(u-t)\log(\frac{u}{2\pi})\,\mathrm{d}u = \frac{1}{2\pi}\log(\frac{T}{2\pi})\cdot(1+O(T^{\varepsilon-1/2}))\int_{t+1/\log_2 T}^{t+H}\log(u-t)\,\mathrm{d}u$$

$$= \frac{1}{2\pi}\log(\frac{T}{2\pi})\cdot\Big(H\log H - H + O(\frac{\log_3 T}{\log_2 T})\Big).$$

By combining the above estimates we obtain

Lemma 2. *Suppose that (7.8) and (7.9) hold. If γ denotes zeros of $M_{Z,f}(t)$, H is given by (7.7) and $|T-t| \le T^{1/2+\varepsilon}$, then*

$$\prod_{t-H\le\gamma\le t+H} |\gamma - t| \;\le\; \exp\Big\{\frac{1}{\pi}\log(\frac{T}{2\pi})\cdot\Big(H\log H - H + O\Big(\frac{\log_3 T}{\log_2 T}\Big)\Big)\Big\}. \tag{7.13}$$

8. The asymptotic formula for the convolution function

In this section we shall prove a sharp asymptotic formula for $M_{Z,f}(t)$, which is given by Theorem 1. This will hold if $f(x)$ belongs to a specific subclass of functions from E_α^0 ($\alpha > 1$ is fixed), and for such $M_{Z,f}(t)$ we may hope that (7.8)–(7.11) will hold. To construct this subclass of functions first of all let let $\varphi(x) \ge 0$ (but $\varphi(x) \not\equiv 0$) belong to E_0^α. Such a choice is possible, since it is readily checked that $f^2(x) \in S_\alpha^\beta$ if $f(x) \in S_\alpha^\beta$, and trivially $f^2(x) \ge 0$. Thus $\varphi(x)$ is of bounded support, so that $\varphi(x) = 0$ for $|x| \ge a$ for some $a > 0$. We normalize $\varphi(x)$ so that $\int_{-\infty}^{\infty} \varphi(x)\,\mathrm{d}x = 1$, and for an arbitrary constant $b > \max(1, a)$ we put

$$\Phi(x) := \int\limits_{x-b}^{x+b} \varphi(t)\,\mathrm{d}t.$$

Then $0 \le \Phi(x) \le 1$, $\Phi(x)$ is even (because $\varphi(x)$ is even) and nonincreasing for $x \ge 0$, and

$$\Phi(x) = \begin{cases} 0 & \text{if} \quad |x| \ge b + a, \\ 1 & \text{if} \quad |x| \le b - a. \end{cases}$$

One can also check that $\varphi(x) \in S_0^\alpha$ implies that $\Phi(x) \in S_0^\alpha$. Namely $|x^k \Phi(x)| \le (b+a)^k$, and for $q \ge 1$ one uses (6.3) (with $k = 0$, $f^{(q)}$ replaced by $\varphi^{(q-1)}$, $(\alpha, \beta) = (0, \alpha)$) to obtain

$$|x^k \Phi^{(q)}(x)| \le (b+a)^k |\Phi^{(q)}(x)| \le (b+a)^k \Big(|\varphi^{(q-1)}(x+b)| + |\varphi^{(q-1)}(x-b)|\Big)$$

$$\le (b+a)^k 2CB^{q-1}(q-1)^{(q-1)\alpha} \le \frac{2C}{B}(b+a)^k B^q q^{q\alpha},$$

hence (6.3) will hold for Φ in place of f, with $A = b + a$ and suitable C. Let

$$f(x) := \int_{-\infty}^{\infty} \Phi(u) \mathrm{e}^{-2\pi i x u}\,\mathrm{d}u = \int_{-\infty}^{\infty} \Phi(u)\cos(2\pi x u)\,\mathrm{d}u.$$

A fundamental property of the class S_α^β (see [9]) is that $\widehat{S_\alpha^\beta} = S_\beta^\alpha$, where in general $\widehat{U} = \{\widehat{f}(x) : f(x) \in U\}$. Thus $f(x) \in S_\alpha^0$, $f(x)$ is even (because $\Phi(x)$ is even), and by the inverse Fourier transform we have $\widehat{f}(x) = \Phi(x)$. The function $f(x)$ is not necessarily nonnegative, but this property is not needed in the sequel.

Henceforth let

$$G = \frac{\delta}{\log(\frac{T}{2\pi})} \qquad (\delta > 0). \tag{8.1}$$

In view of (1.3) it is seen that, on the RH, G is of the order of the average spacing between the zeros of $Z(t)$. If $f(x)$ is as above, then we have

THEOREM 1. *For $|t - T| \le VL$, $L = \log^{\frac12 + \varepsilon} T$, $\log^\varepsilon T \le V \le \frac{T^{\frac14}}{\log T}$, $0 < \delta < 2\pi(b - a)$ and any fixed $N \ge 1$ we have*

$$M_{Z,f}(t) = G\big(Z(t) + O(T^{-N})\big). \tag{8.2}$$

Proof. Observe that the weak error term $O(T^{-1/4})$ in (8.2) follows from (6.10) (with $x = 0$) and (6.10) when we note that

$$\widehat{f}\Big(\frac{G}{2\pi}\log\frac{P}{n}\Big) = \widehat{f}\Big(\frac{\delta}{2\pi}\Big(\frac12 - \frac{\log n}{\log(\frac{T}{2\pi})}\Big)\Big) = 1$$

since by construction $\widehat{f}(x) = 1$ for $|x| < b - a$, and

$$\Big|\frac{\delta}{2\pi}\Big(\frac{1}{2} - \frac{\log n}{\log(\frac{T}{2\pi})}\Big)\Big| \le \frac{\delta}{4\pi} < b - a \qquad \Big(1 \le n \le P = \sqrt{\frac{T}{2\pi}}\Big).$$

Also the hypotheses on t in the formulation of Theorem 1 can be relaxed.

In order to prove (8.2) it will be convenient to work with the real-valued function $\theta(t)$, defined by

$$Z(t) = e^{i\theta(t)}\zeta(\tfrac{1}{2} + it) = \chi^{-1/2}(\tfrac{1}{2} + it)\zeta(\tfrac{1}{2} + it). \tag{8.3}$$

¿From the functional equation (1.2) in the symmetric form

$$\pi^{-s/2}\Gamma(\frac{s}{2})\zeta(s) = \pi^{-(1-s)/2}\Gamma(\frac{1-s}{2})\zeta(1-s)$$

one obtains

$$\chi^{-1/2}(\tfrac{1}{2} + it) = \frac{\pi^{-it/2}\Gamma^{1/2}(\frac{1}{4} + \frac{it}{2})}{\Gamma^{1/2}(\frac{1}{4} - \frac{it}{2})},$$

and consequently

$$\theta(t) = \Im\text{m} \log\Gamma(\tfrac{1}{4} + \tfrac{1}{2}it) - \tfrac{1}{2}t\log\pi. \tag{8.4}$$

We have the explicit representation (see Ch. 3 of [32])

$$\theta(t) = \frac{t}{2}\log\frac{t}{2\pi} - \frac{t}{2} - \frac{\pi}{8} + \Delta(t) \tag{8.5}$$

with ($\psi(x) = x - [x] - \frac{1}{2}$)

$$\Delta(t) := \frac{t}{4}\log(1 + \frac{1}{4t^2}) + \frac{1}{4}\arctan\frac{1}{2t} + \frac{t}{2}\int_0^\infty \frac{\psi(u)\,du}{(u + \frac{1}{4})^2 + t^2}. \tag{8.6}$$

This formula is very useful, since it allows one to evaluate explicitly all the derivatives of $\theta(t)$. For $t \to \infty$ it is seen that $\Delta(t)$ admits an asymptotic expansion in terms of negative powers of t, and from (8.4) and Stirling's formula for the gamma-function it is found that (B_k is the k-th Bernoulli number)

$$\Delta(t) \sim \sum_{n=1}^\infty \frac{(2^{2n} - 1)|B_{2n}|}{2^{2n}(2n-1)2nt^{2n-1}}, \tag{8.7}$$

and the meaning of (8.7) is that, for an arbitrary integer $N \ge 1$, $\Delta(t)$ equals the sum of the first N terms of the series in (8.7), plus the error term which is $O_N(t^{-2N-1})$. In general we shall have, for $k \ge 0$ and suitable constants $c_{k,n}$,

$$\Delta^{(k)}(t) \sim \sum_{n=1}^\infty c_{k,n}t^{1-2n-k}. \tag{8.8}$$

For complex s not equal to the poles of the gamma-factors we have the Riemann-Siegel formula (this is equation (56) of C.L. Siegel [57])

$$\begin{aligned} &\pi^{-s/2}\Gamma(\frac{s}{2})\zeta(s) \\ &= \pi^{-s/2}\Gamma(\frac{s}{2})\int_{0\swarrow 1} \frac{e^{i\pi x^2}x^{-s}}{e^{i\pi x} - e^{-i\pi x}}\,dx + \pi^{(s-1)/2}\Gamma(\frac{1-s}{2})\int_{0\searrow 1} \frac{e^{-i\pi x^2}x^{s-1}}{e^{i\pi x} - e^{-i\pi x}}\,dx. \end{aligned} \tag{8.9}$$

Here $0 \swarrow 1$ (resp. $0 \searrow 1$) denotes a straight line which starts from infinity in the upper complex half-plane, has slope equal to 1 (resp. to -1), and cuts the real axis between 0 and 1. Setting in (8.9) $s = \frac{1}{2} + it$ and using the property (8.4) it follows that

$$Z(t) = 2\,\Re\mathrm{e}\Big(\mathrm{e}^{-i\theta(t)} \int\limits_{0\searrow 1} \frac{\mathrm{e}^{-i\pi z^2} z^{-1/2+it}}{\mathrm{e}^{i\pi z} - \mathrm{e}^{-i\pi z}}\,\mathrm{d}z\Big).$$

As $\Re\mathrm{e}\,(iw) = -\Im\mathrm{m}\, w$, this can be conveniently written as

$$Z(t) = \Im\mathrm{m}\Big(\mathrm{e}^{-i\theta(t)} \int\limits_{0\searrow 1} \mathrm{e}^{-i\pi z^2} z^{-1/2+it} \frac{\mathrm{d}z}{\sin(\pi z)}\Big). \tag{8.10}$$

Since

$$|\sin z|^2 = \sin^2(\Re\mathrm{e}\, z) + \sinh^2(\Im\mathrm{m}\, z),\ z^{-1/2+it} = |z|^{-1/2+it}\mathrm{e}^{-\frac{i}{2}\arg z - t\arg z},$$

and for $z = \eta + u\mathrm{e}^{3\pi i/4}$, u real, $0 < \eta < 1$ we have

$$|\mathrm{e}^{-i\pi z^2}| = \mathrm{e}^{-\pi u^2 + \eta\sqrt{2}\pi u},$$

it follows that the contribution of the portion of the integral in (8.10) for which $|z| \ge \log t$ is $\ll \mathrm{e}^{-\log^2 t}$. Hence

$$Z(t) = \Im\mathrm{m}\Big(\mathrm{e}^{-i\theta(t)} \int\limits_{0\searrow 1, |z|<\log t} \mathrm{e}^{-i\pi z^2} z^{-1/2+it} \frac{\mathrm{d}z}{\sin(\pi z)}\Big) + O(\mathrm{e}^{-\log^2 t}). \tag{8.11}$$

¿From the decay property (6.4) it follows that

$$M_{Z,f}(t) = \int_{-\infty}^{\infty} Z(t+x) f\big(\frac{x}{G}\big)\,\mathrm{d}x = \int\limits_{-\log^{2\alpha-1} t}^{\log^{2\alpha-1} t} Z(t+x) f\big(\frac{x}{G}\big)\,\mathrm{d}x + O(\mathrm{e}^{-c\log^2 t}), \tag{8.12}$$

where c denotes positive, absolute constants which may not be the same ones at each occurrence. Thus from (8.10) and (8.12) we obtain that

$$M_{Z,f}(t) = \Im\mathrm{m}\Big(\int\limits_{0\searrow 1, |z|<\log t} \frac{\mathrm{e}^{-i\pi z^2} z^{-1/2+it}}{\sin(\pi z)} \int\limits_{-\log^{2\alpha-1} t}^{\log^{2\alpha-1} t} \mathrm{e}^{-i\theta(t+x)} z^{ix} f\big(\frac{x}{G}\big)\,\mathrm{d}x\,\mathrm{d}z\Big) + O(\mathrm{e}^{-c\log^2 t}). \tag{8.13}$$

By using Taylor's formula we have

$$\theta(t+x) = \theta(t) + \frac{x}{2}\log\frac{t}{2\pi} + x\Delta'(t) + R(t,x) \tag{8.14}$$

with $\Delta'(t) \ll t^{-2}$ and

$$R(t,x) = \sum_{n=2}^{\infty}\Big(\frac{(-1)^n}{2n(n-1)t^{n-1}} + \frac{\Delta^{(n)}(t)}{n!}\Big)x^n.$$

Now we put

$$\mathrm{e}^{-iR(t,x)} = 1 + S(t,x), \tag{8.15}$$

say, and use (8.5), (8.6), (8.8) and (8.14). We obtain

$$S(t,x) = \sum_{k=1}^{\infty} \frac{(-i)^k R^k(t,x)}{k!} = \sum_{n=2}^{\infty} g_n(t)x^n, \tag{8.16}$$

where each $g_n(t)$ ($\in C^\infty(0,\infty)$) has an asymptotic expansion of the form

$$g_n(t) \sim \sum_{k=0}^{\infty} d_{n,k} t^{-k-[(n+1)/2]} \quad (t \to \infty) \tag{8.17}$$

with suitable constants $d_{n,k}$. From (8.13)-(8.15) we have

$$M_{Z,f}(t) = \Im\mathrm{m}\Big(I_1 + I_2\Big) + O(\mathrm{e}^{-c\log^2 t}), \tag{8.18}$$

where

$$I_1 := \int\limits_{0\searrow 1,|z|<\log t} \frac{\mathrm{e}^{-i\pi z^2} z^{-1/2+it}}{\sin(\pi z)} \int\limits_{-\log^{2\alpha-1} t}^{\log^{2\alpha-1} t} \mathrm{e}^{-i\theta(t)-\frac{ix}{2}\log\frac{t}{2\pi}-ix\Delta'(t)} z^{ix} f\Big(\frac{x}{G}\Big)\,\mathrm{d}x\,\mathrm{d}z, \tag{8.19}$$

$$I_2 := \int\limits_{0\searrow 1,|z|<\log t} \frac{\mathrm{e}^{-i\pi z^2} z^{-1/2+it}}{\sin(\pi z)} \int\limits_{-\log^{2\alpha-1} t}^{\log^{2\alpha-1} t} \mathrm{e}^{-i\theta(t)-\frac{ix}{2}\log\frac{t}{2\pi}-ix\Delta'(t)} z^{ix} f\Big(\frac{x}{G}\Big) S(t,x)\,\mathrm{d}x\,\mathrm{d}z. \tag{8.20}$$

In I_1 we write $z = \mathrm{e}^{i\arg z}|z|$, which gives

$$I_1 := \int\limits_{0\searrow 1,|z|<\log t} \mathrm{e}^{-\theta(t)} \frac{\mathrm{e}^{-i\pi z^2} z^{-1/2+it}}{\sin(\pi z)} h(z)\,\mathrm{d}z + O(\mathrm{e}^{-c\log^2 t}),$$

where

$$\begin{aligned} h(z) &:= \int_{-\infty}^{\infty} \mathrm{e}^{-x\arg z} f\Big(\frac{x}{G}\Big) \exp\Big(ix\log\frac{|z|\mathrm{e}^{-\Delta'(t)}}{\sqrt{t/2\pi}}\Big)\,\mathrm{d}x \\ &= G\int_{-\infty}^{\infty} \mathrm{e}^{-Gy\arg z} f(y) \exp\Big(iyG\log\frac{|z|\mathrm{e}^{-\Delta'(t)}}{\sqrt{t/2\pi}}\Big)\,\mathrm{d}y \\ &= G\sum_{n=0}^{\infty} \frac{(-G\arg z)^n}{n!} \int_{-\infty}^{\infty} y^n f(y) \exp\Big(2i\pi y\cdot\frac{G}{2\pi}\log\frac{|z|\mathrm{e}^{-\Delta'(t)}}{\sqrt{t/2\pi}}\Big)\,\mathrm{d}y. \end{aligned}$$

Change of summation and integration is justified by absolute convergence, since $|z| < \log t, G = \delta/\log(T/2\pi)$, $\Delta'(t) \ll t^{-2}, t \sim T$, and $f(x)$ satisfies (6.4). But

$$\int_{-\infty}^{\infty} f(y) \exp\Big(2i\pi y\cdot\frac{G}{2\pi}\log\frac{|z|\mathrm{e}^{-\Delta'(t)}}{\sqrt{t/2\pi}}\Big)\,\mathrm{d}y = \widehat{f}\Big(\frac{G}{2\pi}\log\frac{|z|\mathrm{e}^{-\Delta'(t)}}{\sqrt{t/2\pi}}\Big) = 1$$

for $\delta < 2\pi(b-a)$, since

$$\Big|\frac{G}{2\pi}\log\frac{|z|\mathrm{e}^{-\Delta'(t)}}{\sqrt{t/2\pi}}\Big| = \frac{\delta}{2\pi\log(\frac{T}{2\pi})}\Big(\log\sqrt{\frac{t}{2\pi}} - \log|z| + \Delta'(t)\Big) = (\frac{\delta}{4\pi} + o(1)) < \frac{\delta}{2\pi} < b-a,$$

and $\widehat{f}(x) = 1$ for $|x| < b - a$. Moreover for $n \ge 1$ and $|x| < b - a$ we have

$$\widehat{f}^{(n)}(x) = (2\pi i)^n \int_{-\infty}^{\infty} y^n e^{2\pi i x y} f(y)\,\mathrm{d}y = 0,$$

hence it follows that

$$\int_{-\infty}^{\infty} y^n f(y) \exp\Bigl(2i\pi y \cdot \frac{G}{2\pi} \log \frac{|z|\mathrm{e}^{-\Delta'(t)}}{\sqrt{t/2\pi}}\Bigr)\,\mathrm{d}y = 0 \quad (n \ge 1, 1 < 2\pi(b-a)).$$

Thus we obtain

$$I_1 = G \int_{0\searrow 1} \frac{\mathrm{e}^{-\theta(t)}\mathrm{e}^{-i\pi z^2} z^{-1/2+it}}{\sin(\pi z)}\,\mathrm{d}z + O(\mathrm{e}^{-c\log^2 t}). \tag{8.21}$$

Similarly from (8.16) and (8.20) we have

$$I_2 = \int_{0\searrow 1, |z|<\log t} \frac{\mathrm{e}^{-i\pi z^2} z^{-1/2+it}}{\sin(\pi z)} \int_{-\log^{2\alpha-1} t}^{\log^{2\alpha-1} t} \mathrm{e}^{-i\theta(t) - \frac{ix}{2}\log\frac{t}{2\pi} - ix\Delta'(t)} z^{ix} f\bigl(\frac{x}{G}\bigr) \sum_{n=2}^{\infty} x^n g_n(t)\,\mathrm{d}x\,\mathrm{d}z$$

$$= \int_{0\searrow 1, |z|<\log t} \frac{\mathrm{e}^{-i\pi z^2} z^{-1/2+it}}{\sin(\pi z)} \int_{-\log^{2\alpha-1} t}^{\log^{2\alpha-1} t} \Bigl(\sum_{n=1}^{N} \frac{P_n(x)}{t^n} + O\bigl(\frac{1+x^{N+2}}{t^{N+1}}\bigr)\Bigr) \mathrm{e}^{-i\theta(t) - \frac{ix}{2}\log\frac{t}{2\pi} - ix\Delta'(t)} z^{ix} f\bigl(\frac{x}{G}\bigr)\,\mathrm{d}x\,\mathrm{d}z$$

$$= \sum_{n=1}^{N} e^{-i\theta(t)} t^{-n} \int_{0\searrow 1, |z|<\log t} \frac{\mathrm{e}^{-i\pi z^2} z^{-1/2+it}}{\sin(\pi z)} \int_{-\log^{2\alpha-1} t}^{\log^{2\alpha-1} t} P_n(x) \mathrm{e}^{-\frac{ix}{2}\log\frac{t}{2\pi} - ix\Delta'(t)} z^{ix} f\bigl(\frac{x}{G}\bigr)\,\mathrm{d}x\,\mathrm{d}z$$

$$+ O\Bigl(\frac{1}{t^{N+1}} \int_{-\log^{2\alpha-1} t}^{\log^{2\alpha-1} t} (1 + x^{N+2}) |f\bigl(\frac{x}{G}\bigr)| \Bigl| \int_{0\searrow 1} \frac{\mathrm{e}^{-i\pi z^2} z^{-1/2+it+ix}}{\sin(\pi z)}\,dz \Bigr|\,\mathrm{d}x\Bigr) + O(\mathrm{e}^{-c\log^2 t}),$$

where each $P_n(x)$ is a polynomial in x of degree $n(\ge 2)$. The integral over z in the error term is similar to the one in (8.10). Hence by the residue theorem we have ($Q = [\sqrt{t/2\pi}\,]$)

$$\int_{0\searrow 1} \frac{\mathrm{e}^{-i\pi z^2} z^{-1/2+it+ix}}{\sin(\pi z)}\,\mathrm{d}z = 2\pi i \sum_{n=1}^{Q} \operatorname*{Res}_{z=n} \frac{\mathrm{e}^{-i\pi z^2} z^{-1/2+it+ix}}{\sin(\pi z)} + \int_{Q\searrow Q+1} \frac{\mathrm{e}^{-i\pi z^2} z^{-1/2+it+ix}}{\sin(\pi z)}\,\mathrm{d}z,$$

similarly as in the derivation of the Riemann-Siegel formula. It follows that

$$\Bigl| \int_{0\searrow 1} \frac{\mathrm{e}^{-i\pi z^2} z^{-1/2+it+ix}}{\sin(\pi z)}\,dz \Bigr| \ll t^{1/4}.$$

Thus analogously as in the case of I_1 we find that, for $n \ge 1$,

$$\int_{-\log^{2\alpha-1} t}^{\log^{2\alpha-1} t} P_n(x) \mathrm{e}^{-i\theta(t) - \frac{ix}{2}\log\frac{t}{2\pi} - ix\Delta'(t)} z^{ix} f\bigl(\frac{x}{G}\bigr)\,\mathrm{d}x = \int_{-\infty}^{\infty} P_n(x) \cdots \mathrm{d}x + O(\mathrm{e}^{-c\log^2 t}) = O(\mathrm{e}^{-c\log^2 t}).$$

Hence it follows that, for any fixed integer $N \geq 1$,

$$I_2 \ll_N T^{-N}. \tag{8.22}$$

Theorem 1 now follows from (8.10) and (8.18)-(8.22), since clearly it suffices to assume that N is an integer. One can generalize Theorem 1 to derivatives of $M_{Z,f}(t)$.

Theorem 1 shows that $Z(t)$ and $M_{Z,f}(t)/G$ differ only by $O(T^{-N})$, for any fixed $N \geq 0$, which is a very small quantity. This certainly supports the belief that, for this particular subclass of functions $f(x)$, (7.8)–(7.11) will be true, but *proving* it may be very hard. On the other hand, nothing precludes the possibility that the error term in Theorem 1, although it is quite small, represents a function possessing many small "spikes" (like $t^{-N}\sin(t^{N+2})$, say). These spikes could introduce many new zeros, thus violating (7.8)–(7.11). Therefore it remains an open question to investigate the distribution of zeros of $M_{Z,f}(t)$ of Theorem 1, and to see to whether there is a possibility that Theorem 1 can be used in settling the truth of the RH.

9. Convolution functions and the RH

In this section we shall discuss the possibility to use convolution functions to disprove the RH, of course assuming that it is false. Let us denote by T_α^β the subclass of S_α^β with $\alpha > 1$ consisting of functions $f(x)$, which are not identically equal to zero, and which satisfy $\int_{-\infty}^{\infty} f(x)\,\mathrm{d}x > 0$. It is clear that T_α^β is non-empty. Our choice for G will be the same one as in (8.1), so that for suitable δ we shall have

$$\widehat{f}\Big(\frac{G}{4\pi}\log\big(\frac{T}{2\pi}\big)\Big) = \widehat{f}\big(\frac{\delta}{4\pi}\big) \gg 1. \tag{9.1}$$

In fact by continuity (9.1) will hold for $|\delta| \leq C_1$, where $C_1 > 0$ is a suitable constant depending only on f, since if $f(x) \in T_\alpha^\beta$, then we have

$$\widehat{f}(0) = \int_{-\infty}^{\infty} f(x)\,\mathrm{d}x > 0. \tag{9.2}$$

Moreover if $f(x) \in S_\alpha^0$, then $\widehat{f}(x) \in S_0^\alpha$ and thus it is of bounded support, and consequently $G \ll 1/\log T$ must hold if the bound in (9.1) is to be satisfied. This choice of $f(x)$ turns out to be better suited for our purposes than the choice made in Section 8, which perhaps would seem more natural in view of Theorem 1. The reason for this is that, if $f(x) \in S_\alpha^\beta$ with parameters A and B, then $\widehat{f}(x) \in S_\beta^\alpha$ with parameters $B+\varepsilon$ and $A+\varepsilon$, respectively (see [9]). But for $f(x)$ as in Section 8 we have $A = a+b$, thus for $\widehat{f}(x)$ we would have (in [9] $\widehat{f}$ is defined without the factor 2π, which would only change the scaling factors) $B = a+b+\varepsilon$, and this value of B would eventually turn out to be too large for our applications. In the present approach we have more flexibility, since only (9.2) is needed. Note that $f(x)$ is not necessarily nonnegative.

Now observe that if we replace $f(x)$ by $f_1(x) := f(Dx)$ for a given $D > 0$, then obviously $f_1(x) \in S_\alpha^0$, and moreover uniformly for $q \geq 0$ we have

$$f_1^{(q)}(x) = D^q f^{(q)}(Dx) \ll (BD)^q \exp(-aD^{1/\alpha}|x|^{1/\alpha}).$$

In other words the constant B in (6.3) or (6.4) is replaced by BD. Take now $D = \eta/B$, where $\eta > 0$ is an arbitrary, but fixed number, and write f for Df_1. If the RH holds, then from (4.5), (6.4) and (6.5) and we have, for k given by (7.12),

$$M_{Z,f}^{(k)}(t) \ll \Big(\frac{\eta}{G}\Big)^k \exp\Big(\frac{B_1\log t}{\log\log t}\Big) \tag{9.3}$$

with a suitable constant $B_1 > 0$.

We shall assume now that the RH holds and that (7.8), (7.11) hold for some $f(x) \in T_\alpha^0$ (for which (9.3) holds, which is implied by the RH), and we shall obtain a contradiction. To this end let $U := T^{1/2+\varepsilon}$, so that we may apply (7.10) or (7.11), $V = T^{1/4}/\log T$, $L = \log^{1/2+\varepsilon} T$. We shall consider the mean value of $|M_{Z,f}(t)|$ over $[T-U, T+U]$ in order to show that, on the average, $|M_{Z,f}(t)|$ is not too small. We have

$$I := \int_{T-U}^{T+U} |M_{Z,f}(t)|\,\mathrm{d}t \ge \sum_{n=1}^{N} \int_{T_n-VL}^{T_n+VL} |M_{Z,f}(t)|\,\mathrm{d}t \tag{9.4}$$
$$\ge \sum_{n=1}^{N} \int_{T_n-VL}^{T_n+VL} |M_{Z,f}(t)| \exp\Bigl(-(T_n-t)^2V^{-2}\Bigr)\,\mathrm{d}t,$$

where $T_n = T - U + (2n-1)VL$, and N is the largest integer for which $T_N + VL \le T + U$, hence $N \gg UV^{-1}L^{-1}$. We use Lemma 1 to bound from below each integral over $[T_n - VL, T_n + VL]$. It follows that

$$I \gg GV \sum_{n=1}^{N} \Bigl(\Bigl|\widehat{f}\bigl(\frac{G}{4\pi}\log\bigl(\frac{T_n}{2\pi}\bigr)\bigr)\Bigr| + O(T^{-1/4})\Bigr) \gg GV(N + O(NT^{-1/4})) \gg GUL^{-1} \tag{9.5}$$

for sufficiently small δ, since (9.1) holds and

$$\widehat{f}\Bigl(\frac{G}{4\pi}\log\bigl(\frac{T_n}{2\pi}\bigr)\Bigr) = \widehat{f}\Bigl(\frac{\delta}{4\pi}\cdot\frac{\log\bigl(\frac{T_n}{2\pi}\bigr)}{\log\bigl(\frac{T}{2\pi}\bigr)}\Bigr) = \widehat{f}\Bigl(\frac{\delta}{4\pi} + O\bigl(\frac{U}{T}\bigr)\Bigr).$$

We have assumed that (7.11) holds, but this implies that (7.10) holds also. Namely it holds unconditionally with $S(t)$ in place of $S_M(t)$. Thus for any fixed integer $m \ge 1$ we have

$$\int_T^{T+U} \bigl(S_M(t+H) - S_M(t-H)\bigr)^{2m}\,\mathrm{d}t$$
$$\ll \int_{T+H}^{T+H+U} \bigl(S_M(t) - S(t)\bigr)^{2m}\,\mathrm{d}t + \int_T^{T+U} \bigl(S(t+H) - S(t-H)\bigr)^{2m}\,\mathrm{d}t + \int_{T-H}^{T-H+U} \bigl(S(t) - S_M(t)\bigr)^{2m}\,\mathrm{d}t$$
$$\ll U(\log\log T)^m \qquad (T^a < U \le T, \tfrac{1}{2} < a \le 1).$$

Let $\mathcal{D}$ be the subset of $[T-U, T+U]$ where

$$|S_M(t+H) - S_M(t-H)| \le \log^{1/2} T \tag{9.6}$$

fails. The bound (7.10) implies that

$$m(\mathcal{D}) \ll U \log^{-C} T \tag{9.7}$$

for any fixed $C > 0$. If we take $C = 10$ in (9.7) and use the Cauchy-Schwarz inequality for integrals we shall have

$$\int_{\mathcal{D}} |M_{Z,f}(t)|\,\mathrm{d}t \le (m(\mathcal{D}))^{1/2} \Bigl(\int_{T-U}^{T+U} M_{Z,f}^2(t)\,\mathrm{d}t\Bigr)^{1/2} \ll GU\log^{-4} T, \tag{9.8}$$

since

$$\int_{T-U}^{T+U} M_{Z,f}^2(t)\,\mathrm{d}t \le \int_{T-U}^{T+U}\int_{-\infty}^{\infty} Z^2(t+x)|f(\frac{x}{G})|\,\mathrm{d}x \int_{-\infty}^{\infty}|f(\frac{y}{G})|\,\mathrm{d}y\,\mathrm{d}t$$

$$\ll G\int_{T-U}^{T+U}\int_{-G\log^{2\alpha-1}T}^{G\log^{2\alpha-1}T} |\zeta(\tfrac{1}{2}+it+ix)|^2|f(\frac{x}{G})|\,\mathrm{d}x\,\mathrm{d}t + G$$

$$\ll G\int_{-G\log^{2\alpha-1}T}^{G\log^{2\alpha-1}T}\Big(\int_{T-2U}^{T+2U}|\zeta(\tfrac{1}{2}+iu)|^2\,\mathrm{d}u\Big)|f(\frac{x}{G})|\,\mathrm{d}x + G \ll G^2U\log T.$$

The last bound easily follows from mean square results on $|\zeta(\frac{1}{2}+it)|$ (see [16]) with the choice $U = T^{\frac{1}{2}+\varepsilon}$. Therefore (9.4) and (9.8) yield

$$GUL^{-1} \ll \int_{\mathcal{D}'} |M_{Z,f}(t)|\,dt, \tag{9.9}$$

where $\mathcal{D}' = [T-U, T+U]\setminus\mathcal{D}$, hence in (9.9) integration is over t for which (9.6) holds. If $t\in\mathcal{D}'$, γ denotes the zeros of $M_{Z,f}(t)$, then from (7.7) and (7.12) we obtain (recall that $\log_r t = \log(\log_{r-1} t)$)

$$k = k(t,T) = \frac{H}{\pi}\log(\frac{T}{2\pi})\cdot\{1+O((\log T)^{\varepsilon-\frac{1}{2}})\},\ \log k = \log H - \log\pi + \log_2(\frac{T}{2\pi}) + O((\log T)^{\varepsilon-1/2}) \tag{9.10}$$

for any given $\varepsilon > 0$. To bound $M_{Z,f}(t)$ we use (7.6), with k given by (9.10), $\tau = \tau(t,k)$, (9.3) and

$$k! = \exp(k\log k - k + O(\log k)).$$

We obtain, denoting by B_j positive absolute constants,

$$GUL^{-1} \ll \int_{\mathcal{D}'}\prod_{|\gamma-t|\le H}|\gamma-t|\frac{|M_{Z,f}^{(k)}(\tau)|}{k!}\,\mathrm{d}t \tag{9.11}$$

$$\ll \exp\Big(\frac{B_2\log T}{\log_2 T}\Big)\int_{\mathcal{D}'}\exp\{k(\log\frac{\eta}{G} - \log k + 1)\}\prod_{|\gamma-t|\le H}|\gamma-t|\,\mathrm{d}t$$

$$\ll \exp\Big(\frac{B_3\log T}{\log_2 T} + \frac{H}{\pi}\log(\frac{T}{2\pi})\cdot\big(\log\frac{\eta}{\delta} + \log_2(\frac{T}{2\pi}) - \log H + \log\pi - \log_2(\frac{T}{2\pi}) + 1\big)\Big)\int_{\mathcal{D}'}\prod_{|\gamma-t|\le H}|\gamma-t|\,\mathrm{d}t.$$

It was in evaluating $k\log k$ that we needed (9.10), since only the bound (7.9) would not suffice (one would actually need the bound $S_M(T) \ll \log T/(\log_2 T)^2$). If the product under the last integral is bounded by (7.13), we obtain

$$GUL^{-1} \ll U\exp\Big(\frac{H}{\pi}\log(\frac{T}{2\pi})\cdot(\log\frac{\eta}{\delta} + B_4)\Big),$$

and thus for $T \ge T_0$

$$1 \le \exp\Big(\frac{H}{\pi}\log(\frac{T}{2\pi})\cdot(\log\frac{\eta}{\delta} + B_5)\Big). \tag{9.12}$$

Now we choose e.g.,

$$\eta = \delta^2,\ \delta = \min(C_1, \mathrm{e}^{-2B_5}),$$

where C_1 is the constant for which (9.1) holds if $|\delta| \le C_1$, so that (9.12) gives

$$1 \le \exp\Big(\frac{-B_5 H}{\pi}\log(\frac{T}{2\pi})\Big),$$

which is a contradiction for $T \ge T_1$. Thus we have proved the following

THEOREM 2. *If* (7.8) *and* (7.11) *hold for suitable* $f(x) \in T_\alpha^0$ *with* G *given by* (8.1), *then the Riemann hypothesis is false.*

Theorem 2 is similar to the result proved also in [23]. Actually the method of proof of Theorem 2 gives more than the assertion of the theorem. Namely it shows that, under the above hypotheses, (4.5) cannot hold for any fixed $A > 0$. Perhaps it should be mentioned that (7.11) is not the only condition which would lead to the disproof of the RH. It would be enough to assume, under the RH, that one had (7.8)–(7.10) for a suitable $f(x)$, or

$$N_M(t) = N(t) + O\Big(\frac{\log T}{(\log\log T)^2}\Big) \tag{9.18}$$

for $t \in [T-U, T+U]$ with a suitable $U(= T^{1/2+\varepsilon}$, but smaller values are possible), to derive a contradiction. The main drawback of this approach is the necessity to impose conditions like (7.8)–(7.10) which can be, for all we know, equally difficult to settle as the assertions which we originaly set out to prove (or disprove). For this reason our results can only be conditional. Even if the RH is false it appears plausible that, as $T \to \infty$, $N_0(T) = (1 + o(1))N(T)$. In other words, regardless of the truth of the RH, almost all complex zeros of $\zeta(s)$ should lie on the critical line. This is the conjecture that the author certainly believes in. No plausible conjectures seem to exist (if the RH is false) regarding the order of $N(T) - N_0(T)$.

REFERENCES

[1] R.J. Anderson and H.M. Stark, *Oscillation theorems*, in LNM's **899**, Springer-Verlag, Berlin-Heidelberg-New York, 1981, 79-106.

[2] F.V. Atkinson, *The mean value of the Riemann zeta-function*, Acta Math. **81**(1949), 353-376.

[3] R. Balasubramanian, *On the frequency of Titchmarsh's phenomenon for $\zeta(s)$ IV*, Hardy-Ramanujan J. **9**(1986), 1-10.

[4] R. Balasubramanian and K. Ramachandra, *On the frequency of Titchmarsh's phenomenon for $\zeta(s)$ III*, Proc. Indian Acad. Sci. Section A **86**(1977), 341-351.

[5] E. Bombieri and D. Hejhal, *Sur les zéros des fonctions zeta d'Epstein*, Comptes Rendus Acad. Sci. Paris **304**(1987), 213-217.

[6] H. Davenport and H. Heilbronn, *On the zeros of certain Dirichlet series I,II*, J. London Math. Soc. **11**(1936), 181-185 and ibid. 307-312.

[7] H.M. Edwards, *Riemann's zeta-function*, Academic Press, New York-London, 1974.

[8] A. Fujii, *On the distribution of the zeros of the Riemann zeta-function in short intervals*, Bull. Amer. Math. Soc. **81**(1975), 139-142.

[9] I.M. Gel'fand and G.E. Shilov, *Generalized functions (vol. 2)*, Academic Press, New York-London, 1968.

[10] A.O. Gel'fond, *The calculus of finite differences (Russian)*, Nauka, Moscow, 1967.

[11] J.L. Hafner and A. Ivić, *On the mean square of the Riemann zeta-function on the critical line*, J. Number Theory **32**(1989), 151-191.

[12] J.L. Hafner and A. Ivić, *On some mean value results for the Riemann zeta-function*, in Proc. International Number Theory Conf. Québec 1987, Walter de Gruyter and Co., Berlin-New York, 348-358.

[13] M.N. Huxley, *Exponential sums and the Riemann zeta-function IV*, Proceedings London Math. Soc. (3)**66**(1993), 1-40.

[14] M.N. Huxley, *A note on exponential sums with a difference*, Bulletin London Math. Soc. **29**(1994), 325-327.

[15] A. Ivić, *Large values of the error term in the divisor problem*, Invent. Math. **71**(1983), 513-520.

[16] A. Ivić, *The Riemann zeta-function*, John Wiley and Sons, New York, 1985 (2nd ed. Dover, 2003).

[17] A. Ivić, *On consecutive zeros of the Riemann zeta-function on the critical line*, Séminaire de Théorie des Nombres, Université de Bordeaux 1986/87, Exposé no. **29**, 14 pp.

[18] A. Ivić, *On a problem connected with zeros of $\zeta(s)$ on the critical line*, Monatshefte Math. **104**(1987), 17-27.

[19] A. Ivić, *Large values of certain number-theoretic error terms*, Acta Arithmetica **56**(1990), 135-159.

[20] A. Ivić, *Mean values of the Riemann zeta-function*, LN's **82**, Tata Institute of Fundamental Research, Bombay, 1991 (distr. by Springer Verlag, Berlin etc.).

[21] A. Ivić, *On a class of convolution functions connected with $\zeta(s)$*, Bulletin CIX Acad. Serbe des Sciences et des Arts, Classe des Sciences mathématiques et naturelles, Math. No **20**(1995), 29-50.

[22] A. Ivić, *On the fourth moment of the Riemann zeta-function*, Publications Inst. Math. (Belgrade) **57(71)**(1995), 101-110.

[23] A. Ivić, *On the distribution of zeros of a class of convolution functions*, Bulletin CXI Acad. Serbe des Sciences et des Arts, Classe des Sciences mathématiques et naturelles, Sciences mathématiques No. **21**(1996), 61-71.

[24] A. Ivić, *On the ternary additive divisor problem and the sixth moment of the zeta-function*, in "Sieve Methods, Exponential Sums, and their Applications in Number Theory" (eds. G.R.H. Greaves, G. Harman, M.N. Huxley), Cambridge University Press (Cambridge, UK), 1996, 205-243.

[25] A. Ivić and M. Jutila, *Gaps between consecutive zeros of the Riemann zeta-function*, Monatshefte Math. **105**(1988), 59-73.

[26] A. Ivić and Y. Motohashi, *A note on the mean value of the zeta and L-functions VII*, Proc. Japan Acad. Ser. A **66**(1990), 150-152.

[27] A. Ivić and Y. Motohashi, *The mean square of the error term for the fourth moment of the zeta-function*, Proc. London Math. Soc. (3)**66**(1994), 309-329.

[28] A. Ivić and Y. Motohashi, *The fourth moment of the Riemann zeta-function*, Journal Number Theory **51**(1995), 16-45.

[29] A. Ivić and H.J.J. te Riele, *On the zeros of the error term for the mean square of* $|\zeta(\frac{1}{2}+it)|$, Math. Comp. **56** No **193**(1991), 303-328.

[30] M. Jutila, *On the value distribution of the zeta-function on the critical line*, Bull. London Math. Soc. **15**(1983), 513-518.

[31] A.A. Karatsuba, *On the zeros of the Davenport-Heilbronn function lying on the critical line (Russian)*, Izv. Akad. Nauk SSSR ser. mat. **54** no. 2 (1990), 303-315.

[32] A.A. Karatsuba and S.M. Voronin, *The Riemann zeta-function*, Walter de Gruyter, Berlin–New York, 1992.

[33] N.V. Kuznetsov, *Sums of Kloosterman sums and the eighth moment of the Riemann zeta-function*, Papers presented at the Ramanujan Colloquium, Bombay 1989, publ. for Tata Institute (Bombay) by Oxford University Press, Oxford, 1989, pp. 57-117.

[34] A.A. Lavrik, *Uniform approximations and zeros of derivatives of Hardy's Z-function in short intervals (Russian)*, Analysis Mathem. **17**(1991), 257-259.

[35] D.H. Lehmer, *On the roots of the Riemann zeta function*, Acta Math. **95**(1956), 291-298.

[36] D.H. Lehmer, *Extended computation of the Riemann zeta-function*, Mathematika **3**(1956), 102-108.

[37] J.E. Littlewood, *Sur la distribution des nombres premiers*, Comptes rendus Académie Sci. (Paris) **158**(1914), 1869-1872.

[38] J. van de Lune, H.J.J. te Riele and D.T. Winter, *On the zeros of the Riemann zeta-function in the critical strip IV*, Math. Comp. **46**(1987), 273-308.

[39] H.L. Montgomery, *Extreme values of the Riemann zeta-function*, Comment. Math. Helv. **52**(1977), 511-518.

[40] Y. Motohashi, *The fourth power mean of the Riemann zeta-function*, in "Proceedings of the Amalfi Conference on Analytic Number Theory 1989", eds. E. Bombieri et al., Università di Salerno, Salerno, 1992, 325-344.

[41] Y. Motohashi, *An explicit formula for the fourth power mean of the Riemann zeta-function*, Acta Math. **170**(1993), 181-220.

[42] Y. Motohashi, *A relation between the Riemann zeta-function and the hyperbolic Laplacian*, Ann. Sc. Norm. Sup. Pisa, Cl. Sci. IV ser. **22**(1995), 299-313.

[43] Y. Motohashi, *The Riemann zeta-function and the non-Euclidean Laplacian*, Sugaku Expositions, AMS **8**(1995), 59-87.

[44] A.M. Odlyzko, *On the distribution of spacings between the zeros of the zeta-function*, Math. Comp. **48**(1987), 273-308.

[45] A.M. Odlyzko, *Analytic computations in number theory*, Proc. Symposia in Applied Math. **48**(1994), 451-463.

[46] A.M. Odlyzko, *The* 10^{20}*-th zero of the Riemann zeta-function and 175 million of its neighbors*, to appear.

[47] A.M. Odlyzko and H.J.J. te Riele, *Disproof of the Mertens conjecture*, J. reine angew. Math. **357**(1985), 138-160.

[48] K. Ramachandra, *Progress towards a conjecture on the mean value of Titchmarsh series*, in "Recent Progress in Analytic Number Theory", symposium Durham 1979 (Vol. 1), Academic Press, London, 1981, 303-318.

[49] K. Ramachandra, *On the mean-value and omega-theorems for the Riemann zeta-function*, LNs 85, Tata Institute of Fundamental Research, Bombay, 1995 (distr. by Springer Verlag, Berlin etc.).
[50] H.J.J. te Riele, *On the sign of the difference* $\pi(x) - \mathrm{li}\, x$, Math. Comp. **48**(1987), 323-328.
[51] H.J.J. te Riele and J. van de Lune, *Computational number theory at CWI in 1970-1994*, CWI Quarterly **7(4)** (1994), 285-335.
[52] B. Riemann, *Über die Anzahl der Primzahlen unter einer gegebener Grösse*, Monats. Preuss. Akad. Wiss. (1859-1860), 671-680.
[53] A. Selberg, *On the zeros of Riemann's zeta-function*, Skr. Norske Vid. Akad. Oslo **10**(1942), 1-59.
[54] A. Selberg, *Contributions to the theory of the Riemann zeta-function*, Arch. Math. Naturvid. **48**(1946) No. 5, 89-155.
[55] A. Selberg, *Harmonic analysis and discontinuous groups in weakly symmetric spaces with applications to Dirichlet series*, J. Indian Math. Soc. **20**(1956), 47-87.
[56] R. Sherman Lehman, *On the difference* $\pi(x) - \mathrm{li}\, x$, Acta Arith, **11**(1966), 397-410.
[57] C.L. Siegel, *Über Riemanns Nachlaß zur analytischen Zahlentheorie*, Quell. Stud. Gesch. Mat. Astr. Physik **2**(1932), 45-80 (also in *Gesammelte Abhandlungen, Band I*, Springer Verlag, Berlin etc., 1966, 275-310).
[58] R. Spira, *Some zeros of the Titchmarsh counterexample*, Math. Comp. **63**(1994), 747-748.
[59] K.-M. Tsang, *Some* Ω*-theorems for the Riemann zeta-function*, Acta Arith. **46**(1986), 369-395.
[60] K.-M. Tsang, *The large values of the Riemann zeta-function*, Mathematika **40**(1993), 203-214.
[61] E.C. Titchmarsh, *The theory of the Riemann zeta-function* (2nd ed.), Clarendon Press, Oxford, 1986.

References added in November 2003:

[62] A. Ivić, *On some results concerning the Riemann Hypothesis*, in "Analytic Number Theory" (Kyoto, 1996) ed. Y. Motohashi, LMS LNS **247**, Cambridge University Press, Cambridge, 1997, 139-167.
[63] Y. Motohashi, Spectral theory of the Riemann zeta-function, Cambridge University Press, 1997.
[64] A. Ivić, *On the multiplicity of zeros of the zeta-function*, Bulletin CXVIII de l'Académie Serbe des Sciences et des Arts - 1999, Classe des Sciences mathématiques et naturelles, Sciences mathématiques No. **24**, 119-131.
[65] E. Bombieri, *Problems of the Millenium: the Riemann Hypothesis*, `http//www.ams.org/claymath/prize_problems/riemann.pdf`, Amer. Math. Soc., Providence, R.I., 2000, 12pp.
[66] J.B. Conrey, *L-functions and random matrices*, in "Mathematics Unlimited" (Part I), B. Engquist and W. Schmid eds., Springer, 2001, 331-352.
[67] J.B. Conrey, D.W. Farmer, J.P. Keating, M.O. Rubinstein and N.C. Snaith, *Integral moments of L-functions*, 2003, 58pp, arXiv:math.NT/0206018, `http://front.math.ucdavis.edu/mat.NT/0206018`.

Aleksandar Ivić
Katedra Matematike RGF-a
Universitet u Beogradu
Đušina 7, 11000 Beograd
Serbia (Yugoslavia)
aivic@matf.bg.ac.yu, aivic@rgf.bg.ac.yu

12

The Experts Speak for Themselves

> *To appreciate the living spirit rather than the dry bones of mathematics, it is necessary to inspect the work of a master at first hand. Textbooks and treatises are an unavoidable evil ... The very crudities of the first attack on a significant problem by a master are more illuminating than all the pretty elegance of the standard texts which has been won at the cost of perhaps centuries of finicky polishing.*
>
> **Eric Temple Bell**

This chapter contains several original papers. These give the most essential sampling of the enormous body of material on the Riemann zeta function, the Riemann hypothesis, and related theory. They give a chronology of milestones in the development of the theory contained in the previous chapters. We begin with Chebyshev's groundbreaking work on $\pi(x)$, continue through Riemann's proposition of the Riemann hypothesis, and end with an ingenious algorithm for primality testing. These papers place the material in historical context and illustrate the motivations for research on and around the Riemann hypothesis. Most papers are preceded by a short biographical note on the author(s) and all are preceded by a short review of the material they present.

12.1 P. L. Chebyshev (1852)

Sur la fonction qui détermine la totalité des nombres premiers inférieurs à une limite donnée

Pafnuty Lvovich Chebyshev was born in Okatovo, Russia, on May 16, 1821. Chebyshev's mother, Agrafena Ivanova Pozniakova, acted as his teacher and taught him basic language skills. His cousin tutored him in French and arithmetic. Chebyshev was schooled at home by various tutors until he entered Moscow University in 1837. In 1847 he was appointed to the University of St. Petersburg. During his career he contributed to several fields of mathematics. Specifically, in the theory of prime numbers he proved Bertrand's postulate in 1850, and came very close to proving the prime number theorem. He was recognized internationally as a preeminent mathematician. He was elected to academies in Russia, France, Germany, Italy, England, Belgium, and Sweden. He was also given honorary positions at every Russian university. Chebyshev died in St. Petersburg in 1894 [119].

This paper is considered the first substantial result toward the prime number theorem. The prime number theorem is the statement that $\pi(x) \sim \frac{x}{\log x}$. Legendre had worked on the prime number theorem and had attempted to give accurate approximations to $\pi(x)$. Chebyshev shows that if any approximation to $\pi(x)$ is within order $\frac{x}{\log^N x}$, for any fixed large $N \in \mathbb{Z}$, then the approximation is $\mathrm{Li}(x)$. Chebyshev considers the expression $\frac{\pi(x)\log x}{x}$, and proves that if $\lim_{x\to\infty} \frac{\pi(x)\log x}{x}$ exists, then the limit is 1.

Sur la fonction qui détermine la totalité des nombres premiers inférieurs à une limite donnée.

§ 1. Legendre, dans sa Théorie des nombres *), propose une formule pour déterminer combien il y a de nombres premiers depuis 1 jusqu'à une limite donnée. Il commence par comparer sa formule avec l'énumération immédiate des nombres premiers faite dans les tables les plus étendues, nommément depuis 10000 jusqu'à 1000000, et l'applique ensuite à la solution de plusieurs questions. Malgré la concordance prononcée de la formule de Legendre avec les tables des nombres premiers, nous nous permettons néanmoins d'élever quelques doutes sur son exactitude, et par conséquent aussi sur les résultats qu'on en a tirés. Nous fondons notre assertion sur un théorème, relatif aux propriétés de la fonction qui détermine combien il y a de nombres premiers inférieurs à une limite donnée, théorème dont on peut déduire plusieurs conséquences curieuses. Nous allons d'abord donner la démonstration du théorème en question, et nous en présenterons ensuite quelques applications.

I-er Théorème.

§ 2. *Si l'on représente par $\varphi(x)$ la totalité des nombres premiers inférieurs à x, par n un entier quelconque, enfin par ρ une quantité > 0, la somme*

$$\sum_{x=2}^{x=\infty} \left[\varphi(x+1) - \varphi(x) - \frac{1}{\log x}\right] \frac{\log^n x}{x^{1+\rho}}$$

jouira de la propriété de s'approcher d'une limite finie, à mesure que ρ converge vers zéro.

*) Tome 2, page 65 (Troisième édition).

— 30 —

Démonstration. Commençons par démontrer que la propriété en question a lieu pour les fonctions que l'on obtient par la différentiation successive des trois expressions

$$\sum \frac{1}{m^{1+\rho}} - \frac{1}{\rho}, \quad \log \rho - \Sigma \log\left(1 - \frac{1}{\mu^{1+\rho}}\right),$$
$$\Sigma \log\left(1 - \frac{1}{\mu^{1+\rho}}\right) + \sum \frac{1}{\mu^{1+\rho}}$$

par rapport à ρ; ici, comme par la suite, la sommation par rapport à m s'étend à tous les entiers depuis $m = 2$ jusqu'à $m = \infty$, et par rapport à μ seulement aux nombres premiers, également depuis $\mu = 2$ jusqu'à $\mu = \infty$.

Considérons la première expression. Il est facile de voir que l'on a

$$\int_0^\infty \frac{e^{-x}}{e^x - 1} x^\rho \, dx = \sum \frac{1}{m^{1+\rho}} \cdot \int_0^\infty e^{-x} x^\rho \, dx,$$

$$\int_0^\infty e^{-x} x^{-1+\rho} \, dx = \frac{1}{\rho} \int_0^\infty e^{-x} x^\rho \, dx,$$

et par conséquent

$$\sum \frac{1}{m^{1+\rho}} - \frac{1}{\rho} = \frac{\int_0^\infty \left(\frac{1}{e^x - 1} - \frac{1}{x}\right) e^{-x} x^\rho \, dx}{\int_0^\infty e^{-x} x^\rho \, dx}.$$

En vertu de cette équation la dérivée d'un ordre quelconque n de $\sum \frac{1}{m^{1+\rho}} - \frac{1}{\rho}$ par rapport à ρ sera égale à une fraction, dont le dénominateur est $\left[\int_0^\infty e^{-x} x^\rho \, dx\right]^{n+1}$ et le numerateur une fonction entière des expressions

$$\int_0^\infty \left(\frac{1}{e^x - 1} - \frac{1}{x}\right) e^{-x} x^\rho \, dx, \quad \int_0^\infty \left(\frac{1}{e^x - 1} - \frac{1}{x}\right) e^{-x} x^\rho \log x \, dx,$$

$$\int_0^\infty \left(\frac{1}{e^x - 1} - \frac{1}{x}\right) e^{-x} x^\rho \log^2 x \, dx, \ldots \int_0^\infty \left(\frac{1}{e^x - 1} - \frac{1}{x}\right) e^{-x} x^\rho \log^n x \, dx,$$

$$\int_0^\infty e^{-x} x^\rho \, dx, \int_0^\infty e^{-x} x^\rho \log x \, dx, \int_0^\infty e^{-x} x^\rho \log^2 x \, dx, \ldots \int_0^\infty e^{-x} x^\rho \log^n x \, dx.$$

Or, une telle fraction, pour $n = 0$ aussi bien que pour $n > 0$, s'approche d'une limite finie à mesure que ρ converge vers zéro; car la limite de l'intégrale $\int_0^\infty e^{-x} x^\rho dx$ pour $\rho = 0$ est 1, et les intégrales

$$\int_0^\infty \left(\frac{1}{e^x-1} - \frac{1}{x}\right) e^{-x} x^\rho dx, \quad \int_0^\infty \left(\frac{1}{e^x-1} - \frac{1}{x}\right) e^{-x} x^\rho \log x\, dx,$$

$$\int_0^\infty \left(\frac{1}{e^x-1} - \frac{1}{x}\right) e^{-x} x^\rho \log^2 x\, dx, \ldots \int_0^\infty \left(\frac{1}{e^x-1} - \frac{1}{x}\right) e^{-x} x^\rho \log^n x\, dx,$$

$$\int_0^\infty e^{-x} x^\rho \log x\, dx, \int_0^\infty e^{-x} x^\rho \log^2 x\, dx, \ldots \int_0^\infty e^{-x} x^\rho \log^n x\, dx$$

pour $\rho = 0$ conservent évidemment des valeurs finies.

Ainsi, il est certain que la fonction $\sum \frac{1}{m^{1+\rho}} - \frac{1}{\rho}$, aussi bien que ses dérivées successives, resteront finies à mesure que ρ convergera vers la limite zéro.

Considérons actuellement la fonction

$$\log \rho - \Sigma \log \left(1 - \frac{1}{\mu^{1+\rho}}\right).$$

On sait que

$$\left[\left(1 - \frac{1}{2^{1+\rho}}\right)\left(1 - \frac{1}{3^{1+\rho}}\right)\left(1 - \frac{1}{5^{1+\rho}}\right) \ldots\ldots\right]^{-1}$$
$$= 1 + \frac{1}{2^{1+\rho}} + \frac{1}{3^{1+\rho}} + \frac{1}{4^{1+\rho}} + \ldots\ldots;$$

d'où l'on tire

$$-\log\left(1 - \frac{1}{2^{1+\rho}}\right) - \log\left(1 - \frac{1}{3^{1+\rho}}\right) - \log\left(1 - \frac{1}{5^{1+\rho}}\right) \ldots$$
$$= \log\left(1 + \frac{1}{2^{1+\rho}} + \frac{1}{3^{1+\rho}} + \frac{1}{4^{1+\rho}} + \ldots\right),$$

équation qui, d'après la notation admise plus haut, peut être écrite de cette manière

$$-\Sigma \log\left(1 - \frac{1}{\mu^{1+\rho}}\right) = \log\left(1 + \sum \frac{1}{m^{1+\rho}}\right);$$

donc

$$\log \rho - \Sigma \log\left(1 - \frac{1}{\mu^{1+\rho}}\right) = \log\left(1 + \sum \frac{1}{m^{1+\rho}}\right)\rho;$$

ou bien

$$\log \rho - \Sigma \log\left(1 - \frac{1}{\mu^{1+\rho}}\right) = \log\left[1 + \rho + \left(\sum \frac{1}{m^{1+\rho}} - \frac{1}{\rho}\right)\rho\right].$$

— 32 —

Cette équation fait voir que toutes les dérivées de

$$\log \rho - \Sigma \log \left(1 - \frac{1}{\mu^{1+\rho}}\right)$$

suivant ρ, s'exprimeront au moyen d'un nombre fini de fractions, dont les dénominateurs seront des puissances entières et positives de

$$1 + \rho + \left(\sum \frac{1}{m^{1+\rho}} - \frac{1}{\rho}\right)\rho,$$

et les numérateurs des fonctions entières de ρ, de l'expression $\sum \frac{1}{m^{1+\rho}} - \frac{1}{\rho}$ et de ses dérivées par rapport à ρ. Or, de telles fractions s'approcheront d'une limite finie à mesure que ρ convergera vers zéro; car l'expression $1+\rho+\left(\sum \frac{1}{m^{1+\rho}} - \frac{1}{\rho}\right)\rho$, qui entre dans les dénominateurs de ces fractions, tendra vers la limite 1 à mesure que ρ s'approchera de zéro, et cela parce-que la différence $\sum \frac{1}{m^{1+\rho}} - \frac{1}{\rho}$, dans cette hypothèse, reste finie comme nous l'avons démontré plus haut. Quant à ce qui concerne les numérateurs, comme ils ne contiennent la différence $\sum \frac{1}{m^{1+\rho}} - \frac{1}{\rho}$ et ses dérivées que sous forme entière, et que ces fonctions tendent vers une limite finie quand ρ converge vers zéro, il en sera de même pour ses numérateurs.

Il nous reste encore à démontrer que la même propriété à lieu relativement aux dérivées de la fonction

$$\Sigma \log \left(1 - \frac{1}{\mu^{1+\rho}}\right) + \sum \frac{1}{\mu^{1+\rho}}.$$

Nous remarquerons d'abord que sa première dérivée sera

$$\sum \frac{1}{\mu^{2+2\rho}} \cdot \frac{\log \mu}{1 - \frac{1}{\mu^{1+\rho}}}.$$

Il est facile de voir par la forme de cette fonction que les dérivées des ordres supérieurs s'exprimeront également au moyen d'un nombre fini des termes tels que

$$\sum \frac{1}{\mu^{2+2\rho}} \cdot \frac{\log^p \mu}{1 - \frac{1}{\mu^{1+\rho}}} \cdot \frac{1}{\mu^s \left(1 - \frac{1}{\mu^{1+\rho}}\right)^r},$$

p, q, r n'étant par inférieurs à zéro. Mais chaque terme de cette nature, pour des valeurs de ρ non-inférieures à zéro, a une valeur finie; en effet, pour $\rho = 0$ et $\rho > 0$, la fonction sous le signe Σ sera une quantité d'un ordre supérieur au premier par rapport à $\frac{1}{\mu}$.

— 33 —

Après nous être convaincu que les dérivées des trois expressions

$$\sum \frac{1}{m^{1+\rho}} - \frac{1}{\rho}, \quad \log \rho - \Sigma \log \left(1 - \frac{1}{\mu^{1+\rho}}\right),$$
$$\Sigma \log \left(1 - \frac{1}{\mu^{1+\rho}}\right) + \sum \frac{1}{\mu^{1+\rho}},$$

pour des valeurs de ρ convergentes vers zéro, tendent vers des limites finies, nous concluons que la même propriété aura également lieu par rapport à l'expression

$$\frac{d^n\left[\Sigma \log \left(1 - \frac{1}{\mu^{1+\rho}}\right) + \sum \frac{1}{\mu^{1+\rho}}\right]}{d\rho^n} - \frac{d^n\left[\log \rho - \Sigma \log \left(1 - \frac{1}{\mu^{1+\rho}}\right)\right]}{d\rho^n} - \frac{d^{n-1}\left(\sum \frac{1}{m^{1+\rho}} - \frac{1}{\rho}\right)}{d\rho^{n-1}},$$

laquelle, après les différentiations effectuées, ce réduira à

$$\pm\left(\sum \frac{\log^n \mu}{\mu^{1+\rho}} - \sum \frac{\log^{n-1} m}{m^{1+\rho}}\right).$$

Ce qui vient d'être dit renferme le théorème énoncé plus haut, car il est facile de remarquer que, d'après notre notation, la différence

$$\sum \frac{\log^n \mu}{\mu^{1+\rho}} - \sum \frac{\log^{n-1} m}{m^{1+\rho}}$$

est identique avec l'expression

$$\sum_{x=2}^{x=\infty} \left[\varphi(x+1) - \varphi(x) - \frac{1}{\log x}\right] \frac{\log^n x}{x^{1+\rho}},$$

ou bien, ce qui revient au même, avec

$$\sum_{x=2}^{x=\infty} \left[\varphi(x+1) - \varphi(x)\right] \frac{\log^n x}{x^{1+\rho}} - \sum_{x=2}^{x=\infty} \frac{\log^{n-1} x}{x^{1+\rho}},$$

Pour le faire voir il n'y a qu'à observer que le premier terme de cette différence est simplement égal à $\sum \frac{\log^n \mu}{\mu^{1+\rho}}$, parceque le facteur $\varphi(x+1) - \varphi(x)$ de $\frac{\log^n x}{x^{1+\rho}}$ se réduit, par la définition même de la fonction φ, à 1 ou à 0 suivant que x est un nombre premier ou un nombre composé. Quant au second terme $\sum_{x=2}^{x=\infty} \frac{\log^{n-1} x}{x^{1+\rho}}$, il se transforme évidemment en $\sum \frac{\log^{n-1} m}{m^{1+\rho}}$ par le changement de x en m.

De cette manière la proposition que nous avions en vue de démontrer, se trouve complètement établie.

— 34 —

§ 3. Le théorème dont on vient de donner la démonstration conduit à plusieurs propriétés curieuses de la fonction qui détermine combien il y a de nombres premiers inférieurs à une limite donnée. Et d'abord observons que la différence

$$\frac{1}{\log x} - \int_x^{x+1} \frac{dx}{\log x},$$

pour x très grand, est une quantité infiniment petite du premier ordre par rapport à $\frac{1}{x}$; par conséquent l'expression

$$\left(\frac{1}{\log x} - \int_x^{x+1} \frac{dx}{\log x}\right) \frac{\log^n x}{x^{1+\rho}},$$

pour x très grand, sera de l'ordre $2+\rho$ relativement à $\frac{1}{x}$; d'après cela, la somme

$$\sum_{x=2}^{x=\infty} \left(\frac{1}{\log x} - \int_x^{x+1} \frac{dx}{\log x}\right) \frac{\log^n x}{x^{1+\rho}}$$

pour des valeurs de ρ non-inférieures à zéro, restera finie. Ajoutant cette somme à l'expression

$$\sum_{x=2}^{x=\infty} \left[\varphi(x+1) - \varphi(x) - \frac{1}{\log x}\right] \frac{\log^n x}{x^{1+\rho}},$$

pour laquelle le théorème I-er a lieu, nous concluons que la valeur de

$$\sum_{x=2}^{x=\infty} \left[\varphi(x+1) - \varphi(x) - \int_x^{x+1} \frac{dx}{\log x}\right] \frac{\log^n x}{x^{1+\rho}} dx$$

restera finie à mesure que ρ convergera vers la limite zéro. De là on tire le théorème suivant:

II-ème Théorème.

La fonction $\varphi(x)$, qui désigne combien il y a de nombres premiers inférieurs à x, satisfera, entre les limites $x=2$ et $x=\infty$, une infinité de fois aux deux inégalités

$$\varphi(x) > \int_2^x \frac{dx}{\log x} - \frac{\alpha x}{\log^n x} \quad \text{et} \quad \varphi(x) < \int_2^x \frac{dx}{\log x} + \frac{\alpha x}{\log^n x},$$

quelque petite que soit la valeur de α, supposée positive, et quelque grand que soit en même temps le nombre n.

Démonstration. Nous nous contenterons de démontrer l'une de ces deux inégalités, parceque l'autre s'établira tout-à-fait de la même manière. Choisissons, par exemple, la suivante:

$$\varphi(x) < \int_2^x \frac{dx}{\log x} + \frac{\alpha x}{\log^n x}. \tag{1}$$

Pour prouver que cette inégalité est satisfaite une infinité de fois, admettons d'abord que le contraire ait lieu, et voyons quelles seront les conséquences de cette hypothèse. Soit a un entier supérieur à e^n et supérieur en même temps au plus grand nombre qui satisfait à l'inégalité (1). Dans cette supposition on aura pour $x > a$ l'inégalité

$$\varphi(x) \geqq \int_2^x \frac{dx}{\log x} + \frac{\alpha x}{\log^n x}, \quad \log x > n,$$

et par conséquent

$$\varphi(x) - \int_2^x \frac{dx}{\log x} \geqq \frac{\alpha x}{\log^n x}, \quad \frac{n}{\log x} < 1. \tag{2}$$

Or, si l'on admettait les inégalités (2), il en résulterait, contrairement à ce qui a été démontré plus haut, que l'expression

$$\sum_{x=2}^{x=\infty} \left[\varphi(x+1) - \varphi(x) - \int_x^{x+1} \frac{dx}{\log x} \right] \frac{\log^n x}{x^{1+\rho}},$$

au lieu de converger vers une limite finie pour des valeurs très petites de ρ, s'approcherait de la limite $+\infty$. En effet, nous pouvons considérer cette expression comme la limite de

$$\sum_{x=2}^{x=s} \left[\varphi(x+1) - \varphi(x) - \int_x^{x+1} \frac{dx}{\log x} \right] \frac{\log^n x}{x^{1+\rho}}$$

pour $s = \infty$. Supposant donc $s > a$, cette quantité peut être mise sous la forme

$$C + \sum_{x=a+1}^{x=s} \left[\varphi(x+1) - \varphi(x) - \int_x^{x+1} \frac{dx}{\log x} \right] \frac{\log^n x}{x^{1+\rho}}, \tag{3}$$

en faisant pour abréger

$$C = \sum_{x=2}^{x=a} \left[\varphi(x+1) - \varphi(x) - \int_x^{x+1} \frac{dx}{\log x} \right] \frac{\log^n x}{x^{1+\rho}},$$

et observant que C désignera une quantité finie pour $\rho = 0$ et $\rho > 0$.

— 36 —

Or, l'expression (3), en vertu de la formule connue

$$\sum_{a+1}^{s} u_x (v_{x+1} - v_x) = u_s v_{s+1} - u_a v_{a+1} - \sum_{a+1}^{s} v_x (u_x - u_{x-1}),$$

et après avoir fait

$$v_x = \varphi(x) - \int_2^x \frac{dx}{\log x}, \quad u_x = \frac{\log^n x}{x^{1+\rho}},$$

se transformera dans la suivante:

$$C - \left[\varphi(a+1) - \int_2^{a+1} \frac{dx}{\log x}\right] \frac{\log^n a}{a^{1+\rho}} + \left[\varphi(s+1) - \int_2^{s+1} \frac{dx}{\log x}\right] \frac{\log^n s}{s^{1+\rho}}$$

$$- \sum_{x=a+1}^{x=s} \left[\varphi(x) - \int_2^x \frac{dx}{\log x}\right] \left[\frac{\log^n x}{x^{1+\rho}} - \frac{\log^n (x-1)}{(x-1)^{1+\rho}}\right],$$

qui, à son tour, en faisant $\theta > 0$ et < 1, pourra s'écrire comme il suit:

$$C - \left[\varphi(a+1) - \int_2^{a+1} \frac{dx}{\log x}\right] \frac{\log^n a}{a^{1+\rho}} + \left[\varphi(s+1) - \int_2^{s+1} \frac{dx}{\log x}\right] \frac{\log^n s}{s^{1+\rho}} +$$

$$+ \sum_{x=a+1}^{x=s} \left[\varphi(x) - \int_2^x \frac{dx}{\log x}\right] \left[1 + \rho - \frac{n}{\log(x-\theta)}\right] \frac{\log^n (x-\theta)}{(x-\theta)^{2+\rho}}.$$

Si l'on représente par F la somme des deux premiers termes de cette expression, et si l'on observe de plus que le troisième est positif en vertu de la condition (2), on sera en droit de conclure que l'expression précédente a une valeur supérieure à

$$F + \sum_{x=a+1}^{x=s} \left[\varphi(x) - \int_2^x \frac{dx}{\log x}\right] \left[1 + \rho - \frac{n}{\log(x-\theta)}\right] \frac{\log^n (x-\theta)}{(x-\theta)^{2+\rho}}.$$

Les mêmes conditions (2) font voir que dans cette expression la fonction sous le signe Σ conservera une valeur positive entre les limites. En outre, on aura entre les limites de la sommation 1°) $1 + \rho - \frac{n}{\log(x-\theta)} > 1 - \frac{n}{\log a}$; car $\rho > 0$, $x \overline{\geqq} a + 1$, $\theta < 1$; 2°) $\varphi(x) - \int_2^x \frac{dx}{\log x} > \frac{\alpha(x-\theta)}{\log^n (x-\theta)}$; car $\varphi(x) - \int_2^x \frac{dx}{\log x} \overline{\geqq} \frac{\alpha x}{\log^n x}$ en vertu de la première des inégalités (2), et en vertu de la seconde la dérivée de $\frac{\alpha x}{\log^n x}$, égale à $\frac{\alpha}{\log^n x}\left(1-\frac{n}{\log x}\right)$, est positive, ce qui donne $\frac{\alpha x}{\log^n x} > \frac{\alpha(x-\theta)}{\log^n (x-\theta)}$. Donc l'expression précédente surpasse la somme

$$F+\sum_{x=a+1}^{x=s} \frac{\alpha(x-\theta)}{\log^n(x-\theta)}\left(1-\frac{n}{\log a}\right)\frac{\log^n(x-\theta)}{(x-\theta)^{2+\rho}},$$

qui, après les réductions, devient

$$F+\alpha\left(1-\frac{n}{\log a}\right)\sum_{x=a+1}^{x=s}\frac{1}{(x-\theta)^{1+\rho}};$$

or, cette dernière expression est évidemment supérieure à celle-ci

$$F+\alpha\left(1-\frac{n}{\log a}\right)\sum_{x=a+1}^{x=s}\frac{1}{x^{1+\rho}},$$

laquelle, pour $s=\infty$, se réduit à

$$F+\alpha\left(1-\frac{n}{\log a}\right)\sum_{x=a+1}^{x=\infty}\frac{1}{x^{1+\rho}},$$

ou à

$$F+\alpha\left(1-\frac{n}{\log a}\right)\frac{\int_0^\infty \frac{e^{-ax}}{e^x-1}\,x^\rho\,dx}{\int_0^\infty e^{-x}x^\rho\,dx}.$$

Il est facile de faire voir que la quantité à laquelle nous sommes parvenus converge vers la limite $+\infty$ pour $\rho=0$. En effet, on a d'abord $\int_0^\infty \frac{e^{-ax}}{e^x-1}\,dx=+\infty$, $\int_0^\infty e^{-x}\,dx=1$, de plus α et $1-\frac{n}{\log a}$ sont toutes deux des quantités positives, la première par hypothèse, et la seconde en vertu de la dernière inégalité (2).

Nous étant assurés de cette manière que, dans l'hypothèse admise, non seulement la somme

$$\sum_{x=a}^{x=\infty}\left[\varphi(x+1)-\varphi(x)-\int_x^{x+1}\frac{dx}{\log x}\right]\frac{\log^n x}{x^{1+\rho}},$$

mais aussi une quantité plus petite qu'elle se réduit à $+\infty$, nous sommes en droit de conclure que l'hypothèse en question est inadmissible, d'où découle de suite la légitimité du théorème II.

— 38 —

§ 4. Il sera facile actuellement, en vertu de la proposition précédente, de démontrer le théorème qui suit:

III-éme Théorème.

L'expression $\frac{x}{\varphi(x)} - \log x$, *pour* $x = \infty$, *ne peut avoir une limite différente de* -1.

Démonstration. Soit L la limite de la différence $\frac{x}{\varphi(x)} - \log x$ pour $x = \infty$. Dans cette hypothése on pourra toujours trouver un nombre N tellement grand que pour $x > N$ la valeur de $\frac{x}{\varphi(x)} - \log x$ sera comprise entre les limites $L - \varepsilon$ et $L + \varepsilon$, ε étant aussi petite qu'on voudra. Ainsi, pour de semblables valeurs de x, et lorsque $\varepsilon > 0$, on aura

$$(4) \qquad \frac{x}{\varphi(x)} - \log x > L - \varepsilon, \quad \frac{x}{\varphi(x)} - \log x < L + \varepsilon.$$

Mais, en vertu du théorème précédent, les inégalités

$$\varphi(x) > \int_2^x \frac{dx}{\log x} - \frac{\alpha x}{\log^n x}, \quad \varphi(x) < \int_2^x \frac{dx}{\log x} + \frac{\alpha x}{\log^n x}$$

sont satisfaites par une infinité de valeurs de x, et par conséquent aussi par des valeurs de x supérieures à N, pour lesquelles les inégalités (4) ont lieu. Or, ces inégalités, combinées avec celles que nous venons d'écrire, conduisent à

$$\frac{x}{\int_2^x \frac{dx}{\log x} - \frac{\alpha x}{\log^n x}} - \log x > L - \varepsilon, \quad \frac{x}{\int_2^x \frac{dx}{\log x} + \frac{\alpha x}{\log^n x}} - \log x < L + \varepsilon;$$

d'où l'on tire

$$L + 1 < \frac{x - (\log x - 1)\left(\int_2^x \frac{dx}{\log x} - \frac{\alpha x}{\log^n x}\right)}{\int_2^x \frac{dx}{\log x} - \frac{\alpha x}{\log^n x}} + \varepsilon,$$

$$L + 1 > \frac{x - (\log x - 1)\left(\int_2^x \frac{dx}{\log x} + \frac{\alpha x}{\log^n x}\right)}{\int_2^x \frac{dx}{\log x} + \frac{\alpha x}{\log^n x}} - \varepsilon.$$

On voit par ces inégalités que la valeur numérique de $L+1$ ne surpasse pas celle de l'une des expressions qui en forment les seconds membres. De plus ε peut devenir aussi petite qu'on voudra dans l'hypothèse de N très grand, et on peut en dire autant de chacune des quantités

$$\frac{x-(\log x-1)\left(\int_2^x \frac{dx}{\log x} \mp \frac{\alpha x}{\log^n x}\right)}{\int_2^x \frac{dx}{\log x} \mp \frac{\alpha x}{\log^n x}};$$

car, pour $x=\infty$, on trouve par les principes du calcul différentiel que leur limite commune est zéro.

Nous étant ainsi convaincus que les limites

$$\frac{x-(\log x-1)\left(\int_2^x \frac{dx}{\log x} \mp \frac{\alpha x}{\log^n x}\right)}{\int_2^x \frac{dx}{\log x} \mp \frac{\alpha x}{\log^n x}} \pm \varepsilon,$$

de la valeur numérique de $L+1$ peuvent être diminuées à volonté, nous sommes en droit de conclure que $L+1=0$, et par conséquent $L=-1$, ce qu'il s'agissait de démontrer.

Ce que nous venons de prouver relativement à la limite de la valeur de $\frac{x}{\varphi(x)}-\log x$, pour $x=\infty$, ne s'accorde pas avec une formule donnée par Legendre pour déterminer approximativement combien il y a de nombres premiers inférieurs à une limite donnée. D'après lui la fonction $\varphi(x)$, pour x très grand, est exprimée avec une approximation suffisante par la formule

$$\varphi(x)=\frac{x}{\log x-1{,}08366},$$

qui donne pour la limite de $\frac{x}{\varphi(x)}-\log x$ le nombre $-1{,}08366$ au lieu de -1.

§ 5. En partant du théorème II on peut déterminer la limite supérieure du dégre de précision avec lequel la fonction, désignée par $\varphi(x)$, peut être remplacée par toute autre fonction donnée $f(x)$. Dans ce qui va suivre nous comparerons la difference $f(x)-\varphi(x)$ avec les expressions

$$\frac{x}{\log x}, \quad \frac{x}{\log^2 x}, \quad \frac{x}{\log^3 x}, \ldots$$

et, pour abréger le discours, nous dirons que A est *une quantité de l'ordre* $\frac{x}{\log^n x}$, quand le rapport de A à $\frac{x}{\log^m x}$ pour $x=\infty$, sera infini pour $m>n$ et zéro pour $m<n$. Cela posé, nous allons démontrer le théorême suivant:

IV-éme Théorème.

Quand l'expression

$$\frac{\log^n x}{x}\left(f(x)-\int_2^x \frac{dx}{\log x}\right),$$

pour $x=\infty$, *a pour limite une quantité finie ou infinie, la fonction* $f(x)$ *ne peut représenter* $\varphi(x)$ *exactement en quantités de l'ordre* $\frac{x}{\log^n x}$ *inclusivement.*

Démonstration. Soit L la limite vers laquelle converge l'expression

$$\frac{\log^n x}{x}\left(f(x)-\int_2^x \frac{dx}{\log x}\right)$$

à mesure que x s'approche de l'infini. Comme L, par hypothèse, est différente de zéro, elle ne pourra être égale qu'à une quantité positive ou négative. Supposons la positive; notre raisonnement s'appliquera sans difficulté au cas de $L<0$.

Si la limite L de l'expression que nous considérons, pour $x=\infty$, est supérieure à zéro, nous pourrons trouver un nombre N assez grand et tel que, pour $x>N$, la valeur de l'expression

$$\frac{\log^n x}{x}\left(f(x)-\int_2^x \frac{dx}{\log x}\right)$$

reste constamment supérieure à une certaine quantité positive l.

Nous aurons donc pour $x>N$

$$(5)\qquad \frac{\log^n x}{x}\left(f(x)-\int_2^x \frac{dx}{\log x}\right)>l.$$

Mais, en vertu du théorême II, quelque petit que soit $\alpha = \frac{l}{2}$, nous aurons pour un nombre infini de valeurs de x l'inégalité

$$(6) \qquad \varphi(x) < \int_2^x \frac{dx}{\log x} + \frac{\alpha x}{\log^n x},$$

qui donne

$$f(x) - \int_2^x \frac{dx}{\log x} < f(x) - \varphi(x) + \frac{\alpha x}{\log^n x};$$

en la multipliant par $\frac{\log^n x}{x}$, et observant que $\alpha = \frac{l}{2}$, on trouve

$$\frac{\log^n x}{x}\left[f(x) - \int_2^x \frac{dx}{\log x}\right] < \frac{\log^n x}{x}[f(x) - \varphi(x)] + \frac{l}{2}$$

ou bien, en vertu de l'négalité (5),

$$\frac{\log^n x}{x}[f(x) - \varphi(x)] > \frac{l}{2}.$$

Or, cette inégalité ayant lieu en même temps que celles marquées par les numéros (5) et (6) pour une infinité de valeurs de x prouve, à cause de $\frac{l}{2} > 0$, que la limite de

$$\frac{\log^n x}{x}[f(x) - \varphi(x)],$$

pour $x = \infty$, ne peut pas être égale à zéro. Si donc cette limite est différente de zéro, la différence $f(x) - \varphi(x)$, d'après la convention établie plus haut, est une quantité de l'ordre $\frac{x}{\log^n x}$ ou d'un ordre inférieur; par conséquent $f(x)$ diffère de $\varphi(x)$ d'une quantité de l'ordre $\frac{x}{\log^n x}$, ou bien d'un ordre inférieur, ce qu'il s'agissait de démontrer.

En nous basant sur ce théorême, nous pouvons faire voir que la formule de Legendre $\frac{x}{\log x - 1{,}08366}$, pour laquelle la limite de l'expression

$$\frac{\log^2 x}{x}\left(\frac{x}{\log x - 1{,}08366} - \int_2^x \frac{dx}{\log x}\right),$$

quand $x = \infty$, est égale à 0,08366, ne peut exprimer $\varphi(x)$ avec un degré de précision allant jusqu'aux quantités de l'ordre $\frac{x}{\log^2 x}$ inclusivement.

On trouve avec la même facilité les valeurs des constantes A et B telles

— 42 —

que la fonction $\frac{x}{A \log x + B}$ puisse représenter $\varphi(x)$ avec une précision poussée aux quantités de l'ordre $\frac{x}{\log^2 x}$ inclusivement. En vertu du théorême précédent de telles valeurs de A et B doivent satisfaire à l'équation

$$\lim. \left[\frac{\log^2 x}{x}\left(\frac{x}{A \log x + B} - \int_2^x \frac{dx}{\log x}\right)\right]_{x=\infty} = 0.$$

Le dévelloppement de $\frac{x}{A \log x + B}$ donne

$$\frac{x}{A \log x + B} = \frac{1}{A} \cdot \frac{x}{\log x} - \frac{B}{A^2} \cdot \frac{x}{\log^2 x} + \frac{B^2}{A^3} \cdot \frac{x}{\log^3 x} - \ldots$$

De plus, intégrant $\int_2^x \frac{dx}{\log x}$ par parties, on trouve

$$\int_2^x \frac{dx}{\log x} = \frac{x}{\log x} + \frac{x}{\log^2 x} + 2\int_2^x \frac{dx}{\log^3 x} + C.$$

En vertu de ce qui vient d'être trouvé l'équation précédente se réduit à

$$\lim. \left\{\frac{\log^2 x}{x}\left(\begin{array}{l}\frac{1}{A} \cdot \frac{x}{\log x} - \frac{B}{A^2} \cdot \frac{x}{\log^2 x} + \frac{B^2}{A^3} \cdot \frac{x}{\log^3 x} - \ldots \\ \ldots - \frac{x}{\log x} - \frac{x}{\log^2 x} - 2\int_2^x \frac{dx}{\log^3 x} - C\end{array}\right)\right\}_{x=\infty} = 0,$$

ou bien

$$\lim. \left\{\begin{array}{l}\left(\frac{1}{A} - 1\right) \log x - \left(\frac{B}{A^2} + 1\right) + \frac{B^2}{A^3} \frac{1}{\log x} - \ldots \\ \ldots - 2\,\frac{\log^2 x}{x}\int_2^x \frac{dx}{\log^3 x} - C\,\frac{\log^2 x}{x}\end{array}\right\}_{x=\infty} = 0.$$

Or, si l'on observe que tous les termes à partir du troisième convergent vers zéro pour des valeurs croissantes de x, on verra immédiatement qu'on ne peut satisfaire à l'équation précédente qu'en faisant $\frac{1}{A} - 1 = 0$, $\frac{B}{A^2} + 1 = 0$. D'où $A = 1$, $B = -1$.

Ainsi, de toutes les fonctions de la forme $\frac{x}{A \log x + B}$ la seule $\frac{x}{\log x - 1}$ peut exprimer $\varphi(x)$ avec une précision poussée aux quantités de l'ordre $\frac{x}{\log^2 x}$ inclusivement.

§ 6. Démontrons actuellement un théorème concernant le choix de la fonction qui détermine, avec un degré de précision requis, la fonction que nous avons représentée par $\varphi(x)$.

V-ème Théorème.

Si la fonction $\varphi(x)$ qui désigne combien il y a de nombres premiers inférieurs à x, peut être représentée algébriquement avec une précision poussée aux quantités de l'ordre $\frac{x}{\log^n x}$ inclusivement au moyen des fonctions x, $\log x$, e^x, alors elle s'exprimera par la formule

$$\frac{x}{\log x} + \frac{1.x}{\log^2 x} + \frac{1.2.x}{\log^3 x} + \ldots + \frac{1.2.3\ldots(n-1)x}{\log^n x}$$

Démonstration. Soit $f(x)$ la fonction qui, contenant sous forme algébrique x, $\log x$, e^x, exprime $\varphi(x)$ exactement jusqu'aux quantités de l'ordre $\frac{x}{\log^n x}$ inclusivement; l'expression

$$\frac{\log^n x}{x}\left[f(x) - \frac{x}{\log x} - \frac{1.x}{\log^2 x} - \frac{1.2.x}{\log^3 x} - \ldots - \frac{1.2.3\ldots(n-1)x}{\log^n x}\right]$$

pour des valeurs croissantes de x, devra converger soit vers zéro, soit vers une limite finie ou infiniment grande. En effet, s'il n'en était pas ainsi, la première dérivée de cette expression changerait de signe une infinité de fois pour des valeurs de x croissantes jusqu'à $+\infty$, ce qui ne peut arriver, comme il est facile de s'en assurer, avec une fonction algébrique de x, $\log x$, e^x *).

Ainsi, on aura nécessairement pour $f(x)$

$$(7)\quad \lim.\left\{\frac{\log^n x}{x}\left(f(x) - \frac{x}{\log x} - \frac{1.x}{\log^2 x} - \frac{1.2x}{\log^3 x} - \ldots - \frac{1.2.3\ldots(n-1)x}{\log^n x}\right)\right\}_{x=\infty} = L.$$

---

*) Il est très facile de s'assurer qu'une fonction algébrique de x, $\log x$, e^x cesse de changer de signe pour une valeur de x surpassant une certaine limite. Si la fonction dont il s'agit est entière, alors son signe dépendra d'un terme de la forme $Kx^{m'}.\log^{m''} x.e^{m'''x}$, pour des valeurs de x plus ou moins considérables, ce terme ne changeant pas de signe pour $x > 1$. Pour toute autre fonction algébrique de x, $\log x$, e^x, que nous représenterons par y, on démontrera la même proposition de la manière suivante: observons d'abord que la fonction y sera la racine de l'équation $u_0 y^m + u_1 y^{m-1} + u_2 y^{m-2} + \ldots + u_{m-1} y + u_m = 0$, $u_0, u_1, u_2, \ldots u_{m-1}, u_m$ étant des fonctions entières de x, $\log x$, e^x; si l'on représente par v la fonction qui résulte de l'élimination de y entre l'équation précédente et sa dérivée, alors les fonctions u_0, u_m et v, comme entières, finiront par ne plus se réduire à zéro ou à changer de signe pour des valeurs de x surpassant une certaine limite; il arrivera donc que y conservera également son signe, car, pour des valeurs de x qui ne réduisent pas v à zéro, l'équation $u_0 y^m + u_1 y^{m-1} + \ldots + u_{m-1} y + u_m = 0$ ne peut avoir de racines égales, et quand les racines sont inégales, le signe de l'une d'elles ne peut changer qu'en supposant que le signe de u_0 ou u_m change. — Cette propriété peut être étendue à beaucoup d'autres fonctions, pour lesquelles, par cette raison, le théorème V ainsi que les conséquences qui s'en déduisent, auront également lieu.

— 44 —

Mais, d'un autre côté, il est facile de s'assurer que

$$\lim.\left[\frac{\log^n x}{x}\left(\frac{x}{\log x}+\frac{1.x}{\log^2 x}+\frac{1.2.x}{\log^3 x}+\ldots+\frac{1.2..(n-1)x}{\log^n x}-\int_2^x\frac{dx}{\log x}\right)\right]_{x=\infty}=0;$$

cette équation ajoutée à la précédente donne

$$\lim.\left[\frac{\log^n x}{x}\left(f(x)-\int_2^x\frac{dx}{\log x}\right)\right]_{x=\infty}=L.$$

Or, comme par hypothèse $f(x)$ représente $\varphi(x)$ exactement jusqu'aux quantités de l'ordre $\frac{x}{\log^n x}$ inclusivement, et que d'après le théorême IV cela ne peut avoir lieu, si la limite de

$$\frac{\log^n x}{x}\left[f(x)-\int_2^x\frac{dx}{\log x}\right],$$

pour $x=\infty$, n'est pas zéro, on aura $L=0$; cela posé, l'équation (7), pour $L=0$, se réduit à

$$\lim.\left\{\frac{\log^n x}{x}\left[f(x)-\frac{x}{\log x}-\frac{1.x}{\log^2 x}-\frac{1.2.x}{\log^3 x}-\ldots-\frac{1.2.3...(n-1)x}{\log^n x}\right]\right\}_{x=\infty}=0,$$

ce qui prouve que la fonction

$$\frac{x}{\log x}+\frac{1.x}{\log^2 x}+\frac{1.2x}{\log^3 x}+\ldots+\frac{1.2.3...(n-1)x}{\log^n x}$$

ne différe pas de $f(x)$ de quantités de l'ordre $\frac{x}{\log^n x}$ et d'ordres inférieurs, et que par conséquent elle peut, aussi bien que $f(x)$, représenter $\varphi(x)$ avec une précision poussée jusqu'aux quantités de l'ordre $\frac{x}{\log^n x}$ inclusivement, ce qu'il s'agissait de démontrer.

D'après le théoréme que nous venons d'établir, nous concluons que si la fonction $\varphi(x)$, qui représente combien il y a de nombres premiers inférieurs à x, peut être exprimée algébriquement au moyen de x, $\log x$, e^x jusqu'aux quantités des ordres $\frac{x}{\log x}, \frac{x}{\log^2 x}, \frac{x}{\log^3 x}, \ldots$ inclusivement, elle devra s'exprimer par

$$\frac{x}{\log x},\ \frac{x}{\log x}+\frac{1.x}{\log^2 x},\ \frac{x}{\log x}+\frac{1.x}{\log^2 x}+\frac{1.2.x}{\log^3 x},\ldots$$

De plus, comme ces sommes ne sont autre chose que les valeurs successives de l'integrale $\int_2^x\frac{dx}{\log x}$, poussées aux quantités des ordres $\frac{x}{\log x}, \frac{x}{\log^2 x}, \frac{x}{\log^3 x}, \ldots$

nous sommes en droit de conclure que dans toutes ces hypothèses l'intégrale $\int_2^x \frac{dx}{\log x}$ exprimera $\varphi(x)$ avec exactitude jusqu'aux quantités de l'ordre pour lequel elle peut encore s'exprimer algébriquement au moyen de x, $\log x$, e^x.

Il est facile de se convaincre par les tables des nombres premiers que l'intégrale $\int_2^x \frac{dx}{\log x}$, pour x très grand, exprime avec assez de précision combien il y a de nombres premiers inférieurs à x. Mais ces tables sont trop peu étendues pour pouvoir décider de la supériorité de la formule $\int_2^x \frac{dx}{\log x}$ sur la formule de Legendre $\frac{x}{\log x - 1{,}08366}$ ou sur toute autre analogue. Dans les limites de ces tables les deux fonctions $\int_2^x \frac{dx}{\log x}$ et $\frac{x}{\log x - 1{,}08366}$ diffèrent peu entr'elles; mais leur différence $\frac{x}{\log x - 1{,}08366} - \int_2^x \frac{dx}{\log x}$, ayant un *minimum* pour $x = e^{\frac{(1{,}08366)^2}{0{,}08366}} = 1247689$, croitra constamment jusqu'à l'infini après cette valeur, et déjà pour $x > 10000000$, aura une valeur considérable. C'est pour des nombres de cette grandeur que l'avantage de l'une des deux formules $\int_2^x \frac{dx}{\log x}$, $\frac{x}{\log x - 1{,}08366}$ devra se manifester. Mais pour effectuer cette vérification il faudrait avoir des tables de nombres premiers beaucoup plus étendues que celle que l'on possède.

§ 7. En adoptant l'intégrale $\int_2^x \frac{dx}{\log x}$ pour la valeur approchée de $\varphi(x)$ nous serons obligés de changer toutes les formules auxquelles Legendre est parvenu en partant de l'hypothsée $\varphi(x) = \frac{x}{\log x - 1{,}08366}$; nos formules ne seront pas plus compliquées que les siennes, et auront sur elles l'avantage d'être plus approchées d'après les théorèmes qui ont été demontrés plus haut.

Appliquons notre formule à la détermination de la somme des deux séries

$$\frac{1}{2} + \frac{1}{3} + \frac{1}{5} + \ldots + \frac{1}{X},$$

$$\left(1 - \frac{1}{2}\right)\left(1 - \frac{1}{3}\right)\left(1 - \frac{1}{5}\right)\ldots\ldots\left(1 - \frac{1}{X}\right)$$

pour X très grand.

Pour déterminer la somme de la première de ces deux séries, nous observerons que, d'après la notation admise plus haut, l'on a

$$\frac{1}{2}+\frac{1}{3}+\frac{1}{5}+\ldots+\frac{1}{X}=\sum_{x=2}^{x=X}\frac{\varphi(x+1)-\varphi(x)}{x},$$

car, $\varphi(x)$ représentant la totalité des nombres premiers inférieurs à x, la différence $\varphi(x+1)-\varphi(x)$ se réduira à zéro toutes les fois que x sera un nombre composé, et à 1, quand x sera premier.

Supposons X très grand, et désignons par λ un nombre inférieur à X, mais assez grand cependant pour que la fonction $\varphi(x)$, entre les limites $x=\lambda$ et $x=X$, puisse être représentée avec une exactitude suffisante par l'intégrale $\int_2^x \frac{dx}{\log x}$. Dans cette hypothèse l'équation précédente pourra s'écrire de cette manière:

$$\frac{1}{2}+\frac{1}{3}+\frac{1}{5}+\ldots+\frac{1}{X}=\sum_{x=2}^{x=\lambda-1}\frac{\varphi(x+1)-\varphi(x)}{x}+\sum_{x=\lambda}^{x=X}\frac{\varphi(x+1)-\varphi(x)}{x}.$$

Remplaçant dans la dernière somme $\varphi(x)$ par $\int_2^x \frac{dx}{\log x}$, on aura

$$\frac{1}{2}+\frac{1}{3}+\frac{1}{5}+\ldots+\frac{1}{X}=\sum_{x=2}^{x=\lambda-1}\frac{\varphi(x+1)-\varphi(x)}{x}+\sum_{x=\lambda}^{x=X}\frac{\int_x^{x+1}\frac{dx}{\log x}}{x}.$$

Or, l'intégrale $\int_x^{x+1}\frac{dx}{\log x}$ peut être représentée par $\frac{1}{\log x}$ avec exactitude jusqu'aux quantités de l'ordre $\frac{1}{x}$; de plus, la somme $\sum_{x=\lambda}^{x=X}\frac{1}{x\log x}$ peut être remplacée par l'intégrale $\int_\lambda^X \frac{dx}{x\log x}$ avec le même degré de précision.

Sous ces conditions l'équation précédente deviendra

$$\frac{1}{2}+\frac{1}{3}+\frac{1}{5}+\ldots+\frac{1}{X}=\sum_{x=2}^{x=\lambda-1}\frac{\varphi(x+1)-\varphi(x)}{x}+\int_\lambda^x \frac{dx}{x\log x},$$

ou bien, effectuant l'intégration,

$$\frac{1}{2}+\frac{1}{3}+\frac{1}{5}+\ldots+\frac{1}{X}=\sum_{x=2}^{x=\lambda-1}\frac{\varphi(x+1)-\varphi(x)}{x}-\log\log\lambda+\log\log X$$

Enfin, si l'on remplace par C la quantité

$$\sum_{x=2}^{x=\lambda-1} \frac{\varphi(x+1)-\varphi(x)}{x} - \log\log\lambda,$$

indépendante de x, on trouvera

$$\frac{1}{2}+\frac{1}{3}+\frac{1}{5}+\ldots+\frac{1}{X}=C+\log\log X. \tag{8}$$

Lorsque l'on aura déterminé la valeur de la constante C, cette équation pourra servir à trouver par approximation la somme de la série

$$\frac{1}{2}+\frac{1}{3}+\frac{1}{5}+\ldots+\frac{1}{X}$$

quand X sera très grand.

La formule que nous venons de trouver est plus simple que celle de Legendre, d'aprés laquelle on a

$$\frac{1}{2}+\frac{1}{3}+\frac{1}{5}+\ldots+\frac{1}{X}=\log(\log X-0{,}08366)+C.$$

Passons maintenant à la détermination du produit

$$\left(1-\frac{1}{2}\right)\left(1-\frac{1}{3}\right)\left(1-\frac{1}{5}\right)\ldots\left(1-\frac{1}{X}\right)=P.$$

Prenant le logarithme des deux membres de cette équation, on aura la formule

$$\log P=\log\left(1-\frac{1}{2}\right)+\log\left(1-\frac{1}{3}\right)+\log\left(1-\frac{1}{5}\right)+\ldots+\log\left(1-\frac{1}{X}\right)$$

qui peut encore s'écrire de la manière suivante:

$$\log P=-\left(\frac{1}{2}+\frac{1}{3}+\frac{1}{5}+\ldots+\frac{1}{X}\right)+\frac{1}{2}+\log\left(1-\frac{1}{2}\right)+$$
$$\frac{1}{3}+\log\left(1-\frac{1}{3}\right)+\frac{1}{5}+\log\left(1-\frac{1}{5}\right)+\ldots+\frac{1}{X}+\log\left(1-\frac{1}{X}\right).$$

Observons actuellement que la série finie

$$\frac{1}{2}+\log\left(1-\frac{1}{2}\right)+\frac{1}{3}+\log\left(1-\frac{1}{3}\right)+\frac{1}{5}+\log\left(1-\frac{1}{5}\right)+\ldots+\frac{1}{X}+\log\left(1-\frac{1}{X}\right),$$

aux quantités de l'ordre $\frac{1}{X}$ près, peut être remplacée par la série infinie

$$\frac{1}{2}+\log\left(1-\frac{1}{2}\right)+\frac{1}{3}+\log\left(1-\frac{1}{3}\right)+\frac{1}{5}+\log\left(1-\frac{1}{5}\right)+\ldots$$

Or, la différence entre ces deux séries est évidemment inférieure à la somme

$$-\frac{1}{X+1}-\log\left(1-\frac{1}{X+1}\right)-\frac{1}{X+2}-\log\left(1-\frac{1}{X+2}\right)-\ldots$$

— 48 —

qui, elle même, est inférieure à l'intégrale

$$\int_X^\infty \left[-\frac{1}{x} - \log\left(1 - \frac{1}{x}\right)\right] dx = 1 + (X - 1)\log\left(1 - \frac{1}{X}\right);$$

de plus, comme la valeur de $1 + (X - 1)\log\left(1 - \frac{1}{X}\right)$ pour X très grand, est une quantité infiniment petite du premier ordre par rapport à $\frac{1}{X}$, nous en concluons que la substitution qui vient d'être indiquée est permise.

D'après ce qui vient d'être dit, si l'on représente par C' la somme de la série infinie

$$\frac{1}{2} + \log\left(1 - \frac{1}{2}\right) + \frac{1}{3} + \log\left(1 - \frac{1}{3}\right) + \frac{1}{5} + \log\left(1 - \frac{1}{5}\right) + \ldots$$

la valeur de $\log P$ s'exprimera, aux quantités de l'ordre de $\frac{1}{X}$ près, par la formule

$$\log P = -\left(\frac{1}{2} + \frac{1}{3} + \frac{1}{5} + \ldots + \frac{1}{X}\right) + C'.$$

Substituant pour

$$\frac{1}{2} + \frac{1}{3} + \frac{1}{5} + \ldots + \frac{1}{X}$$

la valeur (8) trouvée plus haut, on obtiendra

$$\log P = -C - \log\log X + C',$$

d'où

$$P = \frac{e^{C'-C}}{\log X}.$$

Enfin, faisant pour abréger $e^{C'-C} = C_0$, et remplaçant P par le produit

$$\left(1 - \frac{1}{2}\right)\left(1 - \frac{1}{3}\right)\left(1 - \frac{1}{5}\right)\ldots\left(1 - \frac{1}{X}\right),$$

nous aurons

$$\left(1 - \frac{1}{2}\right)\left(1 - \frac{1}{3}\right)\left(1 - \frac{1}{5}\right)\ldots\left(1 - \frac{1}{X}\right) = \frac{C_0}{\log X}.$$

Legendre a trouvé, pour la valeur du même produit, la formule suivante:

$$\left(1 - \frac{1}{2}\right)\left(1 - \frac{1}{3}\right)\left(1 - \frac{1}{5}\right)\ldots\left(1 - \frac{1}{X}\right) = \frac{C_0}{\log X - 0{,}08366}.$$

12.2 B. Riemann (1859)

Ueber die Anzahl der Primzahlen unter einer gegebenen Grösse

Bernhard Riemann was born in 1826 in what is now Germany. The son of a Lutheran minister, Riemann excelled in Hebrew and theology, but had a keen interest in mathematics. He entered the University of Göttingen in 1846 as a student of theology. However, with the permission of his father, he began to study mathematics under Stern and Gauss. In 1847, Riemann moved to Berlin to study under Steiner, Jacobi, Dirichlet, and Eisenstein. In 1851, he submitted his celebrated doctoral thesis on Riemann surfaces to Gauss. In turn, Gauss recommended Riemann for a post at Göttingen. The name Riemann is ubiquitous in modern mathematics. Riemann's work is full of brilliant insight, and mathematicians struggle to this day with the questions he raised. Riemann died in Italy in 1866 [114].

In 1859 Riemann was appointed to the Berlin Academy of Sciences. New members of the academy were required to report on their research, and in complying with this convention Riemann presented his report *Ueber die Anzahl der Primzahlen unter einer gegebenen Grösse* (*On the number of primes less than a given magnitude*). In this paper, Riemann considers the properties of the Riemann zeta function. His results are stated concisely, and they motivated future researchers to supply rigourous proofs. This paper contains the original statement of the Riemann hypothesis.

We have included a digitization of Riemann's original hand-written memoir. It is followed by a translation into English by David R. Wilkins.

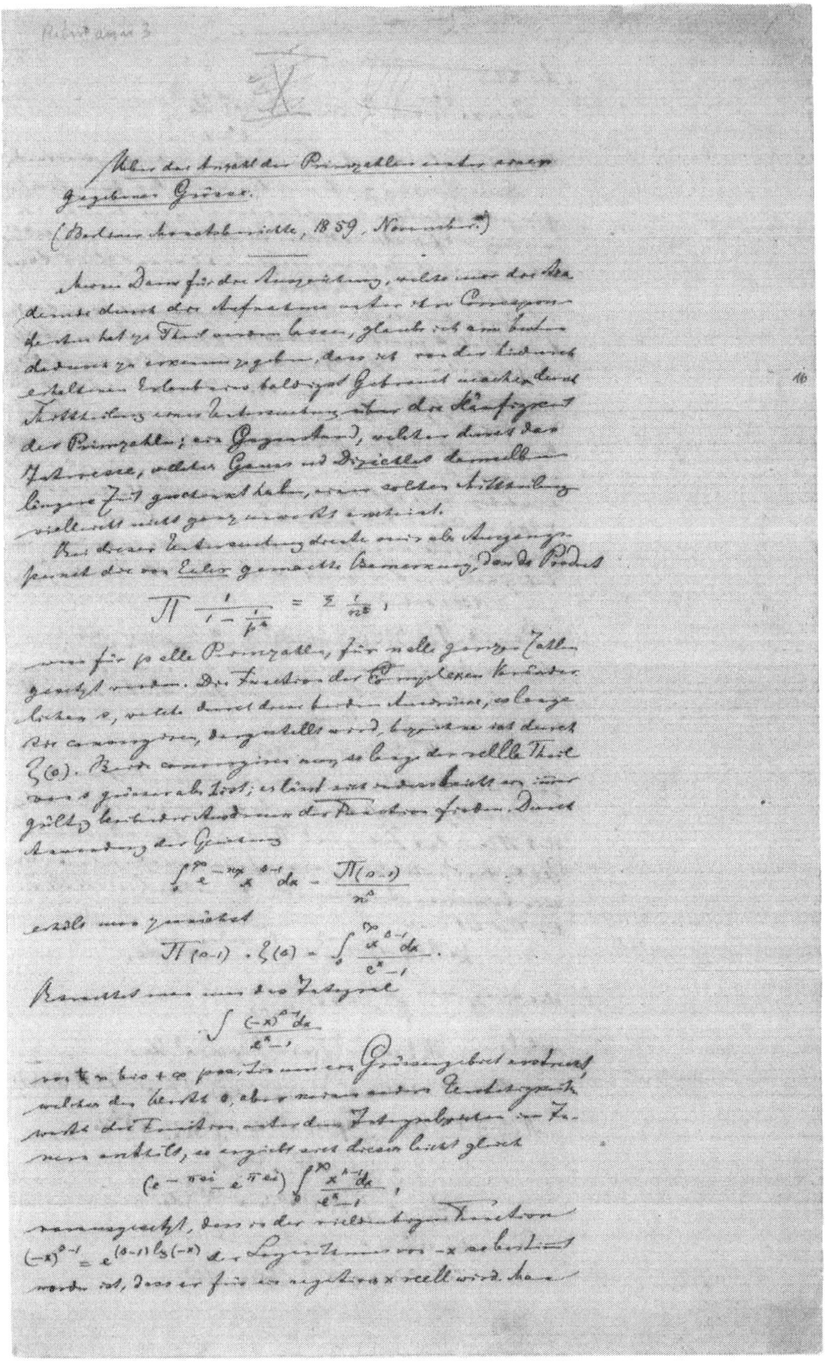

Ueber die Anzahl der Primzahlen unter einer gegebenen Grösse.

(Berliner Monatsberichte, 1859, November.)

Meinen Dank für die Auszeichnung, welche mir die Akademie durch die Aufnahme unter ihre Correspondenten hat zu Theil werden lassen, glaube ich am besten dadurch zu erkennen zu geben, dass ich von der hierdurch erhaltenen Erlaubniss baldigst Gebrauch mache durch Mittheilung einer Untersuchung über die Häufigkeit der Primzahlen; ein Gegenstand, welcher durch das Interesse, welches Gauss und Dirichlet demselben längere Zeit geschenkt haben, einer solchen Mittheilung vielleicht nicht ganz unwerth erscheint.

Bei dieser Untersuchung diente mir als Ausgangspunkt die von Euler gemachte Bemerkung, dass das Product

$$\prod \frac{1}{1-\frac{1}{p^s}} = \sum \frac{1}{n^s},$$

wenn für p alle Primzahlen, für n alle ganzen Zahlen gesetzt werden. Die Function der complexen Veränderlichen s, welche durch diese beiden Ausdrücke, so lange sie convergiren, dargestellt wird, bezeichne ich durch $\zeta(s)$. Beide convergiren nur, so lange der reelle Theil von s grösser als 1 ist; es lässt sich indess leicht ein immer gültig bleibender Ausdruck der Function finden. Durch Anwendung der Gleichung

$$\int_0^\infty e^{-nx} x^{s-1}\,dx = \frac{\Pi(s-1)}{n^s}$$

erhält man zunächst

$$\Pi(s-1)\,\zeta(s) = \int_0^\infty \frac{x^{s-1}\,dx}{e^x-1}.$$

Betrachtet man nun das Integral

$$\int \frac{(-x)^{s-1}\,dx}{e^x-1}$$

von $+\infty$ bis $+\infty$ positiv um ein Grössengebiet erstreckt, welches den Werth 0, aber keinen andern Unstetigkeitswerth der Function unter dem Integralzeichen im Innern enthält, so ergiebt sich dieses leicht gleich

$$(e^{-\pi si} - e^{\pi si})\int_0^\infty \frac{x^{s-1}\,dx}{e^x-1},$$

vorausgesetzt, dass in der vielwerthigen Function $(-x)^{s-1} = e^{(s-1)\log(-x)}$ der Logarithmus von $-x$ so bestimmt worden ist, dass er für ein negatives x reell wird. Man

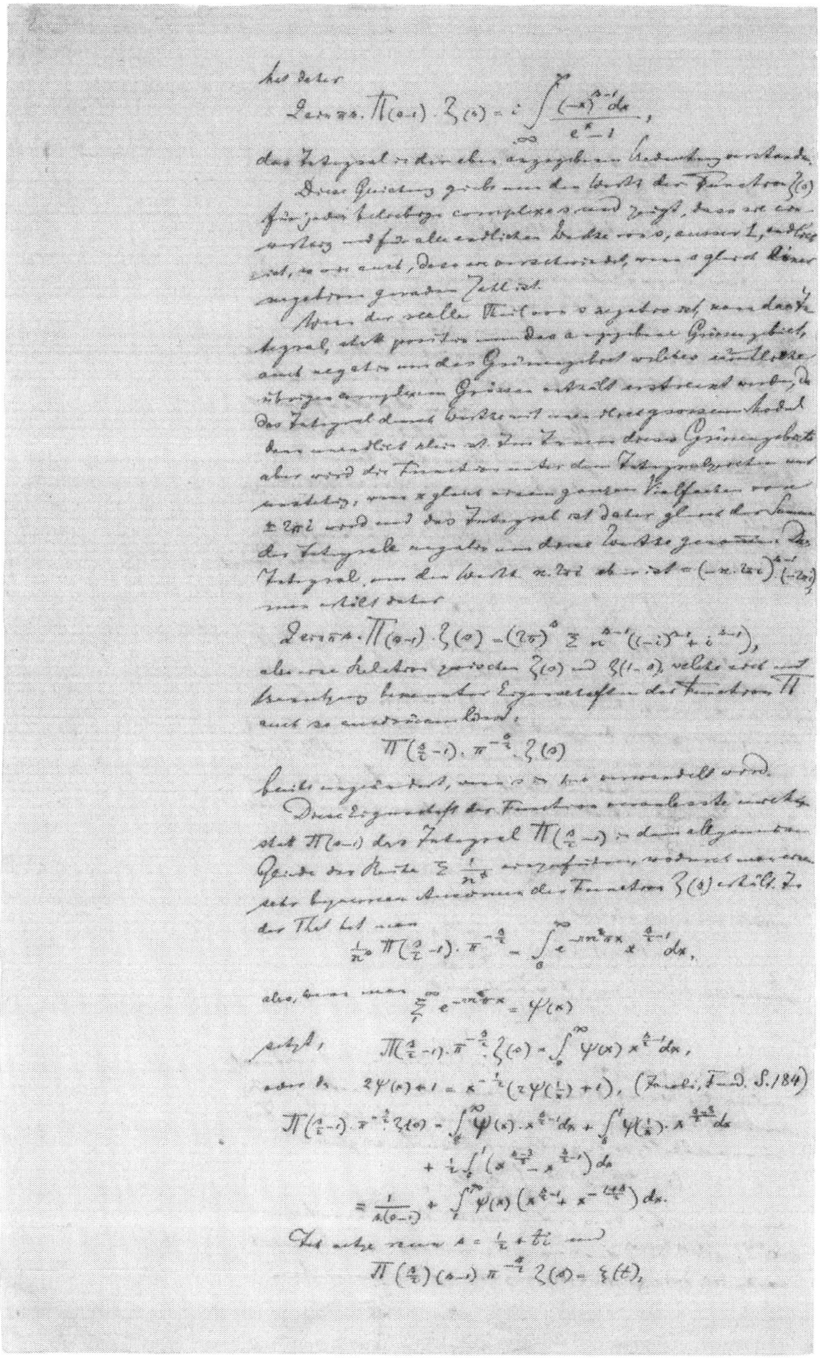

hat daher

$$2\sin \pi s\, \Pi(s-1)\, \zeta(s) = i \int_{\infty}^{\infty} \frac{(-x)^{s-1}\,dx}{e^x - 1},$$

das Integral in der eben angegebenen Bedeutung verstanden.

Diese Gleichung giebt nun den Werth der Function $\zeta(s)$ für jeden beliebigen complexen Werth von s und zeigt, dass sie eindeutig und für alle endlichen Werthe von s, ausser 1, endlich ist, so wie auch, dass sie verschwindet, wenn s gleich einer negativen geraden Zahl ist.

Wenn der reelle Theil von s negativ ist, kann das Integral, statt positiv um das angegebene Grössengebiet, auch negativ um das Grössengebiet welches sämmtliche übrigen complexen Grössen enthält erstreckt werden, da das Integral durch Werthe mit unendlich grossem Modul dann unendlich klein ist. Im Innern dieses Grössengebiets aber wird die Function unter dem Integralzeichen nur unstetig, wenn x gleich einem ganzen Vielfachen von $\pm 2\pi i$ wird und das Integral ist daher gleich der Summe der Integrale negativ um diese Werthe genommen. Das Integral um den Werth $n 2\pi i$ aber ist $= (-n 2\pi i)^{s-1}(-2\pi i)$, man erhält daher

$$2\sin \pi s\, \Pi(s-1)\, \zeta(s) = (2\pi)^s \sum n^{s-1}\left((-i)^{s-1} + i^{s-1}\right),$$

also eine Relation zwischen $\zeta(s)$ und $\zeta(1-s)$, welche sich mit Benutzung bekannter Eigenschaften der Function Π auch so ausdrücken lässt:

$$\Pi\left(\frac{s}{2}-1\right) \pi^{-\frac{s}{2}} \zeta(s)$$

bleibt ungeändert, wenn s in $1-s$ verwandelt wird.

Diese Eigenschaft der Function veranlasste mich statt $\Pi(s-1)$ das Integral $\Pi\left(\frac{s}{2}-1\right)$ in dem allgemeinen Gliede der Reihe $\sum \frac{1}{n^s}$ einzuführen, wodurch man einen sehr bequemen Ausdruck der Function $\zeta(s)$ erhält. In der That hat man

$$\frac{1}{n^s} \Pi\left(\frac{s}{2}-1\right) \pi^{-\frac{s}{2}} = \int_0^{\infty} e^{-nn\pi x} x^{\frac{s}{2}-1}\,dx,$$

also, wenn man

$$\sum_1^{\infty} e^{-nn\pi x} = \psi(x)$$

setzt,

$$\Pi\left(\frac{s}{2}-1\right) \pi^{-\frac{s}{2}} \zeta(s) = \int_0^{\infty} \psi(x)\, x^{\frac{s}{2}-1}\,dx,$$

oder da $2\psi(x) + 1 = x^{-\frac{1}{2}}\left(2\psi\left(\frac{1}{x}\right) + 1\right)$, (Jacobi, Fund. S. 184)

$$\Pi\left(\frac{s}{2}-1\right) \pi^{-\frac{s}{2}} \zeta(s) = \int_1^{\infty} \psi(x)\, x^{\frac{s}{2}-1}\,dx + \int_0^1 \psi\left(\frac{1}{x}\right) x^{\frac{s-3}{2}}\,dx + \frac{1}{2}\int_0^1 \left(x^{\frac{s-3}{2}} - x^{\frac{s}{2}-1}\right) dx$$

$$= \frac{1}{s(s-1)} + \int_1^{\infty} \psi(x)\left(x^{\frac{s}{2}-1} + x^{-\frac{1+s}{2}}\right) dx.$$

Ich setze nun $s = \frac{1}{2} + ti$ und

$$\Pi\left(\frac{s}{2}\right)(s-1)\, \pi^{-\frac{s}{2}} \zeta(s) = \xi(t),$$

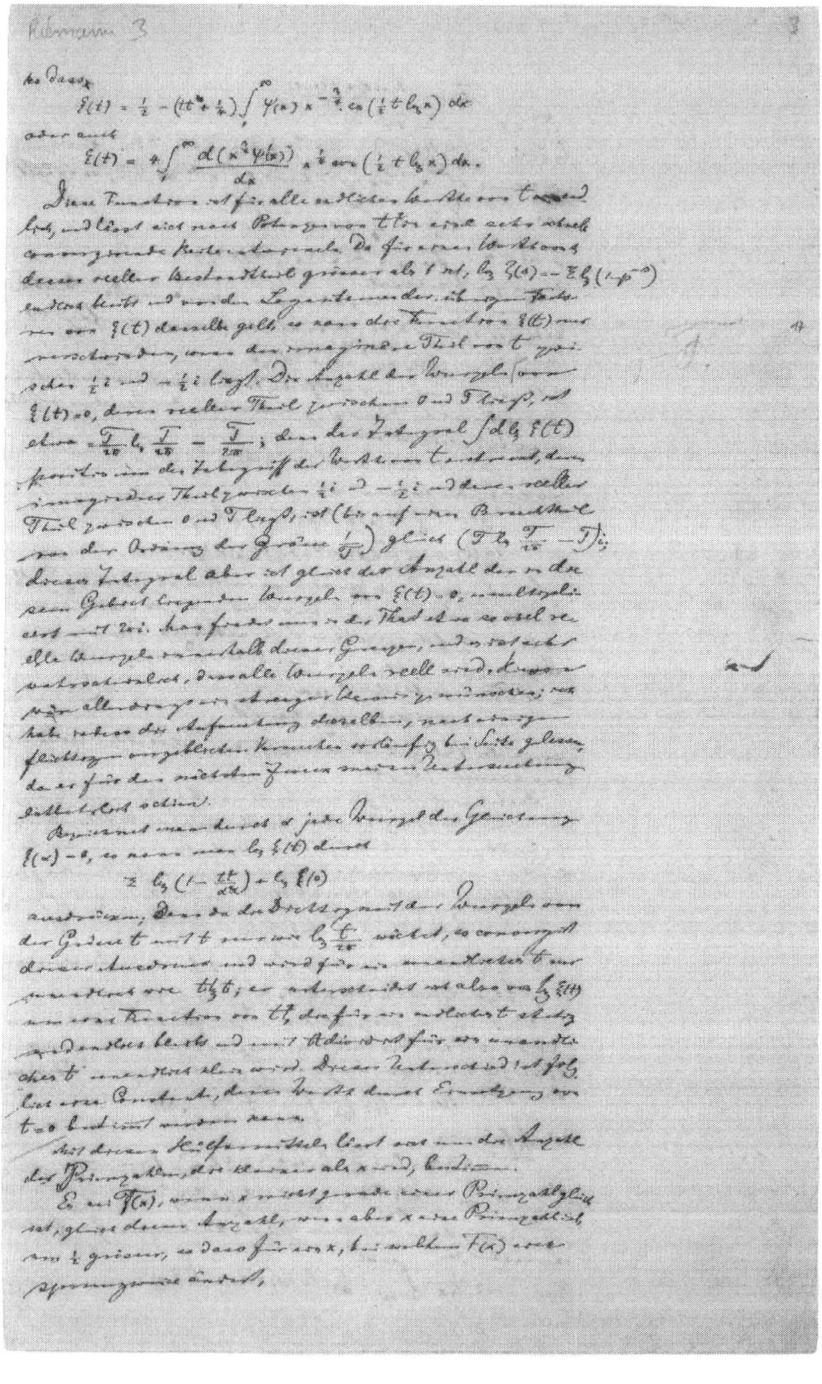

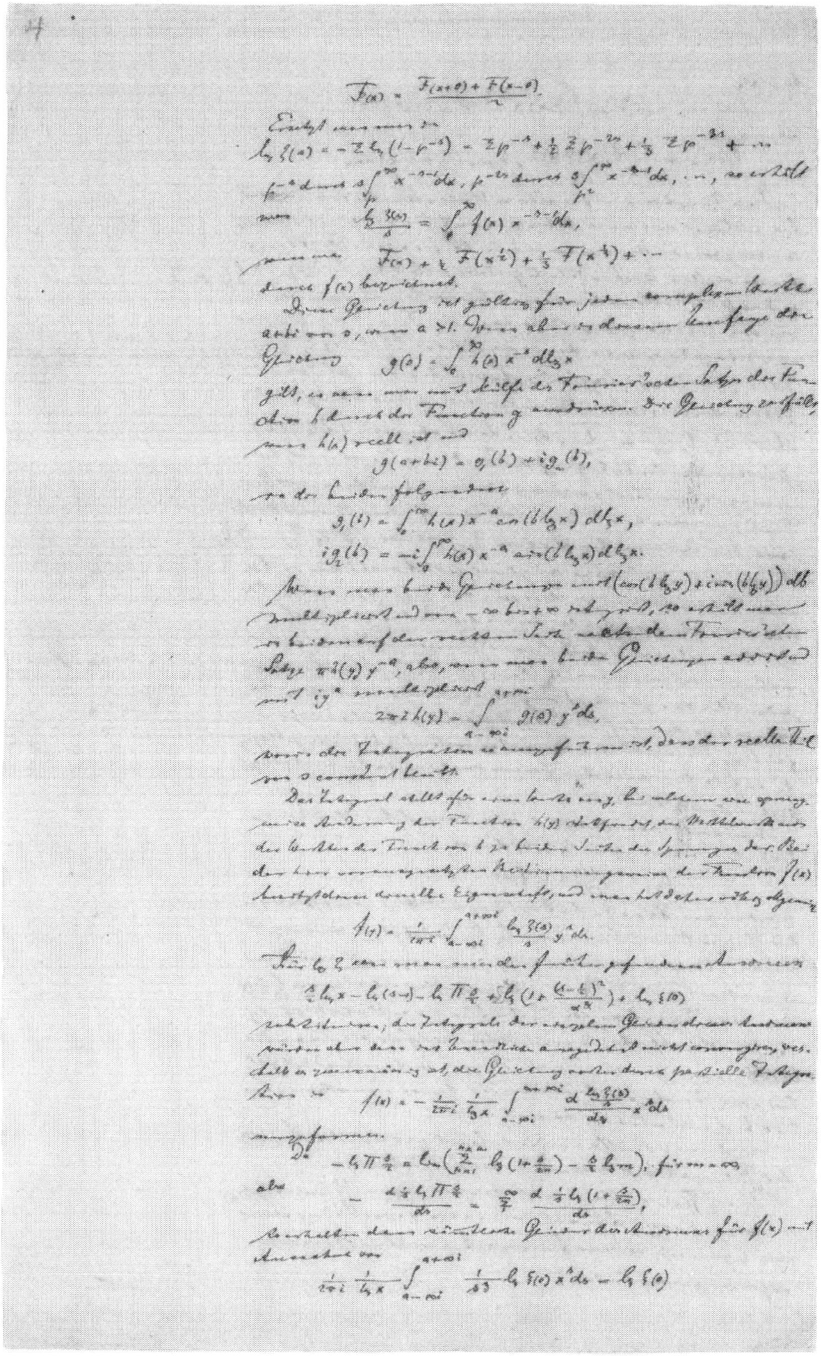
4

$$F(x) = \frac{F(x+0)+F(x-0)}{2}.$$

Ersetzt man nun in

$$\log \zeta(s) = -\Sigma \log(1-p^{-s}) = \Sigma p^{-s} + \tfrac{1}{2}\Sigma p^{-2s} + \tfrac{1}{3}\Sigma p^{-3s} + \cdots$$

p^{-s} durch $s\int_p^\infty x^{-s-1}dx$, p^{-2s} durch $s\int_{p^2}^\infty x^{-s-1}dx$, ..., so erhält man

$$\frac{\log \zeta(s)}{s} = \int_1^\infty f(x)\, x^{-s-1}\, dx,$$

wenn man

$$F(x) + \tfrac{1}{2}F(x^{\frac{1}{2}}) + \tfrac{1}{3}F(x^{\frac{1}{3}}) + \cdots$$

durch $f(x)$ bezeichnet.

Diese Gleichung ist gültig für jeden complexen Werth $a+bi$ von s, wenn $a>1$. Wenn aber in diesem Umfange die Gleichung

$$g(s) = \int_0^\infty h(x)\, x^{-s}\, d\log x$$

gilt, so kann man mit Hülfe des Fourier'schen Satzes die Function h durch die Function g ausdrücken. Die Gleichung zerfällt, wenn $h(x)$ reell ist und

$$g(a+bi) = g_1(b) + i g_2(b),$$

in die beiden folgenden:

$$g_1(b) = \int_0^\infty h(x)\, x^{-a} \cos(b\log x)\, d\log x,$$

$$i g_2(b) = -i\int_0^\infty h(x)\, x^{-a} \sin(b\log x)\, d\log x.$$

Wenn man beide Gleichungen mit $(\cos(b\log y) + i\sin(b\log y))\,db$ multiplicirt und von $-\infty$ bis $+\infty$ integrirt, so erhält man in beiden auf der rechten Seite nach dem Fourier'schen Satze $\pi h(y) y^{-a}$, also, wenn man beide Gleichungen addirt und mit iy^a multiplicirt,

$$2\pi i\, h(y) = \int_{a-\infty i}^{a+\infty i} g(s)\, y^s\, ds,$$

worin die Integration so auszuführen ist, dass der reelle Theil von s constant bleibt.

Das Integral stellt für einen Werth von y, bei welchem eine sprungweise Änderung der Function $h(y)$ stattfindet, den Mittelwerth aus den Werthen der Function h zu beiden Seiten des Sprunges dar. Bei der hier vorausgesetzten Bestimmungsweise der Function $f(x)$ besitzt diese dieselbe Eigenschaft, und man hat daher völlig allgemein

$$f(y) = \frac{1}{2\pi i}\int_{a-\infty i}^{a+\infty i} \frac{\log \zeta(s)}{s}\, y^s\, ds.$$

Für $\log \zeta$ kann man nun den früher gefundenen Ausdruck

$$\frac{s}{2}\log \pi - \log(s-1) - \log \Pi\frac{s}{2} + \Sigma \log\left(1 + \frac{(s-\frac{1}{2})^2}{\alpha\alpha}\right) + \log \xi(0)$$

substituiren; die Integrale der einzelnen Glieder dieses Ausdrucks würden aber dann ins Unendliche ausgedehnt nicht convergiren, weshalb es zweckmässig ist, die Gleichung vorher durch partielle Integration in

$$f(x) = -\frac{1}{2\pi i}\frac{1}{\log x}\int_{a-\infty i}^{a+\infty i} \frac{d\frac{\log \zeta(s)}{s}}{ds}\, x^s\, ds$$

umzuformen.

Da

$$-\log \Pi\frac{s}{2} = \lim\left(\sum_{n=1}^{n=m}\log\left(1+\frac{s}{2n}\right) - \frac{s}{2}\log m\right), \text{ für } m=\infty$$

also

$$-\frac{d\frac{1}{s}\log \Pi\frac{s}{2}}{ds} = \sum_1^\infty \frac{d\,\frac{1}{s}\log\left(1+\frac{s}{2n}\right)}{ds},$$

so erhalten dann sämmtliche Glieder des Ausdrucks für $f(x)$ mit Ausnahme von

$$\frac{1}{2\pi i}\frac{1}{\log x}\int_{a-\infty i}^{a+\infty i} \frac{1}{ss}\log \xi(0)\, x^s\, ds = \log \xi(0)$$

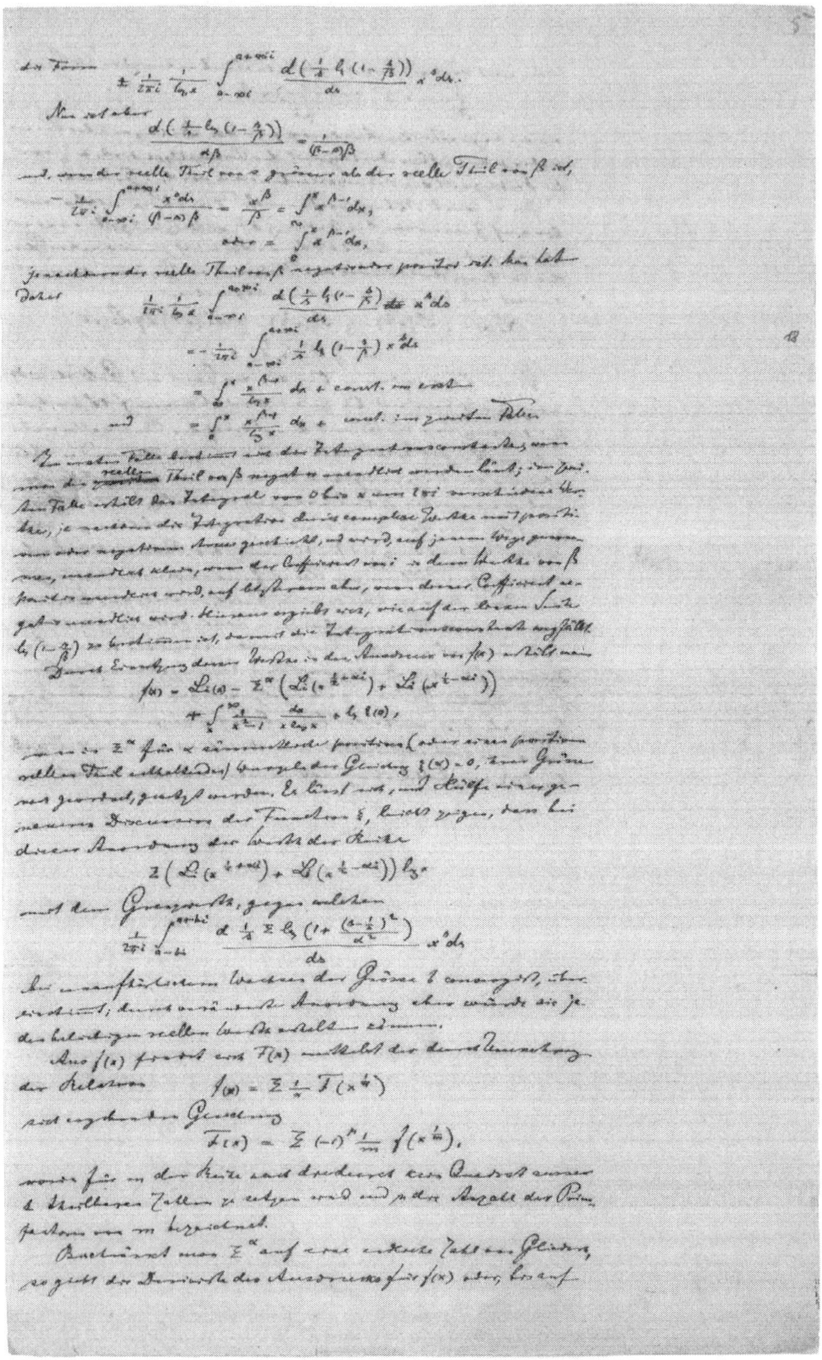

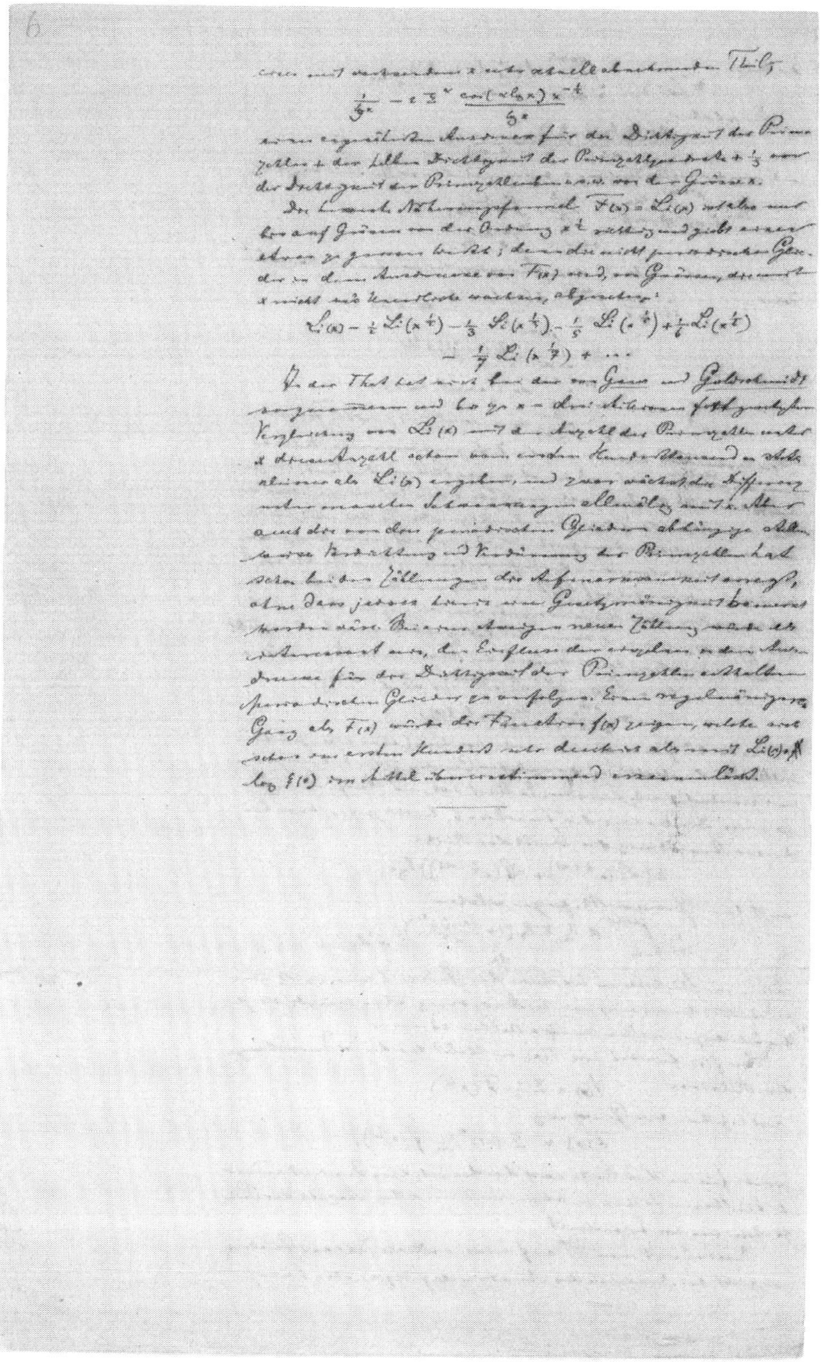

6

eines mit wachsendem x sehr schnell abnehmenden Theils

$$\frac{1}{\log x} - 2\sum^{\alpha} \frac{\cos(\alpha \log x)\, x^{-\frac{1}{2}}}{\log x}$$

einen angenäherten Ausdruck für die Dichtigkeit der Primzahlen + der halben Dichtigkeit der Primzahlquadrate + $\frac{1}{3}$ von der Dichtigkeit der Primzahlcuben u.s.w. von der Grösse x.

Die bekannte Näherungsformel $F(x) = Li(x)$ ist also nur bis auf Grössen von der Ordnung $x^{\frac{1}{2}}$ richtig und giebt einen etwas zu grossen Werth; denn die nicht periodischen Glieder in dem Ausdrucke von $F(x)$ sind, von Grössen, die mit x nicht in's Unendliche wachsen, abgesehen:

$$Li(x) - \tfrac{1}{2} Li(x^{\frac{1}{2}}) - \tfrac{1}{3} Li(x^{\frac{1}{3}}) - \tfrac{1}{5} Li(x^{\frac{1}{5}}) + \tfrac{1}{6} Li(x^{\frac{1}{6}}) - \tfrac{1}{7} Li(x^{\frac{1}{7}}) + \cdots$$

In der That hat sich bei der von Gauss und Goldschmidt vorgenommenen und bis zu x = drei Millionen fortgesetzten Vergleichung von $Li(x)$ mit der Anzahl der Primzahlen unter x diese Anzahl schon im ersten Hunderttausend stets kleiner als $Li(x)$ ergeben, und zwar wächst die Differenz unter manchen Schwankungen allmählich mit x. Aber auch die von den periodischen Gliedern abhängige stellenweise Verdichtung und Verdünnung der Primzahlen hat schon bei den Zählungen die Aufmerksamkeit erregt, ohne dass jedoch hierin eine Gesetzmässigkeit bemerkt worden wäre. Bei einer etwaigen neuen Zählung würde es interessant sein, den Einfluss der einzelnen in dem Ausdrucke für die Dichtigkeit der Primzahlen enthaltenen periodischen Glieder zu verfolgen. Einen regelmässigeren Gang als $F(x)$ würde die Function $f(x)$ zeigen, welche sich schon im ersten Hundert sehr deutlich als mit $Li(x) + \log \xi(0)$ im Mittel übereinstimmend erkennen lässt.

On the Number of Prime Numbers less than a Given Quantity. (Ueber die Anzahl der Primzahlen unter einer gegebenen Grösse.)

Bernhard Riemann
Translated by David R. Wilkins
Preliminary Version: December 1998

[Monatsberichte der Berliner Akademie, November 1859.]

I believe that I can best convey my thanks for the honour which the Academy has to some degree conferred on me, through my admission as one of its correspondents, if I speedily make use of the permission thereby received to communicate an investigation into the accumulation of the prime numbers; a topic which perhaps seems not wholly unworthy of such a communication, given the interest which *Gauss* and *Dirichlet* have themselves shown in it over a lengthy period.

For this investigation my point of departure is provided by the observation of *Euler* that the product

$$\prod \frac{1}{1 - \frac{1}{p^s}} = \sum \frac{1}{n^s},$$

if one substitutes for p all prime numbers, and for n all whole numbers. The function of the complex variable s which is represented by these two expressions, wherever they converge, I denote by $\zeta(s)$. Both expressions converge only when the real part of s is greater than 1; at the same time an expression for the function can easily be found which always remains valid. On making use of the equation

$$\int_0^\infty e^{-nx} x^{s-1}\, dx = \frac{\Pi(s-1)}{n^s}$$

one first sees that

$$\Pi(s-1)\zeta(s) = \int_0^\infty \frac{x^{s-1}\,dx}{e^x - 1}.$$

If one now considers the integral

$$\int \frac{(-x)^{s-1}\,dx}{e^x - 1}$$

from $+\infty$ to $+\infty$ taken in a positive sense around a domain which includes the value 0 but no other point of discontinuity of the integrand in its interior, then this is easily seen to be equal to

$$(e^{-\pi si} - e^{\pi si}) \int_0^\infty \frac{x^{s-1}\,dx}{e^x - 1},$$

provided that, in the many-valued function $(-x)^{s-1} = e^{(s-1)\log(-x)}$, the logarithm of $-x$ is determined so as to be real when x is negative. Hence

$$2\sin \pi s\, \Pi(s-1)\zeta(s) = i \int_\infty^\infty \frac{(-x)^{s-1}\,dx}{e^x - 1},$$

where the integral has the meaning just specified.

This equation now gives the value of the function $\zeta(s)$ for all complex numbers s and shows that this function is one-valued and finite for all finite values of s with the exception of 1, and also that it is zero if s is equal to a negative even integer.

If the real part of s is negative, then, instead of being taken in a positive sense around the specified domain, this integral can also be taken in a negative sense around that domain containing all the remaining complex quantities, since the integral taken though values of infinitely large modulus is then infinitely small. However, in the interior of this domain, the integrand has discontinuities only where x becomes equal to a whole multiple of $\pm 2\pi i$, and the integral is thus equal to the sum of the integrals taken in a negative sense around these values. But the integral around the value $n2\pi i$ is $= (-n2\pi i)^{s-1}(-2\pi i)$, one obtains from this

$$2\sin \pi s\, \Pi(s-1)\zeta(s) = (2\pi)^s \sum n^{s-1}((-i)^{s-1} + i^{s-1}),$$

thus a relation between $\zeta(s)$ and $\zeta(1-s)$, which, through the use of known properties of the function Π, may be expressed as follows:

$$\Pi\left(\frac{s}{2} - 1\right)\pi^{-\frac{s}{2}}\zeta(s)$$

remains unchanged when s is replaced by $1-s$.

This property of the function induced me to introduce, in place of $\Pi(s-1)$, the integral $\Pi\left(\frac{s}{2}-1\right)$ into the general term of the series $\sum \frac{1}{n^s}$, whereby one obtains a very convenient expression for the function $\zeta(s)$. In fact

$$\frac{1}{n^s}\Pi\left(\frac{s}{2}-1\right)\pi^{-\frac{s}{2}} = \int_0^\infty e^{-nn\pi x} x^{\frac{s}{2}-1}\,dx,$$

thus, if one sets

$$\sum_1^\infty e^{-nn\pi x} = \psi(x)$$

then

$$\Pi\left(\frac{s}{2}-1\right)\pi^{-\frac{s}{2}}\zeta(s) = \int_0^\infty \psi(x) x^{\frac{s}{2}-1}\,dx,$$

or since

$$2\psi(x)+1 = x^{-\frac{1}{2}}\left(2\psi\left(\frac{1}{x}\right)+1\right), \quad \text{(Jacobi, Fund. S. 184)}$$

$$\begin{aligned}\Pi\left(\frac{s}{2}-1\right)\pi^{-\frac{s}{2}}\zeta(s) &= \int_1^\infty \psi(x) x^{\frac{s}{2}-1}\,dx + \int_1^\infty \psi\left(\frac{1}{x}\right) x^{\frac{s-3}{2}}\,dx \\ &\quad + \tfrac{1}{2}\int_0^1 \left(x^{\frac{s-3}{2}} - x^{\frac{s}{2}-1}\right)dx \\ &= \frac{1}{s(s-1)} + \int_1^\infty \psi(x)\left(x^{\frac{s}{2}-1} + x^{-\frac{1+s}{2}}\right)dx.\end{aligned}$$

I now set $s = \frac{1}{2} + ti$ and

$$\Pi\left(\frac{s}{2}\right)(s-1)\pi^{-\frac{s}{2}}\zeta(s) = \xi(t),$$

so that

$$\xi(t) = \tfrac{1}{2} - (tt + \tfrac{1}{4})\int_1^\infty \psi(x) x^{-\frac{3}{4}} \cos(\tfrac{1}{2}t\log x)\,dx$$

or, in addition,

$$\xi(t) = 4\int_1^\infty \frac{d(x^{\frac{3}{2}}\psi'(x))}{dx} x^{-\frac{1}{4}} \cos(\tfrac{1}{2}t\log x)\,dx.$$

This function is finite for all finite values of t, and allows itself to be developed in powers of tt as a very rapidly converging series. Since, for a value of s whose real part is greater than 1, $\log\zeta(s) = -\sum\log(1-p^{-s})$ remains finite, and since the same holds for the logarithms of the other factors of $\xi(t)$, it follows that the function $\xi(t)$ can only vanish if the imaginary part of t lies between $\frac{1}{2}i$ and $-\frac{1}{2}i$. The number of roots of $\xi(t) = 0$, whose real parts lie between 0 and T is approximately

$$= \frac{T}{2\pi}\log\frac{T}{2\pi} - \frac{T}{2\pi};$$

because the integral $\int d\log\xi(t)$, taken in a positive sense around the region consisting of the values of t whose imaginary parts lie between $\frac{1}{2}i$ and $-\frac{1}{2}i$ and whose real parts lie between 0 and T, is (up to a fraction of the order of magnitude of the quantity $\frac{1}{T}$) equal to $\left(T\log\frac{T}{2\pi} - T\right)i$; this integral however is equal to the number of roots of $\xi(t) = 0$ lying within in this region, multiplied by $2\pi i$. One now finds indeed approximately this number of real roots within these limits, and it is very probable that all roots are real. Certainly one would wish for a stricter proof here; I have meanwhile temporarily put aside the search for this after some fleeting futile attempts, as it appears unnecessary for the next objective of my investigation.

If one denotes by α all the roots of the equation $\xi(\alpha) = 0$, one can express $\log\xi(t)$ as

$$\sum\log\left(1 - \frac{tt}{\alpha\alpha}\right) + \log\xi(0);$$

for, since the density of the roots of the quantity t grows with t only as $\log\frac{t}{2\pi}$, it follows that this expression converges and becomes for an infinite t only infinite as $t\log t$; thus it differs from $\log\xi(t)$ by a function of tt, that for a finite t remains continuous and finite and, when divided by tt, becomes infinitely small for infinite t. This difference is consequently a constant, whose value can be determined through setting $t = 0$.

With the assistance of these methods, the number of prime numbers that are smaller than x can now be determined.

Let $F(x)$ be equal to this number when x is not exactly equal to a prime number; but let it be greater by $\frac{1}{2}$ when x is a prime number, so that, for any x at which there is a jump in the value in $F(x)$,

$$F(x) = \frac{F(x+0) + F(x-0)}{2}.$$

If in the identity

$$\log\zeta(s) = -\sum\log(1-p^{-s}) = \sum p^{-s} + \tfrac{1}{2}\sum p^{-2s} + \tfrac{1}{3}\sum p^{-3s} + \cdots$$

one now replaces

$$p^{-s} \text{ by } s\int\limits_{p}^{\infty} x^{-s-1}\,ds, \quad p^{-2s} \text{ by } s\int\limits_{p^2}^{\infty} x^{-s-1}\,ds, \ldots,$$

one obtains

$$\frac{\log\zeta(s)}{s} = \int\limits_{1}^{\infty} f(x)x^{-s-1}\,dx,$$

if one denotes

$$F(x) + \tfrac{1}{2}F(x^{\frac{1}{2}}) + \tfrac{1}{3}F(x^{\frac{1}{3}}) + \cdots$$

by $f(x)$.

This equation is valid for each complex value $a + bi$ of s for which $a > 1$. If, though, the equation

$$g(s) = \int\limits_{0}^{\infty} h(x)x^{-s}\,d\log x$$

holds within this range, then, by making use of *Fourier*'s theorem, one can express the function h in terms of the function g. The equation decomposes, if $h(x)$ is real and

$$g(a+bi) = g_1(b) + ig_2(b),$$

into the two following:

$$g_1(b) = \int\limits_{0}^{\infty} h(x)x^{-a}\cos(b\log x)\,d\log x,$$

$$ig_2(b) = -i\int\limits_{0}^{\infty} h(x)x^{-a}\sin(b\log x)\,d\log x.$$

If one multiplies both equations with

$$(\cos(b\log y) + i\sin(b\log y))\,db$$

and integrates them from $-\infty$ to $+\infty$, then one obtains $\pi h(y)y^{-\alpha}$ on the right hand side in both, on account of *Fourier*'s theorems; thus, if one adds both equations and multiplies them by iy^{α}, one obtains

$$2\pi i h(y) = \int\limits_{a-\infty i}^{a+\infty i} g(s)y^{s}\,ds,$$

5

where the integration is carried out so that the real part of s remains constant.

For a value of y at which there is a jump in the value of $h(y)$, the integral takes on the mean of the values of the function h on either side of the jump. From the manner in which the function f was defined, we see that it has the same property, and hence in full generality

$$f(y) = \frac{1}{2\pi i} \int_{a-\infty i}^{a+\infty i} \frac{\log \zeta(s)}{s} y^s \, ds.$$

One can substitute for $\log \zeta$ the expression

$$\frac{s}{2} \log \pi - \log(s-1) - \log \Pi\left(\frac{s}{2}\right) + \sum^{\alpha} \log\left(1 + \frac{(s-\frac{1}{2})^2}{\alpha\alpha}\right) + \log \xi(0)$$

found earlier; however the integrals of the individual terms of this expression do not converge, when extended to infinity, for which reason it is appropriate to convert the previous equation by means of integration by parts into

$$f(x) = -\frac{1}{2\pi i} \frac{1}{\log x} \int_{a-\infty i}^{a+\infty i} \frac{d\frac{\log \zeta(s)}{s}}{ds} x^s \, ds$$

Since

$$-\log \Pi\left(\frac{s}{2}\right) = \lim\left(\sum_{n=1}^{n=m} \log\left(1 + \frac{s}{2n}\right) - \frac{s}{2} \log m\right),$$

for $m = \infty$ and therefore

$$-\frac{d\frac{1}{s} \log \Pi\left(\frac{s}{2}\right)}{ds} = \sum_{1}^{\infty} \frac{d\frac{1}{s} \log\left(1 + \frac{s}{2n}\right)}{ds},$$

it then follows that all the terms of the expression for $f(x)$, with the exception of

$$\frac{1}{2\pi i} \frac{1}{\log x} \int_{a-\infty i}^{a+\infty i} \frac{1}{ss} \log \xi(0) x^s \, ds = \log \xi(0),$$

take the form

$$\pm \frac{1}{2\pi i} \frac{1}{\log x} \int_{a-\infty i}^{a+\infty i} \frac{d\left(\frac{1}{s} \log\left(1 - \frac{s}{\beta}\right)\right)}{ds} x^s \, ds.$$

But now

$$\frac{d\left(\frac{1}{s}\log\left(1-\frac{s}{\beta}\right)\right)}{d\beta} = \frac{1}{(\beta-s)\beta},$$

and, if the real part of s is larger than the real part of β,

$$-\frac{1}{2\pi i}\int_{a-\infty i}^{a+\infty i} \frac{x^s\,ds}{(\beta-s)\beta} = \frac{x^\beta}{\beta} = \int_\infty^x x^{\beta-1}\,dx,$$

or

$$= \int_0^x x^{\beta-1}\,dx,$$

depending on whether the real part of β is negative or positive. One has as a result

$$\frac{1}{2\pi i}\frac{1}{\log x}\int_{a-\infty i}^{a+\infty i} \frac{d\left(\frac{1}{s}\log\left(1-\frac{s}{\beta}\right)\right)}{ds} x^s\,ds$$

$$= -\frac{1}{2\pi i}\int_{a-\infty i}^{a+\infty i} \frac{1}{s}\log\left(1-\frac{s}{\beta}\right) x^s\,ds$$

$$= \int_\infty^x \frac{x^{\beta-1}}{\log x}\,dx + \text{ const.}$$

in the first, and

$$= \int_0^x \frac{x^{\beta-1}}{\log x}\,dx + \text{ const.}$$

in the second case.

In the first case the constant of integration is determined if one lets the real part of β become infinitely negative; in the second case the integral from 0 to x takes on values separated by $2\pi i$, depending on whether the integration is taken through complex values with positive or negative argument, and becomes infinitely small, for the former path, when the coefficient of i in the value of β becomes infinitely positive, but for the latter, when this coefficient becomes infinitely negative. From this it is seen how on the left hand side $\log\left(1-\frac{s}{\beta}\right)$ is to be determined in order that the constants of integration disappear.

Through the insertion of these values in the expression for $f(x)$ one obtains

$$f(x) = Li(x) - \sum^{\alpha}\left(Li\left(x^{\frac{1}{2}+\alpha i}\right) + Li\left(x^{\frac{1}{2}-\alpha i}\right)\right) + \int_x^{\infty} \frac{1}{x^2-1}\frac{dx}{x\log x} + \log\xi(0),$$

if in $\sum^{\alpha}$ one substitutes for α all positive roots (or roots having a positive real part) of the equation $\xi(\alpha) = 0$, ordered by their magnitude. It may easily be shown, by means of a more thorough discussion of the function ξ, that with this ordering of terms the value of the series

$$\sum\left(Li\left(x^{\frac{1}{2}+\alpha i}\right) + Li\left(x^{\frac{1}{2}-\alpha i}\right)\right)\log x$$

agrees with the limiting value to which

$$\frac{1}{2\pi i}\int_{a-bi}^{a+bi} \frac{d\frac{1}{s}\sum\log\left(1+\frac{(s-\frac{1}{2})^2}{\alpha\alpha}\right)}{ds} x^s\, ds$$

converges as the quantity b increases without bound; however when reordered it can take on any arbitrary real value.

From $f(x)$ one obtains $F(x)$ by inversion of the relation

$$f(x) = \sum\frac{1}{n}F\left(x^{\frac{1}{n}}\right),$$

to obtain the equation

$$F(x) = \sum(-1)^{\mu}\frac{1}{m}f\left(x^{\frac{1}{m}}\right),$$

in which one substitutes for m the series consisting of those natural numbers that are not divisible by any square other than 1, and in which μ denotes the number of prime factors of m.

If one restricts $\sum^{\alpha}$ to a finite number of terms, then the derivative of the expression for $f(x)$ or, up to a part diminishing very rapidly with growing x,

$$\frac{1}{\log x} - 2\sum^{\alpha}\frac{\cos(\alpha\log x)x^{-\frac{1}{2}}}{\log x}$$

gives an approximating expression for the density of the prime number + half the density of the squares of the prime numbers + a third of the density of the cubes of the prime numbers etc. at the magnitude x.

The known approximating expression $F(x) = Li(x)$ is therefore valid up to quantities of the order $x^{\frac{1}{2}}$ and gives somewhat too large a value; because the non-periodic terms in the expression for $F(x)$ are, apart from quantities that do not grow infinite with x:

$$Li(x) - \tfrac{1}{2}Li(x^{\frac{1}{2}}) - \tfrac{1}{3}Li(x^{\frac{1}{3}}) - \tfrac{1}{5}Li(x^{\frac{1}{5}}) + \tfrac{1}{6}Li(x^{\frac{1}{6}}) - \tfrac{1}{7}Li(x^{\frac{1}{7}}) + \cdots$$

Indeed, in the comparison of $Li(x)$ with the number of prime numbers less than x, undertaken by *Gauss* and *Goldschmidt* and carried through up to $x =$ three million, this number has shown itself out to be, in the first hundred thousand, always less than $Li(x)$; in fact the difference grows, with many fluctuations, gradually with x. But also the increase and decrease in the density of the primes from place to place that is dependent on the periodic terms has already excited attention, without however any law governing this behaviour having been observed. In any future count it would be interesting to keep track of the influence of the individual periodic terms in the expression for the density of the prime numbers. A more regular behaviour than that of $F(x)$ would be exhibited by the function $f(x)$, which already in the first hundred is seen very distinctly to agree on average with $Li(x) + \log \xi(0)$.

12.3 J. Hadamard (1896)

Sur la distribution des zéros de la fonction $\zeta(s)$ et ses conséquences arithmétiques

Jacques Hadamard was born in France in 1865. Both of his parents were teachers, and Hadamard excelled in all subjects. He entered the École Normale Supérieure in 1884. Under his professors, which included Hermite and Goursat, he began to consider research problems in mathematics. In 1892, Hadamard received his doctorate for his thesis on functions defined by Taylor series. He won several prizes for research in the fields of physics and mathematics. Hadamard was elected president of the French Mathematical Society in 1906. His *Leçons sur le calcul des variations* of 1910 laid the foundations for functional analysis, and introduced the term "functional". In 1912, he was elected to the Academy of Sciences. Throughout his life Hadamard was active politically, and campaigned for peace following the Second World War. His output as a researcher encompasses over 300 papers and books. Hadamard died in Paris in 1963 [117].

This paper gives Hadamard's proof of the prime number theorem. The statement $\pi(x) \sim \mathrm{Li}(x)$ is proven indirectly, by considering the zeros of $\zeta(s)$. Hadamard increases the zero-free region of the Riemann zeta function by proving that $\zeta(\sigma + it) = 0$ implies that $\sigma \neq 1$. This is sometimes referred to as the Hadamard–de la Vallée Poussin Theorem, and from it the desired result follows. An exposition of Hadamard's method in English can be found in Chapter 3 of [155]. This paper, in conjunction with Section 12.4, gives the most important result in the theory of prime numbers since Chebyshev's 1852 result (see Section 12.1).

MÉMOIRES ET COMMUNICATIONS.

SUR LA DISTRIBUTION DES ZÉROS DE LA FONCTION $\zeta(s)$ ET SES CONSÉQUENCES ARITHMÉTIQUES (¹);

Par M. HADAMARD.

I. — *Sur les zéros de la fonction ζ et de quelques fonctions analogues.*

1. La fonction $\zeta(s)$ de Riemann est définie, lorsque la partie réelle de s est plus grande que 1, par l'équation

$$\log\zeta(s) = -\sum_p \log\left(1 - \frac{1}{p^s}\right), \tag{1}$$

où p désigne successivement les différents nombres premiers; les logarithmes sont népériens. Elle est holomorphe dans tout le plan, sauf au point $s = 1$, qui est un pôle simple. Elle ne s'annule pour aucune valeur de s dont la partie réelle soit supérieure à 1, puisque le second membre de l'équation (1) est fini. Mais elle admet une infinité de zéros imaginaires dont la partie réelle est comprise entre 0 et 1. Stieltjes avait démontré, conformément aux prévisions de Riemann, que ces zéros sont tous de la forme

(¹) Les résultats fondamentaux du présent Mémoire ont été communiqués à l'Académie des Sciences, dans la séance du 22 juin 1896.

$\frac{1}{2}+ti$ (le nombre t étant réel); mais sa démonstration n'a jamais été publiée, et il n'a même pas été établi que la fonction ζ n'ait pas de zéros sur la droite (¹) $\Re(s)=1$.

C'est cette dernière conclusion que je me propose de démontrer.

2. Faisons d'abord tendre s vers 1 par valeurs réelles et décroissantes. Le logarithme de $\zeta(s)$, ou, à une quantité finie près, la série

$$S=\sum_p \frac{1}{p^s} \tag{2}$$

augmente indéfiniment comme $-\log(s-1)$.

Remplaçons maintenant s par $s+ti$ et imaginons que le point d'affixe $1+ti$ soit un zéro de ζ. Alors la partie réelle de $\log(s+ti)$, c'est-à-dire (à une quantité finie près) la somme

$$P=\sum_p \frac{1}{p^s}\cos\left(t\log p\right), \tag{3}$$

devra croître indéfiniment par valeurs négatives *comme* $\log(s-1)$, *c'est-à-dire comme* $-S$, lorsque s tendra vers 1 (t restant fixe).

3. Cela posé, soit α un angle que nous supposerons petit; parmi les différents nombres premiers, distinguons deux catégories :

1° Ceux qui satisfont, pour quelque valeur entière de k, à la double inégalité

$$\frac{(2k+1)\pi-\alpha}{t}\leqq \log p\leqq \frac{(2k+1)\pi+\alpha}{t}. \tag{4}$$

Les parties des sommes S_n et P_n [c'est-à-dire des séries (2) et (3) bornées à leurs n premiers termes] correspondant à cette première catégorie de nombres premiers seront désignées par S'_n et P'_n.

2° Les nombres premiers restants, c'est-à-dire ceux qui ne

(¹) $\Re(s)$ désigne, comme d'habitude, la partie réelle de s.

vérifient la double inégalité (4) pour aucune valeur de k, donneront, dans les sommes S_n et P_n, les parties S''_n et P''_n.

Considérons le rapport $\rho_n = \frac{S'_n}{S_n}$, lequel est compris entre o et 1 : lorsque n augmentera indéfiniment, ce rapport aura soit une limite, soit des limites d'oscillation. *Si $\zeta(1+ti)$ était nul, cette ou ces limites devraient tendre vers* 1 *avec s*. Autrement dit, ρ étant un nombre quelconque plus petit que 1, on pourrait faire correspondre à toute valeur réelle de s supérieure à 1, mais suffisamment voisine de 1, une valeur de n à partir de laquelle on aurait

$$\rho_n > \rho. \tag{5}$$

On peut, en effet, écrire évidemment :

$$P'_n \geqq - S'_n \geqq - \rho_n S_n,$$
$$P''_n \geqq - S''_n \cos\alpha \geqq - (1-\rho_n) S_n \cos\alpha$$

(les inégalités ayant leur sens algébrique). Si donc on avait

$$\rho_n \leqq \rho,$$

il en résulterait

$$P_n = - \theta S_n$$

où $\theta = \rho + (1-\rho)\cos\alpha$ est un nombre fixe plus petit que 1 ; et si cela avait lieu pour une infinité de valeurs de n, on pourrait passer à la limite et écrire

$$P \geqq - \theta S,$$

ce qui serait en contradiction avec l'hypothèse $\zeta(1+ti) = 0$, ainsi qu'il a été remarqué au numéro précédent.

L'égalité $\zeta(1+ti) = 0$ exige donc bien que la ou les limites de ρ_n tendent vers 1 avec s.

4. Changeons alors t en $2t$, dans la série (3) et soit Q la nouvelle série ainsi obtenue : les termes qui formaient, dans la série (3), les sommes $P'_n, P''_n, P_n = P'_n + P''_n$ donneront, dans cette nouvelle série, respectivement les sommes $Q'_n, Q''_n, Q_n = Q'_n + Q''_n$ et l'on aura, cette fois,

$$Q'_n \geqq S'_n \cos 2\alpha \geqq \rho_n S_n \cos 2\alpha.$$
$$Q''_n \geqq - S''_n \geqq - (1-\rho_n) S_n$$

et, par conséquent,

$$Q_n \geqq S_n[\rho_n \cos 2\alpha - (1-\rho_n)]:$$

d'où, moyennant l'inégalité (5) supposée vérifiée pour n suffisamment grand,

$$Q_n \geqq \theta' S_n,$$

θ' désignant le nombre $\rho \cos 2\alpha - (1-\rho)$, lequel est positif si nous avons pris $1 > \rho > \frac{1}{1+\cos 2\alpha}$.

Or ceci donnerait $Q \geqq \theta' S$ et, par suite, Q augmenterait indéfiniment par valeurs positives; de sorte que le point d'affixe $1+2ti$ serait un infini de $\zeta(s)$: ce que nous savons n'avoir pas lieu.

L'impossibilité de l'hypothèse $\zeta(1+ti)=0$ est donc mise en évidence.

5. Il est remarquable que cette démonstration ne repose que sur les propriétés les plus simples de $\zeta(s)$: nous nous sommes, en effet, exclusivement servi des remarques suivantes : 1° le logarithme de notre fonction est développable en série de la forme $\Sigma a_n e^{-\lambda_n s}$, les nombres a_n étant tous positifs; 2° la fonction est uniforme sur la droite qui limite la convergence de cette série et ne présente sur cette droite qu'un seul pôle simple.

Toute fonction satisfaisant à ces conditions sera donc différente de 0 sur la droite limite.

Ainsi, dans la démonstration précédente, c'était uniquement pour simplifier l'écriture que nous avons réduit le second membre de l'équation (1) à la série S : la démonstration se serait également appliquée au développement complet de $\log \zeta(s)$. De même, les nombres premiers ayant été distribués d'une façon quelconque en deux catégories, les nombres de la première catégorie étant désignés par p', ceux de la deuxième par p'', si la fonction représentée (lorsque la partie réelle de s est supérieure à 1) par le produit infini

$$f(s) = \frac{1}{\prod_{p'}\left(1-\frac{1}{p'^s}\right)\prod_{p''}\left(1+\frac{1}{p''^s}\right)} \tag{6}$$

est holomorphe sur la droite limite $\mathcal{R}(s) = 1$, elle est différente de 0 sur cette droite ([1]).

En effet, le logarithme du produit

$$f(s)\zeta(s) = \frac{1}{\prod\limits_{p'}\left(1 - \frac{1}{p'^s}\right)^2 \prod\limits_{p''}\left(1 - \frac{1}{p''^{2s}}\right)}$$

est représenté par une série $\sum a_n e^{-\lambda_n s}$ à coefficients positifs ; ce produit satisfait donc aux conditions ci-dessus indiquées.

Ce cas est, par exemple, celui de la fonction de Schlömilch

$$(7) \qquad \sum \frac{(-1)^n}{(2n+1)^s} = \prod_p \frac{1}{1 + \frac{(-1)^{\frac{p+1}{2}}}{p^s}}.$$

6. Plus généralement, nous allons étendre la proposition qui précède aux séries introduites en Arithmétique par Dirichlet, et dont nous devons tout d'abord rappeler, en les complétant sur certains points, les principales propriétés.

Ces séries appartiennent à la catégorie des séries de la forme $\sum \frac{a_n}{n^s}$ *périodiques,* c'est-à-dire dont les coefficients a_n se reproduisent de k en k. De telles séries sont évidemment des combinaisons linéaires des k fonctions

$$\xi_1(s) = \frac{1}{1^s} + \frac{1}{(k+1)^s} + \frac{1}{(2k+1)^s} + \ldots,$$

$$\xi_2(s) = \frac{1}{2^s} + \frac{1}{(k+2)^s} + \frac{1}{(2k+2)^s} + \ldots,$$

$$\ldots\ldots\ldots\ldots\ldots\ldots\ldots\ldots\ldots,$$

$$\xi_k(s) = \frac{1}{k^s} + \frac{1}{(k+k)^s} + \frac{1}{(2k+k)^s} + \ldots,$$

étudiées par MM. Hurwitz ([2]) et Cahen ([3]). Ces fonctions sont

---

([1]) Sauf peut-être au point $s = 1$; mais cette circonstance ne se présentera pas dans la suite.

([2]) *Zeitschrift für Mathematik und Physik*, t. XXVII, p. 86-102; 1882.

([3]) *Thèse de Doctorat*, 1894, et *Annales de l'École Normale supérieure*, 3e série, t. XI.

uniformes dans tout le plan, avec le seul pôle simple $s = 1$ et le résidu correspondant $\frac{1}{k}$, ainsi qu'il résulte de l'expression

$$(8) \qquad \xi_r(s) = \frac{i}{2\pi}\Gamma(1-s)\int (-x)^{s-1}\frac{e^{(k-r)x}}{e^{kx}-1}\,dx,$$

l'intégrale étant prise le long d'un contour C partant de $+\infty$ et y revenant après avoir tourné dans le sens trigonométrique autour de l'origine, et $-x$ étant considéré comme ayant (pour x réel et positif) l'argument $-i\pi$ dans la première partie du chemin d'intégration et, par suite, l'argument $+i\pi$ dans la seconde.

L'intégrale qui figure dans la formule précédente est une fonction entière de s, et les théorèmes généraux donnés dans mon Mémoire *Sur les propriétés des fonctions entières* (¹) permettent d'en déterminer le genre. A cet effet, on peut, par exemple, diviser le contour C en deux parties : l'une C′ partant du point $x = 1$ et y revenant après circulation autour de l'origine; l'autre C″ comprenant les deux traits de 1 à $+\infty$; les intégrales prises suivant ces deux traits ne diffèrent entre elles et de l'intégrale

$$(9) \qquad \int_1^\infty x^{s-1}\frac{e^{(k-r)x}}{e^{kx}-1}\,dx$$

que par les facteurs exponentiels $e^{-i\pi s}$ pour la première, $e^{i\pi s}$ pour la seconde. Or, le coefficient de s^n dans l'intégrale (9), qui a pour valeur

$$\frac{1}{n!}\int_1^\infty (\log x)^n \frac{e^{(k-r)x}}{e^{kx}-1}\frac{dx}{x},$$

est (puisque r est un entier plus grand que 0) au plus comparable au coefficient correspondant de la fonction

$$Q(s) = \int_1^\infty x^{s-1}e^{-x}\,dx,$$

qui intervient dans l'étude de la fonction Γ et dont l'ordre de

(¹) *Journal de M. Jordan*, 4ᵉ série, t. IX; 1893.

grandeur pour s infini est celui de Γ. Quant à l'intégrale prise le long de C', le coefficient de s^n, qui a pour valeur

$$\frac{1}{n!}\int_{C'}(\log x)^n \frac{e^{(k-r)x}}{e^{kx}-1}\,dx,$$

y est au plus de l'ordre de $\frac{K^n}{1.2\ldots n}$, en désignant par K le module maximum de $\log x$ sur le contour en question. On voit donc que la fonction considérée est de genre 1 : le nombre de zéros de cette fonction, compris dans le cercle de rayon R, est de l'ordre de R log R.

7. Lorsqu'on change s en $1-s$, les nouvelles valeurs des fonctions ξ s'expriment en fonction des anciennes par les relations établies par M. Hurwitz ([1]) et que l'on peut prendre sous la forme ([2])

$$(10)\quad \left\{\begin{aligned} &\frac{1}{\Gamma(1-s)}\sum_{l=1}^{k}\sigma^{lr}\xi_l(s)\\ &\quad=\left(\frac{2\pi}{k}\right)^{s-1}\left[e^{(s-1)\frac{i\pi}{2}}\xi_{k-r}(1-s)+e^{-(s-1)\frac{i\pi}{2}}\xi_r(1-s)\right]\\ &\qquad\qquad (r=1,2,\ldots,k),\end{aligned}\right.$$

où σ désigne $e^{\frac{2i\pi}{k}}$

8. Pour définir ses séries, Dirichlet ([3]) part de la décomposition du nombre k en facteurs premiers

$$(11)\qquad k = 2^\lambda p^\varpi p'^{\varpi'}\ldots \qquad (\lambda \geqq 0;\ \varpi, \varpi', \ldots > 0),$$

et, à tout entier n premier avec k, fait correspondre les indices

$$\alpha,\ \beta,\ \gamma,\ \gamma',\ \ldots.$$

([1]) Hurwitz, *loc. cit.*, p. 93.

([2]) Cahen, *loc. cit.*, nos 47, 53.

([3]) *Abhandlungen der Berl. Acad.*, 1837; traduit par Terquem, *Journal de Liouville*, 1re série, t. IV; 1839. Nous nous conformons aux notations employées dans les *Vorlesungen über Zahlentheorie,* éditées par Dedekind, édition de 1863, supplément VI.

définis par les congruences

$$(12) \quad \begin{cases} n \equiv (-1)^\alpha 5^\beta & (\text{mod } 2^\lambda), \\ n \equiv g^\gamma & (\text{mod } p^\varpi), \\ n \equiv g'^{\gamma'} & (\text{mod } p'^{\varpi'}), \\ \ldots\ldots\ldots & \ldots\ldots\ldots, \end{cases}$$

où g, g', ... sont des racines primitives pour les modules respectifs p^ϖ, $p'^{\varpi'}$, ... Les nombres α et β sont ainsi définis aux modules a et b près : les nombres a et b ayant tous deux la valeur 1, si $\lambda = 0, 1$, et prenant les valeurs $a = 2$, $b = \frac{1}{2}\varphi(2^\lambda)$, si $\lambda \geqq 2$. Pareillement, les nombres γ, γ', ... sont définis relativement aux modules

$$c = \varphi(p^\varpi), \qquad c' = \varphi(p'^{\varpi'}), \qquad \ldots,$$

où φ est la fonction bien connue qui exprime combien il y a de nombres premiers à un entier donné et inférieurs à lui.

Réciproquement, la connaissance des indices α, β, γ, γ', ... fait connaître le nombre n, au module k près. Autrement dit, aux $\varphi(k)$ valeurs de n premières avec k et incongrues entre elles suivant le module k correspondent, d'une façon univoque, les

$$a\, b\, c\, c' \ldots = \varphi(k)$$

systèmes de valeurs de α, β, γ, γ', ... incongrus entre eux suivant les modules a, b, c, c', ...

Désignant par θ, η, ω, ω', ... respectivement une racine $a^{\text{ième}}$, une racine $b^{\text{ème}}$, une racine $c^{\text{ième}}$, une racine $c'^{\text{ième}}$, ... de l'unité, autrement dit posant

$$(13) \quad \begin{cases} \theta = \pm 1, \\ \eta = e^{\frac{2i\pi\mu}{b}}, \\ \varpi = e^{\frac{2i\pi\tau}{c}}, \\ \varpi' = e^{\frac{2i\pi\tau'}{c'}}, \\ \ldots\ldots\ldots \end{cases}$$

Dirichlet introduit la fonction

$$\psi_\nu(n) = \begin{cases} 0, \text{ si } n \text{ n'est pas premier avec } k, \\ \theta^\alpha \eta^\beta \omega^\gamma \omega'^{\gamma'} \ldots, \text{ si } n \text{ est premier avec } k, \end{cases}$$

— 207 —

$\alpha, \beta, \gamma, \gamma', \ldots$ étant les indices de n [l'indice ν a pour but de distinguer les unes des autres les $\varphi(k)$ fonctions ψ correspondant aux différents choix possibles des racines $\theta, \eta, \omega, \omega', \ldots$].

Il forme ensuite la série (périodique au sens indiqué ci-dessus)

$$(14) \qquad L_\nu(s) = \sum_{n=1}^{\infty} \frac{\psi_\nu(n)}{n^s} = \sum_{r=1}^{k} \xi_r(s)\,\psi_\nu(r) \qquad [\nu = 1, 2, \ldots, \varphi(k)],$$

égale au produit infini

$$(15) \qquad L_\nu(s) = \prod \frac{1}{1 - \dfrac{\psi_\nu(q)}{q^s}},$$

dans lequel q doit être remplacée successivement par tous les nombres premiers.

Les séries L_ν se répartissent en trois catégories : la première comprend une seule série L_1, celle qui correspond à

$$\theta = \eta = \omega = \omega' = \ldots = 1;$$

la seconde comprend toutes les séries L, pour lesquelles les nombres $\theta, \eta, \ldots$ sont égaux à $+1$ ou à -1 (à l'exception de L_1); la troisième, les séries correspondant aux cas où l'un au moins de ces nombres est imaginaire. Ces dernières sont conjuguées deux à deux; la série

$$L_\nu(s) = \sum \xi_r(s)\,\psi_\nu(r),$$

déduite des racines $\theta, \eta, \omega, \omega', \ldots$, est conjuguée de la série

$$L_{\nu'}(s) = \sum \frac{\xi_r(s)}{\psi_\nu(r)},$$

déduite des racines $\frac{1}{\theta}, \frac{1}{\eta}, \frac{1}{\omega}, \frac{1}{\omega'}, \ldots$.

La série L_1 admet, comme seule singularité, le pôle simple $s = 1$. Quant aux autres séries L, elles sont holomorphes dans tout le plan $\left[\text{parce que la somme } \frac{1}{k}\sum_r \psi_\nu(r) \text{ des résidus au point } s = 1 \text{ est nulle}\right]$.

Dirichlet démontre qu'elles sont toutes différentes de 0 pour $s = 1$.

9. De la relation générale (10), M. Hurwitz a pu déduire que certaines séries de seconde catégorie se reproduisent, à un facteur près, par le changement de s en $(1 - s)$ à la façon de la fonction ζ.

Cette proposition est un cas particulier d'un théorème démontré par M. Lipschitz ([1]), et qui est le suivant : *La série* $L_\nu(s)$ *est (à un facteur exponentiel et trigonométrique près, analogue à celui qui se rencontre dans la formule relative à la fonction* ζ*) changée en sa conjuguée par le changement de* s *en* $1 - s$*, sous les conditions suivantes :*

1° $\lambda \geqq 3$, μ *impair;*

2° $\tau \neq p - 1$, *si* $\varpi = 1$; τ *non divisible par* p, *si* $\varpi > 1$;

3° $\tau' \neq p' - 1$, *si* $\varpi' = 1$; τ' *non divisible par* p', *si* $\varpi' > 1$; ...

Ce théorème nous fournit un renseignement important sur la distribution des zéros de $L_\nu(s)$. Puisque cette fonction n'a aucun zéro imaginaire dont la partie réelle soit plus grande que s, elle n'en a non plus aucun dont la partie réelle soit négative : les zéros imaginaires sont compris dans la même bande que ceux de $\zeta(s)$. Ils sont même, comme ceux de $\zeta(s)$, disposés symétriquement par rapport à la droite $\mathcal{R}(s) = \frac{1}{2}$, puisqu'à tout zéro α correspond un zéro α' (différent ou non du premier), tel que α et $1 - \alpha'$ soient imaginaires conjugués.

Toutefois, cette conclusion n'est pas encore démontrée dans les cas où la relation de Lipschitz ne s'applique pas; mais on ramène ces cas aux autres par les remarques suivantes :

1° Si une racine ω, par exemple, est égale à 1, on aura

$$L_\nu(s) = [1 - \psi'_\nu(p)p^{-s}]L'_\nu(s),$$

la série L'_ν étant composée en partant du nombre k supposé débarrassé du facteur p^ϖ. La même circonstance se produit pour le facteur 2 lorsque l'exposant λ est égal à 1 ;

2° Si l'entier τ est divisible par p^h, la série peut se composer en partant de l'entier k, divisé par p^h, la racine primitive g de p^ϖ étant une racine primitive de $p^{\varpi - h}$. La nouvelle valeur de τ ne con-

([1]) *Journal de Crelle*. t. 105, p. 127-157.

tiendra plus p en facteur. Il en est de même pour le facteur 2 lorsque l'entier μ est pair, et aussi lorsque $\lambda = 2$, $\theta = 1$.

3° Le raisonnement de l'auteur est encore valable pour $\lambda = 2$, $\theta = -1$, en prenant pour l'expression (1) $\left(\theta, \psi; e^{\frac{2ri\pi}{2^\lambda}}\right)$ la valeur $e^{\frac{2ri\pi}{4}} + \theta e^{\frac{-2ri\pi}{4}}$.

Notre conclusion est donc établie pour toutes les séries L_ν. On pourrait dès lors développer, sur la distribution des zéros de L_ν, une théorie analogue à celle de M. von Mangoldt (2). La seule remarque sur laquelle se fonde cet auteur, outre les propriétés communes à $\zeta(s)$ et aux séries L_ν, est que l'argument de $\zeta(s)$ reste fini lorsque le point d'affixe s décrit la droite $\mathfrak{R}(s) = a > 1$. Or cette propriété appartient également aux fonctions L_ν. On pourrait donc compléter l'analyse présentée à cet égard (3) par Piltz.

10. L'équation fondamentale utilisée par Dirichlet pour la démonstration de son théorème, est

$$(16) \qquad \sum_\nu \frac{\log L_\nu(s)}{\psi_\nu(m)} = \varphi(k)\left(\sum \frac{1}{q^s} + \frac{1}{2}\sum' \frac{1}{q^{2s}} + \frac{1}{3}\sum'' \frac{1}{q^{3s}} + \ldots\right),$$

où m est un entier quelconque premier avec k et où les signes $\sum, \sum', \sum'', \ldots$, s'étendent, le premier aux nombres premiers q tels que $q \equiv m \pmod{k}$, le second aux nombres premiers q tels que $q^2 \equiv m \pmod{k}$, etc. Pour $m = 1$, ceci donne

$$\log \prod_\nu L_\nu(s) = \varphi(k)\left(\sum \frac{1}{q^s} + \frac{1}{2}\sum' \frac{1}{q^{2s}} + \frac{1}{3}\sum'' \frac{1}{q^{3s}} + \ldots\right).$$

Donc *les séries de Dirichlet n'ont aucun zéro sur la droite* $\mathfrak{R}(s) = 1$, car la fonction $\prod_\nu L_\nu(s)$ satisfait aux conditions énumérées au n° 5.

(1) *Loc. cit.*, p. 144, formule (9). M. Lipschitz désigne par la lettre ψ la quantité que nous nommons τ_i.
(2) *Journal de Crelle*, t. 114.
(3) *Habilitationschrift*, Iéna, 1884.

II. — *Conséquences arithmétiques.*

11. Nous sommes bien loin, comme on le voit, d'avoir démontré l'assertation de Riemann-Stieltjes; nous n'avons même pas pu exclure l'hypothèse d'une infinité de zéros de $\zeta(s)$ s'approchant indéfiniment de la droite limite. Cependant le résultat auquel nous sommes parvenu suffit, à lui seul, pour démontrer les principales conséquences arithmétiques que l'on a, jusqu'ici, essayé de tirer des propriétés de $\zeta(s)$.

Tout d'abord on peut remarquer que l'équation

$$\sum_p \frac{1}{p^s} = -\log(s-1) + \text{quantité finie}$$

fournit déjà quelques renseignements sur la distribution des nombres premiers. Soit, en effet, a un nombre plus grand que 1, et désignons par N_λ le nombre des nombres premiers compris entre a^λ et $a^{\lambda+1}$. Le premier membre de l'équation précédente est compris entre $\sum\limits_\lambda \frac{N_\lambda}{a^{\lambda s}}$ et $\sum\limits_\lambda \frac{N_\lambda}{a^{(\lambda+1)s}}$. En posant $\frac{1}{a^{s-1}} = x$ et remarquant que $s - 1 = \frac{\log\frac{1}{x}}{\log a}$ peut être ici remplacé par $1 - x$, on peut écrire, à une quantité finie près, pour x plus petit que 1, mais tendant vers 1 :

$$\sum \frac{N_\lambda}{a^\lambda} x^\lambda > \log(1-x) > \frac{x}{a} \sum \frac{N_\lambda}{a^\lambda} x^\lambda,$$

d'où l'on déduit que, ε étant un nombre positif aussi petit qu'on veut, on aura une infinité de fois

$$N_\lambda > \frac{(1-\varepsilon)a^\lambda}{\lambda}$$

et une infinité de fois

$$N_\lambda < \frac{(1+\varepsilon)a^{\lambda+1}}{\lambda+1},$$

conclusions analogues à celles que donne, par exemple, M. Poincaré dans son *Mémoire sur l'extension aux nombres premiers complexes des inégalités de M. Tchebicheff* ([1]), et qui suffiraient, comme elles, à établir que si le rapport d'un nombre x à la somme des logarithmes des nombres premiers plus petits que lui a une limite, cette limite ne peut être que 1.

D'autres inégalités pourraient sans doute être tirées de ce fait que, quel que soit le nombre réel t différent de 0, la quantité $\sum \frac{1}{p^s}\cos(t\log p)$ reste finie lorsque s tend vers 1.

12. Dans son Mémoire précédemment cité, M. Cahen présente une démonstration du théorème énoncé par Halphen : *La somme des logarithmes des nombres premiers inférieurs à x est asymptotique à x.* Toutefois son raisonnement dépend de la proposition de Stieltjes sur la réalité des racines de $\zeta(\frac{1}{2}+ti)=0$. Nous allons voir qu'en modifiant légèrement l'analyse de l'auteur on peut établir le même résultat en toute rigueur.

A cet effet, au lieu de partir de l'intégrale $\frac{1}{2i\pi}\int_{a-\infty i}^{a+\infty i}\frac{x^z}{z}dz$, égale à 1 ou à 0 suivant que x est plus grand ou plus petit que 1, nous considérerons l'intégrale plus générale

$$J_\mu = \frac{1}{2i\pi}\int_{a-\infty i}^{a+\infty i}\frac{x^z}{z^\mu}dz.$$

Dans cette intégrale, comme dans la première, x est une quantité positive ainsi que a; μ est positif.

Lorsque μ est un entier, cette intégrale s'évalue par les mêmes méthodes que J, ou s'en déduit par une intégration par parties, déduite de l'identité

$$\frac{1}{z^\mu} = \frac{(-1)^{\mu-1}}{\Gamma(\mu)}\frac{d^{\mu-1}}{dz^{\mu-1}}\left(\frac{1}{z}\right).$$

La partie tout intégrée disparaît à l'infini et il vient

$$(17)\qquad J_\mu = \begin{cases} 0, \text{ si } x<1 \\ \dfrac{1}{\Gamma(\mu)}\log^{\mu-1}x, \text{ si } x>1. \end{cases}$$

([1]) *Journal de M. Jordan*, 4e série, t. VIII; 1892.

La même formule peut se démontrer pour μ non entier, auquel cas il est entendu que z^μ doit recevoir la détermination qui est réelle et positive pour $z = 0$. Pour $x < 1$, on intégrera le long d'un rectangle ayant un de ses côtés sur la droite $\mathcal{R}(z) = a$ et situé dans la région $\mathcal{R}(z) > a$, le second côté du rectangle augmentant indéfiniment comme la puissance $\mu'^{\text{ième}}(0 < \mu' < \mu)$ du premier. Le résultat est alors évident.

Pour $x > 1$, on commencera par supposer $\mu < 1$. On intégrera alors le long d'un contour ABCDEFGHA (*fig.* 1) composé encore d'un rectangle ayant un côté AB sur la droite $\mathcal{R}(z) = a$, mais situé dans la région $\mathcal{R}(z) < a$ et interrompu sur son côté DA par un lacet qui va à l'origine et en revient en suivant la partie négative de l'axe imaginaire. Si le côté BC augmente indéfiniment comme la puissance $\mu'^{\text{ième}}(0 < \mu' < \mu)$ de AB, l'intégrale prise le long des côtés qui s'éloignent à l'infini disparaît et il reste

$$J_\mu = \frac{1}{2i\pi} \lim \left(\int_{HG} + \int_{FE} \right).$$

Or, sur le chemin HG, l'argument de z est $-\frac{i\pi}{2}$, et, sur le

Fig. 1.

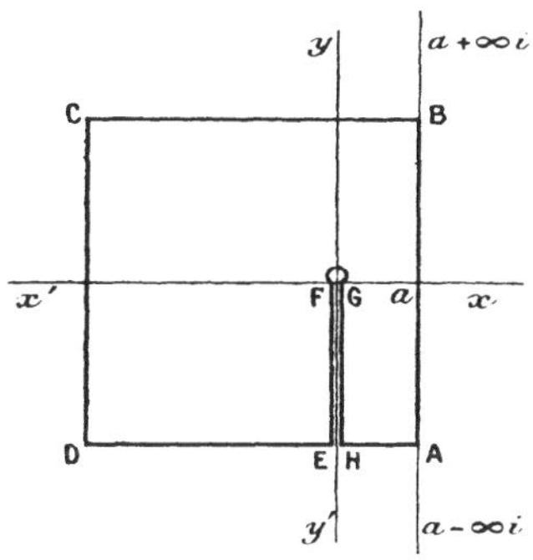

chemin FE, $\frac{3i\pi}{2}$. Il vient donc bien

$$J_\mu = \frac{\left(e^{\frac{\mu i\pi}{2}} - e^{-\frac{3\mu i\pi}{2}}\right)}{2\pi} \int_0^\infty \frac{x^{-it}\,dt}{t^\mu}$$

$$= -\frac{e^{-\frac{\mu i\pi}{2}} \sin \mu\pi}{i\pi} \int_0^\infty \frac{\cos(t \log x) - i \sin(t \log x)}{t^\mu}\,dt = \frac{\log^{\mu-1} x}{\Gamma(\mu)}.$$

Cette formule, établie pour $\mu < 1$, s'étendra au cas de $\mu > 1$ par une intégration par parties déduite de l'identité

$$\frac{1}{z^{\mu+m}} = \frac{(-1)^m \Gamma(\mu)}{\Gamma(\mu+m)} \frac{d^m}{dz^m} \frac{1}{z^\mu}.$$

13. Parallèlement à la voie suivie par M. Cahen, nous appliquerons la formule (17) à l'intégrale

$$(18) \qquad \psi_\mu(x) = -\frac{1}{2i\pi}\int_{a-\infty i}^{a+\infty i} \frac{x^z}{z^\mu} \frac{\zeta'(z)}{\zeta(z)} dz,$$

où a est un nombre quelconque plus grand que 1. En vertu du développement

$$\frac{\zeta'(z)}{\zeta(z)} = -\sum_p \log p \left(\frac{1}{p^s} + \frac{1}{p^{2s}} + \dots\right),$$

notre formule donne

$$(19) \quad \psi_\mu(x) = \frac{1}{\Gamma(\mu)}\left(\sum \log p \log^{\mu-1}\frac{x}{p} + \sum{}' \log p \log^{\mu-1}\frac{x}{p^2} + \dots\right),$$

le signe $\sum$ s'étendant aux nombres premiers plus petits que x, le signe $\sum'$ aux nombres premiers plus petits que $x^{\frac{1}{2}}$, etc.

14. L'avantage que nous trouvons à prendre $\mu > 1$ réside dans la convergence de la série $\frac{1}{|\alpha|^\mu}$, où α désigne successivement les zéros de $\zeta(z)$, convergence sur laquelle reposent, comme nous allons le voir, les raisonnements qui vont suivre.

Dans ces conditions, en effet, nous pouvons séparer de l'ensemble des racines α un nombre M de ces quantités assez grand pour que la somme $\sum \frac{1}{|\alpha|^\mu}$, étendue aux racines restantes, soit plus petite qu'un nombre positif quelconque ε. Aucun des α n'ayant sa partie réelle égale à 1, nous pourrons (*fig.* 2) tracer une parallèle CD à l'axe imaginaire, laissant à sa droite la parallèle $\mathcal{R}(z) = 1$ et à sa gauche les M premières racines α. Des points C, D de cette droite, nous ferons partir des parallèles CE, DF à l'axe réel, parallèles comprenant entre elles les M racines en question, ne passant par aucune autre racine, et que nous prolongerons jusqu'à rencontre en E, F respectivement avec deux droites OEG, OFH issues de l'origine et situées respectivement dans les deux angles formés par la partie négative de l'axe réel avec les deux directions de l'axe imaginaire. Enfin nous fermerons le contour d'intégration ABGECDFHA (*fig.* 2) par deux paral-

Fig. 2.

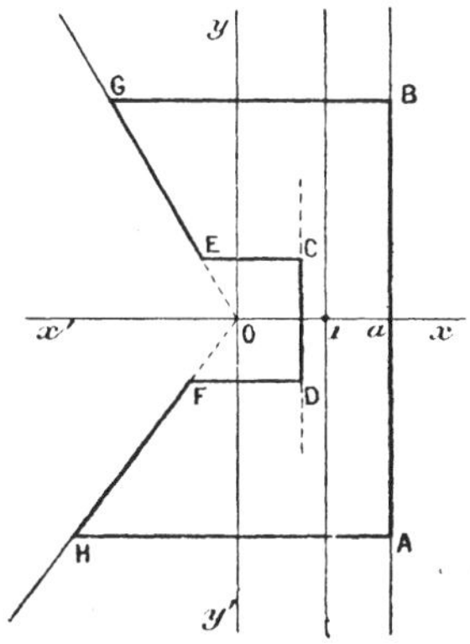

lèles variables BG, AH à l'axe réel (parallèles comprenant, bien entendu, CE et DF entre elles), rejoignant en A, B la droite $\mathcal{R}(z) = a$.

15. Je dis, en premier lieu, que l'on peut éloigner les parallèles BG, AH à l'infini, de telle façon que la partie de l'intégrale ψ_μ relative à ces droites tende vers zéro.

On peut suivre pour cela une marche analogue à celle qui est exposée dans mon Mémoire sur les propriétés des fonctions entières (¹). La méthode qui va suivre diffère légèrement de celle-là; elle me paraît plus avantageuse.

Soit A un nombre plus grand que l'unité. Traçons des parallèles à l'axe réel à des distances de cet axe représentées par A^3, A'^6, ..., $A^{3\lambda}$, ... Le nombre (²) des racines α, dont les coordonnées sont comprises entre $A^{3\lambda}$ et $A^{3\lambda+3}$ est au plus égal à $K\lambda A^{3\lambda}$, le nombre K

(¹) *Loc. cit.*, n° 29 et suiv.

(²) Chaque racine est, bien entendu, comptée un nombre de fois égal à son ordre de multiplicité.

étant fini ([1]), et il en sera de même *a fortiori* de l'intervalle $(A^{3\lambda+1}, A^{3\lambda+2})$; de sorte que si l'on range les racines α par ordre de coefficients de i croissants, il en existera au moins deux consécutives pour lesquelles les coefficients de i différeront d'une quantité supérieure à $\frac{A^{3\lambda+2} - A^{3\lambda+1}}{K\lambda A^{3\lambda}} = \frac{A(A-1)}{K\lambda}$.

Nous tracerons, à égale distance de ces deux racines, une parallèle à l'axe réel dont l'ordonnée sera désignée par z_0, et cette ordonnée aura, avec celle de toute racine α, une différence supérieure à $\frac{A(A-1)}{2K\lambda}$.

Or on a

$$(20)\quad \left\{\begin{aligned}\frac{\zeta'(z)}{\zeta(z)} &= \sum_\alpha \left(\frac{1}{z-\alpha} + \frac{1}{\alpha}\right) - \sum_\beta \left(\frac{1}{z-\beta} + \frac{1}{\beta}\right) - \frac{1}{z} + C \\ &= \sum_\alpha \frac{z}{\alpha(z-\alpha)} - \sum_\beta \frac{z}{\beta(z-\beta)} - \frac{1}{z} + C,\end{aligned}\right.$$

les α désignant les zéros, les β les pôles (réels et négatifs) de ζ, et C étant une constante. Lorsque z varie sur le segment BG de la parallèle d'ordonnée z_0, le rapport $\frac{z-\beta}{\beta}$ reste supérieur à un nombre fixe, indépendant de β, et il en est de même pour le rapport $\frac{z-\alpha}{\alpha}$, si l'ordonnée de α est extérieure à l'intervalle $(A^{3\lambda}, A^{3\lambda+3})$. Les parties correspondantes du second membre de l'équation (20) donnent donc le produit de z par une somme finie $\left(\text{puisque les sommes} \sum \frac{1}{\alpha^2}, \sum \frac{1}{\beta^2} \text{ sont finies}\right)$.

Quant aux termes correspondant aux racines α comprises entre les parallèles d'ordonnées $A^{3\lambda}$ et $A^{3\lambda+3}$, elles donneront, d'après ce qui a été dit plus haut, une somme moindre que $K\lambda A^{3\lambda} \frac{2K\lambda A^3}{A(A-1)}$, quantité de la forme $K' z_0 \log z_0$ (où K' est un nouveau nombre fini).

([1]) Il est clair qu'on peut se dispenser des précautions que nous prenons ici en utilisant les résultats obtenus par M. von Mangoldt sur la distribution des quantités α; la méthode du texte a l'avantage de s'appliquer chaque fois qu'on connaît le genre de la fonction étudiée.

On aura donc

$$\left|\frac{\zeta'(z)}{\zeta(z)}\right| < K' z_0 \log z_0 ;$$

d'où en reportant dans notre intégrale

$$\left|\int_{BG} \frac{x^z}{z^\mu} \frac{\zeta'(z)}{\zeta(z)} dz\right| < \frac{K' \log z_0}{z_0^{\mu-1}} \int |x^z| \, dz,$$

quantité infiniment petite pour z_0 infini.

16. L'intégrale prise le long de la droite indéfinie AB peut donc être remplacée par l'intégrale prise le long du contour indéfini HFDCEG, augmentée de la somme des résidus relatifs au pôle $z = 1$ et aux zéros α non compris entre les parallèles CE, DF.

Le résidu relatif au pôle $z = 1$ est $-x$.

Les résidus relatifs aux zéros de α non compris entre CE et DF ont une somme moindre que εx, où ε a été choisi aussi petit qu'on veut, et cela indépendamment de x.

Quant à l'intégrale prise le long du contour HFDCBG, elle est infiniment petite relativement à x. Cela est évident pour la partie finie FDCE, où il suffit de remarquer que $\frac{1}{z^\mu} \frac{\zeta'(z)}{\zeta(z)}$ est fini. Sur les parties infinies EG, FH, les rapports $\left|\frac{z-\alpha}{\alpha}\right|$, $\left|\frac{z-\beta}{\beta}\right|$ sont supérieurs à un nombre fixe, et, par conséquent, la quantité $\left|\frac{1}{z} \frac{\zeta'(z)}{\zeta(z)}\right|$ est finie. L'intégrale sur un de ces chemins est donc moindre que $K \int \left|\frac{x^z}{z^{\mu-1}}\right| |dz|$ (le nombre K étant fini), c'est-à-dire qu'une quantité finie, décroissante quand x croît.

$\psi_\mu(x)$ est donc asymptotique à x, car, pour rendre la différence $[x - \psi_\mu(x)]$ moindre que ηx, il suffira de choisir $\varepsilon < \frac{\eta}{2}$, puis x assez grand pour que l'intégrale $\int_{HFDCEG}$ soit inférieure à $\frac{\eta}{2} x$.

17. Dans l'expression (19) de $\psi_\mu(x)$, nous ferons abstraction des termes compris sous les signes $\sum$ autres que le premier. Le nombre de ces signes est, en effet, moindre que $\frac{\log x}{\log 2}$, et la plus grande des sommes correspondantes est la première, inférieure

— 217 —

elle-même à $\log\Gamma\left(1+x^{\frac{1}{2}}\right)\log^{\mu-1}x$, par conséquent (à un facteur fini près) à $x^{\frac{1}{2}}\log^{\mu}x$. Nous négligeons donc une quantité moindre que $x^{\frac{1}{2}}\log^{\mu+1}x$; et le résultat obtenu ci-dessus peut s'énoncer ainsi : *la somme* $\frac{1}{\Gamma(\mu)}\sum\log p\log^{\mu-1}\frac{x}{p}$, *étendue aux nombres premiers inférieurs à* x, *est asymptotique à* x.

Ce résultat (où il est entendu que nous devons supposer $\mu > 1$) diffère de l'énoncé d'Halphen : *la somme des logarithmes des nombres premiers inférieurs à* x *est asymptotique à* x. Nous allons voir qu'il le comprend comme cas particulier.

18. Pour cela, prenons $\mu = 2$, ce qui donne

$$\sum_0^x \log p \log\frac{x}{p} = x(1+\eta),$$

η étant (pour x assez grand) inférieur en valeur absolue à tel nombre qu'on voudra.

Dans cette relation, changeons x en $x(1+h)$ et retranchons membre à membre : il vient

$$\sum_0^x \log p\log(1+h) + \sum_x^{x(1+h)} \log p \log\frac{x(1+h)}{p} = x(h+\eta),$$

égalité dans laquelle le signe $\sum\limits_\alpha^\beta F(p)$ désigne la somme des valeurs de la fonction F pour les nombres premiers compris entre α et β.

Pour les nombres premiers qui figurent sous le second signe $\sum$, la quantité $\frac{x(1+h)}{p}$ est comprise entre 1 et $1+h$: on peut donc écrire, en divisant par $\log(1+h)$,

$$(21) \qquad \sum_0^x \log p < x\frac{(h+\eta)}{\log(1+h)}$$

$$\sum_0^{x(1+h)} \log p > x\frac{(h+\eta)}{\log(1+h)}.$$

Dans cette dernière, nous changerons x en $\frac{x}{1+h}$: elle deviendra

$$\sum_0^x \log p > x \frac{h+\eta}{(1+h)\log(1+h)}. \tag{22}$$

Les formules (21) et (22) démontrent l'énoncé d'Halphen. On voit, en effet, que $\sum_0^x \log p$ sera compris entre $x(1+\rho)$ et $x(1-\rho)$, si l'on a choisi h tel que

$$1-\frac{\rho}{2} < \frac{h}{(1+h)\log(1+h)} < \frac{h}{\log(1+h)} < 1+\frac{\rho}{2},$$

puis x assez grand pour que $\eta < \frac{\rho}{2}\log(1+h)$.

19. Les résultats qui précèdent s'étendent d'eux-mêmes aux séries de Dirichlet. On considérera l'intégrale

$$-\frac{1}{2i\pi}\int_{AB}\left[\sum_\nu \frac{1}{\psi_\nu(m)}\frac{L'_\nu(z)}{L_\nu(z)}\right]\frac{x^z}{z^\mu}dz, \tag{23}$$

où μ est un nombre plus grand que 1, les autres lettres ayant le même sens que dans les n^os **8-10**. Cette intégrale représente, à une quantité près infiniment petite relativement à x, le produit de $\varphi(k)$ par la somme des logarithmes des nombres premiers q congrus à m, suivant le module k et plus petits que x, multipliés respectivement par les valeurs correspondantes de $\log^{\mu-1}\frac{x}{q}$.

Or on peut raisonner sur cette intégrale exactement comme nous l'avons fait sur l'intégrale (18), car les propriétés de $\zeta(s)$, que nous avons utilisées et qui sont relatives à la distribution des zéros et au genre, ont été démontrées pour les séries de Dirichlet. La quantité qui figure sous le signe $\int$ dans l'intégrale (23) a pour pôle simple $z=1$, pôle de $\frac{L'_1(z)}{L_1(z)}$, et le résidu correspondant $-\frac{x}{\psi_1(m)}=-x$; les résidus relatifs aux autres pôles donnent une somme qu'on peut considérer comme négligeable vis-à-vis de x, ainsi que l'intégrale prise le long du contour GECDFH, comme il a été expliqué.

Donc l'intégrale (23) est asymptotique à x. En suivant la même marche qu'au numéro précédent nous reconnaissons que *la somme des logarithmes des nombres premiers inférieurs à x et compris dans une progression arithmétique déterminée de raison k est asymptotique à* $\frac{x}{\varphi(k)}$.

L'équation générale

$$\frac{1}{\Gamma(\mu)}\sum_{0}^{x} \log q \log^{\mu-1}\frac{x}{q} = \frac{x}{\varphi(k)}(1+\rho)$$

qui, comme nous venons de le voir, comprend la relation correspondant à $\mu = 1$, ne paraît pas pouvoir se déduire inversement de celle-ci; il serait intéressant de rechercher quels renseignements cette équation fournit sur l'ordre de grandeur de ρ, c'est-à-dire de l'erreur commise en remplaçant $\sum_{0}^{x} \log q$ par sa valeur asymptotique.

20. En terminant, je signalerai l'application possible de la même méthode aux séries de Weber (¹) et de Meyer (²), par lesquelles on étend le théorème de Dirichlet sur la progression arithmétique aux formes quadratiques. Une fois démontré que ces séries sont uniformes, la relation (³), analogue à celle donnée précédemment au nº 10, prouvera qu'elles ne s'annulent pas sur la droite $\mathfrak{R}(s) = 1$.

Dans le cas où le déterminant est négatif, et où l'on considère la forme quadratique seule (sans faire intervenir de progression arithmétique), une formule donnée par Weber (⁴) fournit la démonstration demandée; en même temps, elle fait connaître le genre de ces séries et fournit la relation correspondant au changement de s en $1 - s$. Les méthodes exposées dans le présent Mémoire sont donc dès à présent applicables à ce cas particulier.

(¹) *Math. Annalen*, t. XX, p. 301.

(²) *Journal de Crelle*, t. 103, p. 98; cf. Bachmann, *Analytische Zahlentheorie*, Ch. X; Leipzig, Teubner, 1894.

(³) Bachmann, *loc. cit.*, p. 291, ligne 6, formule (34).

(⁴) Bachmann, *loc. cit.*, p. 302, ligne 4.

Nota. — Pendant la correction des épreuves, je reçois communication des recherches que M. de la Vallée-Poussin consacre au même sujet dans les *Annales de la Société scientifique de Bruxelles* (¹). Nos raisonnements, trouvés d'une façon indépendante, ont quelques points communs : il est remarquable, en particulier, de constater que M. de la Vallée-Poussin a été conduit, lui aussi, à employer comme intermédiaire le fait que la fonction ζ n'a pas de racine de la forme $1 + ti$, quoique les procédés de démonstration soient tout à fait différents. Je crois qu'on ne refusera pas à ma méthode l'avantage de la simplicité.

Les critiques, adressées par M. de la Vallée-Poussin aux démonstrations fondées sur l'emploi de l'intégrale $\int_{a-\infty i}^{a+\infty i} x^z \frac{dz}{z}$, n'intéressent point la nôtre, fondée sur l'intégrale

$$\int_{a-\infty i}^{a+\infty i} \frac{x^z\,dz}{z^\mu} \quad (\mu > 1),$$

grâce au fait que cette dernière garde un sens, même lorsqu'on remplace chaque élément par son module.

(¹) Tome XX, 2ᵉ Partie: 1896.

12.4 C. de la Vallée Poussin (1899)

Sur la fonction $\zeta(s)$ de Riemann et le nombre des nombres premiers inférieurs a une limite donnée

Charles de la Vallée Poussin was born in 1866 in Belgium. Though he came from a family of diverse intellectual background, he was originally interested in becoming a Jesuit priest. He attended the Jesuit College at Mons, but eventually found the experience unappealing. He changed the course of his studies and obtained a diploma in engineering. While at the University of Leuven, de la Vallée Poussin became inspired to study mathematics under Louis-Philippe Gilbert. In 1891, he was appointed Gilbert's assistant. Following Gilbert's death one year later, de la Vallée Poussin was elected to his chair at Louvain. De la Vallée Poussin's early work focused on analysis; however, his most famous result was in the theory of prime numbers. In 1896, he proved the long-standing prime number theorem independently of Jacques Hadamard. His only other works in number theory are two papers on the Riemann zeta function, published in 1916. His major work was his *Cours d'Analyse*. Written in two volumes, it provides a comprehensive introduction to analysis, both for the beginner and the specialist. De la Vallée Poussin was elected to academies in Belgium, Spain, Italy, America, and France. In 1928, the King of Belgium conferred on de la Vallée Poussin the title of Baron. He died in Louvain, Belgium, in 1962 [115].

This paper details de la Vallée Poussin's historic proof of the prime number theorem. The author proves that $\pi(x) \sim \mathrm{Li}(x)$ by considering the zeros of $\zeta(\sigma + it)$. This paper contains a proof of the fact that $\zeta(\sigma + it) = 0$ implies $\sigma \neq 1$. From this fact, sometimes referred to as the Hadamard–de la Vallée Poussin Theorem, the desired result is deduced. The basics of de la Vallée Poussin's approach are explained in English in Chapter 3 of [155].

SUR LA FONCTION $\zeta(s)$ DE RIEMANN ET LE NOMBRE DES NOMBRES PREMIERS INFÉRIEURS A UNE LIMITE DONNÉE

PAR

Ch.-J. DE LA VALLÉE POUSSIN
Correspondant de l'Académie royale de Belgique

(Présenté à la Classe des sciences dans la séance du 4 juin 1898.)

INTRODUCTION

RELATIVE A L'OBJET DU MÉMOIRE (*).

Je me suis occupé à plusieurs reprises, dans les *Annales de la Société scientifique de Bruxelles,* de la théorie des nombres premiers.

La fonction de $\zeta(s)$ de Riemann joue dans cette théorie un rôle fondamental. J'ai démontré, pour la première fois, dans mes *Recherches sur la théorie des nombres premiers* (**), que la fonction $\zeta(s)$ n'a pas de racines de la forme $1 + \beta i$. M. Hadamard a également, avant d'avoir eu connaissance de mes recherches, trouvé le même théorème par une voie plus simple. L'importance de ce théorème est considérable, par le nombre des conséquences asymptotiques que l'on peut en déduire.

(*) Voir le rapport de M. Mansion sur ce Mémoire (*Bulletin de l'Académie royale de Belgique,* juillet 1898). — Depuis que ce Mémoire a été soumis à l'appréciation de l'Académie, nous avons refait tous les calculs numériques en y apportant plus de précision. Nous avons aussi perfectionné notre travail sur plusieurs points de détail, et nous avons tenu compte des conseils de M. Mansion. C'est ce qui explique les légères divergences que l'on remarquera entre le présent Mémoire et l'analyse que M. Mansion en a faite.

(**) *Annales de la Société scientifique de Bruxelles*, 1896.

(4)

Celle qui a peut-être le plus d'intérêt par le nombre considérable de travaux auxquels elle a donné lieu, s'exprime par le théorème suivant :

Le nombre des nombres premiers inférieurs à x *a pour expression asymptotique, lorsque* x *est grand,*

$$Li(x) = \lim_{\varepsilon=0} \int_0^{1-\varepsilon} \frac{dy}{ly} + \int_{1+\varepsilon}^{x} \frac{dy}{ly}$$

avec une erreur qui devient infiniment petite par rapport à Li (x) *quand* x *tend vers l'infini.*

La démonstration de ce théorème a été publiée pour la première fois dans un article de M. von Mangoldt (*).

On trouve aussi dans cet article des renseignements historiques qu'il est intéressant de reproduire (**).

Le pressentiment du théorème précédent a été d'abord exprimé par Dirichlet en 1838 (***), puis par Gauss en 1849 (IV). C'est Tchebychev qui a le premier donné deux limites certaines où l'on peut renfermer le nombre des nombres premiers (V). Mais cet intervalle, quand x croît indéfiniment, n'est pas une fraction infiniment petite de sa limite supérieure, car il reste égal au $\frac{1}{10^e}$ au moins de cette limite. Sylvester, dans

(*) *Ueber eine Anwendung der Riemann'sche Formel für die Anzahl der Primzahlen unter einer gegebenen Grenze.* (JOURN. F. D. REINE U. ANGEW. MATH., Bd 119.)

(**) On lira aussi avec intérêt, à ce point de vue, le savant rapport de M. Mansion sur notre Mémoire.

(***) LEJEUNE-DIRICHLET. *Werke*, Bd I, 2e note, p. 372.

(IV) Lettre à ENCKE. *Werke*, Bd II, pp. 444-447.

(V) *Journal de mathématiques*, t. XVII, 1852, p. 389.

(5)

son article sur le travail de Tchebychev (*) n'a pas réussi à pousser plus loin l'approximation. On voyait donc bien que $Li(x)$ représentait approximativement le nombre des nombres premiers $< x$, mais on n'avait pas encore pu démontrer que l'approximation surpassât $\frac{1}{10^n}$. Le travail de Riemann, qui devait finalement conduire à la solution du problème, était resté, jusque dans ces derniers temps, compliqué de difficultés qu'on ne savait surmonter. C'est la résolution au moins partielle de ces difficultés qui a permis d'établir le théorème rappelé ci-dessus et dont l'importance ne peut échapper, puisqu'il exprime que le logarithme intégral est, au sens mathématique du mot, l'expression asymptotique du nombre des nombres premiers $< x$, c'est-à-dire que l'erreur commise est infiniment petite par rapport à lui.

Lorsque l'on cherche à exprimer $Li(x)$ sous forme finie, on trouve une suite de termes de la forme

$$\frac{x}{lx} + \frac{x}{(lx)^2} + \frac{2x}{(lx)^3} + \frac{2.3x}{(lx)^4} + \ldots$$

Le premier terme est la valeur principale de $Li(x)$ et fournit donc aussi une expression asymptotique du nombre des nombres premiers $< x$. C'est ce que j'ai montré directement dans une note à la fin de l'article de M. von Mangoldt cité plus haut.

Le seul fait que $\zeta(s)$ n'a pas de racines de la forme $1 + \beta i$ ne permet pas de décider laquelle des deux expressions $\frac{lx}{x}$ ou $Li(x)$ est l'expression asymptotique la plus exacte.

(*) *On Tchebycheffs' theorem of the totalyty of the primnumbers comprised within given limits.* (AMERICAN JOURNAL OF MATHEMATICS, vol. IV, 1881, pp. 230-247.)

Pour pouvoir trancher la question, il faut savoir trouver une limite supérieure < 1 des parties réelles α des racines imaginaires $\alpha + \beta i$ de $\zeta(s)$. C'est ce qui n'était fait ni dans mon Mémoire ni dans celui de M. Hadamard. C'est ce qui est fait dans celui-ci, où nous donnons une limite inférieure de $1 - \alpha$. Cette limite fournie par le théorème du n° 30 est très petite, il est vrai, dépend de β et tend vers zéro quand β augmente. Elle n'en a pas moins une très grande valeur. Elle nous permettra de démontrer que $Li(x)$ représente le nombre $F(x)$ des nombres premiers $< x$ avec une erreur qui ne peut surpasser une quantité de la forme.

$$a \frac{x}{lx} e^{-b\sqrt{lx}},$$

où a et b sont des nombres fixes, et que, par conséquent, *le logarithme intégral est une expression asymptotique de* $F(x)$ *plus exacte que toutes ses expressions possibles sous forme finie.*

La méthode que nous allons suivre s'étend d'elle-même aux nombres premiers de la progression arithmétique. Nous y reviendrons plus tard.

SUR LA FONCTION $\zeta(s)$ DE RIEMANN ET LE NOMBRE DES NOMBRES PREMIERS INFÉRIEURS A UNE LIMITE DONNÉE

PREMIÈRE PARTIE.

SUR LES ZÉROS DE $\zeta(s)$.

CHAPITRE PREMIER.

CALCULS PRÉLIMINAIRES.

§ 1. *Rappel de quelques formules connues* (*).

1. La forme $\zeta(s)$ est définie par les relations

$$\zeta(s) = \sum_{n=1}^{\infty} \frac{1}{n^s} = \prod \left(1 - \frac{1}{p^s}\right)^{-1}, \tag{1}$$

la somme s'étendant à tous les entiers et le produit infini à tous les nombres premiers.

2. La fonction $\zeta(s)$ a des zéros sur l'axe réel qui sont les pôles de $\Gamma\left(\frac{s}{2}+1\right)$, mais elle a en outre une infinité de racines imaginaires dont la partie réelle est comprise entre 0 et 1. Nous désignerons, en général, ces racines imaginaires par

$$\rho = \alpha + \beta i \qquad (0 < \alpha < 1).$$

(*) Voir mes *Recherches sur la théorie des nombres premiers*, 1re partie.

3. Ces racines jouissent des propriétés suivantes :

1° Elles sont conjuguées deux à deux ;

2° A toute racine ρ correspond une racine $1-\rho$;

3° Si la partie réelle α de ρ n'est pas $\frac{1}{2}$, les racines ρ et $1-\rho$ forment avec leurs conjuguées un système de quatre racines distinctes :

$$\begin{array}{cc} \alpha+\beta i & \alpha-\beta i \\ 1-\alpha+\beta i & 1-\alpha-\beta i. \end{array}$$

4° M. Van Mangoldt a démontré que la valeur absolue de β est toujours supérieure à 12 (*).

4. La dérivée logarithmique de $\zeta(s)$ vérifie la relation

$$(2) \qquad \frac{\zeta'(s)}{\zeta(s)} = \frac{1}{2}\log \pi - \frac{1}{s-1} - \mathrm{D}\log.\Gamma\left(\frac{s}{2}+1\right) + \sum_{\rho}\frac{1}{s-\rho},$$

où les racines ρ doivent être rangées par ordre de modules croissants.

On en tire, en remplaçant $\frac{\zeta'(s)}{\zeta(s)}$ par sa valeur obtenue par la différentiation du produit infini (1),

$$(3) \qquad \sum_{\rho}\frac{1}{s-\rho} = \frac{1}{s-1} - \frac{1}{2}\log \pi + \mathrm{D}\log.\Gamma\left(\frac{s}{2}+1\right) - \sum\frac{lp}{p^s-1}.$$

Cette formule va jouer un rôle important dans notre étude. Nous allons d'abord en déduire une conséquence utile.

5. *Cas particuliers de la formule* (3). J'ai démontré dans mon Mémoire sur la fonction $\zeta(s)$, au n° 54, que l'on a

$$\lim_{s=1}\left(\frac{\zeta' s}{\zeta s} + \frac{1}{s-1}\right) = \mathrm{C},$$

(*) *Zu Riemann's Abhandlung : Ueber die Anzahl der Primzahlen unter einer gegebenen Grösse.* (JOURN. FÜR DIE REINE UND ANGEW. MATHEMATIK, Bd. 114.)

(9)

où C désigne la constante d'Euler. L'équation (2) donne donc, pour $s = 1$,

$$\sum \frac{1}{1-\rho} = -\frac{1}{2}\log\pi + \frac{1}{2}\frac{\Gamma'\left(\frac{3}{2}\right)}{\Gamma\left(\frac{3}{2}\right)} + C.$$

On a d'ailleurs, par une formule bien connue de Gauss,

$$\frac{\Gamma'\left(\frac{3}{2}\right)}{\Gamma\left(\frac{3}{2}\right)} + C = \int_0^1 \frac{1-x^{\frac{1}{2}}}{1-x}dx = 2\int_0^1 \frac{x\,dx}{1+x} = 2(1-l2),$$

de sorte que l'équation précédente donnera, eu égard à la propriété (2°) des racines ρ (n° 3),

$$(4)\ \ldots\ \sum \frac{1}{\rho} = \sum \frac{1}{1-\rho} = \frac{C}{2} + 1 - l2 - \frac{1}{2}l\pi.$$

Comme on a

$$l2 = 0,69314\ 71805\ 60,$$
$$\frac{1}{2}l\pi = 0,57236\ 49429\ 24,$$
$$\frac{1}{2}C = 0,28860\ 78324\ 51,$$

on en déduit

$$(4^{\text{bis}})\ \ldots\ \sum \frac{1}{\rho} = \sum \frac{1}{1-\rho} = 0,02309\ 57089\ 67\ldots$$

Les racines $\rho = \alpha + \beta i$ sont conjuguées deux à deux, de sorte que les parties imaginaires se détruisent dans les sommes précédentes. Par conséquent, la formule (4) peut s'écrire aussi

$$(5)\qquad \sum \frac{\alpha}{\alpha^2+\beta^2} = \sum \frac{1-\alpha}{(1-\alpha)^2+\beta^2} = 0,02309\ 57089\ 67\ldots$$

La formule (4) entraîne aussi comme conséquence, qui peut être utile,

$$(6)\qquad \sum \frac{1}{\rho} + \sum \frac{1}{1-\rho} = \sum \frac{1}{\rho(1-\rho)} = 0,04619\ 14179\ 34\ldots$$

§ 2. *Sur une expression approchée de* Dlog Γ(a).

6. Schaar a établi, pour a réel et positif, la formule (*)

$$\log \Gamma(a) = \left(a - \frac{1}{2}\right)\log a - a + \log\sqrt{2\pi} + \frac{1}{\pi}\int_0^{-\infty} \frac{a \log(1-e^{2\pi x})\,dx}{a^2+x^2}.$$

On obtient, en différentiant,

$$\frac{\Gamma'(a)}{\Gamma(a)} = \log a - \frac{1}{2a} + \frac{1}{\pi}\int_0^{-\infty} \frac{x^2-a^2}{(x^2+a^2)^2}\log(1-e^{2\pi x})dx,$$

puis, par le changement de la variable x en ax,

$$(7)\qquad \frac{\Gamma'(a)}{\Gamma(a)} = \log a - \frac{1}{2a} + \frac{1}{\pi a}\int_0^{-\infty} \frac{x^2-1}{(x^2+1)^2}\log(1-e^{2\pi a x})dx.$$

Cette formule vient d'être établie pour a réel et >0; mais comme ses deux membres représentent des fonctions synec-

(*) MEYER, *Bestimmte Integrale,* § 52. (Leipzig, 1871), et LIMBOURG, *Théorie de la fonction* Γ. (MÉMOIRE ACAD. DE BELGIQUE, t. XII.)

(11)

tiques de la variable imaginaire a pourvu que la partie réelle de a soit > 0, l'équation (7) subsiste aussi sous cette seule condition.

Soit donc

$$a = u + ti,$$

où u et t sont des variables réelles, la première positive; nous allons chercher une limite supérieure de l'intégrale de l'équation (7).

On a, pour $x < 0$,

$$\text{mod}\log(1 - e^{-2\pi a x}) < -\log(1 - e^{-2\pi u x});$$

par conséquent,

$$\text{mod}\int_0^{-\infty} \frac{x^2 - 1}{(x^2 + 1)^2}\log(1 - e^{2\pi a x})dx < \int_0^{-\infty}\log(1 - e^{2\pi u x})dx.$$

Par la substitution $e^{2\pi u x} = z$, cette dernière intégrale devient

$$\frac{1}{2\pi u}\int_1^0 \frac{l(1-z)}{z}dz = \frac{1}{2\pi u}\int_0^1\left(1 + \frac{z}{2} + \frac{z^2}{3} + \ldots\right)dz$$

$$= \frac{1}{2\pi u}\left(1 + \frac{1}{2^2} + \frac{1}{3^2} + \ldots\right) = \frac{\pi}{12u}.$$

Désignons donc par θ une quantité de module inférieur à l'unité; pour $a = u + ti$ et $u > 0$, l'équation (7) pourra se mettre sous la forme

$$(8) \quad \ldots\ldots \quad \frac{\Gamma'(a)}{\Gamma(a)} = \log a - \frac{1}{2a} + \frac{\theta}{12au}.$$

7. Cette équation va nous servir à trouver une expression approchée de la partie réelle du premier membre.

Convenons de désigner la partie réelle d'une quantité complexe en faisant précéder celle-ci de la caractéristique $\mathfrak{R}$. On a, pour $a = u + ti$,

$$\mathfrak{R} \log a = \frac{1}{2} \log(u^2 + t^2) = \log |t| + \frac{1}{2} \log\left(1 + \frac{u^2}{t^2}\right).$$

Si l'on suppose $u^2 < t^2$, on aura, en désignant par θ_1 une quantité > 0 et < 1,

$$\mathfrak{R} \log a = \log |t| + \frac{\theta_1}{2} \frac{u^2}{t^2}.$$

On a ensuite, eu égard à la signification de θ,

$$\mathfrak{R} \frac{\theta}{12au} = \frac{\theta}{12u\sqrt{u^2 + t^2}} = \frac{\theta_2}{12ut} \quad (-1 < \theta_2 < 1)$$

$$\mathfrak{R} \frac{1}{2a} = \frac{u}{2(u^2 + t^2)}.$$

Donc l'équation (8) donnera, pour $t^2 > u^2$,

$$(9)\ldots \begin{cases} \mathfrak{R} \dfrac{\Gamma'(u + ti)}{\Gamma(u + ti)} = \log |t| - \dfrac{u}{2(u^2 + t^2)} + \theta\left(\dfrac{1}{12u|t|} + \dfrac{u^2}{2t^2}\right) \\ \qquad (-1 < \theta < 1). \end{cases}$$

§ 3. *Évaluation de* $\Sigma \frac{1}{\alpha^2 + \beta^2}$.

8. Considérons de nouveau des sommes étendues à toutes les racines imaginaires $\rho = \alpha + \beta i$ de $\zeta(s)$.

On a identiquement

$$\sum \frac{1 - \alpha}{(1 - \alpha)^2 + \beta^2} = \sum \frac{1 - \alpha}{\alpha^2 + \beta^2} + \sum \frac{(1 - \alpha)(2\alpha - 1)}{(\alpha^2 + \beta^2)(\overline{1 - \alpha}^2 + \beta^2)}.$$

(13)

Donc, en ajoutant membre à membre avec

$$\sum \frac{1-\alpha}{(1-\alpha)^2+\beta^2} = \sum \frac{\alpha}{\alpha^2+\beta^2},$$

il vient

$$2\sum \frac{1-\alpha}{(1-\alpha)^2+\beta^2} = \sum \frac{1}{\alpha^2+\beta^2} + \sum \frac{(1-\alpha)(2\alpha-1)}{(\alpha^2+\beta^2)(\overline{1-\alpha}^2+\beta^2)}.$$

D'ailleurs la dernière somme ne devant pas changer par la permutation de α en $1-\alpha$, est aussi égale à

$$\frac{1}{2}\sum \frac{(1-\alpha)(2\alpha-1)+\alpha(1-2\alpha)}{(\alpha^2+\beta^2)(\overline{1-\alpha}^2+\beta^2)} = -\frac{1}{2}\sum \frac{(2\alpha-1)^2}{(\alpha^2+\beta^2)(\overline{1-\alpha}^2+\beta^2)},$$

et l'on trouve, par conséquent,

$$(10)\quad \sum \frac{1}{\alpha^2+\beta^2} = 2\sum \frac{\alpha}{\alpha^2+\beta^2} + \frac{1}{2}\sum \frac{(2\alpha-1)^2}{(\alpha^2+\beta^2)(\overline{1-\alpha}^2+\beta^2)}.$$

D'ailleurs, puisque β est toujours > 12 en valeur absolue et $(2\alpha-1)^2 < 1$, cette dernière somme est inférieure

$$\frac{1}{2}\frac{1}{12^2}\sum \frac{1}{\alpha^2+\beta^2} = \frac{1}{288}\sum \frac{1}{\alpha^2+\beta^2}$$

et l'équation (10) donne

$$(11).\quad \sum \frac{1}{\alpha^2+\beta^2} < \frac{2}{1-\frac{1}{288}}\sum \frac{\alpha}{\alpha^2+\beta^2} < \frac{0{,}046192}{1-\frac{1}{288}},$$

en vertu de l'équation (5).

§ 4. *Évaluation de* $\Sigma \frac{1}{(u-\rho)(1-\rho)}$.

9. Nous désignons dans ce qui suit par u une quantité réelle et positive > 1.

On a, ρ étant égal à $\alpha + \beta i$,

$$(12)\quad \begin{cases} -\sum \dfrac{1}{(u-\rho)(1-\rho)} = \sum \dfrac{\beta^2 - (u-\alpha)(1-\alpha)}{[(u-\alpha)^2+\beta^2][1-\alpha)^2+\beta^2]} \\ = \sum \dfrac{1}{(1-\alpha)^2+\beta^2} - \sum \dfrac{(u-\alpha)(u-2\alpha+1)}{[(u-\alpha)^2+\beta^2][(1-\alpha)^2+\beta^2]}. \end{cases}$$

Dans cette dernière somme, on a identiquement

$$\frac{1}{(u-\alpha)^2+\beta^2} = \frac{1}{\alpha^2+\beta^2} + \frac{(2\alpha-u)u}{[(u-\alpha)^2+\beta^2][\alpha^2+\beta^2]},$$

et par conséquent, en substituant cette valeur,

$$\sum \frac{(u-\alpha)(u-2\alpha+1)}{[(u-\alpha)^2+\beta^2][(1-\alpha)^2+\beta^2]} = \sum \frac{(u-\alpha)(u-2\alpha+1)}{(\alpha^2+\beta^2)(\overline{1-\alpha}^2+\beta^2)}$$
$$+ u\sum \frac{(u-\alpha)(2\alpha-u)(u-2\alpha+1)}{[(u-\alpha)^2+\beta^2][\alpha^2+\beta^2][(1-\alpha)^2+\beta^2]}.$$

Si 2α est $< u$, on a, au dernier numérateur,

$$0 > (u-\alpha)(2\alpha-u)(u-2\alpha+1) > -u^2(u+1).$$

Si $2\alpha > u$ et $\alpha < u$, on a

$$0 < (u-\alpha).(2\alpha-u)[1-(2\alpha-u)] < (u-\alpha)\frac{1}{4} < \frac{u}{4}.$$

On peut donc poser, pour $u > 1$, puisque $\beta^2 > 144$,

$$(13)\quad \begin{cases} \sum \dfrac{(u-\alpha)(u-2\alpha+1)}{[(u-\alpha)^2+\beta^2][(1-\alpha)^2+\beta^2]} = \sum \dfrac{(u-\alpha)(u-2\alpha+1)}{(\alpha^2+\beta^2)(\overline{1-\alpha}^2+\beta^2)} \\ \qquad\qquad + \tau \sum \dfrac{1}{(\alpha^2+\beta^2)(\overline{1-\alpha}^2+\beta^2)} \\ \qquad \left(-\dfrac{u^3(u+1)}{144} < \tau < \dfrac{u^2}{4.144}\right). \end{cases}$$

(15)

Au second membre de l'équation (13) ci-dessus, la première somme ne doit pas changer, si l'on remplace α par $1-\alpha$.

On peut donc y remplacer les numérateurs par

$$\frac{1}{2}[(u-\alpha)(u-2\alpha+1)+(u-1+\alpha)(u+2\alpha-1)]=\frac{2u^2-u+(2\alpha-1)^2}{2}.$$

Faisons cette substitution dans (13), puis portons la valeur trouvée dans (12) et changeons α en $1-\alpha$; il vient

$$-\sum\frac{1}{(u-\rho)(1-\rho)}=\sum\frac{1}{\alpha^2+\beta^2}-\left(u^2-\frac{u}{2}+\tau\right)\sum\frac{1}{(\alpha^2+\beta^2)(\overline{1-\alpha}^2+\beta^2)}$$
$$-\frac{1}{2}\sum\frac{(2\alpha-1)^2}{(\alpha^2+\beta^2)(\overline{1-\alpha}^2+\beta^2)}.$$

Remplaçons encore $\Sigma\frac{1}{\alpha^2+\beta^2}$ par sa valeur (10), nous trouverons l'expression définitive

$$(14)\left\{\begin{array}{l}-\sum\frac{1}{(u-\rho)(1-\rho)}=2\sum\frac{\alpha}{\alpha^2+\beta^2}-\left(u^2-\frac{u}{2}+\tau\right)\sum\frac{1}{(\alpha^2+\beta^2)(\overline{1-\alpha}^2+\beta^2)}\\ \qquad\left(-\frac{u^3(u+1)}{144}<\tau<\frac{u^2}{4.144}\right).\end{array}\right.$$

10. Dans le second membre de la formule (14), le premier terme est connu, sa valeur 0,0461914... est donnée par l'équation (5). La seconde somme est très petite et sa valeur pourrait se calculer avec une très grande approximation en posant $u=1$ dans la formule (14), mais nous pouvons nous dispenser de faire ce calcul. La formule (14) représente donc, avec une grande approximation, vu la petitesse de τ, la manière dont varie son premier membre quand $u>1$ varie dans le voisinage de l'unité. C'est à ce point de vue que nous allons l'utiliser. Si l'on remarque que, dans la formule (14), on a par (11)

$$\sum\frac{1}{(\alpha^2+\beta^2)(\overline{1-\alpha}^2+\beta^2)}<\frac{1}{12^2}\sum\frac{1}{\alpha^2+\beta^2}<\frac{0{,}046192}{144-\frac{1}{2}},$$

cette formule pourra aussi s'écrire

$$(15). \quad \left\{ \begin{aligned} & -\sum \frac{1}{(u-\rho)(1-\rho)} = 2\sum \frac{\alpha}{\alpha^2+\beta^2} - \sigma \\ & \sigma = \left(u^2 - \frac{u}{2} + \tau\right) \sum \frac{1}{(\alpha^2+\beta^2)(\overline{1-\alpha}^2+\beta^2)} \\ & < \frac{0{,}046192}{144 - \frac{1}{2}} \left(u^2 - \frac{u}{2} + \tau\right). \end{aligned} \right.$$

§ 5. *Évaluation approchée de* $\dfrac{\Gamma'\left(\frac{u}{2}+1\right)}{\Gamma\left(\frac{u}{2}+1\right)}$.

11. Nous allons encore indiquer la manière de faire ce calcul quand u est réel > 1 et voisin de l'unité.

On a, par la formule de Gauss,

$$\frac{\Gamma'(a)}{\Gamma(a)} + C = \int_0^1 \frac{1 - x^{a-1}}{1-x}\,dx,$$

l'équation

$$\frac{\Gamma'\left(\frac{3}{2}\right)}{\Gamma\left(\frac{3}{2}\right)} - \frac{\Gamma'\left(\frac{u}{2}+1\right)}{\Gamma\left(\frac{u}{2}+1\right)} = \int_0^1 \frac{x^{\frac{u}{2}} - x^{\frac{1}{2}}}{1-x}\,dx = 2\int_0^1 \frac{x^{u-1}-1}{1-x^2}\,x^2dx$$

$$= 2\int_0^1 (x^{u-1}-1)dx \sum_{n=1}^{\infty} x^{2n} = 2\sum_{n=1}^{\infty}\left(\frac{1}{2n+u} - \frac{1}{2n+1}\right)$$

$$= -2(u-1)\sum_{n=1}^{\infty} \frac{1}{(2n+1)^2 + (2n+1)(u-1)}$$

$$= -2(u-1)\sum_{n=1}^{\infty}\frac{1}{(2n+1)^2} + 2(u-1)^2\sum_{n=1}^{\infty}\frac{1}{(2n+1)^3} - \ldots$$

On a d'ailleurs (*)

$$\sum \frac{1}{(2n+1)^2} = \left(1 - \frac{1}{4}\right) \sum \frac{1}{n^2} - 1 = 0{,}23370\ 05501$$

$$\sum \frac{1}{(2n+1)^3} = \left(1 - \frac{1}{8}\right) \sum \frac{1}{n^3} - 1 = 0{,}05179\ 97903$$

$$\sum \frac{1}{(2n+1)^4} = \left(1 - \frac{1}{16}\right) \sum \frac{1}{n^4} - 1 = 0{,}01467\ 80316$$

$$\sum \frac{1}{(2n+1)^5} = \left(1 - \frac{1}{32}\right) \sum \frac{1}{n^5} - 1 = 0{,}00452\ 37628$$

$$\sum \frac{1}{(2n+1)^6} = \left(1 = \frac{1}{64}\right) \sum \frac{1}{n^6} - 1 = 0{,}00144\ 70766$$

$$\sum \frac{1}{(2n+1)^7} = \left(1 - \frac{1}{128}\right) \sum \frac{1}{n^7} - 1 = 0{,}00047\ 15487$$

$$\sum \frac{1}{(2n+1)^8} = \left(1 - \frac{1}{256}\right) \sum \frac{1}{n^8} - 1 = 0{,}00015\ 51790$$

$$\sum \frac{1}{(2n+1)^9} = \left(1 - \frac{1}{512}\right) \sum \frac{1}{n^9} - 1 = 0{,}00005\ 13452$$

En substituant ces valeurs dans la formule précédente, on obtient la formule propre au calcul que nous voulions établir.

§ 6. *Évaluation approchée de* $\frac{\zeta' u}{\zeta u}$.

12. Nous avons besoin d'évaluer approximativement $\frac{\zeta' u}{\zeta u}$ pour u réel et voisin de l'unité par excès. Ce calcul est devenu facile, moyennant les résultats qui précèdent.

Changeons s en u, puis s en 1 dans l'équation (2) et sous-

(*) Les sommes $\Sigma n^{-2} = s_2$, $\Sigma n^{-3} = s_3, \ldots$ ont été calculées par Legendre. (Voir Lacroix, *Traité du calcul différentiel et du calcul intégral*, t. III, seconde édition, p. 149.)

trayons membre à membre les deux équations ainsi obtenues, nous trouverons

$$\frac{\zeta' u}{\zeta u} + \frac{1}{u-1} = C - \frac{1}{2}\left[\frac{\Gamma'\left(\frac{u}{2}+1\right)}{\Gamma\left(\frac{u}{2}+1\right)} - \frac{\Gamma'\left(\frac{5}{2}\right)}{\Gamma\left(\frac{5}{2}\right)}\right]$$

$$-(u-1)\sum \frac{1}{(u-\rho)(1-\rho)}.$$

Donc, par la formule (15) et par celle du paragraphe précédent, il viendra

$$(16)\quad \begin{cases} \dfrac{\zeta' u}{\zeta u} + \dfrac{1}{u-1} = 0{,}57721\,5665 \qquad - 0{,}18750\,915\,(u-1) \\ \quad + 0{,}05179\,979(u-1)^2 - 0{,}01467\,803\,(u-1)^3 \\ \quad + 0{,}00452\,576\,(u-1)^4 - 0{,}00144\,708(u-1)^5 \\ \quad + 0{,}00047\,155(u-1)^6 - 0{,}00015\,518\,(u-1)^7 \\ \quad + 0{,}00005\,135(u-1)^8 - \ldots\ldots\ldots \\ \quad - (u-1)\sigma, \end{cases}$$

en posant, pour abréger, comme au n° 10,

$$(17)\ \ldots\quad \sigma = \left(u^2 - \frac{u}{2} + \tau\right)\sum \frac{1}{(\alpha^2+\beta^2)(\overline{1-\alpha}^2+\beta^2)}.$$

Nous allons faire deux applications de cette formule.

13. *Première application*. Posons d'abord $u - 1 = \frac{1}{2}$; l'équation (16) donnera, en changeant les signes,

$$-\frac{\zeta'\left(\frac{3}{2}\right)}{\zeta\left(\frac{3}{2}\right)} - 2 = -0{,}57721\,566 + 0{,}09375\,456 + \frac{\sigma}{2}$$

$$- 0{,}01294\,995 + 0{,}00183\,475$$

$$- 0{,}00028\,279 + 0{,}00004\,522$$

$$- 0{,}00000\,737 + 0{,}00000\,121$$

$$- 0{,}00000\,020 + \ldots.$$

On trouve une limite inférieure du second membre en bornant la série, qui a ses termes alternativement positifs et négatifs, aux termes écrits ci-dessus, dont le dernier est négatif, et en négligeant le terme complémentaire $\frac{\sigma}{\sqrt{2}}$, qui est positif. On obtient ainsi

$$(18)\ldots\ldots \quad -\frac{\zeta'\left(\frac{5}{2}\right)}{\zeta\left(\frac{5}{2}\right)} > 1{,}50517\,98$$

14. *Seconde application.* Nous allons supposer que $u - 1$, encore positif, est $< \frac{1}{12}$. Dans ce cas, comme le montre la formule (15), le dernier terme $(u-1)\sigma$ de la formule (16) est inférieur en valeur absolue à

$$\frac{1}{12}\frac{0{,}046192}{143{,}5}\left(u^2 - \frac{u}{2} + \tau\right) < 0{,}0000178,$$

cette valeur s'obtenant en remplaçant $\left(u^2 - \frac{u}{2} + \tau\right)$ par la quantité supérieure $\frac{2}{3}$.

On a donc par la formule (16), pour $u - 1 < \frac{1}{12}$, en arrêtant la série à son quatrième terme qui est négatif,

$$\begin{aligned}\frac{\zeta' u}{\zeta u} + \frac{1}{u-1} &> 0{,}57721\,56 - \frac{0{,}18751}{12} + \frac{0{,}05179}{(12)^2} - \frac{0{,}01468}{(12)^3} \\ &\quad - 0{,}00001\,78 \\ &> 0{,}56192\,29\end{aligned}$$

§ 7. *La partie réelle α des racines $\alpha + \beta i$ ne peut différer de $\frac{1}{2}$ que si $|\beta|$ est > 28* (*).

15. Reprenons la formule (3), qui peut s'écrire

$$(19)\quad \sum \frac{1}{s-\rho} = \frac{1}{s-1} - \frac{1}{2}l\pi + \frac{1}{2}\frac{\Gamma'\left(\frac{s}{2}+1\right)}{\Gamma\left(\frac{s}{2}+1\right)} - \sum \frac{lp}{p^{ms}},$$

(*) Dans le manuscrit soumis à l'examen des commissaires, nous avions

la dernière somme s'étendant à toute les puissances p^m des nombres premiers.

Nous aurons, dans ce paragraphe, à considérer une série d'inégalités. Pour en simplifier l'écriture, nous conviendrons que lorsque nous écrirons des inégalités entre quantités imaginaires, ces inégalités se rapporteront toujours aux parties réelles des deux membres.

Soit, comme dans les paragraphes précédents, $s = u + ti$, où u et t sont réels; on aura d'abord

$$\sum \frac{lp}{p^{mu}} + \sum \frac{lp}{p^{ms}} > 0,$$

car cette inégalité revient, par la convention qui précède, à

$$\sum \frac{lp}{p^{mu}} [1 + \cos(mtlp)] > 0,$$

qui est évidente, tous les termes de la somme étant positifs.

Changeons s en u dans la formule (19) et ajoutons membre à membre; on aura, en vertu de l'inégalité précédente,

$$(20)\ldots \left\{ \begin{aligned} &\sum \frac{1}{s-\rho} + \sum \frac{1}{u-\rho} < \frac{1}{s-1} + \frac{1}{u-1} \\ &\qquad - l\pi + \frac{1}{2}\left[\frac{\Gamma'\left(\frac{s}{2}+1\right)}{\Gamma\left(\frac{s}{2}+1\right)} + \frac{\Gamma'\left(\frac{u}{2}+1\right)}{\Gamma\left(\frac{u}{2}+1\right)} \right]. \end{aligned} \right.$$

16. C'est sur cette inégalité (20) que nous allons raisonner pour trouver une limite inférieure de β quand α diffère de $\frac{1}{2}$. Pour cela nous supposerons que $\alpha + \beta i$ est une racine de ζs où α diffère de $\frac{1}{2}$ et où $\beta > 0$, et nous poserons, dans la for-

déduit la limite inférieure de β d'une formule que nous établirons plus loin. (Voir le Rapport de M. Mansion.) La méthode nouvelle exposée dans ce paragraphe fournit une limite un peu plus élevée.

mule (20), $u = 2$ et $t = \beta$, donc $s = 2 + \beta i$, Nous allons transformer, dans cette hypothèse, les différents termes qui s'y trouvent.

Nous avons d'abord à étudier la somme $\Sigma \frac{1}{s-\rho}$. Tous les termes de cette somme ont leur partie réelle positive, car celle de s est > 1 et celle d $\alpha < 1$. Nous diminuerons donc la partie réelle de la somme, en limitant celle-ci à un certain nombre de ses termes.

Nous bornerons cette somme aux deux racines $\alpha + \beta i$ et $1 - \alpha + \beta i$ qui existent toujours et sont différentes, α étant supposé différent de $\frac{1}{2}$ (n° 3, 3°). On obtient ainsi, pour $s = 2 + \beta i$,

$$\text{(21)} \qquad \sum \frac{1}{s-\rho} > \frac{1}{2-\alpha} + \frac{1}{1+\alpha} = \frac{3}{(2-\alpha)(1+\alpha)} > \frac{4}{3},$$

ce minimum ayant lieu pour $\alpha = \frac{1}{2}$.

Passons à l'évaluation du second terme de la formule (20). On peut l'écrire comme suit :

$$\sum \frac{1}{u-\rho} = \sum \frac{1}{1-\rho} - (u-1) \sum \frac{1}{(u-\rho)(1-\rho)}$$
$$= \sum \frac{\alpha}{\alpha^2+\beta^2} - (u-1) \sum \frac{1}{(u-\rho)(1-\rho)}$$

et, par les formules (15) et (14),

$$\left| \begin{aligned} &\sum \frac{1}{u-\rho} = (2u-1) \sum \frac{\alpha}{\alpha^2+\beta^2} - (u-1)\sigma \\ &\sigma < \frac{0,046192}{143,5}\left(u^2 - \frac{u}{2} + \tau\right) \\ &\tau < \frac{u^2}{4.144}. \end{aligned} \right.$$

Dans le cas actuel, $u = 2$ et il vient

$$\sum \frac{1}{2-\rho} > 3 \sum \frac{\alpha}{\alpha^2+\beta^2} - \frac{0,046192}{143,5}\left(3 + \frac{1}{144}\right).$$

(22)

Ce dernier terme négatif est inférieur à 0,000968 en valeur absolue; le précédent se calcule par la formule (5). Il vient ainsi

$$(22) \ldots\ldots \sum \frac{1}{2-\rho} > 0{,}068319.$$

Nous arrivons maintenant au dernier terme de la formule (20).

Pour $u = 2$, on a par la formule de Gauss, déjà utilisée au n° 11,

$$(23) \ldots \frac{\Gamma'\left(\frac{u}{2}+1\right)}{\Gamma\left(\frac{u}{2}+1\right)} = \frac{\Gamma'(2)}{\Gamma(2)} = 1 - C.$$

Substituons les valeurs (21), (22) et (23) dans l'inégalité (20), où l'on a $u = 2$, $s = 2 + \beta i$; il viendra *a fortiori*, à cause du sens des inégalités 21 et 22,

$$\frac{4}{3} + 0{,}068319 < \frac{1}{1+\beta^2} + 1 - l\pi + \frac{1}{2}\left[\frac{\Gamma'\left(2+\frac{\beta}{2}i\right)}{\Gamma\left(2+\frac{\beta}{2}i\right)} + 1 - C\right]$$

On en tire, en multipliant par 2 et en effectuant les calculs numériques,

$$2{,}66998 < \frac{\Gamma'\left(2+\frac{\beta i}{2}\right)}{\Gamma\left(2+\frac{\beta i}{2}\right)} + \frac{2}{1+\beta^2}.$$

Remplaçons encore $\Gamma' : \Gamma$ par son expression tirée de la formule (9); nous aurons encore *a fortiori*, β étant positif,

$$2{,}66998 < \log\frac{\beta}{2} + \frac{1}{12\beta} + \frac{8}{\beta^2} - \frac{4}{16+\beta^2} + \frac{2}{1+\beta^2}$$

(23)

d'où, en observant que $\frac{4}{16+\beta^2} > \frac{4}{\beta^2} - 4\frac{16}{\beta^4}$ et $\frac{2}{1+\beta^2} < \frac{2}{\beta^2}$,

$$(24) \quad . \quad . \quad \log\frac{\beta}{2} > 2{,}66998 - \frac{1}{12\beta} - \frac{1}{\beta^2}\left(6 + \frac{64}{\beta^2}\right).$$

Comme on sait déjà que β est > 12, la dernière parenthèse est inférieure à 6,5; donc

$$\log\frac{\beta}{2} > 2{,}66998 - \frac{7{,}5}{144} > 2{,}617$$
$$\beta > 27{,}38.$$

On peut donc supposer maintenant $\beta > 27{,}38$ dans la formule (24). Dans ce cas, la parenthèse de cette formule est inférieure à 6,1 et l'on a

$$\log\frac{\beta}{2} > 2{,}66998 - \frac{100{,}58}{12(27{,}38)^2} > 2{,}65879$$
$$\beta > 28{,}558;$$

de là le théorème suivant :

17. Théorème. *La fonction* $\zeta(s)$ *n'a pas de racine imaginaire* $\alpha + \beta i$ *où* α *diffère de* $\frac{1}{2}$, *à moins que* β *ne soit plus grand que* 28,558 *en valeur absolue.*

CHAPITRE II.

Recherche d'une limite supérieure de la partie réelle des racines ρ.

§ 1er. *Démonstration d'une inégalité fondamentale.*

18. Posons $s = u + ti$. Nous supposerons, dans tout ce chapitre, que u est une quantité réelle, voisine de l'unité par excès, et que t est réel, positif et > 12.

L'équation (3), eu égard à l'identité

$$\frac{\zeta'(s)}{\zeta(s)} + \sum \frac{lp}{p^s - 1} = 0,$$

peut évidemment s'écrire

$$(25) \quad . \quad . \quad \sum_\rho \frac{1}{s-\rho} = \frac{1}{s-1} - \frac{1}{2}\log\pi + \frac{1}{2}\frac{\Gamma'\left(\frac{s}{2}+1\right)}{\Gamma\left(\frac{s}{2}+1\right)}$$

$$- \sum \left[\frac{lp}{p^u - 1} + \frac{lp}{p^s - 1}\right] - \frac{\zeta' u}{\zeta u}.$$

On a

$$\sum \frac{lp}{p^s - 1} = \sum \frac{lp}{p^{ms}} = \sum \frac{lp}{p^{m\varkappa}}[\cos mtlp + i \sin mtlp],$$

où la première somme s'étend à tous les nombres premiers successifs et les deux suivantes à toutes les puissances p^m des nombres premiers. On aura donc

$$\Re \sum \left[\frac{lp}{p^u - 1} + \frac{lp}{p^s - 1}\right] = \sum \frac{lp}{p^{m\varkappa}}(1 + \cos mtlp)$$

et la somme du second membre s'étend à toutes les puissances p^m des nombres premiers.

Remarquons maintenant qu'on a l'équation

$$\frac{1}{2}(1 - \cos mtlp)(1 + \cos mtlp) = \frac{1 - \cos 2mtlp}{4};$$

il viendra, en multipliant respectivement les termes de la dernière somme (qui sont tous positifs) par $\frac{1-\cos mtlp}{2}$, qui est > 0 et < 1,

$$\Re \sum \left[\frac{lp}{p^u - 1} + \frac{lp}{p^s - 1}\right] > \frac{1}{4} \sum \frac{lp}{p^{m\varkappa}}(1 - \cos 2mtlp).$$

Afin d'abréger l'écriture, nous allons convenir une fois pour toutes que, lorsque nous écrirons des inégalités entre quantités imaginaires, ces inégalités se rapporteront aux parties réelles des deux membres seulement.

Moyennant cette convention, l'équation (25) nous donnera par l'inégalité précédente

$$(26). \quad \left\{ \begin{aligned} &\sum \frac{1}{s-\rho} < \frac{1}{s-1} - \frac{1}{2}\log\pi + \frac{1}{2}\frac{\Gamma'\left(\frac{s}{2}+1\right)}{\Gamma\left(\frac{s}{2}+1\right)} \\ &\qquad - \frac{1}{4}\sum \frac{lp}{p^{mu}}(1-\cos 2mtlp) - \frac{\zeta' u}{\zeta u}. \end{aligned} \right.$$

Posons maintenant $s' = u + 2ti$.

On peut écrire une équation analogue à (25) en changeant le signe des deux termes en u qui se détruisent et en changeant s en s' dans cette équation; ce sera

$$\sum \frac{1}{s'-\rho} = \frac{1}{s'-1} - \frac{1}{2}\log\pi + \frac{1}{2}\frac{\Gamma'\left(\frac{s'}{2}+1\right)}{\Gamma\left(\frac{s'}{2}+1\right)} + \sum\left[\frac{lp}{p^{u}-1} - \frac{lp}{p^{s'}-1}\right] + \frac{\zeta' u}{\zeta u}.$$

Divisons cette équation par 4 et ajoutons-la membre à membre à l'inégalité précédente. Les parties réelles des sommes Σ se détruiront au second membre et nous trouverons ($s = u + ti$, $s' = u + 2ti$)

$$(27) \left\{ \begin{aligned} &\sum \frac{1}{s-\rho} + \frac{1}{4}\sum \frac{1}{s'-\rho} < -\frac{5}{8}\log\pi + \frac{1}{s-1} + \frac{1}{2}\frac{\Gamma'\left(\frac{s}{2}+1\right)}{\Gamma\left(\frac{s}{2}+1\right)} \\ &\qquad - \frac{3}{4}\frac{\zeta' u}{\zeta u} + \frac{1}{4}\frac{1}{s'-1} + \frac{1}{8}\frac{\Gamma'\left(\frac{s'}{2}+1\right)}{\Gamma\left(\frac{s'}{2}+1\right)}. \end{aligned} \right.$$

Cette inégalité qui a lieu entre les parties réelles des deux membres est l'inégalité fondamentale sur laquelle se base notre démonstration.

§ 2. *Limite supérieure de* ($1 - \alpha$), *qui résulte immédiatement de l'inégalité fondamentale* (27).

19. La démonstration que nous poursuivons résulte déjà de la relation (27). On remarque en effet que, si u est > 1, tous les termes du premier membre ont leur partie réelle positive, puisque ρ a la sienne < 1. L'inégalité (27) subsistera donc *a fortiori* si l'on borne le premier membre au seul terme de la première somme relatif à la racine particulière $\rho = \alpha + \beta i$. Si l'on fait alors $t = \beta$, ce terme unique se réduit à

$$\frac{1}{u - \alpha}.$$

Faisons maintenant tendre u vers l'unité; il n'y a qu'un seul terme au second membre de (27) qui puisse croître indéfiniment, c'est le premier de la seconde ligne, qui devient infini comme

$$\frac{3}{4} \frac{1}{u - 1}.$$

Donc α est < 1, sans quoi le premier membre de (27) finirait par surpasser le second. Mais on voit facilement que l'on peut obtenir une limite inférieure de la différence $1 - \alpha$.

Nous ne nous contenterons pas de la démonstration générale qui résulte de la formule (27), parce qu'il y a moyen d'abaisser davantage la limite supérieure de α, comme on le verra au § 4. Nous allons cependant exposer en détail cette démonstration, parce qu'elle est plus simple que celle que nous exposerons ensuite et rendra l'intelligence de celle-ci plus facile. D'ailleurs, les calculs que nous allons faire ne nous seront pas inutiles.

(27)

20. Substituons au second membre de la formule (27) les valeurs suivantes, tirées de la formule (9) :

$$\mathfrak{R}\frac{1}{2}\frac{\Gamma'\left(\frac{u+ti}{2}+1\right)}{\Gamma\left(\frac{u+ti}{2}+1\right)}=\frac{1}{2}\left[\log\frac{t}{2}-\frac{2(u+2)}{(u+2)^2+t^2}+\theta\left(\frac{1}{3(u+2)t}+\frac{(u+2)^2}{2t^2}\right)\right]$$

$$\mathfrak{R}\frac{1}{8}\frac{\Gamma'\left(\frac{u+2ti}{2}+1\right)}{\Gamma\left(\frac{u+2ti}{2}+1\right)}=\frac{1}{8}\left[\log t-\frac{2(u+2)}{(u+2)^2+4t^2}+\theta\left(\frac{1}{6(u+2)t}+\frac{(u+2)^2}{8t^2}\right)\right]$$

la formule (27) deviendra

$$(28)\quad \sum\frac{1}{s-\rho}+\frac{1}{4}\sum\frac{1}{s'-\rho}<\frac{5}{8}\log\frac{t}{\pi}-\frac{t2}{2}-\frac{3}{4}\frac{\zeta' u}{\zeta u}+\varphi(u,t)$$

où $\varphi(u, t)$ désigne l'expression, évanouissante avec $1 : t$,

$$(29)\ldots\left\{\begin{aligned}\varphi=&\frac{u-1}{(u-1)^2+t^2}+\frac{1}{4}\frac{u-1}{(u-1)^2+4t^2}-\frac{u+2}{(u+2)^2+t^2}\\&-\frac{1}{4}\frac{u+2}{(u+2)^2+4t^2}+\theta\left[\frac{3}{16(u+2)t}+\frac{17}{64}\left(\frac{u+2}{t}\right)^2\right].\end{aligned}\right.$$

Dans cette formule, θ est compris entre -1 et $+1$.

On peut aussi mettre $\varphi(u, t)$ sous la forme

$$(30)\ldots\left\{\begin{aligned}\varphi(u,t)=&-3\frac{t^2-(u-1)(u+2)}{[(u-1)^2+t^2][(u+2)^2+t^2]}\\&-\frac{3}{4}\frac{4t^2-(u-1)(u+2)}{[(u-1)^2+4t^2][(u+2)^2+4t^2]}\\&+\theta\left[\frac{3}{16}\frac{1}{(u+2)t}+\frac{17}{64}\left(\frac{u+2}{t}\right)^2\right].\end{aligned}\right.$$

Comme t est > 12, les deux premiers termes du second membre qui sont négatifs l'emportent certainement de beaucoup sur le second des termes entre crochets si $u - 1$ est petit; et l'on a, si φ est positif,

$$(31). \quad . \quad . \quad . \quad . \quad \varphi(u,t) < \frac{5}{16}\frac{1}{(u+2)t} < \frac{1}{16t},$$

car, u étant > 1, $u + 2$ sera > 3.

21. Nous allons utiliser immédiatement ce résultat. La formule (28) peut s'écrire, en abrégé,

$$(32) \quad \sum \frac{1}{s-\rho} + \frac{1}{4}\sum \frac{1}{s'-\rho} < \frac{3}{4}\frac{1}{u-1} + \frac{5}{8}\log t - m,$$

en posant

$$m = \frac{5}{8}\log\pi + \frac{l2}{2} + \frac{5}{4}\left(\frac{\zeta' u}{\zeta u} + \frac{1}{u-1}\right) - \varphi.$$

On a maintenant, pour $u - 1 \overline{\overline{<}} \frac{1}{12}$,

$$\left.\begin{aligned} \frac{5}{8}\log\pi &> 0{,}71545 \\ \frac{l2}{2} &> 0{,}34657 \\ \frac{3}{4}\left(\frac{\zeta' u}{\zeta u} + \frac{1}{u-1}\right) &> 0{,}42144 \end{aligned}\right\} = 1{,}48346...,$$

la dernière inégalité se déduisant de la formule finale du n° **14**. Donc

$$m > 1{,}4834 - \frac{1}{16t}.$$

22. Soit maintenant $\rho = \alpha + \beta i$ une racine de $\zeta(s)$; bornons le premier membre de la formule (32) au seul terme de la première somme relatif à cette racine et faisons $t = \beta$; il viendra *a fortiori*

$$(33). \quad . \quad . \quad . \quad \frac{1}{u-\alpha} < \frac{3}{4}\frac{1}{u-1} + h\log\beta - m,$$

en posant, en abrégé,

$$(34). \quad . \quad . \quad . \quad . \quad h = \frac{5}{8} \text{ et } m = 1{,}4834 - \frac{1}{16\beta}.$$

On tire alors de la formule (33)

$$u - \alpha > \frac{1}{\frac{3}{4}\frac{1}{u-1} + (h\log\beta - m)}$$

et, comme $1 - \alpha = u - \alpha - (u-1)$,

$$1 - \alpha > \frac{\frac{1}{4} - (u-1)(h\log\beta - m)}{\frac{3}{4}\frac{1}{u-1} + (h\log\beta - m)}.$$

Posons, pour simplifier, x pouvant être choisi arbitrairement avec u,

$$(35) \quad (u-1)(h\log\beta - m) = x, \quad \text{d'où} \quad u - 1 = \frac{x}{h\log\beta - m};$$

il viendra

$$1 - \alpha > \frac{1}{h\log\beta - m}\cdot\frac{x - 4x^2}{3 + 4x}.$$

Le maximum de cette fraction en x a lieu pour

$$x = \frac{2\sqrt{3} - 3}{4}, \quad \text{d'où} \quad \frac{x - 4x^2}{3 + 4x} = \frac{7}{4} - \sqrt{3}.$$

On en conclut, par conséquent, comme $\sqrt{3} = 1{,}73205$,

$$(36). \quad . \quad 1 - \alpha > \left(\frac{7}{4} - \sqrt{3}\right)\frac{1}{h\log\beta - m} = \frac{0{,}01795}{h\log\beta - m}.$$

$$(37). \quad . \quad u - 1 = \frac{2\sqrt{3} - 3}{4(h\log\beta - m)} = \frac{0{,}11602}{h\log\beta - m}.$$

23. Si l'on remplace dans ces expressions h et m par leurs valeurs (34), il vient

$$(38) \quad . \; . \; . \quad 1-\alpha > \frac{0{,}02872}{\log\beta - 2{,}3734 + \dfrac{1}{10\beta}}$$

$$(39) \quad . \; . \; . \quad u-1 = \frac{0{,}18563}{\log\beta - 2{,}3734 + \dfrac{1}{10\beta}}.$$

La formule (38) donne la limite inférieure de $1-\alpha$ que nous voulions obtenir. Mais, pour qu'on puisse l'appliquer, il faut, comme nous l'avons supposé, que $u-1$ soit $< \frac{1}{12}$. Cela aura lieu, d'après la formule (39), si

$$\log\beta - 2{,}3734 > 12 . 0{,}18563 = 2{,}2275$$

$$\log\beta > 4{,}6009,$$

et, par conséquent,

$$\beta > 99{,}56.$$

Remarque — Si l'on substitue cette valeur limite de β et de son logarithme dans la formule (38), il vient

$$1-\alpha > \frac{0{,}0287}{2{,}228} > 0{,}0128.$$

Si β est $< 99{,}56$ ou 100 environ, on ne peut plus appliquer la formule (38), mais la valeur de $1-\alpha$ sera au moins égale à cette limite 0,0128 que fournit la formule pour $\beta = 100$. En effet, on peut toujours supposer $u - 1 = \frac{1}{12}$ dans la formule (33); alors, β étant < 100, il est clair que cette formule donne pour $u-\alpha$ et, par suite, pour $1-\alpha$ une limite au moins égale à celle que l'on obtient pour $\beta = 100$, savoir 0,0128.

§ 3. *Transformation de l'inégalité fondamentale* (27).

24. *Objet de ce paragraphe.* Dans la démonstration faite au paragraphe précédent, on a négligé complètement les sommes qui figurent au premier membre de la formule (26), sauf un seulement de leurs termes. Nous allons reprendre la démonstration en nous proposant de tenir compte autant que possible de ces termes.

A cet effet, nous distinguerons deux espèces de sommes Σ étendues aux racines ρ. Nous désignerons par Σ' des sommes étendues à toutes les racines dont la partie réelle α est $\overline{\gtrless} \frac{1}{2}$ et par Σ'' des sommes étendues à toutes les racines dont la partie réelle est $> \frac{1}{2}$.

25. *Recherche d'une inégalité relative aux sommes* Σ'. Nous allons d'abord chercher une limite inférieure des sommes Σ' que nous venons de définir. A cet effet, soit u' une quantité réelle $> u$; on a (u étant lui-même > 1)

$$\frac{u-\alpha}{(u-\alpha)^2+(t-\beta)^2} > \frac{u-\alpha}{u'-\alpha}\,\frac{u'-\alpha}{(u'-\alpha)^2+(t-\beta)^2}.$$

Mais $\frac{u-\alpha}{u'-\alpha}$, où $u' > u > \alpha$, diminue quand α augmente et acquiert sa plus petite valeur $\frac{2u-1}{2u'-1} > \frac{1}{2u'-1}$ quand α atteint le maximum $\frac{1}{2}$ qui lui est assigné. Il vient donc, en se rappelant que les inégalités se rapportent aux parties réelles des deux membres seulement,

$$(40)\qquad \sum{}'\frac{1}{s-\rho} > \frac{1}{2u'-1}\sum{}'\frac{u'-\alpha}{(u'-\alpha)^2+(t-\beta)^2}.$$

On a ensuite, pour $0 \overline{\gtrless} \alpha \overline{\gtrless} \frac{1}{2}$,

$$\frac{\dfrac{u'-(1-\alpha)}{(u'-1+\alpha)^2+(t-\beta)^2}}{\dfrac{u'-\alpha}{(u'-\alpha)^2+(t-\beta)^2}}$$

$$= \frac{u'-1+\alpha}{u'-\alpha}\,\frac{(u'-\alpha)^2+(t-\beta)^2}{(u'-1+\alpha)^2+(t-\beta)^2} < \frac{u'-\alpha}{u'-1+\alpha} < \frac{u'}{u'-1},$$

et l'on en déduit, par l'addition de l'unité aux deux membres,

$$\frac{u'-(1-\alpha)}{(u'-1+\alpha)^2+(t-\beta)^2}+\frac{u'-\alpha}{(u'-\alpha)^2+(t-\beta)^2}<\frac{2u'-1}{u'-1}\,\frac{u'-\alpha}{(u'-\alpha)^2+(t-\beta)^2}.$$

L'inégalité (40) devient donc *a fortiori*

$$\sum{}'\frac{1}{s-\rho}>\frac{u'-1}{(2u'-1)^2}\sum{}'\left[\frac{u'-\alpha}{(u'-\alpha)^2+(t-\beta)^2}+\frac{(u'-1+\alpha)}{(u'-1+\alpha)^2+(t-\beta)^2}\right].$$

Or, la somme au second membre contient maintenant tous les termes réels de la somme non accentuée

$$\sum\frac{1}{u'+ti-\rho},$$

étendue à toutes les racines ρ, et même deux fois ceux où $\alpha=\frac{1}{2}$ s'il y en a. Donc on aura *a fortiori*

$$(41)\quad\ldots\quad\sum{}'\frac{1}{s-\rho}>\frac{u'-1}{(2u'-1)^2}\sum\frac{1}{u'+ti-\rho}.$$

Nous allons maintenant choisir u' de manière à rendre maximum le coefficient de la somme du second membre. La dérivée fournit l'équation $2u'-1=4(u'-1)$, d'où l'on tire

$$(42).\quad\ldots\ldots\quad u'=\frac{3}{2},\qquad\frac{u'-1}{(2u'-1)^2}=\frac{1}{8},$$

de sorte que l'inégalité (41) devient

$$\sum{}'\frac{1}{s-\rho}>\frac{1}{8}\sum\frac{1}{u'+ti-\rho}.$$

Ajoutons à cette inégalité celle qui s'en déduit par le changement de t en $2t$ et par conséquent de s en s' (n° 18); il viendra

$$\sum{}'\frac{1}{s-\rho}+\frac{1}{4}\sum{}'\frac{1}{s'-\rho}>\frac{1}{8}\left[\sum\frac{1}{u'+ti-\rho}+\frac{1}{4}\sum\frac{1}{u'+2ti-\rho}\right].$$

Dans le cas particulier où $u-1 \leqq \frac{1}{12}$ et où s a la même partie imaginaire $ti = \beta i$ que l'une des racines ρ, on peut encore ajouter $\frac{23}{39}$ au second membre de cette inégalité. En effet, celle-ci a été obtenue par la sommation d'une suite de relations de la forme

$$\frac{1}{s-\rho} > \frac{1}{8}\left[\frac{u'-\alpha}{(u'-\alpha)^2+(t-\beta)^2} + \frac{u'-1+\alpha}{(u'-1+\alpha)^2+(t-\beta)^2}\right]$$

où $u' = \frac{3}{2}$. Considérons celle où $t = \beta$ et, par suite, $s - \rho = u - \alpha$; la différence entre ses deux membres sera minimum si u reçoit sa plus grande valeur $\frac{13}{12}$ et α sa plus petite valeur 0. Ce minimum est $\frac{23}{39}$ et c'est *a fortiori* une limite inférieure de la différence entre les deux membres de l'inégalité obtenue après la sommation.

Dans la suite, nous supposerons toujours que s a la même partie imaginaire qu'une des racines ρ, de sorte qu'on aura l'inégalité

$$(45)\quad \left\{\begin{aligned} \sum{}' \frac{1}{s-\rho} + \frac{1}{4}\sum{}' \frac{1}{s'-\rho} > \frac{1}{8}\Bigg[\sum \frac{1}{u'+ti-\rho} \\ + \frac{1}{4}\sum \frac{1}{u'+2ti-\rho}\Bigg] + \frac{23}{39} \end{aligned}\right.$$

26. ***Recherche d'une inégalité relative aux sommes*** Σ''. Les sommes Σ se composent des sommes Σ' et Σ'' (n° 24); on aura donc en vertu de l'inégalité précédente

$$\begin{aligned} \sum{}'' \frac{1}{s-\rho} + \frac{1}{4}\sum{}'' \frac{1}{s'-\rho} < \left[\sum \frac{1}{s-\rho} + \frac{1}{4}\sum \frac{1}{s'-\rho}\right] \\ - \frac{1}{8}\left[\sum \frac{1}{u'+ti-\rho} + \frac{1}{4}\sum \frac{1}{u'+2ti-\rho}\right] - \frac{23}{39}. \end{aligned}$$

Remplaçons au second membre la première quantité entre crochets par le second membre de l'inégalité (28) et la seconde quantité entre crochets par le même second membre de (28), où l'on change seulement u en u'. Je dis que l'on aura *a fortiori*

$$(44)\quad \left\{ \begin{aligned} \sum'' \frac{1}{s-\rho} + \frac{1}{4} \sum'' \frac{1}{s'-\rho} < \frac{7}{16}\left(\frac{5}{4}\log\frac{t}{\pi} - l2\right) - \frac{23}{39} \\ - \frac{3}{4}\frac{\zeta' u}{\zeta u} + \varphi(u, t) \\ + \frac{5}{32}\frac{\zeta' u'}{\zeta u'} - \frac{1}{8}\varphi(u', t). \end{aligned} \right.$$

Comme nous avons remplacé les deux crochets par des quantités plus grandes en vertu de l'inégalité (28), nous devons encore justifier que leur différence a été augmentée en même temps. Pour cela, il suffira de montrer que la différence des deux membres de l'inégalité (28) diminue quand u augmente et est, par suite, moindre pour u' que pour u. La différence entre les deux membres de (28) est la même que celle des deux membres de (27) ou de (26). Celle-ci, d'après la manière dont la formule a été établie au n° (18), a pour valeur

$$\sum \frac{lp}{p^{mu}}(1 + \cos mtlp) - \frac{1}{4}\sum \frac{lp}{p^{mu}}(1 - \cos 2mtlp)$$

et, en la mettant sous la forme

$$\frac{1}{2}\sum \frac{lp}{p^{mu}}(1 + \cos mtlp)^2,$$

on reconnaît immédiatement qu'elle diminue quand u augmente. L'inégalité (44) est donc justifiée.

Reportons-nous encore au raisonnement que l'on a fait au n° 20 pour passer de la formule (30) à la formule (31). On peut reproduire sur la différence des termes correspondants de $\varphi(u,t)$ et de $\frac{1}{8}\varphi(u',t)$ le raisonnement que l'on a fait sur chacun des termes de $\varphi(u, t)$. Il vient ainsi

$$\varphi(u, t) - \frac{1}{8}\varphi(u', t) < \frac{1}{16t} + \frac{1}{8}\frac{1}{16t}.$$

(35)

La formule (44) peut donc se mettre sous la forme suivante, analogue à celle de la formule (32),

$$(45) \qquad \sum{}'' \frac{1}{s-\rho} + \frac{1}{4}\sum{}'' \frac{1}{s'-\rho} < \frac{3}{4}\frac{1}{u-1} + \frac{35}{64}\log t - m,$$

en posant, pour abréger,

$$(46).\ . \quad \begin{cases} m = \dfrac{3}{4}\left(\dfrac{1}{u-1} + \dfrac{\zeta' u}{\zeta u}\right) + \dfrac{7}{16}\left(\dfrac{5}{4} l\pi + l2\right) \\ \quad - \dfrac{3}{32}\dfrac{\zeta' u'}{\zeta u'} + \dfrac{23}{39} - \dfrac{9}{8}\dfrac{1}{16t}. \end{cases}$$

On a dans cette expression $u - 1 < \frac{1}{12}$ et $u' = \frac{3}{2}$.

§ 4. *Limite inférieure de* (1 — α) *qui résulte de la formule* (45).

27. La valeur de m donnée par (46) se calcule au moyen des résultats obtenus aux numéros 13 et 14. On a, $u - 1$ étant $\overline{\overline{<}}\ \frac{1}{12}$ et u' égal à $\frac{3}{2}$,

$$\frac{3}{4}\left(\frac{1}{u-1} + \frac{\zeta' u}{\zeta u}\right) > 0{,}42144\,22$$

$$-\frac{3}{32}\frac{\zeta' u'}{\zeta u'} > 0{,}14111\,06$$

$$\frac{7}{16}\left(\frac{5}{4} l\pi + l2\right) = 0{,}92927\,60.$$

On peut donc faire

$$(47)\ .\ .\ .\ .\ .\ . \qquad m = 2{,}08157\,23 - \frac{9}{8}\frac{1}{16t}.$$

28. Limitons maintenant le premier membre de la formule (45) à une seule racine $\alpha + \beta i$ et posons $t = \beta$. Il viendra *a fortiori*

$$(48)\ldots\quad \begin{cases} \dfrac{1}{u-\alpha} < \dfrac{3}{4}\dfrac{1}{u-1} + h\log\beta - m \\ h = \dfrac{35}{64} \\ m = 2{,}08157\,23 - \dfrac{9}{128\beta}. \end{cases}$$

29. Cette formule a exactement la même forme que la formule (33) sur laquelle on a raisonné au n° 22; il n'y a de changé que les valeurs de h et de m. Cette formule se transforme donc exactement comme celle du n° 22 et l'on trouve les formules correspondant à (36) et (37) :

$$(49)\qquad 1-\alpha > \left(\frac{7}{4} - \sqrt{3}\right)\frac{1}{h\log\beta - m} = \frac{0{,}01794\,92}{h\log\beta - m}$$

$$(50)\qquad u - 1 = \frac{2\sqrt{3}-3}{4(h\log\beta - m)} = \frac{0{,}11602\,54}{h\log\beta - m}.$$

En remplaçant maintenant h et m par leurs valeurs écrites plus haut, ces formules peuvent aussi prendre la forme

$$(51)\ldots\ldots\quad 1-\alpha > \frac{p}{\log\beta - \dfrac{m}{h}}$$

en posant, en abrégé,

$$(52)\ldots\ldots\quad \begin{cases} p = 0{,}03282\,14 \\ \dfrac{m}{h} = 3{,}80630\,36 - \dfrac{9}{70\beta}. \end{cases}$$

(37)

30. Pour que ces formules soient applicables, il faut, comme le suppose la démonstration, que la valeur de $(u - 1)$ donnée par (50) soit inférieure à $\frac{1}{12}$. Il faut donc que l'on ait

$$h \log \beta - m > 12.0{,}116025 = 1{,}5923.$$

Cela aura lieu, m étant $< 2{,}08157\,23$ (form. 48), si

$$h \log \beta > 3{,}473877,$$

$$\log \beta > 6{,}35223$$

et, *a fortiori*, si $\beta > 574$.

Dans ce cas, on a $9 : 70\,\beta < 0{,}000224$; la formule (52) donne

$$\frac{m}{h} > 3{,}806079$$

et, en substituant cette valeur dans la formule (51), on obtient le théorème suivant :

31. Théorème. *A partir de $\beta > 574$, on a, entre les parties réelles et imaginaires d'une racine $\alpha + \beta i$ de $\zeta(s)$, la relation*

$$(53) \quad \ldots\ldots \quad 1 - \alpha > \frac{0{,}03282\,14}{\log \beta - 3{,}806}.$$

Si β est < 574, la valeur de $1 - \alpha$ sera au moins égale à celle que fournit cette relation pour $\beta = 574$.

La dernière partie du théorème reste seule à démontrer, mais il suffit pour cela de reproduire la remarque qui termine le n° 23.

32. La formule (53) peut encore s'écrire sous une autre forme, qui est plus avantageuse au point de vue des démonstrations ultérieures.

(38)

Comme 3,806079 est le logarithme naturel de 44,9737, la formule s'écrira aussi sous la forme plus condensée

$$(54) \quad \ldots\ldots \quad 1 - \alpha > \frac{p}{l\beta - ln},$$

où l'on a

$$\begin{cases} p = 0{,}0328214 \\ n = 44{,}9737 \end{cases}$$

C'est cette formule (54) que nous allons appliquer maintenant dans la seconde partie du mémoire (*).

(*) En établissant la formule (54), nous n'avons pas encore tiré tout le parti possible de l'artifice utilisé au § 3. Nous indiquerons dans une note à la fin du Mémoire le moyen d'augmenter légèrement les valeurs de p et de n. Ces nouvelles valeurs seront applicables également dans la seconde partie du Mémoire.

DEUXIÈME PARTIE.

Lois asymptotiques relatives aux nombres premiers.

CHAPITRE III.

ÉVALUATION DE QUELQUES SOMMES OU FIGURENT LES RACINES ρ.

§ 1. *Évaluation de la somme* $\sum\limits_{\beta > b} \frac{y^{\alpha-1}}{\beta^2}$.

33. La somme indiquée dans le titre de ce paragraphe est supposée s'étendre à toutes les racines $\rho = \alpha + \beta i$ pour lesquelles β est positif et $> b$. Nous supposerons que b est assez grand pour que l'on puisse appliquer l'inégalité (54) du n° 32 dans tous les termes de cette somme.

34. Rappelons d'abord un théorème établi par M. *von Mangoldt* et que nous aurons à appliquer (*) :
Si h *et* k *sont deux nombres réels vérifiant la condition*

$$\operatorname{tg} 1 = 1{,}5541 \overline{\overline{<}} k \overline{\overline{<}} h - 4,$$

le nombre des racines $\alpha + \beta$i, *pour lesquelles* β *est compris entre* h — k *et* h + k, *est inférieur à* klh.

35. Pour utiliser ce théorème, écrivons la somme à évaluer sous la forme suivante :

$$\sum \frac{1}{\beta^{1+\varepsilon}} \cdot \frac{y^{\alpha-1}}{\beta^{1-\varepsilon}},$$

où $0 < \varepsilon < 1$.

(*) *Zu Riemann's Abhandlung : Ueber die Anzahl der Primzahlen unter einer gegebenen Grösse.* JOURN. F. DIE R. U. A. MATH., B. CXIV, p. 265.

On a, par la formule (54) de la première partie,

$$\frac{y^{\alpha-1}}{\beta^{1-\varepsilon}} < \frac{1}{\beta^{1-\varepsilon}} e^{-\frac{ply}{l\beta \; ln}}.$$

La dérivée par rapport à β de la fonction qui est au second membre est, à un facteur positif près,

$$ply - (1-\varepsilon)\left(l \cdot \frac{\beta}{n}\right)^2;$$

cette fonction deviendra donc maximum pour

$$(1). \quad . \quad l\frac{\beta}{n} = \sqrt{\frac{ply}{1-\varepsilon}}, \quad \text{d'où} \quad \beta = ne^{\sqrt{\frac{ply}{1-\varepsilon}}};$$

et ce maximum sera

$$\frac{1}{n^{1-\varepsilon}} e^{-2\sqrt{(1-\varepsilon)ply}}.$$

On a donc

$$(2) \quad . \quad . \quad . \quad . \quad . \quad \sum_{\beta > b} \frac{y^{\alpha-1}}{\beta^2} = \frac{e^{-2\sqrt{(1-\varepsilon)ply}}}{n^{1-\varepsilon}} \sum_{\beta > b} \frac{1}{\beta^{1+\varepsilon}}.$$

36. Tout revient donc maintenant à l'évaluation de cette dernière somme.

A cet effet, désignons par Σ_r^s une somme étendue aux valeurs de β comprises entre r et s; on aura

$$\sum_{\beta > b} \frac{1}{\beta^{1+\varepsilon}} = \sum_{b}^{b+2k} + \sum_{b+2k}^{b+4k} + \sum_{b+4k}^{b+6k} + \ldots$$

et, par le théorème de M. von Mangoldt (n° 34),

$$(5) . \quad . \quad \sum_{\beta > b} \frac{1}{\beta^{1+\varepsilon}} < k\left[\frac{l(b+k)}{b^{1+\varepsilon}} + \frac{l(b+3k)}{(b+2k)^{1+\varepsilon}} + \ldots\right].$$

Mais on a d'abord, par la formule $l(a+x) < la + \frac{x}{a}$,

$$\frac{l(b+3k)}{(b+2k)^{1+\varepsilon}} < \frac{l(b+2k)}{(b+2k)^{1+\varepsilon}} + \frac{k}{(b+2k)^{2+\varepsilon}};$$

(41)

ensuite, en intégrant par parties, on a

$$\frac{1}{2k}\int_0^{2k}\frac{l(b+x)dx}{(b+x)^{1+\varepsilon}}=\frac{l(b+2k)}{(b+2k)^{1+\varepsilon}}+(1+\varepsilon)\frac{1}{2k}\int_0^{2k}\frac{xl(b+x)dx}{(b+x)^{2+\varepsilon}}$$
$$-\frac{1}{2k}\int_0^{2k}\frac{xdx}{(b+x)^{2+\varepsilon}}.$$

Si donc b est assez grand pour qu'on ait

$$(1+\varepsilon)\int_0^{2k}\frac{xl(b+x)dx}{(b+x)^{2+\varepsilon}}>\int_0^{2k}\frac{xdx}{(b+x)^{2+\varepsilon}}+\frac{2k^2}{(b+2k)^{2+\varepsilon}},$$

ce qui aura certainement lieu si $lb>2$ (*) et ce que nous supposons (n° 33), il viendra

$$\frac{l(b+3k)}{(b+2k)^{1+\varepsilon}}<\frac{1}{2k}\int_0^{2k}\frac{l(b+x)dx}{(b+x)^{1+\varepsilon}}.$$

Comme cette relation s'applique aussi aux termes suivants de la formule (3), il viendra, le premier terme seul au second membre n'étant pas transformé dans cette formule,

$$(4)\quad .\ .\ .\quad \sum_{\beta>b}\frac{1}{\beta^{1+\varepsilon}}<\frac{kl(b+k)}{b^{1+\varepsilon}}+\frac{1}{2}\int_b^{\infty}\frac{lxdx}{x^{1+\varepsilon}}.$$

On trouve, en effectuant l'intégration,

$$\int_b^{\infty}\frac{lxdx}{x^{1+\varepsilon}}=\frac{1}{\varepsilon}\frac{lb}{b^{\varepsilon}}+\frac{1}{\varepsilon}\int_b^{\infty}\frac{dx}{x^{1+\varepsilon}}=\frac{1}{\varepsilon b^{\varepsilon}}\left[lb+\frac{1}{\varepsilon}\right];$$

(*) On déduit, en effet, de $lb>1+1$

$$\int_0^{2k}\frac{xl(b+x)dx}{(b+x)^{2+\varepsilon}}>\int_0^{2k}\frac{xdx}{(b+x)^{2+\varepsilon}}+\int_0^{2k}\frac{xdx}{(b+2k)^{2+\varepsilon}}$$
$$>\int^{2k}\frac{xdx}{(b+x)^{2+\varepsilon}}+\frac{2k^2}{(b+2k)^{2+\varepsilon}}.$$

et, en substituant dans (4),

$$(5). \quad . \quad . \quad . \quad \sum_{\beta>b} \frac{1}{\beta^{1+\varepsilon}} < \frac{kl(b+k)}{b^{1+\varepsilon}} + \frac{1}{2\varepsilon b^{\varepsilon}} \left[lb + \frac{1}{\varepsilon} \right].$$

37. Si l'on porte cette valeur dans (2), il vient

$$(6) \qquad \sum_{\beta>b} \frac{y^{\alpha-1}}{\beta^2} < \frac{1}{n^{1-\varepsilon} b^{\varepsilon}} \left[\frac{kl(b+k)}{b} + \frac{lb}{2\varepsilon} + \frac{1}{2\varepsilon^2} \right] e^{-2\sqrt{(1-\varepsilon)ply}}.$$

Cette formule, où l'on donnera à k sa plus plus petite valeur $k = 1,5541$ (n° 34), renferme encore un paramètre arbitraire $\varepsilon > 0$ et < 1, que l'on peut particulariser de différentes manières.

C'est ce que nous allons faire au paragraphe suivant.

§ 2. *Cas particuliers de la formule* (6) *qui précède.*

38. *Premier cas particulier.* Si l'on désire obtenir la formule la plus avantageuse au point de vue asymptotique, c'est-à-dire pour les valeurs indéfiniment croissantes de y, il faut poser

$$\varepsilon = \frac{1}{\sqrt{ply}},$$

ce qui rend minimum la valeur principale du second membre de (6). Si l'on observe alors que l'on a

$$\sqrt{1-\varepsilon} = \frac{1-\varepsilon}{\sqrt{1-\varepsilon}} > (1-\varepsilon)\left(1+\frac{\varepsilon}{2}\right) = 1 - \frac{\varepsilon}{2} - \frac{\varepsilon^2}{2}$$

d'où

$$\sqrt{(1-\varepsilon)ply} > \sqrt{ply} - \frac{1}{2} - \frac{1}{2\sqrt{ply}},$$

la formule (6) prendra la forme

$$(1) \quad . \quad . \quad . \quad . \quad . \quad . \quad \sum_{\beta>b} \frac{y^{\alpha-1}}{\beta^2} < A\, ply\, e^{-2\sqrt{ply}},$$

(43)

où A est une quantité qui reste finie. On peut faire, b étant $> n$,

$$(2) \ldots \quad A = \frac{e^{1+\frac{1}{\sqrt{ply}}}}{2n}\left[1 + \frac{lb}{\sqrt{ply}} + 2\frac{kl(b+k)}{bply}\right].$$

Comme la valeur de A n'est pas très petite, la formule (1) ne devient avantageuse que pour de très grandes valeurs de y. On peut obtenir d'autres formules également utiles, en donnant à ε une valeur fixe. En voici un exemple.

39. *Deuxième cas particulier.* On obtient une formule simple en donnant à ε la valeur $\frac{3}{4}$. Si, en outre, pour déterminer complètement la formule (6), nous y posons $b = 5^4 = 625$, nous obtiendrons

$$(3). \ldots \quad \sum_{\beta > 625} \frac{y^{\alpha-1}}{\beta^2} < 0{,}01611\, e^{\sqrt{ply}}.$$

Car on trouve, tous calculs faits,

$$\frac{1}{\sqrt[4]{n}.125}\left[\frac{kl(5^4+k)}{625} + \frac{8l5}{3} + \frac{8}{9}\right] < 0{,}01611.$$

Rappelons, à ce propos, les valeurs de p et de n (n° 32)

$$p = 0{,}03282, \qquad n = 44{,}9737.$$

§ 3. *Évaluation de la somme* $\sum \frac{y^{\alpha-1}}{\alpha^2+\beta^2}$ *ou* σ^2.

40. Nous supposerons dans le paragraphe actuel que cette somme s'étend à toutes les racines $\alpha + \beta i$ sans distinction.

Nous pouvons déjà la partager en deux parties, la première Σ' étendue aux racines $\alpha + \beta i$ où $\alpha \leqq \frac{1}{2}$, la seconde Σ'' étendue aux racines où $\alpha > \frac{1}{2}$.

On a donc

$$(1) \ldots \quad \sum \frac{y^{\alpha-1}}{\alpha^2+\beta^2} = \sum{}' \frac{y^{\alpha-1}}{\alpha^2+\beta^2} + \sum{}'' \frac{y^{\alpha-1}}{\alpha^2+\beta^2}.$$

41. La somme Σ' a comme limite supérieure la valeur obtenue en supposant que $\alpha = \frac{1}{2}$ pour toutes les racines. On a donc

$$(2). \quad . \quad . \quad . \quad \sum' \frac{y^{\alpha-1}}{\alpha^2+\beta^2} \leqq \frac{2}{\sqrt{y}} \sum \frac{\alpha}{\alpha^2+\beta^2} \leqq \frac{0,0462}{\sqrt{y}}$$

par la formule (5) du n° 5 de la première partie du mémoire.

42. Passons à la somme Σ''. Nous pouvons aussi la partager en deux parties

$$\sum_{\text{mod}\,\beta<b}'' \frac{y^{\alpha-1}}{\alpha^2+\beta^2} + \sum_{\text{mod}\,\beta>b}'' \frac{y^{\alpha-1}}{\alpha^2+\beta^2}.$$

On a identiquement

$$\sum'' \frac{1}{\alpha^2+\beta^2} = \sum'' \frac{\alpha}{\alpha^2+\beta^2} + \sum'' \frac{1-\alpha}{(1-\alpha)^2+\beta^2}$$
$$+ \sum'' \frac{1-\alpha}{\alpha^2+\beta^2} - \sum'' \frac{1-\alpha}{(1-\alpha)^2+\beta^2}.$$

La différence écrite en seconde ligne est négative, puisque $1-\alpha < \alpha$; ensuite la somme écrite en première ligne ne peut surpasser la somme non accentuée $\Sigma \frac{\alpha}{\alpha^2+\beta^2}$; il vient donc

$$\sum_{\text{mod}\,\beta<b}'' \frac{1}{\alpha^2+\beta^2} < \sum \frac{\alpha}{\alpha^2+\beta^2} < 0,0231.$$

Enfin, comme, pour $\beta < b$, on a $\alpha - 1 < -\frac{p}{lb-ln}$, il vient

$$(3). \quad . \quad . \quad . \quad . \quad \sum_{\text{mod}\,\beta<b}'' \frac{y^{\alpha-1}}{\alpha^2+\beta^2} < 0,0231\, e^{-\frac{p\,ly}{lb-ln}}.$$

Considérons maintenant la seconde partie

$$\sum_{\text{mod}\,\beta>b}'' \frac{y^{\alpha-1}}{\alpha^2+\beta^2}.$$

Celle-ci est inférieure à

$$\sum_{\alpha:\text{ud},\beta>b}'' \frac{y^{\alpha-1}}{\beta^2}.$$

Les formules des deux paragraphes précédents nous donnent une limite supérieure de cette somme. Elles nous donnent, en effet, une limite de

$$\sum_{\beta>b} \frac{y^{\alpha-1}}{\beta^2}.$$

Dans cette dernière somme, on ne tient pas compte d'une moitié des racines qui correspondent aux valeurs négatives de β; mais, dans la somme précédente, il y a aussi la moitié au moins des racines qui ne figurent pas, celles où $\alpha \overline{\overline{<}} \frac{1}{2}$. Les racines négligées dans les deux cas se correspondent deux à deux pour les mêmes valeurs de β^2, nous pouvons donc faire usage des mêmes limites.

43. En substituant ces valeurs, on trouve, suivant celles des deux formules (1) ou (3) du paragraphe précédent que l'on utilise,

$$(4) \qquad \sum \frac{y^{\alpha-1}}{\alpha^2+\beta^2} < \frac{0,0462}{\sqrt{y}} + 0,0231\, e^{-\frac{ply}{lb-ln}} + Aplye^{-2\sqrt{ply}},$$

ou bien

$$(5). \quad . \quad . \sum \frac{y^{\alpha-1}}{\alpha^2+\beta^2} < \frac{0,0462}{\sqrt{y}} + \frac{0,0231}{y^{0,0124}} + 0,01611\, e^{\sqrt{ply}},$$

car dans cette dernière formule $b = 625$, $lb = 6,43775$.

On a d'ailleurs dans les deux formules

$$p = 0,03282, \qquad n = 44,9757$$
$$ln = 3,806079$$

et l'expression de A se trouve au nº 38.

44. L'examen de la formule (4) permet d'écrire aussi immédiatement la formule, commode au point de vue des conséquences asymptotiques,

$$(6). \quad \ldots\ldots \quad \sum \frac{y^{\alpha-1}}{\alpha^2+\beta^2} < Bply e^{-2\sqrt{ply}},$$

où B est une quantité qui reste finie quand y tend vers l'infini et dont l'expression complète, assez longue, s'obtient en remplaçant A par sa valeur (2) au n° 38 dans la formule

$$B = A + 0{,}0231\, e^{-\frac{ply}{16-\pi}+2\sqrt{ply}} + \frac{0{,}0462}{\sqrt{y}} e^{2\sqrt{ply}}.$$

Les formules (4), (5) et (6) montrent que la somme que nous étudions tend vers zéro quand y tend vers l'infini. Mais, à cause de la petitesse de p, cette décroissance est extrêmement lente et ne devient considérable qu'à partir des valeurs de y ayant au moins une centaine de chiffres. C'est, en tout cas, tout ce que l'on peut conclure de ces formules.

45. La somme que nous étudions ici reviendra souvent dans la suite. Pour abréger l'écriture, nous représenterons sa racine carrée par σ, ou, en d'autres termes, nous poserons

$$(7) \quad \ldots\ldots \quad \sigma^2 = \sum \frac{y^{\alpha-1}}{\alpha^2+\beta^2}.$$

La quantité σ est donc une quantité qui tend vers zéro quand y tend vers l'infini, et les différentes formules de ce paragraphe permettent d'en déterminer approximativement la valeur quand y est donné.

En particulier, la formule (6) donnera

$$(8) \quad \ldots\ldots \quad \sigma < \sqrt{B}\sqrt{ply}\, e^{-\sqrt{ply}},$$

où B est une quantité qui reste inférieure à une limite fixe que l'on pourrait facilement assigner.

(47)

De là découle le théorème suivant :

46. *L'expression désignée par σ est infiniment petite en même temps que 1 : y et elle est d'un ordre de petitesse au moins égal à celui de l'expression*

$$\sqrt{\mathrm{p}ly}\, e^{-\sqrt{\mathrm{p}ly}},$$

où p = 0,03282..

CHAPITRE IV.

FORMULES ASYMPTOTIQUES.

§ 1. *Évaluation de* $\sum\limits_{p^m<y} lp$.

47. J'ai établi, dans la première partie de mon *Mémoire sur la théorie des nombres premiers* (ANNALES DE LA SOCIÉTÉ SCIENTIFIQUE DE BRUXELLES, t. XX), n° 52, la formule

$$(1) \quad \left\{ \begin{aligned} \sum_{p^m<y} \frac{lp}{p^m} - \frac{1}{y}\sum_{p^m<y} lp = ly - C - 1 + \frac{1}{y}\frac{\zeta'(o)}{\zeta(o)} \\ - \sum_{\rho} \frac{y^{\rho-1}}{\rho(\rho-1)} - \sum_{m} \frac{y^{-2m-1}}{2m(2m+1)}. \end{aligned} \right.$$

Dans cette formule, les sommes du premier membre sont étendues à toutes les puissances des nombres premiers qui sont inférieures à y.

En d'autres termes, on a posé, en abrégé,

$$\sum_{p^m<y} lp = \sum_{p<y} lp + \sum_{p<y^{\frac{1}{2}}} lp + \cdots$$

$$\sum_{p^m<y} \frac{lp}{p^m} = \sum_{p<y} \frac{lp}{p} + \sum_{p<y^{\frac{1}{2}}} \frac{lp}{p^2} + \cdots$$

On peut aussi remplacer dans le second membre $\frac{\zeta'o}{\zeta o}$ par sa valeur connue

$$\frac{\zeta'o}{\zeta o} = l(2\pi).$$

48. Multiplions (1) par dy et intégrons de 1 à x ; il viendra

$$x\sum_{p^m<x}\frac{lp}{p^m}-\sum_{p^m<x}lp-\int_1^x\frac{dy}{y}\sum_{p<y}lp=x(lx-1)+1-(C+1)(x-1)$$

$$+\ lx\frac{\zeta'o}{\zeta o}-\sum\frac{x^\rho-1}{\rho^2(\rho-1)}+\sum\frac{x^{-2m}-1}{4m^2(2m+1)}.$$

Mais on a par (1)

$$x\sum\frac{lp}{p^m}-\sum lp=xlx-(1+C)x+\frac{\zeta'o}{\zeta o}-\sum\frac{x^\rho}{\rho(\rho-1)}-\sum\frac{x^{-2m}}{2m(2m+1)}.$$

De cette équation soustrayons la précédente ; il vient

$$\int_1^x\frac{dy}{y}\sum_{p^m<y}lp=x-1-(1+C)+\frac{\zeta'o}{\zeta o}-\sum\frac{1}{\rho^2(1-\rho)}+\sum\frac{1}{4m^2(2m+1)}$$

$$-\ lx\frac{\zeta o}{\zeta o}-\sum\frac{x^\rho}{\rho^2}-\sum\frac{x^{-2m}}{4m^2}.$$

49. Remplaçons dans cette équation x par $(1+k)\,x$, où k est une constante positive quelconque, et soustrayons membre à membre ; nous trouvons, en divisant par k,

$$(2)\quad \left\{\begin{aligned}\frac{1}{k}\int_x^{(1+k)x}\frac{dy}{y}\sum_{p^m<y}lp &= x-\frac{l(1+k)}{k}\frac{\zeta'o}{\zeta o}\\ &-\sum\frac{(1+k)^\rho-1}{k}\frac{x^\rho}{\rho^2}\\ &-\sum\frac{(1+k)^{-2m}-1}{k}\frac{x^{-2m}}{4m^2}.\end{aligned}\right.$$

50. La somme qui est sous le signe d'intégration au premier membre étant constante ou croissante, on a

$$\frac{1}{k}\sum_{p^m<x}lp\int_x^{(1+k)x}\frac{dx}{x}\overline{\lessgtr}\frac{1}{k}\int_x^{(1+k)x}\frac{dx}{x}\sum lp\overline{\lessgtr}\frac{1}{k}\sum_{p^m<(1+k)x}lp\int_x^{(1+k)x}\frac{dx}{x},$$

c'est-à-dire

$$(3)\qquad \frac{l(1+k)}{k}\sum_{p^m<x} lp \leqq \frac{1}{k}\int_x^{(1+k)x}\frac{dx}{x}\sum lp \leqq \frac{l(1+k)}{k}\sum_{p^m<(1+k)x} l.p$$

La première de ces inégalités donne par (2)

$$(4)\quad \left\{\begin{aligned} \frac{l(1+k)}{k}\sum_{p^m<x} lp \leqq x &- \frac{l(1+k)}{k}\frac{\zeta'0}{\zeta 0} \\ &- \sum\frac{(1+k)^{\rho}-1}{k}\frac{x^{\rho}}{\rho^2} - \sum\frac{(1+k)^{-2m}-1}{k}\frac{x^{-2m}}{4m^2}. \end{aligned}\right.$$

Changeons dans (2) et (3) x en $\frac{x}{1+k}$, nous aurons de même, par la deuxième inégalité (3),

$$(5)\quad \left\{\begin{aligned} \frac{l(1+k)}{k}\sum_{p^m<x} lp \geqq \frac{x}{1+k} &- \frac{l(1+k)}{k}\frac{\zeta'0}{\zeta 0} \\ &- \sum\frac{1-(1+k)^{-\rho}}{k}\frac{x^{\rho}}{\rho^2} - \sum\frac{1-(1+k)^{+2m}}{k}\frac{x^{-2m}}{4m^2}. \end{aligned}\right.$$

Ces inégalités (4) et (5) peuvent aussi s'écrire *a fortiori*

$$(6)\quad \left\{\begin{aligned} \sum_{p^m<x} lp \leqq \frac{kx}{l(1+k)} - \frac{\zeta'0}{\zeta 0} &+ \sum\frac{(1+k)^{\alpha}+1}{l(1+k)}\left|\frac{x^{\rho}}{\rho^2}\right| \\ &- \sum\frac{(1+k)^{-2m}-1}{l(1+k)}\frac{x^{-2m}}{4m^2} \\ \sum_{p^m<x} lp \geqq \frac{kx}{(1+k)l(1+k)} - \frac{\zeta'0}{\zeta 0} &- \sum\frac{(1+k)^{-\alpha}+1}{l(1+k)}\left|\frac{x^{\rho}}{\rho^2}\right| \\ &+ \sum\frac{(1+k)^{2m}-1}{l(1+k)}\frac{x^{-2m}}{4m^2}. \end{aligned}\right.$$

51. On est ainsi conduit à prendre, pour valeur approchée du premier membre, la demi-somme des seconds. L'erreur commise Δ sera inférieure à leur demi-différence.

Posons en abrégé

$$(7). \quad . \quad . \quad . \quad . \quad \frac{1}{2}\frac{k(2+k)}{(1+k)l(1+k)} = 1 + w;$$

on trouvera

$$(8) \quad \begin{cases} \sum_{p^m < x} lp = (1+w)x - \frac{\zeta' o}{\zeta o} + \sum \frac{(1+k)^\alpha - (1+k)^{-\alpha}}{2l(1+k)} \left| \frac{x^\rho}{\rho^2} \right| \\ \qquad + \sum \frac{(1+k)^{2m} - (1+k)^{-2m}}{2l(1+k)} \frac{x^{-2m}}{4m^2} + \Delta \end{cases}$$

et l'on aura

$$(9) \quad \text{mod } \Delta < \frac{k^2 x}{2(1+k)l(1+k)} + \sum \frac{2+(1+k)^\alpha + (1+k)^{-\alpha}}{2l(1+k)} \left| \frac{x^\rho}{\rho^2} \right|$$
$$+ \sum \frac{2-(1+k)^{+2m} - (1+k)^{-2m}}{2l(1+k)} \frac{x^{-2m}}{4m^2}$$

52. Observons que l'on a toujours

$$2+(1+k)^\alpha + (1+k)^{-\alpha}) < 2+(1+k)+(1+k)^{-1} < 4 + \frac{k^2}{1+k};$$

on voit que nous aurons *a fortiori*, en négligeant une somme négative,

$$\text{mod } \Delta < \frac{x}{2}\left[\frac{k^2}{(1+k)l(1+k)}\left(1+\sum \frac{x^{\alpha-1}}{\alpha^2+\beta^2}\right) + \frac{4}{l(1+k)}\sum \frac{x^{\alpha-1}}{\alpha^2+\beta^2}\right].$$

On a posé, en abrégé,

$$\sum \frac{x^{\alpha-1}}{\alpha^2+\beta^2} = \sigma^2,$$

de sorte que l'on aura

$$\text{mod } \Delta < \frac{x}{2}\left[\frac{k^2}{(1+k)l(1+k)}(1+\sigma^2) + \frac{4\sigma^2}{l(1+k)}\right].$$

(51)

Ensuite, comme on a, par la formule (7),

$$\frac{1}{l(1+k)} = \frac{1+k}{k}\,\frac{1+w}{1+\frac{k}{2}},$$

il vient

$$\operatorname{mod}\Delta < \frac{x}{2}\left[k(1+\sigma^2) + \frac{1+k}{k}4\sigma^2\right]\frac{1+w}{1+\frac{k}{2}}$$

$$\operatorname{mod}\Delta < \frac{x}{2}\left[k + \left(\frac{4}{k} + 4 + k\right)\sigma^2\right]\frac{1+w}{1+\frac{k}{2}}.$$

Donnons à k la valeur

(10). . . . . $k = 2\sigma$;

il viendra

(11) . . . . $\operatorname{mod}\Delta < 2x\left[\sigma + \sigma^2 + \frac{1}{2}\sigma^3\right]\frac{1+w}{1+\frac{k}{2}}.$

L'évaluation de Δ est donc ramenée par cette formule à celle de la somme σ^2 qui a été étudiée au paragraphe 4 du chapitre précédent et à celle de w que nous allons réduire aussi à celle de σ.

53. La formule (8) peut se mettre sous la forme

(12). . . . . $\sum_{p^m<x} lp = (1+\varepsilon)x - \frac{\zeta' o}{\zeta o}$

en posant

(13). . $\begin{cases} \varepsilon = w + \sum \frac{(1+k)^{\alpha} - (1+k)^{\alpha}}{2l(1+k)}\left|\frac{x^{\rho-1}}{\rho^2}\right| \\ \quad + \sum \frac{(1+k)^{2m} - (1+k)^{-2m}}{2l(1+k)}\frac{x^{-2m-1}}{4m^2} + \frac{\Delta}{x} \end{cases}$

et w est donné par la formule (7).

54. En vertu de l'inégalité

$$(1 + k)^{\alpha} - (1 + k)^{-\alpha} < (1 + k) - (1 + k)^{-1} = \frac{2k + k^2}{1 + k},$$

on a

$$\sum \frac{(1 + k)^{\alpha} - (1 + k)^{-\alpha}}{2l(1 + k)} \left| \frac{x^{\rho - 1}}{\rho^2} \right| < \frac{2k + k^2}{2(1 + k)l(1 + k)} \sum \frac{x^{\alpha - 1}}{\alpha^2 + \beta^2}.$$

Opérons comme au n° 52 ; faisons la substitution

$$\sum \frac{x^{\alpha - 1}}{\alpha^2 + \beta^2} = \sigma^2,$$

remplaçons $(2k + k^2) : 2(1 + k)l(1 + k)$ par sa valeur $1 + w$; on aura

$$(14) \qquad \sum \frac{(1 + k)^{\alpha} - (1 + k)^{-\alpha}}{2l(1 + k)} \left| \frac{x^{\rho - 1}}{\rho^2} \right| < (1 + w)\sigma^2.$$

Considérons maintenant la quantité w qui est donnée par la formule (7) ; elle est positive et l'on a, par (7),

$$w = \frac{k + \frac{k^2}{2}}{k + \frac{k^2}{1.2} - \frac{k^3}{2.3} + \cdots} - 1 = \frac{\frac{k^2}{2.3} - \frac{k^3}{3.4} + \frac{k^4}{4.5} - \cdots}{1 + \frac{k}{1.2} - \frac{k^2}{2.3} + \cdots}$$

donc

$$w < \frac{k^2}{6} \frac{1 - \frac{k}{2} + \frac{3k^2}{10}}{1 + \frac{k}{2} - \frac{k^2}{6}} < \frac{k^2}{6}(1 - k + k^2)$$

et, k étant égal à 2σ,

$$(15) \; . \; . \; . \; . \; . \quad w < \frac{2}{3}\sigma^2(1 - 2\sigma + 4\sigma^2)$$

(53)

La dernière somme

$$\sum \frac{(1+k)^{2m} - (1+k)^{-2m}}{2l(1+k)} \frac{x^{-2m-1}}{4m^2}$$

tend avec une grande rapidité vers zéro quand x augmente.

En effet, elle est inférieure à

$$\frac{1}{2l(1+k)} \cdot \frac{1}{x} \cdot \sum \frac{1}{4m^2} \left(\frac{1+k}{x}\right)^{2m} < \frac{k}{2l(1+k)} \cdot \frac{1}{kx} \cdot \left(\frac{1+k}{x}\right)^2 \frac{\pi^2}{24}.$$

Mais on a $k = 2\sigma$ et, en remplaçant σ par sa valeur, on en tire

$$kx = 2\sqrt{x}\sqrt{\sum \frac{x^\alpha}{\alpha^2+\beta^2}} > 2\sqrt{x}\sqrt{\sum \frac{1}{\alpha^2+\beta^2}} > 0{,}4\sqrt{x},$$

par la formule (11) de la première partie du Mémoire. La somme en question sera donc encore inférieure à

$$\frac{\pi^2 k}{48l(1+k)} \frac{1}{0{,}4\sqrt{x}} \left(\frac{1+k}{x}\right)^2 < \frac{\pi^2}{19{,}2} \frac{(1+k)^3}{x^2\sqrt{x}} < \frac{1}{x^2\sqrt{x}}$$

et l'on aura, en définitive,

$$(16) \quad . \ . \quad \sum \frac{1+k)^{2m} - (1+k)^{-2m}}{2l(1+k)} \frac{x^{-2m-1}}{4m^2} < \frac{1}{x^2\sqrt{x}}.$$

En substituant ces différentes valeurs 11, 14 et 16 dans 13, on trouve

$$\text{mod}\ \varepsilon < w + (1+w)\left(2\sigma + \sigma^2 + \frac{\sigma^3}{1+\sigma}\right) + \frac{1}{x^2\sqrt{x}}$$

$$< w + (1+w)(2\sigma + \sigma^2) + \sigma^3 + \frac{1}{x^2\sqrt{x}};$$

puis, par (15), il vient *a fortiori*

$$(17) \qquad \text{mod}\ \varepsilon < 2\sigma + \frac{5}{3}\sigma^2 + \sigma^3 + \frac{8}{3}\sigma^4 + \frac{1}{x^2\sqrt{x}}.$$

L'évaluation de ε se ramène donc enfin à celle de la somme $\Sigma \frac{x^{\alpha} \, 1}{\alpha^2+\beta^2} = \sigma^2$ que nous avons faite au chapitre précédent.

55. En rapprochant la formule (17) de l'énoncé du théorème du n° 46 du chapitre précédent et en observant que y est remplacé ici par x, on obtient le théorème suivant :

Si l'on pose

$$\frac{1}{x} \sum_{p^m<x} lp = 1 + \eta,$$

la quantité η *tendra vers zéro quand* x *tendra vers l'infini, et cette quantité sera un infiniment petit d'un ordre de petitesse au au moins égal à celui de l'expression*

$$\sqrt{\mathrm{p}lx}\, e^{-\sqrt{\mathrm{p}lx}},$$

où $\mathrm{p} = 0{,}03282\ldots$

56. Si l'on remarque que l'on a

$$\sum_{p^m<y} lp - 2 \sum_{p^m<y^{\frac{1}{2}}} lp = \sum_{p<y} - \sum_{p<y^{\frac{1}{2}}} + \sum_{p<y^{\frac{1}{3}}} - \cdots < \sum_{p<y} lp,$$

on reconnaîtra immédiatement que la différence entre $\sum\limits_{p<y} lp$ et $\sum\limits_{p^m<y} lp$ est de l'ordre de $\sqrt{y}$. On peut donc énoncer cet autre théorème analogue au précédent :

Le rapport

$$\frac{1}{x} \sum_{p<x} lp,$$

où la somme s'étend aux logarithmes des nombres premiers $< x$, *est de la forme*

$$1 + \eta_1,$$

où η_1 *est infiniment petit avec* $\frac{1}{x}$ *et d'un ordre de petitesse au moins égal à celui de*

$$\sqrt{\mathrm{p}lx}\, e^{-\sqrt{\mathrm{p}lx}},$$

où l'on a encore $\mathrm{p} = 0{,}03282\ldots$

(55)

Ce théorème entraîne aussi le suivant :

57. *On peut assigner un nombre fixe* H, *indépendant de* x, *et tel qu'il y ait toujours au moins un nombre premier compris entre*

$$x \quad \text{et} \quad x + \mathrm{H}\sqrt{plx}\, e^{-\sqrt{plx}}x.$$

Les calculs qui précèdent permettraient facilement de trouver une valeur de ce nombre H, mais, comme elle serait assez grande, son calcul ne présenterait pas grand intérêt.

§ 2. *Évaluation de* $\sum\limits_{p^m < x} \frac{lp}{p^m}$.

58. Remplaçons dans la formule (1) du paragraphe précédent Σlp par sa valeur tirée de la formule (12) et changeons y en x; il viendra

$$\sum_{p^m < x} \frac{lp}{p^m} = lx - \mathrm{C} + \varepsilon - \sum_\rho \frac{x^{\rho-1}}{\rho(\rho-1)} - \sum_m \frac{x^{-2m-1}}{2m(2m+1)}.$$

L'évaluation de ε se fait par la formule (17) du paragraphe précédent. Il reste donc à évaluer

$$\sum_\rho \frac{x^{\rho-1}}{\rho(\rho-1)}.$$

Cette somme est inférieure en valeur absolue à

$$\sum \frac{x^{\alpha-1}}{\sqrt{\alpha^2+\beta^2}\sqrt{(1-\alpha)^2+\beta^2}} = \sum \frac{x^{\alpha-1}}{(\alpha^2+\beta^2)\sqrt{1+\dfrac{1-2\alpha}{\alpha^2+\beta^2}}}.$$

Les termes de cette dernière somme sont inférieurs à $\frac{x^{\alpha-1}}{\alpha^2+\beta^2}$, sauf si $\alpha > \frac{1}{2}$, et, dans ce cas, β étant > 28, ils sont inférieurs à

$$\frac{x^{\alpha-1}}{(\alpha^2+\beta^2)\sqrt{1-\dfrac{1}{(28)^2}}} < 1{,}00064\,\frac{x^{\alpha-1}}{\alpha^2+\beta^2},$$

Il vient donc

$$\text{mod} \sum \frac{x^{\rho-1}}{\rho(\rho-1)} < 1{,}00064\, \sigma^2.$$

Nous sommes donc ramenés encore une fois aux évaluations du § 4 du chapitre III.

On déduit de là et du théorème du n° 46 le théorème suivant :

59. *Si l'on pose*

$$\sum_{p^m < x} \frac{lp}{p^m} = lx - C + \eta_2,$$

la quantité η_2 *tendra vers zéro quand* x *tendra vers l'infini et elle sera un infiniment petit d'un ordre au moins aussi élevé que celui de l'expression*

$$\sqrt{plx}\, e^{-\sqrt{plx}},$$

où p = 0,03282.

60. On peut établir un théorème analogue relativement à la somme

$$\sum_{p<x} \frac{lp}{p-1},$$

qui s'étend aux nombres premiers $< x$. En effet, les deux sommes

$$\sum_{p<x} \frac{lp}{p-1} \quad \text{et} \quad \sum_{p^m<x} \frac{lp}{p^m}$$

ont une différence égale à

$$\sum_{p>x^{\frac{1}{2}}}^{p<x} \frac{lp}{p^2} + \sum_{p>x^{\frac{1}{3}}}^{p<x} \frac{lp}{p^3} + \sum_{p>x^{\frac{1}{4}}}^{p<x} \frac{lp}{p^4} + \cdots$$

et par suite inférieure à

$$lx \left[\sum_{n>x^{\frac{1}{2}}} \frac{1}{n^2} + \sum_{n>x^{\frac{1}{3}}} \frac{1}{n^3} + \sum_{n>x^{\frac{1}{4}}} \frac{1}{n^4} + \cdots \right]$$

(57)

où n est maintenant un entier quelconque. Le premier des termes entre crochets est le plus grand et le nombre de ces termes est égal au degré de la plus grande puissance de 2 contenue dans x, et ne surpassera pas $\frac{lx}{l2}$. La somme ci-dessus sera donc évidemment inférieure à

$$\frac{(lx)^2}{l2}\sum_{n>\sqrt{x}}\frac{1}{n^2}<\frac{(lx)^2}{\sqrt{x}\,l2}$$

et tendra vers zéro plus rapidement que la limite assignée à n dans le théorème précédent. Donc la première des sommes (4) vérifie ce théorème comme la seconde.

D'où le théorème :

61. *Si l'on pose l'équation*

$$\sum_{p<x}\frac{lp}{p-1}=lx-C+\eta_5,$$

la quantité η_5 *tendra vers zéro avec* $\frac{1}{x}$ *et elle sera d'un ordre de petitesse au moins égal à celui de la fonction*

$$\sqrt{\mathrm{p}lx}\,e^{-\sqrt{\mathrm{p}lx}},$$

où $\mathrm{p}=0{,}03282\ldots$

CHAPITRE V.

NOMBRE DES NOMBRES PREMIERS INFÉRIEURS A UNE LIMITE DONNÉE.

62. Pour simplifier l'écriture, nous désignerons par $F(x)$ la fonction qui exprime combien il y a de nombres premiers $< x$ et nous définirons une autre fonction $f(x)$ par l'équation

$$(1).\quad\ldots\quad f(x)=F(x)+\frac{1}{2}F(x^{\frac{1}{2}})+\frac{1}{3}F(x^{\frac{1}{3}})+\cdots$$

63. Nous allons utiliser une formule que j'ai démontrée dans mes *Recherches sur la théorie des nombres premiers* au n° 28 de la première partie, et que voici :

$$(2)\quad \begin{cases} \sum_{p^m<x} \frac{lp}{p^{mu}} - \frac{1}{x^u}\sum_{p^m<x} lp = \frac{1}{x^u}\frac{\zeta' o}{\zeta o} - \frac{\zeta' u}{\zeta u} + \left(\frac{1}{1-u} - 1\right)x^{1-u} \\ \qquad - \sum_\rho \frac{u x^{\rho-u}}{\rho(\rho-u)} - \sum \frac{u x^{-2m-u}}{2m(2m+u)}. \end{cases}$$

On a, dans cette relation,

$$\frac{\zeta' o}{\zeta o} = l(2\pi),$$

et l'on peut, si l'on veut, substituer cette valeur dans l'équation.

Le variable u est quelconque, réelle ou imaginaire.

Multiplions l'équation par du et intégrons de u à ∞ ; il viendra

$$\begin{aligned} &\sum_{p^m<x} \frac{1}{m}\frac{1}{p^{mu}} - \frac{1}{x^u lx}\sum_{p^m<x} lp \\ &= \frac{1}{x^u lx}\frac{\zeta' o}{\zeta o} - \frac{x^{1-u}}{lx} + \int_u^\infty \left(\frac{x^{1-u}}{1-u} - \frac{\zeta' u}{\zeta u}\right) du \\ &- \sum_\rho \int_u^\infty \frac{u x^{\rho-u} du}{\rho(\rho-u)} - \sum_m \int_u^\infty \frac{u x^{-2m-u} du}{2m(2m+u)}. \end{aligned}$$

Posons $u = 0$ dans cette équation et supposons que l'intégration se fasse le long de l'axe réel. Les fonctions sous le signe resteront finies et continues dans chaque intégrale et l'on aura

$$(4)\quad \begin{cases} f(x) - \frac{1}{lx}\sum_{p^m<x} lp = \frac{1}{lx}\frac{\zeta' o}{\zeta o} - \frac{x}{lx} + \int_0^\infty \left(\frac{x^{1-u}}{1-u} - \frac{\zeta' u}{\zeta u}\right) du \\ \qquad - \sum_\rho \int_0^\infty \frac{u x^{\rho-u} du}{\rho(\rho-u)} - \sum_m \int_0^\infty \frac{u x^{-2m-u} du}{2m(2m+u)}. \end{cases}$$

(59)

64. Nous allons maintenant transformer ces différentes intégrales. En introduisant des termes qui se détruisent, on peut écrire l'équation suivante, dans laquelle les fonctions restent continues sous les signes d'intégration :

$$\int_0^{\infty}\left(\frac{x^{1-u}}{1-u}-\frac{\zeta'u}{\zeta u}\right)du = \lim_{\varepsilon=0}\int_0^{1-\varepsilon}+\int_{1+\varepsilon}^{\infty}\frac{x^{1-u}du}{1-u}$$
$$-\int_0^2\left(\frac{\zeta'u}{\zeta u}+\frac{1}{u-1}\right)du-\int_2^{\infty}\frac{\zeta'u}{\zeta u}du$$
$$+\lim_{\varepsilon=0}\int_0^{1-\varepsilon}+\int_{1+\varepsilon}^{2}\frac{du}{u-1}+\int_{1-\varepsilon}^{1+\varepsilon}\frac{x^{1-u}-1}{1-u}du.$$

Le terme écrit en dernière ligne est nul ; celui qui est écrit sur la ligne précédente a pour valeur

$$-[\log(u-1)\zeta u]_0^2-[\log\zeta u]_2^{\infty}=\log(-\zeta o)=-l2,$$

car la relation (voir mon Mémoire déjà cité n° 4)

$$\zeta(1-s)=\frac{2}{(2\pi)^s}\cos\frac{s\pi}{2}\Gamma(s)\zeta(s)$$

donne, pour $s=1$,

$$\zeta(o)=\lim_{s=1}\frac{1}{\pi}\frac{\cos\frac{s\pi}{2}}{s-1}=-\frac{1}{2}.$$

Il vient donc

$$\int_0^{\infty}\left(\frac{x^{1-u}}{1-u}-\frac{\zeta'u}{\zeta u}\right)du=-l2+\lim_{\varepsilon=0}\int_0^{1-\varepsilon}+\int_{1+\varepsilon}^{\infty}\frac{x^{1-u}du}{1-u}.$$

Mais, en faisant le changement de variables

$$1-u=\frac{lt}{lx},$$

on trouve

$$\int_0^{1-\varepsilon} \frac{x^{1-u}du}{1-u} = \int_{x^{\varepsilon}}^{x} \frac{dt}{lt}, \quad \int_{1+\varepsilon}^{\infty} \frac{x^{1-u}du}{1-u} = \int_0^{x^{-\varepsilon}} \frac{dt}{lt}.$$

Par suite,

$$(5)\ldots \left\{ \begin{array}{l} \displaystyle\int_0^{\infty} \left(\frac{x^{1-u}}{1-u} - \frac{\zeta' u}{\zeta u} \right) du = -l2 + \mathrm{Li}(x), \\ \displaystyle\mathrm{Li}(x) = \lim_{\varepsilon=0} \int_0^{1-\varepsilon} + \int_{1+\varepsilon}^{x} \frac{dt}{lt}. \end{array} \right.$$

Passons aux intégrales suivantes de la formule (4). On a, par décomposition puis intégration par parties,

$$(6) \left\{ \begin{array}{l} \displaystyle\int_0^{\infty} \frac{ux^{\rho-u}dx}{\rho(\rho-u)} = -\frac{1}{\rho}\int_0^{\infty} x^{\rho-u}du + \int_0^{\infty} \frac{x^{\rho-u}du}{\rho-u} \\ \displaystyle\qquad = -\frac{x^{\rho}}{\rho lx} + \frac{x^{\rho}}{\rho lx} + \frac{1}{lx}\int_0^{\infty} \frac{x^{\rho-u}}{(\rho-u)^2} \\ \displaystyle\qquad = \frac{1}{lx}\int_0^{\infty} \frac{x^{\rho-u}du}{(\rho-u)^2}. \end{array} \right.$$

De même

$$(7)\ldots \int_0^{\infty} \frac{ux^{-2m-u}du}{2m(2m+u)} = \frac{1}{lx}\int_0^{\infty} \frac{x^{-2m-u}du}{(2m+u)^2}.$$

Substituons les valeurs (5), (6) et (7) dans (4); il vient

$$(8) \left\{ \begin{array}{l} \displaystyle f(x) = \mathrm{Li}(x) - l2 + \frac{1}{lx}\left[\sum_{p^m < x} lp - x - \frac{\zeta' o}{\zeta o} \right] \\ \displaystyle\qquad - \frac{1}{lx}\left[\sum_{\rho} \int_0^{\infty} \frac{x^{\rho-u}du}{(\rho-u)^2} + \sum_m \int_0^{\infty} \frac{x^{-2m-u}du}{(2m+u)^2} \right]. \end{array} \right.$$

(61)

Cette équation est celle qui paraît donner l'expression la plus commode de la fonction $f(x)$.

65. Si l'on remplace dans (8) le premier crochet par sa valeur déduite de la formule (12) du chapitre précédent, il vient

$$(9)\quad \left\{ \begin{aligned} f(x) &= \mathrm{Li}(x) - l2 \\ &+ \frac{1}{lx}\left[\varepsilon x - \sum \int_0^{\infty} \frac{x^{\rho-u}du}{(\rho-u)^2} - \sum \int_0^{\infty} \frac{x^{-2m-u}du}{(2m+u)^2}\right]. \end{aligned} \right.$$

La valeur de ε est donnée par la formule (13), et la formule (17) du chapitre précédent en donne une limite supérieure propre au calcul. Comme ε tend vers zéro avec $\frac{1}{x}$, la formule précédente montre déjà que $Li(x)$ est la valeur principale de $F(x)$.

Il reste encore à évaluer les deux autres sommes qui figurent dans la formule (9). On a, par intégration par parties,

$$\int_0^{\infty} \frac{x^{\rho-u}du}{(\rho-u)^2} = \frac{x^\rho}{\rho^2 lx} + \frac{2}{lx}\int_0^{\infty} \frac{x^{\rho-u}du}{(\rho-u)^3};$$

ensuite

$$\begin{aligned} \mathrm{mod}\,.\int_0^{\infty} \frac{x^{\rho-u}du}{(\rho-u)^3} &< \int_0^{\infty} \frac{x^{\alpha-u}du}{\beta^3} = \frac{x^\alpha}{\beta^3 lx} \\ &< \frac{1}{lx}\left(\frac{\alpha^2+\beta^2}{\beta^3}\right)\frac{x^\alpha}{\alpha^2+\beta^2}. \end{aligned}$$

il vient donc

$$(10)\quad \mathrm{mod}\int_0^{\infty} \frac{x^{\rho-u}du}{(\rho-u)^2} < \left[\frac{1}{lx} + \frac{2}{(lx)^2}\left(\frac{1}{12} + \frac{1}{(12)^3}\right)\right]\sum \frac{x^\alpha}{\alpha^2+\beta^2}.$$

Enfin, on a

$$(11)\quad \sum_m \int_0^{\infty} \frac{x^{-2m-u}du}{(2m+u)^2} < \frac{1}{4x^2}\left(\sum \frac{1}{m^2}\right)\int_0^{\infty} x^{-u}du < \frac{1}{2x^2 lx}.$$

En se servant de ces relations (10) et (11) et de la formule (17) du chapitre précédent qui donne une limite supérieure de ε, on voit que l'on peut poser

$$(12) \quad . \quad . \quad . \quad . \quad . \quad f(x) = \mathrm{L}i(x) - l2 + \tau \frac{x}{lx},$$

où l'on a

$$(13) \quad |\tau| < 2\sigma + \left(\frac{5}{3} + \frac{1}{lx} + \frac{0{,}168}{(lx)^2}\right)\sigma^2 + \sigma^3 + \frac{8}{3}\sigma^4 + \frac{3}{2x^2}.$$

L'évaluation de τ est donc encore une fois ramenée à celle de la somme

$$\sum \frac{x^{\alpha-1}}{\alpha^2 + \beta^2} = \sigma^2$$

qui a fait l'objet du § 4 du chapitre 3.

66. Reportons-nous à la formule (6) de ce même paragraphe; comme le premier terme de la valeur de τ dans (13) donne aussi la valeur principale de τ, on a le théorème suivant :

Si l'on pose

$$f(x) = \mathrm{L}i(x) - l2 + \tau \frac{x}{lx},$$

la quantité τ sera infiniment petite avec $\frac{1}{x}$, *et elle sera d'un ordre de petitesse au moins égal à celui de l'expression*

$$\sqrt{plx}\, e^{-\sqrt{plx}},$$

où p = 0,03282.

67. Observons maintenant que la relation

$$f(x) = \mathrm{F}(x) + \frac{1}{2}\mathrm{F}(x^{\frac{1}{2}}) + \frac{1}{3}\mathrm{F}(x^{\frac{1}{3}}) + \cdots$$

(63)

donne

$$f(x) - f(x^{\frac{1}{2}}) = F(x) - \frac{1}{2}F(x^{\frac{1}{2}}) + \frac{1}{3}F(x^{\frac{1}{3}}) - \cdots < F(x);$$

on aura les inégalités

$$f(x) > F(x) > f(x) - f(\sqrt{x}),$$

et, comme $f(\sqrt{x})$ est $< \sqrt{x}$, on voit que le théorème précédent s'applique aussi à la fonction F (x), qui exprime combien il y a de nombres premiers inférieurs à x.

Nous énoncerons ce théorème comme il suit :

Si l'on désigne par F (x) *la fonction qui exprime combien il y a de nombres premiers* < x, F (x) *aura pour valeur asymptotique* Li (x) *quand* x *tendra vers l'infini. De plus, la différence entre ces deux fonctions ne pourra pas être d'un ordre de grandeur supérieur à celui de la fonction*

$$\frac{x}{lx}\sqrt{plx}\, e^{-\sqrt{plx}},$$

où p = 0,03282.

Ce théorème est celui auquel nous avons fait allusion dans l'introduction en insistant sur son importance.

CHAPITRE VI.

SUR LA CONVERGENCE DE LA SÉRIE $\sum \frac{\mu(k)}{k}$.

68. Définissons, avec M. *Mertens* (*), la fonction numérique $\mu(k)$ d'un paramètre entier k de la manière suivante :

$\mu(k) = 1$ pour $k = 1$,

$= 0$ si k a un facteur carré autre que 1,

$= -1$ si k est formé d'un nombre impair et

$= +1$ si k est formé d'un nombre pair de facteurs premiers différents.

(*) *Ueber einige asymtotische Gesetze der Zahlentheorie,* JOURNAL F. D. R. U. A. MATH. B. 77, 1874, p. 289.

D'après cette définition, on a, s étant une variable réelle ou imaginaire dont la partie réelle est supérieure à l'unité,

$$\sum_{k=1}^{\infty} \frac{\mu(k)}{k^s} = \prod\left(1 - \frac{1}{p^s}\right) = \frac{1}{\zeta(s)}.$$

Si la série, qui est au premier membre, converge pour $s = 1$, il résulte de cette relation que sa valeur sera 0, mais le point délicat est de prouver la convergence.

69. *Euler* paraît être le premier qui ait considéré cette série, et il a affirmé le résultat auquel nous venons de faire allusion, mais sans l'établir rigoureusement (*).

Quoique de nombreux mathématiciens, Mertens, Stieltjes, Gram, etc., se soient occupés de cette fonction, la démonstration du théorème énoncé par Euler n'a été établie que dans ces tout derniers temps par M. *von Mangoldt* (**).

La démonstration de M. von Mangoldt est assez détournée et d'ailleurs elle se borne strictement à établir la convergence de la série, sans fournir aucune approximation de la rapidité de la convergence. Nous pensons donc qu'il y a quelque intérêt à revenir sur cette question, d'abord pour présenter une démonstration plus directe et plus simple, ensuite pour donner une limite supérieure du reste de cette série, ce qui n'a pas encore été fait jusqu'à présent.

70. Différentions l'équation

$$\frac{1}{\zeta(s)} = \sum_{m>1}^{\infty} \frac{\mu(k)}{k^s},$$

où $\Re(s) > 1$; il vient, en changeant les signes,

$$\frac{\zeta'(s)}{\zeta(s)^2} = \frac{\zeta' s}{\zeta s} \cdot \frac{1}{\zeta s} = \sum_{m=1}^{\infty} \frac{\mu(k) l k}{k^s}.$$

(*) *Introductio in analysis infinitorum*, t. I. Lausanne, 1748, Cap. XV, n° 277.

(**) *Sitzungsberichte der K. P. Akademie der Wiss. zu Berlin*, 22 Juli 1897. (Voir aussi cet article pour les renseignements bibliographiques.)

Faisons les substitutions

$$\frac{\zeta'(s)}{\zeta(s)} = -\sum \frac{lp}{p^s - 1} = -\sum \frac{lp}{p^{ms}} \quad \text{et} \quad \frac{1}{\zeta(s)} = \sum \frac{\mu(k)}{k^s};$$

il viendra

$$\sum \frac{\mu(k)}{k^s} \sum \frac{lp}{p^{ms}} = -\sum \frac{\mu(k) lk}{k^s}.$$

Si l'on effectue les multiplications au premier membre et réduit les termes semblables, on devra trouver les mêmes termes dans les deux membres, parce que des exponentielles différentes ne sont pas du même ordre de grandeur pour s infini. Supposons implicitement ces multiplications effectuées, et égalons la somme des termes des deux membres dans lesquels la base de l'exponentielle est $< x$. Ce résultat peut s'écrire :

$$\sum_{k<x} \left[\frac{\mu(k)}{k^s} \sum_{p^m < \frac{x}{k}} \frac{lp}{p^{ms}} \right] = -\sum_{k<x} \frac{\mu(k) lk}{k^s}.$$

et, pour $s = 1$,

$$(1) \quad \ldots\ldots \quad \sum_{k<x} \frac{\mu(k)}{k} \sum_{p^m < \frac{x}{k}} \frac{lp}{p^m} = -\sum_{k<x} \frac{\mu(k) lk}{k}.$$

La somme relative à p^m a été étudiée au § 2 du chapitre IV, et l'on peut poser, par le théorème du n° (59) de ce paragraphe,

$$\sum_{p^m < \frac{x}{k}} \frac{lp}{p^m} = lx - lk - C + \eta\left(\frac{x}{k}\right).$$

Substituons cette valeur dans l'équation précédente et supprimons les termes qui se détruisent; il restera simplement

$$(2) \quad \ldots \quad (lx - C) \sum_{k<x} \frac{\mu(k)}{k} + \sum_{k<x} \frac{\mu(k)}{k} \eta\left(\frac{x}{k}\right) = 0.$$

Telle est l'équation fondamentale dont nous allons déduire les conséquences que nous avons en vue. A cet effet, nous allons utiliser les formules que nous avons établies pour l'approximation de η.

71. Il résulte immédiatement du théorème du n° (59) que l'on peut poser

$$\operatorname{mod} \eta\left(\frac{x}{k}\right) < be^{-a\sqrt{l\frac{x}{k}}},$$

où a et b désignent deux nombres positifs fixes, indépendants de x et de k, et auxquels les résultats que nous avons obtenus dans les chapitres antérieurs permettent d'attribuer des valeurs précises. En particulier, on peut donner à a toute valeur $< \sqrt{p} = 0{,}03282$ et la valeur de b en résultera.

Cela fait, la seconde somme de l'équation (2) sera certainement inférieure en valeur absolue à

$$b \sum_{k<x} \frac{1}{k} e^{-a\sqrt{l\frac{x}{k}}},$$

et nous allons montrer qu'elle ne peut surpasser un nombre que nous allons assigner.

Si l'on remarque que la fonction à sommer, ayant pour dérivée, à un facteur positif près

$$a - 2\sqrt{l\frac{x}{k}},$$

est d'abord décroissante, pour devenir ensuite constamment croissante quand k tend vers x, il est clair que l'on peut poser, le premier terme étant < 1, et le dernier $< \frac{1}{x}$,

$$\sum_{k<x} \frac{1}{k} e^{-a\sqrt{l\frac{x}{k}}} < 1 + \int_1^x \frac{dk}{k} e^{-a\sqrt{l\frac{x}{k}}} + \frac{1}{x}.$$

(67)

Par les changements de variables $k = \frac{x}{t}$, puis $lt = z^2$, on a

$$\int_1^x \frac{dk}{k} e^{-a\sqrt{l\frac{x}{k}}} = \int_1^x \frac{dt}{t} e^{-a\sqrt{lt}} = 2\int_1^{\sqrt{lx}} z e^{-az} dz$$

$$= 2\,\frac{1 - e^{-a\sqrt{lx}}(a\sqrt{lx} + 1)}{a^2} < \frac{2}{a^2},$$

et, par conséquent

$$b \sum_{k<x} \frac{1}{k} e^{-a\sqrt{l\frac{x}{k}}} < b\left(1 + \frac{2}{a^2} + \frac{1}{x}\right) < 2b\left(1 + \frac{1}{a^2}\right).$$

Le dernier membre de cette inégalité est un nombre fixe h auquel nous sommes en état d'assigner une valeur précise et qui est aussi la limite supérieure de la seconde somme de l'équation (2). Cette équation nous donne donc immédiatement

$$\text{mod} \sum_{k<x} \frac{\mu(k)}{k} < \frac{h}{lx - C}.$$

De là le théorème suivant :

THÉORÈME. *La somme*

$$\sum_{k<x} \frac{\mu(k)}{k}$$

tend vers zéro quand* x *tend vers l'infini, et sa valeur absolue reste inférieure à une expression de la forme

$$\frac{h}{lx},$$

où* h *est un nombre fixe, auquel nos calculs antérieurs permettent d'assigner une valeur précise.

Ce théorème résout la question que nous nous étions proposée.

NOTE COMPLÉMENTAIRE DE LA PREMIÈRE PARTIE.

73. *Objet de cette note.* Nous nous proposons, dans cette note, d'apporter un léger perfectionnement aux calculs numériques de la première partie. L'analyse des paragraphes 3 et 4 du chapitre II, qui conduit à la relation

$$1-\alpha > \frac{p}{l\beta - ln},$$

a son point de départ dans les inégalités du n° (25). Nous avons remarqué qu'on peut les remplacer par d'autres un peu plus avantageuses. Nous allons les établir, et nous reproduirons sur elles tous nos raisonnements antérieurs; les calculs numériques seuls seront changés et nous trouverons pour p et n des valeurs un peu plus élevées que celles du n° (31).

74. *Inégalité relative aux sommes* Σ' (voir n° 25). En supposant $u' > u > 1$, nous avons trouvé au n° 25 :

$$\frac{u-\alpha}{(u-\alpha)^2+(t-\beta)^2} > \frac{u-\alpha}{u'-\alpha}\,\frac{u'-\alpha}{(u'-\alpha)^2+(t-\beta)^2},$$

$$\frac{u'-1+\alpha}{(u'-1+\alpha)^2+(t-\beta)^2} : \frac{u'-\alpha}{(u'-\alpha)^2+(t-\beta)^2} < \frac{u'-\alpha}{u'-1+\alpha}.$$

On tire de la seconde de ces inégalités

$$\frac{u'-1+\alpha}{(u'-1+\alpha)^2+(t-\beta)^2} + \frac{u'-\alpha}{(u'-\alpha)^2+(t-\beta)^2} < \frac{2u'-1}{u'-1+\alpha}\,\frac{u'-\alpha}{(u'-\alpha)^2+(t-\beta)^2};$$

puis, par la première,

$$(1)\qquad \begin{cases} \dfrac{u-\alpha}{(u-\alpha)^2+(t-\beta)^2} > \dfrac{u-\alpha}{u'-\alpha}\,\dfrac{u'-1+\alpha}{2u'-1}\left[\dfrac{u'-\alpha}{(u'-\alpha)^2+(t-\beta)^2}\right. \\ \qquad\qquad\left. + \dfrac{u'-1+\alpha}{(u'-1+\alpha)^2+(t-\beta)^2}\right]. \end{cases}$$

(69)

Je dis que si α est compris entre 0 et $\frac{1}{2}$, u et u' entre 1 et 2, on aura

$$(2) \quad . \quad . \quad . \quad . \quad \frac{u-\alpha}{u'-\alpha}\,\frac{u'-1+\alpha}{2u'-1} \geqq \frac{u'-1}{u'(2u'-1)},$$

ce minimum étant atteint pour ($u = 1$, $\alpha = 0$). En effet, la différence des deux membres de (2) a le signe de la quantité

$$u'(u-\alpha)(u'-1+\alpha) - (u'-\alpha)(u'-1),$$

qui peut s'écrire aussi

$$u'(u'-1)(u-1) + \alpha[u'(u-\alpha) - (u'-1)^2].$$

Comme u et u' sont > 1, le premier terme est positif; le second l'est aussi, car α étant $\overline{\gtrless} \frac{1}{2}$, donc $u - \alpha > \frac{1}{2}$, ce terme surpasse

$$\alpha\left[\frac{u'}{2} - (u'-1)^2\right] = \alpha\left(u' + \frac{1}{2}\right)\left(1 - \frac{u'}{2}\right),$$

qui est positif pour u' compris entre 1 et 2.

Choisissons maintenant u' de manière à rendre maximum le second membre de la formule (2). En annulant sa dérivée, on trouve

$$(3) \qquad u' = 1 + \frac{1}{\sqrt{2}}, \quad \text{d'où} \quad \frac{u'-1}{u'(2u'-1)} = 3 - 2\sqrt{2}.$$

Cette valeur de u' étant comprise entre 1 et 2, peut se mettre dans (2), et il vient par (1)

$$\frac{u-\alpha}{(u-\alpha)^2 + (l-\beta)^2} > (3 - 2\sqrt{2})\left[\frac{u'-\alpha}{(u'-\alpha)^2 + (l-\beta)^2} + \frac{u'-1+\alpha}{(u'-1+\alpha)^2 + (l-\beta)^2}\right].$$

On peut sommer toutes les inégalités analogues pour toutes les racines $\alpha + \beta i$ où $\alpha \overline{\overline{<}} \frac{1}{2}$; soient Σ' une somme étendue à ces racines particulières et Σ une somme étendue à toutes les racines sans distinction; on aura *a fortiori* (l'inégalité ayant lieu entre les parties réelles des deux membres)

$$\sum' \frac{1}{s-\rho} > (3-2\sqrt{2}) \sum \frac{1}{u'+ti-\rho}.$$

Ajoutons à cette inégalité celle qui s'en déduit par le changement de t en $2t$ et, par suite (n° 18), de s en s'; il vient

$$\sum' \frac{1}{s-\rho} + \frac{1}{4}\sum' \frac{1}{s'-\rho} > (3-2\sqrt{2})\left[\sum \frac{1}{u'+ti-\rho} + \frac{1}{4}\sum \frac{1}{u'+2ti-\rho}\right].$$

Comme on supposera dorénavant que s a la même partie imaginaire $ti = \beta i$ que l'une des racines ρ, on peut encore ajouter au second membre la différence des deux membres de (4) pour cette racine, savoir

$$\frac{1}{u-\alpha} - (3-2\sqrt{2})\left[\frac{1}{u'-\alpha} + \frac{1}{u'-1+\alpha}\right].$$

Cette différence, où $u' = 1 + \frac{1}{\sqrt{2}}$ et où nous ferons $u - 1 \overline{\overline{<}} \frac{1}{12}$, surpasse la quantité

$$\frac{12}{13} - 2(3-2\sqrt{2}).$$

Nous trouvons donc, sous ces conditions, l'inégalité

$$(5)\quad \ldots\quad \left\{\begin{aligned} &\sum' \frac{1}{s-\rho} + \frac{1}{4}\sum' \frac{1}{s'-\rho} > (3-2\sqrt{2})\left[\sum \frac{1}{u'+ti-\rho}\right.\\ &\qquad\left. + \frac{1}{4}\sum \frac{1}{u'+2ti-\rho}\right] + \frac{12}{13} - 2(3-2\sqrt{2}), \end{aligned}\right.$$

qui correspond à la relation (42) du n° 25.

(71)

75. ***Inégalité relative aux sommes*** Σ'' (voir n° 26). Désignons par Σ'' une somme étendue aux racines ρ, où $\alpha > \frac{1}{2}$. On aura $\Sigma'' = \Sigma - \Sigma'$; par suite, en vertu de la relation (5),

$$\sum'' \frac{1}{s-\rho} + \frac{1}{4}\sum'' \frac{1}{s'-\rho} < \left[\sum \frac{1}{s-\rho} + \frac{1}{4}\sum \frac{1}{s'-\rho}\right]$$
$$-(3-2\sqrt{2})\left[\sum \frac{1}{u'+ti-\rho} + \frac{1}{4}\sum \frac{1}{u'+2ti-\rho}\right]$$
$$+2(3-2\sqrt{2}) - \frac{12}{13}.$$

Comme on l'a expliqué au n° 26, on peut remplacer respectivement ces deux crochets par leurs limites supérieures tirées de la relation (28) au n° 19. Ceci donne la formule

$$(6)\ldots \left\{ \begin{aligned} &\sum'' \frac{1}{s-\rho} + \frac{1}{4}\sum'' \frac{1}{s'-\rho} < (\sqrt{2}-1)\left(\frac{5}{4}\log\frac{t}{\pi} - l(2) - \frac{3\zeta'u}{4\zeta u}\right.\\ &\qquad \left. + \varphi(u,t) + (3-2\sqrt{2})\left[\frac{3\zeta'u'}{4\zeta u'} - \varphi(u',t) + 2\right]\right) - \frac{12}{13}, \end{aligned} \right.$$

qui correspond à la relation (44) au n° 26.

On voit, par le raisonnement de ce même numéro, que l'on a

$$\varphi(u,t) - (3-2\sqrt{2})\varphi(u',t) < \frac{1+(3-2\sqrt{2})}{16t} < \frac{2-\sqrt{2}}{8t},$$

de telle sorte qu'on a la formule, analogue à (45),

$$(7)\ldots \sum'' \frac{1}{s-\rho} + \frac{1}{4}\sum'' \frac{1}{s'-\rho} < \frac{3}{4}\frac{1}{u-1} + h\log t - m,$$

en posant, en abrégé,

$$(8)\ldots\ldots h = \frac{5}{4}(\sqrt{2}-1) = 0{,}51776\,69\ldots$$

$$(9) \quad \begin{cases} m = \frac{3}{4}\left(\frac{\zeta' u}{\zeta u} + \frac{1}{u-1}\right) + (\sqrt{2}-1)\left(\frac{5}{4} l\pi + l2\right) \\ \quad - (5 - 2\sqrt{2})\left(\frac{3}{4}\frac{\zeta' u'}{\zeta u'} + 2\right) + \frac{12}{13} - \frac{2-\sqrt{2}}{8t}. \end{cases}$$

76. *Évaluation de* m. Dans cette valeur de m, nous supposerons $u - 1 \gtreqless \frac{1}{12}$, par conséquent (n° 14)

$$\frac{5}{4}\left(\frac{\zeta' u}{\zeta u} + \frac{1}{u-1}\right) > 0{,}42144\,22.$$

On a ensuite

$$-\frac{3}{4}\frac{\zeta' u'}{\zeta u'} > 0{,}71096\,44;$$

car, $u' - 1$ étant égal à $\frac{\sqrt{2}}{2} = \frac{1{,}41421\,356}{2}$, la formule (16) du n° 12 donne

$$\begin{aligned} \frac{\zeta' u'}{\zeta u'} + \sqrt{2} &< 0{,}57721\,566 - 0{,}09575\,456\sqrt{2} \\ &+ 0{,}02589\,989 - 0{,}00366\,951\sqrt{2} \\ &+ 0{,}00113\,094 - 0{,}00018\,088\sqrt{2} \\ &+ 0{,}00005\,894 - 0{,}00000\,969\sqrt{2} \\ &+ 0{,}00000\,321 \\ &< 0{,}60430\,864 - 0{,}09761\,464\sqrt{2} \end{aligned}$$

d'où $-\frac{\zeta' u'}{\zeta u'} > 0{,}94795\,28\ldots$

Enfin, comme on a

$$\frac{5}{4} l\pi + l2 = 2{,}12405\,953\ldots,$$

on voit, par la substitution de ces diverses valeurs dans (9), que l'on peut poser *a fortiori* dans (7)

$$(10) \ldots\ldots \quad m = 2{,}00316\,98 - \frac{2-\sqrt{2}}{8t}.$$

77. ***Limite inférieure de*** $1-\alpha$. On peut maintenant donner à h et à m les valeurs (8) et (10) ci-dessus dans les formules (49) et (50) du n° 29, savoir

$$1-\alpha > \frac{7-4\sqrt{3}}{4(h\log\beta - m)} \quad \text{et} \quad u-1 = \frac{2\sqrt{3}-3}{4(h\log\beta - m)}.$$

Celles-ci donnent ainsi les relations

$$(11) \qquad 1-\alpha > \frac{p}{\log\beta - \frac{m}{h}} \quad \text{et} \quad u-1 = \frac{q}{\log\beta - \frac{m}{h}},$$

où l'on a

$$p = \frac{7-4\sqrt{3}}{5(\sqrt{2}-1)} = 0{,}034666..$$

$$q = \frac{2\sqrt{3}-3}{5(\sqrt{2}-1)} = 0{,}224088...$$

$$\frac{m}{h} = 3{,}86886\,45 - \frac{0{,}024264}{\beta}.$$

Pour que la formule (11) soit applicable, il faut que β soit assez grand pour que $u-1$ soit $> \frac{1}{12}$.

On voit que cette condition aura lieu, si

$$\log\beta > \frac{m}{h} + 12q \qquad \text{ou} \qquad > 6{,}5579...,$$

et, *a fortiori*, si $\beta > 705$.

On a, dans ce cas,

$$\frac{m}{h} > 3{,}868829 = \log(47{,}886\,..)$$

et l'on peut énoncer le théorème suivant, que l'on peut substituer à celui du n° 31 :

78. Théorème. *A partir de* $\beta \gtreqless 705$, *on a, entre les parties réelles et imaginaires d'une racine* $\alpha + \beta i$ *de* $\zeta(s)$, *la relation*

$$(12) \ldots\ldots\ldots \quad 1 - \alpha > \frac{p}{l\beta - ln},$$

où p *et* n *ont les valeurs déterminées* *

$$p = 0{,}034666\ldots, \qquad n = 47{,}886\ldots$$

Pour $\beta < 705$, *la valeur de* $1 - \alpha$ *sera supérieure à la limite* 0,0128 *que donne cette formule pour* $\beta = 705$.

79. *Remarque.* Les valeurs précédentes de p et de n sont aussi applicables dans toutes les formules et dans tous les théorèmes de la seconde partie du Mémoire. Il n'y a d'exception que pour les formules du n° 39 (parce que l'on n'y suppose pas $\beta > 705$).

* Cette valeur de p (sauf une erreur de calcul sur la 5e décimale) se trouvait déjà dans le manuscrit soumis à l'examen de M. Mansion, et on la trouvera dans son rapport, mais avec une valeur beaucoup moins élevée de n.

12.5 G. H. Hardy (1914)

Sur les zéros de la fonction $\zeta(s)$ de Riemann

Godfrey Harold Hardy was born in England in 1877 to Isaac and Sophia Hardy. He performed exceptionally in school, but did not develop a passion for mathematics during his youth. Instead, he was drawn to the subject as a way to assert his intellectual superiority over his peers. He won scholarships to Winchester College in 1889, and to Trinity College, Cambridge, in 1896. He was elected fellow of Trinity College in 1900. Much of his best work was done in collaboration with Littlewood and Ramanujan. Hardy also worked with Titchmarsh, Ingham, Landau, and Pólya. He died in Cambridgeshire, England, in 1947 [116].

In this paper Hardy proves that there are infinitely many zeros of $\zeta(s)$ on the critical line. This is the first appearance of such a result. It gives a valuable piece of heuristic evidence and is requisite to the truth of the Riemann hypothesis.

ANALYSE MATHÉMATIQUE. — *Sur les zéros de la fonction $\zeta(s)$ de Riemann.* Note de **M. G.-H. Hardy.**

1. MM. H. Bohr et E. Landau ont donné tout récemment ([1]) la démonstration que la plupart des zéros complexes de $\zeta(s)$ sont situés, quel que soit δ positif, dans le domaine $\frac{1}{2} - \delta < \sigma < \frac{1}{2} + \delta$. Je me propose maintenant de démontrer que, *parmi les zéros de $\zeta(s)$, il y en a une infinité sur la droite $\sigma = \frac{1}{2}$* ([2]).

Je pars d'une formule connue de M. Cahen ([3]), savoir

$$e^{-y} = \frac{1}{2\pi i}\int_{k-i\infty}^{k+i\infty} \Gamma(u)\, y^{-u}\, du \qquad [\Re(y) > 0,\ k > 0];$$

d'où l'on déduit immédiatement

$$1 + \frac{1}{\pi i}\int_{k-i\infty}^{k+i\infty} \Gamma(u)\, y^{-u}\, \zeta(2u)\, du = 1 + 2\sum_{1}^{\infty} e^{-n^2 y} \qquad \left(k > \frac{1}{2}\right).$$

Je prends maintenant pour chemin d'intégration la droite $\sigma = \frac{1}{4}$. En faisant application du théorème de Cauchy et des formules de Riemann

$$\frac{s(s-1)}{2}\Gamma\left(\frac{s}{2}\right)\pi^{-\frac{s}{2}}\zeta(s) = \xi(s) = \xi\left(\frac{1}{2} + ti\right) = \Xi(t),$$

où $\Xi(t)$ est réelle pour t réel, on est conduit à l'équation

$$(1) \qquad 1 + \sqrt{\frac{\pi}{y}} - \frac{2}{\pi}\int_{-\infty}^{\infty}\left(\frac{\pi}{y}\right)^{\frac{1}{4}+ti}\frac{\Xi(2t)}{\frac{1}{4} + 4t^2}\, dt = 1 + 2\sum_{1}^{\infty} e^{-n^2 y}.$$

Dans cette équation, je pose $y = \pi e^{i\alpha}$, où $-\frac{1}{2}\pi < \alpha < \frac{1}{2}\pi$, et

$$e^{-y} = e^{-\pi\cos\alpha - i\pi\sin\alpha} = e^{\pi i \tau} = q = \rho e^{i\Phi};$$

([1]) *Comptes rendus*, 12 janvier 1914.

([2]) J'ai communiqué déjà ce résultat à la Société mathématique de Londres (séance du 12 mars 1914).

([3]) Thèse (*Annales École Normale supérieure*, 1894, p. 99). Cette formule a été retrouvée par M. Mellin (*Acta Soc. Fennicæ*, t. XX, n° 7, 1895, p. 6) qui en a fait des applications intéressantes.

1914, 1 *Comptes Rendus*, 158, 1012–14.

1013

et j'obtiens la formule

$$(2)\qquad \int_0^\infty \frac{(e^{\alpha t}+e^{-\alpha t})\,\Xi(2t)}{\frac{1}{4}+4t^2}\,dt = \pi\cos\frac{1}{4}\alpha - \frac{1}{2}\pi e^{\frac{1}{2}i\alpha}\,F(q),$$

où

$$F(q) = 1 + 2\sum_1^\infty q^{n^2} = \vartheta_3(0,\tau).$$

Enfin, en différentiant $2p$ fois par rapport à α, on a

$$(3)\qquad \int_0^\infty \frac{(e^{\alpha t}+e^{-\alpha t})\,t^{2p}\,\Xi(2t)}{\frac{1}{4}+4t^2}\,dt = \frac{(-1)^p\pi}{4^{2p}}\cos\frac{1}{4}\alpha - \left(\frac{d}{d\alpha}\right)^{2p}\left[\frac{1}{2}\pi e^{\frac{1}{2}i\alpha}F(q)\right]$$

2. Je vais me servir maintenant d'un lemme tiré de la théorie des fonctions elliptiques. Je suppose que α tende vers $\frac{1}{2}\pi$, de sorte que q tende vers -1 suivant un chemin tangent au rayon $\Phi = \pi$. Cela étant, je dis que le dernier terme de l'équation (3) tend, quel que soit p, vers la limite zéro. Pour cela, il suffit évidemment que toutes les fonctions

$$\left(\frac{d}{dy}\right)^{2p} F(q) = 2\sum_1^\infty n^{4p} q^{n^2}$$

tendent vers zéro. Mais cette dernière proposition se déduit comme corollaire des théorèmes généraux qu'on doit à MM. Bohr et Marcel Riesz, au sujet de la sommabilité des séries de Dirichlet.

La série

$$1^{-s} + 0 + 0 - 4^{-s} + 0 + 0 + 0 + 0 + 9^{-s} + 0 + \ldots,$$

convergente pour $\sigma > 0$, représente la fonction

$$(1 - 2^{1-2s})\,\zeta(2s),$$

régulière dans tout le plan et d'ordre fini dans tout demi-plan $\sigma > \sigma_0$. La série est donc sommable, pour toute valeur de s, par les moyennes de Cesàro d'ordre assez élevé; et pour s entier négatif, elle a la somme

$$(1 - 2^{1-2s})\,\zeta(2s) = 0.$$

3. Il s'ensuit que, quand α tend vers $\frac{1}{2}\pi$, l'intégrale (3) tend vers la limite $\frac{(-1)^p\pi}{4^{2p}}\cos\frac{1}{8}\pi$. Supposons maintenant que $\Xi(2t)$ garde un signe

pour $t > T > 1$, par exemple le signe positif. Alors on a, par un théorème connu,

(4) $$\int_0^\infty \frac{\left(e^{\frac{1}{2}\pi t} + e^{-\frac{1}{2}\pi t}\right) t^{2p}\, \Xi(2t)}{\frac{1}{4} + 4t^2}\, dt = \frac{(-1)^p \pi}{4^{2p}} \cos\frac{1}{8}\pi.$$

Soit p impair. On a

(5) $$\int_T^\infty < -\int_0^T < KT^{2p},$$

où K est indépendant de p. Mais cela est impossible. Il y a en effet, d'après notre hypothèse, un nombre δ positif tel que $\Xi(2t) > \delta$, pour $2T < t < 2T + 1$. Donc

(6) $$\int_T^\infty > \int_{2T}^{2T+1} > \delta K_1 (2T)^{2p},$$

où K_1, comme K, est positif et ne dépend nullement de p. Enfin, des inégalités (5) et (6) je tire

$$\delta K_1 2^{2p} < K;$$

donc, pour p assez grand, une contradiction.

COMMENTS

See comments following 1921, 2.

12.6 G. H. Hardy (1915)

Prime Numbers

> *The Theory of Numbers has always been regarded as one of the most obviously useless branches of Pure Mathematics. The accusation is one against which there is no valid defence; and it is never more just than when directed against the parts of the theory which are more particularly concerned with primes. A science is said to be useful if its development tends to accentuate the existing inequalities in the distribution of wealth, or more directly promotes the destruction of human life. The theory of prime numbers satisfies no such criteria. Those who pursue it will, if they are wise, make no attempt to justify their interest in a subject so trivial and so remote, and will console themselves with the thought that the greatest mathematicians of all ages have found in it a mysterious attraction impossible to resist.*
>
> **G. H. Hardy**

Hardy's abilities as a mathematician were complemented by his literary abilities. He is respected not only for his research, but for his ability to communicate mathematics clearly and beautifully. His book, A Mathematician's Apology, is considered one of the best accounts of the life of a creative artist.

In this paper, Hardy gives a brief exposition on the prime number theorem. He begins by presenting a brief history of results in the field of prime number theory. The bulk of the paper discusses the application of analysis to number theory and in particular to the asymptotic distribution of the prime numbers. Hardy gives Chebyshev's results. They appear in Section 12.1. He then proceeds chronologically to give Riemann's innovations, that appear in Section 12.2. Finally, Hardy gives an outline of the proofs given by Hadamard and de la Vallée Poussin, see Sections 12.3 and 12.4, respectively.

MANCHESTER, 1915.] [BRITISH ASSOCIATION.

Prime Numbers

By G. H. HARDY, F.R.S.

(Ordered by the General Committee to be printed *in extenso*.)

THE Theory of Numbers has always been regarded as one of the most obviously useless branches of Pure Mathematics. The accusation is one against which there is no valid defence; and it is never more just than when directed against the parts of the theory which are more particularly concerned with primes. A science is said to be useful if its development tends to accentuate the existing inequalities in the distribution of wealth, or more directly promotes the destruction of human life. The theory of prime numbers satisfies no such criteria. Those who pursue it will, if they are wise, make no attempt to justify their interest in a subject so trivial and so remote, and will console themselves with the thought that the greatest mathematicians of all ages have found in it a mysterious attraction impossible to resist.

The foundations of the theory were laid by Euclid. Among Euclid's theorems two in particular are of fundamental importance. The first (Euc. vii. 24) is that *if a and b are both prime to c, then ab is also prime to c.* This theorem is the basis of the whole theory of the factorisation of numbers, systematised later by Euler and by Gauss, and in particular of the theorem that *every number can be expressed in one and only one way as a product of primes.* The second theorem (Euc. ix. 20) is that *the number of primes is infinite*: to this theorem I shall return in a moment.

In modern times the theory has developed in two different directions. In the first place there is what may be called roughly the theory of *individual* or *isolated* primes, a theory which it is difficult to define precisely, but of which a general idea may be formed by considering a few of its characteristic problems. How can we determine whether a given number is prime? what conditions are necessary and what sufficient? Can we define forms which represent prime numbers only? Are there infinitely many pairs of primes which differ by 2? Is (as Goldbach asserted) every even number the sum of two primes? This theory I shall dismiss very briefly. We know a number of very beautiful theorems of this character. I need only mention Wilson's theorem, Fermat's theorem, and the extensions of the latter by Lucas. But on the whole the record of research in this direction is a record of failure. The difficulties are too great for the methods of analysis at our command, and the problems remain unsolved.

Very different results are revealed when we turn to the second principal branch of the modern theory, the theory of the *average or asymptotic distribution of primes.* This theory (though one of its most famous problems is still unsolved) is in some ways almost complete, and certainly represents one of the most remarkable triumphs of modern analysis. The theory centres round one theorem, the *Primzahlsatz* or *Prime Number Theorem*; and it is to the history of this theorem, which may almost be said to embody the history of the whole subject, that I shall devote the remainder of this lecture.*

The problem may be stated crudely as follows: *How many primes*

* A full account of the history of the theorem will be found in Landau's *Handbuch der Lehre von der Verteilung der Primzahlen* (Teubner, 1909).

are there less than a given number x? More precisely, let $\pi(x)$ denote the number of primes * not exceeding x: then *what is the order of magnitude of $\pi(x)$?* The Prime Number Theorem provides a complete answer to this last question. It asserts that

$$\pi(x) \sim \frac{x}{\log x},$$

that is to say, that $\pi(x)$ *and* $x/(\log x)$ *are asymptotically equivalent*, or that their ratio tends to 1 when x tends to infinity.

The first step towards the proof of this theorem was made by Euclid, when he proved that the number of primes is infinite, or that

$$\pi(x) \to \infty.$$

Euclid's proof is classical, and can hardly be repeated too often. If the number of primes is finite, let them be 2, 3, 5, . . ., P. The number 2. 3. 5. . . . P + 1 is not divisible by any of 2, 3, 5, ..., P. It is therefore prime itself, or divisible by some prime greater than P; and either alternative contradicts the hypothesis that P is the greatest prime. It is worth remarking that Euclid's reasoning may be used to prove rather more, viz. that the order of $\pi(x)$ is at least as great as that of $\log \log x$.†

The next advances were made by Euler, probably about 1740. It was Euler to whom we owe the introduction into analysis of the Zeta-function, the function on whose properties, as later research has shown, the whole theory depends.

Let $s = \sigma + i\,t$. Then the function $\zeta(s)$ is defined, when $\sigma > 1$, by the equations

$$\zeta(s) = \sum n^{-s} = 1^{-s} + 2^{-s} + 3^{-s} + \ldots;$$

and Euler's fundamental contribution to the theory is the formula

$$\zeta(s) = \Pi\left(\frac{1}{1-p^{-s}}\right),$$

where the product extends over all prime values of p. Euler, it is true, considered $\zeta(s)$ as a function of a *real* variable only. But his formula at once indicates the existence of a deep-lying connection between the theory of $\zeta(s)$ and the theory of primes.

Euler deduced from his formula that the series $\sum p^{-s}$, obviously convergent when $s > 1$, is divergent when $s = 1$; and from this it is easy to deduce important consequences as to the order of $\pi(x)$. It is evident that $\pi(x) < x$, so that the order of $\pi(x)$ certainly does not exceed that of x, or, in the notation which is usual now, $\pi(x) = O(x)$.‡ It is an easy corollary of Euler's result that the order of $\pi(x)$ is *not very much less than that of x*; that, for example, $\pi(x) \neq O(x^a)$ for any value of a less than 1; or again, more precisely, that

$$\pi(x) \neq O\left\{\frac{x}{(\log x)^{1+a}}\right\}$$

for any value of a greater than 1.

* It proves most convenient not to count 1 as a prime.

† This was pointed out to me by Prof. H. Bohr of Copenhagen.

‡ $f = O(\phi)$ means that the absolute value of f is less than a constant multiple of ϕ: thus $\sin x = O(1)$, $100\,x = O(x)$.

It is also easy to prove that the order of $\pi(x)$ is *definitely less than that of* x, or that, as we should express it now, $\pi(x) = o(x)$.* This theorem, when read in conjunction with those which precede, is, I think, enough to suggest the Prime Number Theorem as a very plausible conjecture, or at any rate to suggest that the true order is that of $x/(\log x)$. The theorem was in fact conjectured first by Gauss (1793) and by Legendre (1798); and it is in Legendre's *Essai sur la théorie des nombres* that the conjecture first appears in print.

In this state the problem remained for fifty years, until the publication (1849–1852) of the researches of the Russian mathematician Tschebyschef. I have no time to speak of Tschebyschef's work as fully as it deserves, but his chief results, in so far as they bear directly on the problem now before us, were as follows :—

(1) Tschebyschef showed that the problem is simplified if we take as fundamental not the function $\pi(x)$ itself, but the closely related function

$$\theta(x) = \sum_{p \leq x} \log p$$

(the sum of the logarithms of all primes not exceeding x). He showed that the order of $\theta(x)$ is the same as that of $\pi(x) \log x$, and that the Prime Number Theorem itself is equivalent to the theorem that

$$\theta(x) \sim x.$$

(2) He showed that $\theta(x)$ is actually of order x, and $\pi(x)$ of order $x/(\log x)$, in fact that positive constants A and B exist such that

$$A \frac{x}{\log x} < \pi(x) < B \frac{x}{\log x}.$$

(3) He showed that *if* $\theta(x)/x$ *tends to a limit*, then that limit must be unity.

What Tschebyschef could not prove is that the limit does in fact exist, and, as he failed to prove this, he failed to prove the Prime Number Theorem. And about Tschebyschef's methods (interesting as they are), I shall say nothing ; for later research has shown that it was the essential inadequacy of his methods which was responsible for his failure, and that the theorem lies deeper in analysis than any of the ideas on which he relied.

The next great step was taken by Riemann in 1859, and it is in Riemann's famous memoir *Ueber die Anzahl der Primzahlen unter einer gegebenen Grösse* that we first find the ideas upon which the theory has now been shown really to rest. Riemann did not prove the Prime Number Theorem: it is remarkable, indeed, that he never mentions it. His object was a different one, that of finding an explicit expression for $\pi(x)$, or rather for another closely associated function, as a sum of an infinite series. But it was Riemann who first recognised that, if we are to solve any of these problems, we must study the Zeta-function as a function of the *complex* variable $s = \sigma + it$, and in particular study the distribution of its zeros.

* $f = o(\phi)$ means that $f/\phi \to 0$. Thus $\sin x = o(x)$. This theorem also was stated by Euler, but without satisfactory proof.

Riemann proved

(1) that $\zeta(s)$ is an analytic function of s, regular all over the plane except for a simple pole at the point 1;

(2) that $\zeta(s)$ satisfies the functional equation

$$\zeta(1-s) = 2(2\pi)^{-s} \cos \tfrac{1}{2} s\pi \, \Gamma(s)\zeta(s);$$

(3) that $\zeta(s)$ has zeros at the points $-2, -4, -6, \ldots$, and no other zeros *except possibly complex zeros whose real parts lie between* 0 *and* 1 *inclusive.*

To these propositions he added certain others of which he could produce no satisfactory proof. In particular he asserted that there is in fact an infinity of complex zeros, all naturally situated in the 'critical strip' $0 \leq \sigma < 1$; an assertion now known to be correct. Finally he asserted that it was 'sehr wahrscheinlich' that all these zeros have the real part $\frac{1}{2}$: the notorious 'Riemann hypothesis', unsettled to this day.

We come now to the time when, a hundred years after the conjectures of Gauss and Legendre, the theorem was finally proved. The way was opened by the work of Hadamard on integral transcendental functions. In 1893 Hadamard proved that the complex zeros of Riemann actually exist; and in 1896 he and de la Vallée-Poussin proved independently that *none of them have the real part* 1, and deduced a proof of the Prime Number Theorem.

It is not possible for me now to give an adequate account of the intricate and difficult reasoning by which these theorems are established. But the general ideas which underlie the proofs are, I think, such as should be intelligible to any mathematician.

In the first place Euler's formula shows that $\log \zeta(s)$ behaves, throughout the half-plane $\sigma > 1$, much like the series $\sum p^{-s}$. But $\zeta(s)$ has a simple pole for $s = 1$, and so the sum of the series $\sum p^{-1-\delta}$ tends logarithmically to $+\infty$ when $\delta \to 0$ through positive values. Suppose now that (if possible) $\zeta(1 + ti) = 0$. Then the real part of $\log \zeta(1 + \delta + ti)$, and therefore the real part of the series $\sum p^{-1-\delta-ti}$, tends, also logarithmically, to $-\infty$ when $\delta \to 0$. It follows that the series

$$\sum p^{-1-\delta}, \quad -\sum p^{-1-\delta} \cos (t \log p)$$

tend to $+\infty$ *with equal rapidity* when $\delta \to 0$. As the first series is a series of positive terms, while the signs of the terms in the second series change with a certain regularity, it is natural to suppose that our last conclusion is impossible; and this is in fact not particularly difficult to prove.

I come now to the proof of the Prime Number Theorem itself. If we differentiate Euler's formula logarithmically, we obtain

$$\frac{\zeta'(s)}{\zeta(s)} = \sum \left(\frac{\log p}{p^s} + \frac{\log p}{p^{2s}} + \ldots \right) = \sum_{p,m} \frac{\log p}{p^{ms}};$$

or (1)

$$-\frac{\zeta'(s)}{\zeta(s)} = \sum \frac{\Lambda(n)}{n^s}$$

where p assumes all prime values, m and n all positive integral values, and $\Lambda(n)$ is equal to log p if n is of the form p^m and to zero otherwise. Let

$$\psi(x) = \sum_{n \leq x} \Lambda(n)$$

Then $\psi(x)$ is, for our present purpose, equivalent to $\theta(x)$: it is easy to show that the difference between the two functions is of order $\sqrt{x}$. We have therefore to prove that $\psi(x) \sim x$.

The series on the right-hand side of the equation (1) is what is called a 'Dirichlet's series'; and the theory of such series resembles the more familiar theory of Taylor's series in one very important respect. *We can express the coefficients by contour integrals in which the function represented by the series appears under the sign of integration.* In particular we can show that

$$(2) \qquad \psi(x) = -\frac{1}{2\pi i}\int \frac{\zeta'(s)}{\zeta(s)}\frac{x^s}{s}\,ds,$$

where the path of integration is a line parallel to the imaginary axis and passing to the right of the point $s=1$.

The general idea of the proof is now easy enough to grasp. Every element of the integral (2) is of order x^σ, where $\sigma > 1$: we can therefore draw no *direct* conclusion as to the behaviour of $\psi(x)$ when x is large. But it is at once suggested that we should try to make use of Cauchy's theorem. The subject of integration has a simple pole at the point 1, corresponding to the pole of $\zeta(s)$ itself, and the residue at the pole is precisely x; and there are no other singularities on the line $\sigma = 1$, since $\zeta(s)$, as we have seen, has no poles *or zeros* on that line. Suppose then that we can move the path of integration across to the left of the line, introducing the appropriate correction due to the pole. Plainly we shall then have an expression for $\psi(x) - x$ in the form of an integral *in which every element is of order less than that of* x. And if we can show that the same is true of the integral itself, we shall have proved that $\psi(x) \sim x$, that is to say, we shall have proved the Prime Number Theorem. It will be observed that, if $\zeta(s)$ had zeros whose real part is equal to 1, then the result would be definitely false, since there would be additional residues of order x. It thus becomes clear why the older attempts to prove the theorem, without using the theory of functions of a complex variable, were unsuccessful.

The arguments which I have advanced are not exact: I have merely put forward a chain of reasoning which seems likely to lead to the desired result. The achievement of Hadamard and de la Vallée-Poussin was to replace these plausibilities by rigorous proofs. It might be difficult for me to make clear to you how great this achievement was. Some branches of pure mathematics have the pleasant characteristic that what seems plausible at first sight is generally true. In this theory anyone can make plausible conjectures, and they are almost always false. Nothing short of absolute rigour counts; and it is for this reason that the Analytic Theory of Numbers, while hardly a subject for an amateur, provides the finest possible discipline in accurate reasoning for anyone who will make a real effort to understand its results.

COMMENTS

See 1918, 1 (§ 2.1), and comments thereon.

12.7 G. H. Hardy and J. E. Littlewood (1915)

New Proofs of the Prime-Number Theorem and Similar Theorems

> *I believe this to be false. There is no evidence whatever for it (unless one counts that it is always nice when any function has only real roots). One should not believe things for which there is no evidence. In the spirit of this anthology I should also record my feeling that there is no imaginable reason why it should be true. Titchmarsh devised a method, of considerable theoretical interest, for calculating the zeros. The method reveals that for a zero to be off the critical line a remarkable number of 'coincidences' have to happen. I have discussed the matter with several people who know the problem in relation to electronic calculation; they are all agreed that the chance of finding a zero off the line in a lifetime's calculation is millions to one against. It looks then as if we may never know.*
> *It is true that the existence of an infinity of L-functions raising the same problems creates a remarkable situation. Nonetheless life would be more comfortable if one could believe firmly that the hypothesis is false [101].*
>
> **John E. Littlewood**

John Edensor Littlewood was born in England in 1885 to Edward and Sylvia Littlewood. When Littlewood was seven, he and his family moved to South Africa so his father could take a position as the headmaster of a new school. His father, a mathematician by trade, recognized his son's talents and their failure to be realized in Africa. Littlewood was sent back to England at the age of fifteen to study at St. Paul's school, in London. In 1902, he won a scholarship to Cambridge and entered Trinity College in 1903. In 1906, he began his research under the direction of E. W. Barnes. After solving the first problem assigned him by Barnes, Littlewood was assigned the Riemann hypothesis. His legendary 35-year collaboration with G. H. Hardy started in 1911, and its products include the present paper [118]. For biographical information on G. H. Hardy, see Sections 12.5 and 12.6.

In this paper Hardy and Littlewood develop new analytic methods to prove the prime number theorem. The tools they develop are general and concern Dirichlet series. To apply these methods to the prime number theorem one can consider any of

$$M(x) = o(x), \quad \sum_{n \leq x} \Lambda(n) \sim x, \quad \sum_{n=1}^{\infty} \frac{\mu(n)}{n} = 0.$$

Each of these statements can be shown to be equivalent to the prime number theorem without employing the theory of functions of a complex variable.

NEW PROOFS OF THE PRIME-NUMBER THEOREM AND SIMILAR THEOREMS.

By G. H. HARDY and J. E. LITTLEWOOD.

1. OUR object in writing this paper is to give a short sketch of a new method which we have found for the proof of certain fundamental theorems in the Analytic Theory of Numbers. A fuller account of our researches will be published elsewhere.*

Our method depends upon the use of the formula

$$e^{-y} = \frac{1}{2\pi i} \int_{\kappa - i\infty}^{\kappa + i\infty} \Gamma(s)\, y^{-s}\, ds, \dagger \tag{1.1}$$

where κ and the real part of y are positive, and y^{-s} has its principal value, in connection with the "Tauberian" theorems proved by us in a recent paper in the *Messenger of Mathematics*.‡ The theorems which we have principally in view are those expressed by the formulæ

$$M(x) = o(x), \tag{1.21}$$

$$\Sigma \frac{\mu(n)}{n} = 0, \tag{1.22}$$

$$\psi(x) \sim x, \tag{1.23}$$

* As part of a memoir, "Contributions to the theory of the Riemann Zeta-function and the theory of the distribution of primes", to appear in the *Acta Mathematica*.

† This formula was first given by Cahen (*Annales de l'École Normale Supérieure*, vol. xi, p. 75). It was found independently by Mellin (*Acta Societatis Fennicae*, vol. xx., No. 7, p. 6), to whom the first rigorous proof is due.

‡ Vol. xliii., p. 134.

1915, 1 (with J. E. Littlewood) *Quarterly Journal of Mathematics*, 46, 215–19.

where $M(x)$, $\mu(n)$, and $\psi(x)$ have their usual meanings. All these theorems are known to be equivalent to* the "Prime-Number Theorem"

(1.24) $$\Pi(x) \sim \frac{x}{\log x}.$$

They will appear here as particular cases of general theorems concerning Dirichlet's series.

2. We begin by stating the following theorem, which is equivalent to Theorems D, E, and F of our paper in the *Messenger of Mathematics* already quoted.

THEOREM A. *Suppose that*

(i) $\lambda_1, \lambda_2, \lambda_3, \ldots$ *is a sequence of numbers satisfying the conditions* $\lambda_1 > 0$, $\lambda_n > \lambda_{n-1}$, $\lambda_n \to \infty$, $\lambda_n/\lambda_{n-1} \to 1$;

(ii) $\alpha \geq 0$;

(iii) a_n *is real and satisfies one or other of the conditions*

$$a_n > -K\lambda_n^{\alpha-1}(\lambda_n - \lambda_{n-1}), \quad a_n < K\lambda_n^{\alpha-1}(\lambda_n - \lambda_{n-1}),$$

or is complex and of the form

$$O\{\lambda_n^{\alpha-1}(\lambda_n - \lambda_{n-1})\};$$

(iv) *the series* $$f(y) = \Sigma a_n e^{-\lambda_n y}$$

is convergent for $y > 0$, *and*

$$f(y) \sim Ay^{-\alpha}$$

as y *tends to zero. Then*

$$A_n = a_1 + a_2 + \ldots a_n \sim \frac{A\lambda_n^{\alpha}}{\Gamma(1+\alpha)}$$

as n *tends to infinity.*†

3. We now prove

THEOREM B. *If*

(i) *the series* $\Sigma a_n \lambda_n^{-s}$ *is absolutely convergent for* $\sigma > \sigma_0 > 0$;

(ii) *the function* $F(s)$ *defined by the series is regular for* $\sigma > c$, *where* $0 < c \leq \sigma_0$, *and continuous throughout any finite part of the plane for which* $\sigma \geq c$;

* By this we mean that, from any one of them, the rest can be deduced by elementary reasoning which involves no appeal to the theory of functions of a complex variable.

† When $A = 0$, the last two formulæ must be interpreted as $f(y) = o(y^{-\alpha})$ and $A_n = o(\lambda_n^{\alpha})$ respectively.

(iii) $$F(s) = O(e^{C|t|}),$$

where $C < \frac{1}{2}\pi$, *uniformly for* $\sigma \geq c$:

then the series $$f(y) = \Sigma a_n e^{-\lambda_n y}$$

is convergent for all positive values of y, *and*

$$f(y) = o(y^{-c})$$

as y *tends to zero.*

Suppose that $y > 0$ and $\kappa > \sigma_0$. Then

$$e^{-\lambda_n y} = \frac{1}{2\pi i}\int_{\kappa - i\infty}^{\kappa + i\infty} \Gamma(s)(\lambda_n y)^{-s} ds; \tag{3.1}$$

and it is easy to see that we may multiply by a_n and sum with respect to n. We thus obtain

$$f(y) = \frac{1}{2\pi i}\int_{\kappa - i\infty}^{\kappa + i\infty} \Gamma(s) F(s) y^{-s} ds. \tag{3.2}$$

An application of Cauchy's theorem enables us to replace this equation by

$$\begin{aligned} f(y) &= \frac{1}{2\pi i}\int_{c - i\infty}^{c + i\infty} \Gamma(s) F(s) y^{-s} ds \\ &= \frac{y^{-c}}{2\pi}\int_{-\infty}^{\infty} \Gamma(c + it) F(c + it) y^{-it} dt \\ &= \frac{y^{-c}}{2\pi}\int_{-\infty}^{\infty} \Phi(c + it) y^{-it} dt, \end{aligned} \tag{3.3}$$

say. But the integral

$$\int_{-\infty}^{\infty} |\Phi(c + it)| dt$$

is convergent, so that the integrals

$$\int_{-\infty}^{\infty} \Phi(c + it) \begin{matrix}\cos \\ \sin\end{matrix} (t \log y) dt,$$

tend to zero as $\log y$ tends, positively or negatively, to infinity.* The theorem now follows from (3.3).

We have supposed λ_n subject to the conditions (i) of Theorem A. The condition

$$\lambda_n / \lambda_{n-1} \rightarrow 1$$

is unnecessary here, but is necessary in the next theorem.

* Cf. Landau, *Prace Matematyczno-Fizycznch*, vol. xxi., pp. 173 *et seq.*

Combining Theorems A and B, we obtain

THEOREM C. *If the conditions of Theorem B are satisfied, and a_n is real and satisfies one or other of the inequalities*

$$a_n > -K\lambda_n^{c-1}(\lambda_n - \lambda_{n-1}), \quad a_n < K\lambda_n^{c-1}(\lambda_n - \lambda_{n-1}),$$

or is complex and of the form

$$O\{\lambda_n^{c-1}(\lambda_n - \lambda_{n-1})\},$$

then
$$A_n = a_1 + a_1 + \ldots + a_n = o(\lambda_n^c).$$

Suppose in particular that

$$\lambda_n = n, \quad a_n = \mu(n), \quad c = 1.$$

Then
$$F(s) = \Sigma \frac{\mu(n)}{n^s} = \frac{1}{\zeta(s)},$$

and all the conditions of Theorem C are satisfied. It follows that

$$M(n) = \mu(1) + \mu(2) + \ldots + \mu(n) = o(n),$$

i.e. that (1.21) is true.

4. In order to obtain a direct proof of (1.22) we must modify Theorem C in such a way that it shall apply to the case in which $c = 0$.

THEOREM D. *Suppose that*

(i) *the conditions of Theorem C are satisfied, except that $c = 0$;*

(ii) *$F(s)$ is regular for $s = 0$.*

Then the series Σa_n is convergent, and has the sum $F(0)$.

The proof is very much the same as that of Theorem C. Suppose that $F(s)$ is regular for $|s| \leq \delta$. Then, instead of (3.3), we have

$$f(y) = F(0) + \frac{1}{2\pi i}\int \Gamma(s) F(s) y^{-s} ds, \tag{4.1}$$

the contour of integration consisting of the parts $(-i\infty, -i\delta)$ and $(i\delta, i\infty)$ of the imaginary axis, and a semicircle γ described on and to the left of the line $(-i\delta, i\delta)$. We show, substantially as in the proof of Theorem C, that the rectilinear parts of the integral tend to zero. Also

$$\int_\gamma \Gamma(s) F(s) y^{-s} ds = \frac{\Gamma(i\delta) F(i\delta) y^{-i\delta} - \Gamma(-i\delta) F(-i\delta) y^{i\delta}}{\log(1/y)}$$
$$- \frac{1}{\log(1/y)} \int_\gamma y^{-s} \frac{d}{ds}\{\Gamma(s) F(s)\} ds$$
$$= O\left\{\frac{1}{\log(1/y)}\right\} = o(1).$$

Thus
$$f(y) \to F(0)$$
as $y \to 0$, and so
$$\Sigma a_n = F(0).$$

The conditions of this theorem are satisfied, for example, when

$$\lambda_n = n, \quad a_n = \frac{\mu(n)}{n}, \quad c = 0,$$

$$F(s) = \frac{1}{\zeta(s+1)}, \quad F(0) = 0;$$

and (1.22) is a corollary.

5. In order to prove (1.23), a slightly different modification of Theorem C is required.

THEOREM E. *Suppose that the conditions of Theorem C are satisfied, except that $F(s)$ is analytic near $s=c$, and has there a simple pole with residue g. Then*

$$A_n = a_1 + a_2 + \ldots + a_n \sim (g/c)\,\lambda_n^c.$$

The formula (3.3) is in this case replaced by

$$(5.1) \quad f(y) = g\,\Gamma(c)\,y^{-c} + \frac{1}{2\pi i}\int \Gamma(s)\,F(s)\,y^{-s}\,ds,$$

the path of integration being a modification of the line $\sigma = c$ similar to that of the imaginary axis used in the proof of Theorem D. Practically the same argument as was used in the last proof gives the result

$$f(y) \sim g\,\Gamma(c)\,y^{-c},$$

and the theorem follows immediately. If we take

$$\lambda_n = n, \quad a_n = \Lambda(n), \quad c = 1, \quad F(s) = -\zeta'(s)/\zeta(s), \quad g = 1,$$

we obtain (1.23).

We may add in conclusion that the truth of Theorem B does not really depend on the condition $C < \frac{1}{2}\pi$, which may be removed by a modification of the argument. This is, however, immaterial for the applications which we have had in view.

COMMENTS

See 1918, 1 (§ 2.1), and comments thereon.

12.8 A. Weil (1941)

On the Riemann hypothesis in Function-Fields

André Weil was born in France in 1906. He became interested in mathematics at a very young age and studied at the École Normale Supérieure in Paris. Weil received his doctoral degree from the University of Paris under the supervision of Hadamard. His thesis developed ideas on the theory of algebraic curves. Life during the Second World War was difficult for Weil, who was a conscientious objector. After the war he moved to the United States, where he became a member of the Institute for Advanced Study at Princeton. He retired from the institute in 1976 and became a professor emeritus.

Weil's research focused on the areas of number theory, algebraic geometry, and group theory. He laid the foundations for abstract algebraic geometry, as well as for the theory of abelian varieties. His work led to two Fields Medals (Deligne in 1978 for solving the Weil conjecture, and Yau in 1982 for work in three–dimensional algebraic geometry). Weil was an honorary member of the London Mathematical Society, a fellow of the Royal Society of London, and a member of both the Academy of Sciences in Paris and the National Academy of Sciences in the United States [113].

In this paper, Weil gives a proof of the Riemann hypothesis for function fields. This proof builds on Weil's earlier work, presented in [169].

invariant subgroup of order $1 + 3k$. Since s_1s_2 is of order 3 we have $(s_1s_2)^3 = (s_1s_2^{-1}\, s_1s_2)^3 = 1$. It is easy to prove that $s_1s_2s_1$ generates a cyclic group whose generators are commutative with their conjugates under s_1 and under s_2. That is, if a group contains a maximal subgroup of order 3 it contains an invariant abelian subgroup of index 3 which is either cyclic or of type 1^m. The groups which contain maximal subgroups of order 2 or of order 3 may therefore be regarded as determined, but those which contain maximal subgroups of order 2 are naturally considerably simpler than those which contain maximal subgroups of order 3.

[1] Cf. these PROCEEDINGS **27**, 212–216 (1941).

ON THE RIEMANN HYPOTHESIS IN FUNCTION-FIELDS

BY ANDRÉ WEIL

NEW SCHOOL FOR SOCIAL RESEARCH

Communicated June 11, 1941

A year ago[1] I sketched the outline of a new theory of algebraic functions of one variable over a finite field of constants, which may suitably be described as transcendental, in view of its close analogy with that portion of the classical theory of algebraic curves which depends upon the use of Abelian integrals of the first kind and of Jacobi's inversion theorem; and I indicated how this led to the solution of two outstanding problems, viz., the proof of the Riemann hypothesis for such fields, and the proof that Artin's non-abelian L-functions on such fields are polynomials. I have now found that my proof of the two last-mentioned results is independent of this "transcendental" theory, and depends only upon the algebraic theory of correspondences on algebraic curves, as due to Severi.[2]

Γ being a non-singular projective model of an algebraic curve over an algebraically closed field of constants, the variety of ordered couples (P, Q) of points on Γ has the non-singular model $\Gamma \times \Gamma$ (in a bi-projective space); correspondences are divisors on this model, additively written; they form a module $\mathfrak{C}$ on the ring Z of rational integers. Let Γ_A, Γ_B', Δ, respectively, be the loci of points (P, A), (B, P) and (P, P), on $\Gamma \times \Gamma$, A and B being fixed on Γ and P a generic point of Γ (in the precise sense defined by van der Waerden[3]). The intersection number (C, D) of C and D being defined by standard processes[4] for irreducible, non-coinciding correspondences C, D, will be defined for any C and D which have no irreducible component in common, by the condition of being linear both in C and in D. The degrees $r(C)$, $s(C)$, and the coincidence number $f(C)$

of C are defined as its intersection numbers with a generic Γ_A, with a generic Γ_B' and with Δ, respectively. Transformation $(P, Q) \rightarrow (Q, P)$, which is a birational, one-to-one involution of $\Gamma \times \Gamma$ into itself, transforms a correspondence C into a correspondence which will be denoted by C'. To every rational function φ on $\Gamma \times \Gamma$ is attached a divisor (the divisor of its curves of zeros and poles), i.e., a corresponding C_φ; and $(C_\varphi, D) = 0$ whenever it is defined. Let $\mathfrak{C}_0$ be the module of all correspondences of the form $C_\varphi + \Sigma a_i\Gamma_{A_i} + \Sigma b_j\Gamma_{B_j}'$: if $C \in \mathfrak{C}_0$, then $f(C) = r(C) + s(C)$ (whenever $f(C)$ is defined), since this is true for $C = C_\varphi$ (both sides then being 0) and for $C = \Gamma_A$, $C = \Gamma_B'$. Put $\mathfrak{R} = \mathfrak{C} - \mathfrak{C}_0$; an element γ in $\mathfrak{R}$ is a co-set in $\mathfrak{C}$ modulo $\mathfrak{C}_0$, and it can be shown that there always is a C in this co-set for which $f(C)$, $r(C)$, $s(C)$ are defined, so that, putting $\sigma(\gamma) = r(C) + s(C) - f(C)$, $\sigma(\gamma)$ is defined as a linear function over $\mathfrak{R}$. If γ_0 in $\mathfrak{R}$ corresponds to Δ in $\mathfrak{C}$, $\sigma(\gamma_0) = 2g$, where g is the genus of Γ. If $C \in \mathfrak{C}_0$, we have $C' \in \mathfrak{C}_0$, and so the reciprocal relation $(C \rightleftarrows C')$ in $\mathfrak{C}$ induces a similar relation $(\gamma \rightleftarrows \gamma')$ in $\mathfrak{R}$; and $\sigma(\gamma') = \sigma(\gamma)$.

The product of correspondences now being defined essentially as in Severi, $\mathfrak{C}$ becomes a ring, with Δ as unit-element; $\mathfrak{C}_0$ is two-sided ideal in $\mathfrak{C}$, so that $\mathfrak{R}$ also can be defined as ring, with unit-element γ_0: we write $\gamma_0 = 1$. It is found that the degrees of $C \cdot D$ are $r(C) \cdot r(D)$, $s(C) \cdot s(D)$; and that (C, D) is the same as $(C \cdot D', \Delta) = f(C \cdot D')$, and hence is equal to $r(C) \cdot s(D) + r(D) \cdot s(C) - \sigma(\gamma \cdot \delta')$ if γ, δ are the elements of $\mathfrak{R}$ which correspond to C, D. It follows that $\sigma(\gamma\delta) = \sigma(\delta\gamma)$. From $\sigma(n \cdot 1) = 2ng$ it follows that ring $\mathfrak{R}$ has characteristic 0.

Defining the complementary correspondence to C as in Severi,[5] it is found that the generic complementary correspondence to C is irreducible and has degrees (g, m), where $2m = \sigma(\gamma\gamma')$: it follows that $\sigma(\gamma\gamma') \geqslant 0$, and that $\sigma(\gamma\gamma') = 0$ only if $\gamma = 0$.

Now, let k (as in my note[1]) be the Galois field with q elements; k_n its algebraic extension of degree n; $\bar{k}$ its algebraic closure. Let $K = k(x, y)$ be a separable algebraic extension of $k(x)$; $K_n = k_n(x, y)$; $\bar{K} = \bar{k}(x, y)$. Then $(x, y) \rightarrow (x^q, y^q)$ defines a correspondence I, of degrees 1, q, on a non-singular model Γ of field $\bar{K}$; I^n is the correspondence, of degrees 1, q^n, defined by $(x, y) \rightarrow (x^{q^n}, y^{q^n})$; and $I \cdot I' = q \cdot \Delta$ (more generally, any correspondence C, such that $r(C) = 1$, satisfies $C \cdot C' = s(C) \cdot \Delta$). Let ι be the element of $\mathfrak{R}$ which corresponds to I; we have $\iota \cdot \iota' = q$. The intersections of I^n with Δ, which are all found to be of multiplicity 1, are those points of Δ which have coördinates in k_n, so that the number ν_n of such points (i.e., of points on Γ with coördinates in k_n) is $f(I^n) = 1 + q^n - \sigma(\iota^n)$. But the numerator of the zeta-function $\zeta_K(s)$ of K is[6] a polynomial $P(u) = u^{2g} - (1 + q - \nu_1)u^{2g-1} + \ldots$, of degree $2g$, in $u = q^s$; and, putting $P(u) = \prod_i (u - \alpha_i)$, the numerator of the zeta-function of K_n is

$$P_n(u^n) = \prod_i (u^n - \alpha_i^n) = u^{2ng} - (1 + q^n - \nu_n)u^{n(2g-1)} + \ldots,$$

so that we find that $\Sigma\alpha_i^n = 1 + q^n - \nu_n = \sigma(\iota^n)$.

Now, let $F(x) = \Sigma a_\mu x^\mu$ be any polynomial with coefficients in Z; apply $\sigma(\gamma\gamma') \geqslant 0$ to $\gamma = F(\iota)$; using $\iota\cdot\iota' = q$, $\sigma(1) = 2g$, and $\sigma(\iota'^n) = \sigma(\iota^n) = \Sigma\alpha_i^n$, we find

$$\sum_\mu a_\mu^2\cdot 2g + 2\sum_{\mu<\nu} a_\mu a_\nu q^\mu\cdot\sum_i \alpha_i^{\nu-\mu} \geqslant 0.$$

The functional equation of $\zeta_K(s)$ implies that to every root α_i of $P(u)$ there is a root $\alpha_i' = q/\alpha_i$, so that our inequality can be written as $\Sigma F(\alpha_i)F(\alpha_i') \geqslant 0$. The left-hand side of this is a quadratic form in the coefficients of F, which is $\geqslant 0$ for all integral values, and therefore also for all real values, of these coefficients. If α_1 is such that $\alpha_1\bar{\alpha}_1 = |\alpha_1|^2 \neq q$, i.e., $\alpha_1' \neq \bar{\alpha}_1$, suppose, first, that $\alpha_1' \neq \alpha_1$; put $\alpha_2 = \alpha_1'$; a polynomial $F(x)$ can be found, with real-valued coefficients, which vanishes for all roots of $P(u)$ except $\alpha_1, \alpha_2, \bar{\alpha}_1, \bar{\alpha}_2$, and takes prescribed values z_1, z_2 at α_1, α_2: then $\Sigma F(\alpha_i)F(\alpha_i') = z_1z_2 + \bar{z}_1\bar{z}_2$, which becomes < 0 for suitable z_1, z_2: this contradicts our previous inequality. We reason similarly if $\alpha_1' = \alpha_1$. Thus all roots of $P(u)$ must satisfy $|\alpha_i|^2 = q$, which is the Riemann hypothesis for ζ_K.

A detailed account of this theory, including the application to Artin's L-functions, and of the "transcendental" theory as outlined in my previous note, is being prepared for publication.

[1] Sur les fonctions algébriques à corps de constantes fini, C.R.t. 210 (1940), p. 592.

[2] F. Severi, *Trattato di Geometria algebrica*, vol. 1, pt. 1, Bologna, Zanichelli 1926, chapter VI. It should be observed that Severi's treatment, although undoubtedly containing all the essential elements for the solution of the problems it purports to solve, is meant to cover only the classical case where the field of constants is that of complex numbers, and doubts may be raised as to its applicability to more general cases, especially to characteristic $p \neq 0$. A rewriting of the whole theory, covering such cases, is therefore a necessary preliminary to the applications we have in view.

[3] A generic point of an irreducible variety of dimension n is a point, the coördinates of which satisfy the equations of the variety and generate a field of degree of transcendency n over the field of constants. Cf. B. L. van der Waerden, *Einführung in die algebraische Geometrie*, Berlin, Springer 1939, chap. IV, § 29.

[4] B. L. van der Waerden, "Zur algebraischen Geometrie XIII," *Math. Ann.*, **115**, 359, and XIV, *Ibid.* 619.

[5] Loc. cit.,[2] chap. VI, No. 75 (pp. 228–229) and No. 84 (pp. 259–267). Severi's treatment can be somewhat clarified and simplified at this point, as will be shown elsewhere.

[6] H. Hasse, "Über die Kongruenzzetafunktionen," *Sitz.-ber. d. Preuss. Akad. d. Wiss.* **1934**, 250.

12.9 P. Turán (1948)

On some approximative Dirichlet-Polynomials in the Theory of the Zeta-Function of Riemann

Pál Turán was born in 1910 in Hungary. He demonstrated a talent for mathematics early in his schooling. Turán attended Pázmány Péter University in Budapest, where he met Pál Erdős, with whom he would collaborate extensively. He completed his Ph.D. under Fejér in 1935 and published his thesis, *On the Number of Prime Divisors of Integers.* Turán found it extremely difficult to find employment owing to the widespread anti-semitism of the period. He spent the war in forced labour camps, which not only saved his life, but according to Turán, gave him time to develop his mathematical ideas. During his lifetime, he published approximately 150 papers in several mathematical disciplines. Among his several honours are the Kossuth Prize from the Hungarian government (which he won on two occasions), the Szele Prize from the János Bolyai Mathematical Society, and membership in the Hungarian Academy of Sciences [121].

In this paper Turán considers partial sums of the Riemann zeta function, $\sum_{k=1}^{n} k^{-s}$. He proves the condition that if there is an n_0 such that for $n > n_0$ the partial sums do not vanish in the half plane $\Re(s) > 1$, then the Riemann hypothesis follows. Hugh Montgomery later proved that this condition cannot lead to a proof of the hypothesis, since the partial sums of the zeta function have zeros in the region indicated.

DET KGL. DANSKE VIDENSKABERNES SELSKAB
MATEMATISK-FYSISKE MEDDELELSER, BIND XXIV, NR. 17

ON SOME APPROXIMATIVE DIRICHLET-POLYNOMIALS IN THE THEORY OF THE ZETA-FUNCTION OF RIEMANN

BY

PAUL TURÁN

KØBENHAVN
I KOMMISSION HOS EJNAR MUNKSGAARD
1948

1. The zeta-function of Riemann is defined in the complex $s = \sigma + it$ plane for the half-plane $\sigma > 1$ by

$$\zeta(s) = \frac{1}{1^s} + \frac{1}{2^s} + \cdots + \frac{1}{n^s} + \cdots. \tag{1.1}$$

Here is valid also the product-representation of Euler

$$\zeta(s) = \prod_p \frac{1}{1 - \frac{1}{p^s}},$$

where p runs through the consecutive primes. From this representation it follows clearly that

$$\zeta(s) \neq 0 \qquad \text{for } \sigma > 1. \tag{1.2}$$

As is well-known, the function $\zeta(s)$ is regular in the whole plane except at $s = 1$, where there is a pole of the first order. It is also well-known that the distribution of its roots is of fundamental importance in the theory of numbers. We know from the functional equation

$$\pi^{-\frac{s}{2}} \Gamma\left(\frac{s}{2}\right) \zeta(s) = \pi^{-\frac{1-s}{2}} \Gamma\left(\frac{1-s}{2}\right) \zeta(1-s) \tag{1.3}$$

that in the half-plane $\sigma < 0$ the only zeros are $s = -2, -4, -6, \ldots$ and that there are an infinite number of roots $\varrho = \sigma_\varrho + it_\varrho$, the so called "non trivial roots", such that

$$0 < \sigma_\varrho < 1. \tag{1.4}$$

The famous hypothesis of Riemann, unproved so far, states that these all lie on the line $\sigma = \frac{1}{2}$. Using the fact obvious from the functional-equation (1.3) that they are symmetrical with respect to $s = \frac{1}{2}$ we can express the content of this hypothesis in the form that

$$\zeta(s) \neq 0 \qquad \text{for} \quad \sigma > \frac{1}{2}. \tag{1.5}$$

No one has yet been able to prove even the existence of a ϑ with $\frac{1}{2} \leqq \vartheta < 1$ such that

$$\zeta(s) \neq 0 \qquad \text{for} \quad \sigma > \vartheta. \tag{1.6}$$

2. Next we consider the partial-sums

$$U_n(s) = \frac{1}{1^s} + \frac{1}{2^s} + \cdots + \frac{1}{n^s} \tag{2.1}$$

of the series (1.1). They obviously converge to $\zeta(s)$ for $\sigma > 1$. We ask whether these partial-sums share with $\zeta(s)$ the property of being non-vanishing in the half-plane $\sigma > 1$. We have found the somewhat striking

Theorem I. If there is an n_0 such that for $n > n_0$ the partial-sums $U_n(s)$ do not vanish in the half-plane $\sigma > 1$, then Riemann's conjecture (1.5) is true.[1)]

More generally

Theorem II. If there are positive numbers n_0 and K such that for $n > n_0$ the partial-sum $U_n(s)$ does not vanish in the half-plane

$$\sigma \geqq 1 + \frac{K}{\sqrt{n}}, \tag{2.2}$$

then Riemann's hypothesis (1.5) is true.[2)]

Still more generally

Theorem III. If there are positive numbers n_0, K and ϑ satisfying

1) This elegant form of the theorem is due to Prof. B. Jessen; my original form was more awkward.

2) This theorem is due to my pupil Mr. P. Ungár who observed that the method of proof of theorem I furnishes at the same time the proof of theorem II.

$$\frac{1}{2} \leqq \vartheta < 1 \tag{2.3}$$

such that for $n > n_0$ the sum $U_n(s)$ does not vanish in the half-plane

$$\sigma \geqq 1 + \frac{K}{n^{1-\vartheta}}, \tag{2.4}$$

then $\zeta(s) \neq 0$ in the half-plane $\sigma > \vartheta$.

A further not uninteresting generalisation is given by

Theorem IV. If there are positive n_0, K, K_1 and ϑ satisfying (2.3) such that for $n > n_0$ the polynomial $U_n(s)$ omits in the half-plane (2.4) a real value c_n with[1)]

$$-\frac{K_1}{n^{1-\vartheta}} \leqq c_n \leqq \frac{K_1}{n^{1-\vartheta}},$$

then $\zeta(s) \neq 0$ for $\sigma > \vartheta$.

3. All these theorems admit a further generalisation which asserts that these theorems remain true even if there is an infinity of exceptional n's provided that there are "not too many". We state explicitly only the analogue of theorem II.

Theorem V. If there is a positive K such that—denoting by $a(x)$ the number of n-values not exceeding x for which $U_n(s)$ has zeros in the half-plane $\sigma \geqq 1 + \frac{K}{\sqrt{n}}$—we have

$$\lim_{x \to \infty} \frac{a(x)}{\log x} = 0, \tag{3.1}$$

then Riemann's hypothesis (1.5) is true.

Such connection between Riemann's hypothesis and the roots of the partial-sums seems not to have been observed so far. The very interesting question whether, supposing Riemann's hypothesis to be true, we can deduce consequences on the roots of the sections, remains open.

On the basis of theorem III we have an interesting situation for the roots of the partial-sums $U_n(s)$. If Riemann's hypothesis

1) The stronger statement that the omitted value c_n must satisfy only $|c_n| \leqq K_1 n^{\vartheta - 1}$ we cannot prove.

(1.5) is not true, or more exactly $\sup \sigma_\varrho = \Theta > \frac{1}{2}$, then there is an infinity of n's such that $U_n(s)$ vanishes in the half-plane $\sigma > 1$ and even in the half-plane $\sigma > 1 + n^{\Theta-1-\varepsilon}$, where ε is an arbitrarily small preassigned number. But if Riemann's hypothesis (1.5) is true, then, curiously, the method fails and nothing can be said about the roots of $U_n(s)$ this way.

4. What can actually be said about the roots of $U_n(s)$? According to a theorem of K. Knopp[1] every point of the line $\sigma = 1$ is a condensation-point for the zeros of $U_n(s)$. But in an interesting way this condensation happens at least for $|t| \geqq \tau_0$, where τ_0 is a sufficiently large numerical constant[2] only from the left.

More exactly we can prove

Theorem VI. There exist numerical τ_0 and K_2 such that $U_n(s)$ does not vanish for

$$\tau_0 \leqq |t| \leqq e^{K_2 \log n \log\log n}, \qquad \sigma \geqq 1, \qquad n > n_0. \tag{4.1}$$

Further $U_n(s)$ does not vanish in the half-plane

$$\sigma \geqq 1 + 2\frac{\log\log n}{\log n}, \qquad n > n_0. \tag{4.2}$$

In the estimation (4.1) of the domain of non-vanishing we could replace $\log n \log\log n$ by $\log^k n$ with a suitable $k > 1$, using estimations of Vinogradoff instead of estimations of Weyl.

The first part of theorem VI shows the indicated behaviour of the roots of $U_n(s)$; but to prove only this for all sufficiently large t we could use a more elementary reasoning. We write $U_n(s)$ in the form

$$U_n(s) = \zeta(s) - r_n(s), \qquad r_n(s) = \sum_{\nu > n}{}' \nu^{-s}. \tag{4.3}$$

In what follows we denote by $K_3, \ldots$ positive quantities, whose dependence upon eventual parameters will be indicated explicitly; if no such dependence is mentioned they denote numerical constants. If

1) See the paper of R. Jentzsch: Untersuchungen zur Theorie der Folgen analytischer Functionen. Acta Math. 41 (1918), p. 219—251, in particular p. 236.

2) This probably also holds with $\tau_0 = 0$.

$$1 < \sigma \leqq 2, \qquad |t| \geqq 4, \tag{4.4}$$

then

$$\left|(\nu+1)^{1-s} - \nu^{1-s} - \frac{1-s}{\nu^s}\right| \leqq \left|\left(1+\frac{1}{\nu}\right)^{1-s} - 1 - \frac{1-s}{\nu}\right| < \frac{K_3 t^2}{\nu^2},$$

and summing over $\nu > n$

$$|r_n(s)| < 2\frac{n^{1-\sigma}}{|t|} + \frac{2K_3|t|}{n}.$$

This is true for any s in the domain (4.4) and obviously for $\sigma \geqq 1$, $|t| \geqq 4$; hence for $n \geqq t^2$

$$|r_n(s)| \leqq \frac{K_4}{|t|}. \tag{4.5}$$

Since for a suitable positive K_5 we have[1)] for $\sigma \geqq 1$, $|t| \geqq 4$

$$\frac{1}{|\zeta(s)|} < K_5 \log|t|, \tag{4.6}$$

it follows from this, (4.5) and (4.3), that for $\sigma \geqq 1$, $|t| \geqq K_6$, $n \geqq t^2$

$$\left|U_n(s)\right| > \frac{1}{K_5 \log|t|} - \frac{K_4}{|t|} > 0. \qquad Q.\ e.\ d.$$

We do not know so far of a single $U_n(s)$ vanishing in the half-plane $\sigma > 1$. Beyond the obvious fact that $U_n(s) \neq 0$ there for $n \leqq 3$, we know only from a remark of Prof. B. Jessen that $U_4(s)$ as well as $U_5(s)$ does not vanish in the half-plane $\sigma \geqq 1$. For the set of values of $U_4(s)$ coincides "essentially" with that of the function

$$g_4(\varphi, \psi, \sigma) = 1 + \frac{1}{2^\sigma}e^{i\varphi} + \frac{1}{3^\sigma}e^{i\psi} + \frac{1}{4^\sigma}e^{2i\varphi}$$

and

$$\Re g_4(\varphi, \psi, \sigma) = 1 + \frac{1}{2^\sigma}\cos\varphi + \frac{1}{3^\sigma}\cos\psi + \frac{1}{4^\sigma}\cos 2\varphi,$$

so that for fixed $\sigma = \sigma_0 \geqq 1$

1) T. H. Gronwall: Sur la fonction $\zeta(s)$ de Riemann au voisinage de $\sigma = 1$. Rend. Circ. Mat. di Pal. T. XXXV, 1913, p. 95—102.

$$\Re g_4(\varphi, \psi, \sigma_0) \geqq 1 - \frac{1}{3^{\sigma_0}} + \min_{\varphi}\left(\frac{1}{2^{\sigma_0}} \cos\varphi + \frac{1}{4^{\sigma_0}} \cos 2\varphi\right) =$$

$$= 1 - \frac{1}{3^{\sigma_0}} - \frac{1}{8} - \frac{1}{4^{\sigma_0}}$$

and for $\sigma \geqq 1$

$$\Re U_4(s) \geqq \frac{7}{24}.$$

Similarly we have

$$\Re U_5(s) \geqq \frac{7}{24} - \frac{1}{5} = \frac{11}{120}.$$

5. We can prove all the corresponding theorems on replacing $U_n(s)$ with the Cesaro-means[1)] of the series (1.1)

$$C_n(s) = \sum_{\nu \leqq n}' \frac{n - \nu + 1}{n + 1} \cdot \nu^{-s}. \tag{5.1}$$

To see what alterations are necessary in the proofs we shall treat explicitly only

Theorem VII. If there exist positive K and n_0 so that the polynomial $C_n(s)$ does not vanish in the half-plane

$$\sigma \geqq 1 + \frac{K}{\sqrt{n}},$$

then Riemann's hypothesis (1.5) is true.

Though the numerical evidences that the polynomials $C_n(s)$ do not vanish in the half-plane $\sigma \geqq 1$ are more numerous (e. g. the non-vanishing of $C_6(s)$ in this half-plane follows quite trivially), we cannot enlarge the domain of non-vanishing (4.1) of theorem VI for the Cesaro-means. For the Riesz-means

$$R_n(s) = \sum_{\nu \leqq n}' \left(1 - \frac{\log \nu}{\log n}\right) \nu^{-s},$$

for which the analogous theorems would have a somewhat enhanced interest because of the fact that they converge to $\zeta(s)$ in the closed half-plane $\sigma \geqq 1$, our method fails in principle.

1) Analogous theorems hold for Cesaro-means of higher order.

6. Another interesting series for the zeta-function is

$$\frac{1}{1^s} - \frac{1}{2^s} + \frac{1}{3^s} - \cdots + \frac{(-1)^{n+1}}{n^s} + \cdots, \tag{6.1}$$

which represents in the half-plane $\sigma > 0$ the function

$$\left(1 - \frac{2}{2^s}\right) \zeta(s). \tag{6.2}$$

This function vanishes in the half-plane $\sigma > 0$ at the points $s = \varrho$ as well as at

$$s = w_k = 1 + \frac{2k\pi i}{\log 2} \tag{6.3}$$

$$k = \pm 1, \pm 2, \cdots.$$

Hence from the well-known theorem of Hurwitz it follows that the partial-sums

$$V_n(s) = \sum_{m \leqq n} (-1)^{m+1} m^{-s} \tag{6.4}$$

have roots "near" to w_k if n is sufficiently large, and we prove that these occur infinitely often in the half-plane $\sigma > 1$. Hence the analogue of theorem I is meaningless; but the analogues of theorems II, III, IV and V are true. We shall prove only

Theorem VIII. If there exist positive n_0 and K such that for $n > n_0$ the partial-sums $V_n(s)$ do not vanish in the half-plane $\sigma \geqq 1 + \frac{K}{\sqrt{n}}$, then Riemann's conjecture (1.5) is true,

and

Theorem IX. There is an infinity of values of n for which $V_n(s)$ vanishes in the half-plane $\sigma > 1$.

This will follow from the fact that those roots of $V_{2n}(s)$ which converge for a fixed k to w_k may be expressed asymptotically as

$$w_k + \frac{1}{4 \log 2 \cdot \zeta'(w_k)} \cdot n^{-1 + \frac{2k\pi i}{\log 2}}.$$

7. Returning to the partial-sums $U_n(s)$ we have mentioned the fact that every point of the line $\sigma = 1$ is a clustering point of the zeros of the polynomials $U_n(s)$. Are these all the clustering points? The answer, as we can prove easily, is affirmative. For

$$(\nu+1)^{1-s} - \nu^{1-s} - (1-s)\,\nu^{-s} = \nu^{1-s}\left(\left(1+\frac{1}{\nu}\right)^{1-s} - 1 - \frac{1-s}{\nu}\right) =$$

$$= \nu^{1-s}\int_0^{1/\nu} (1-s)(-s)(1+r)^{-s-1}\left(\frac{1}{\nu} - r\right) dr$$

and summing over ν

$$\left.\begin{aligned} |(n+1)^{1-s} - 1 - (1-s)\,U_n(s)| &\leq |s||s-1| \sum_{1 \leq \nu \leq n} \nu^{1-\sigma}\frac{1}{\nu^2} \max_{0 \leq y \leq \frac{1}{\nu}} (1+y)^{-\sigma-1} \leq \\ &\leq |s||s-1| \sum_{\nu=1}^{\infty} \nu^{-1-\sigma}. \end{aligned}\right\} \quad (7.1)$$

If a point $s^* = \sigma^* + it^*$ in the strip $\varepsilon \leq \sigma \leq 1 - \varepsilon \left(0 < \varepsilon < \frac{1}{4}\right)$ could be an accumulation-point for the zeros of $U_n(s)$ then we have for an infinity of values of n that $U_n(s)$ vanishes in the domain

$$\frac{\varepsilon}{2} \leq \sigma \leq 1 - \frac{\varepsilon}{2}, \qquad |t| \leq (t^* + 1). \qquad (7.2)$$

But from (7.1) it follows that in the domain (7.2)

$$|(n+1)^{1-s} - 1 - (1-s)\,U_n(s)| \leq \frac{4|s||s-1|}{\varepsilon}$$

$$|1-s||U_n(s)| \geq (n+1)^{\frac{\varepsilon}{2}} - \frac{4|s||s-1|}{\varepsilon} - 1 > 0$$

if $n > n_2 = n_2(t^*, \varepsilon)$, which is a contradiction. For the half-plane $\sigma < \varepsilon$ the proof runs similarly. Analogous statements hold for the Cesaro-means $C_n(s)$ and the Riesz-means $R_n(s)$. Similarly we can show that the complete set of accumulation points of zeros of $V_n(s)$ consists of the points of the line $\sigma = 0$, the non trivial roots ϱ of the zeta-function and the points w_k of (6.3).

8. These results suggest interesting further questions. Let there be given the series

$$e^{-\lambda_1 s} + a_2 e^{-\lambda_2 s} + \cdots + a_n e^{-\lambda_n s} + \cdots, \qquad 0 \leqq \lambda_1 < \lambda_2 < \cdots \to \infty, \quad (8.1)$$

which is convergent for $\sigma > 0$. We denote by H the clustering set of the zeros of its partial-sums. Is it always true that H consists of the zeros in $\sigma > 0$ of the function $f(s)$ defined by the series (8.1) and of the points of the line $\sigma = 0$? We can show that the set H can consist of the whole half-plane $\sigma \leqq 0$; by this we give the answer to a question raised by L. Fejér. Let $r_1, r_2, \cdots$ be the set of all positive rational numbers, arranged in such a way that every fixed one occurs infinitely often; we consider the product

$$g(s) = \prod_{\nu=1}^{\infty} \left[1 - \left(e^{-s-r_\nu}\right)^{2^\nu}\right]. \quad (8.2)$$

Since the product

$$\prod_{\nu=1}^{\infty} \left[1 + e^{-2^\nu \sigma}\right]$$

is convergent for $\sigma > 0$ the product (8.2) can be expressed in the form (8.1) convergent for $\sigma > 0$. We observe that because of the rapid growth of the numbers 2^ν every partial-product is at the same time a partial-sum; hence all the numbers

$$s = -r_\nu + \frac{2 l \pi i}{2^\nu} \qquad \begin{aligned} l &= 0, \pm 1, \pm 2, \cdots \\ \nu &= 1, 2, \cdots \end{aligned}$$

are roots of certain partial-sums. Since every fixed r_μ occurs infinitely often, every point of the line $\sigma = -r_\mu$ is a clustering point of such roots and so are all points of the half-plane[1] $\sigma \leqq 0$.

Somewhat more peculiar is the behaviour of the zeros of the sections of the series

$$1 + e^{-(s+10)} + e^{-2s} + e^{-3(s+10)} + \cdots + e^{-2ns} + e^{-(2n+1)(s+10)} + \cdots,$$

as P. Erdös remarked. As is easily shown, the set H consists

[1] Putting $e^{-s} = z$, we obtain a power series, regular for $|z| < 1$ with the property, that the roots of the partial-sums cluster to every point in $|z| \geqq 1$.

in this case of the lines $\sigma = 0$ and $\sigma = -10$. Probably one can prescribe in the half-plane $\sigma < 0$ the cluster set H of a Dirichlet-series regular in $\sigma > 0$.

9. The theorems I—V raise the question whether the zeros of the partial-sums of a series (8.1) convergent for $\sigma > 0$ can cluster to the points of the line $\sigma = 0$ only from the left side. That this is, indeed, possible is shown for example by series of the type

$$\sum_{\nu=1}^{\infty} a_\nu e^{-\nu s}, \tag{9.1}$$

where the a_ν's are positive and tend monotonously to 0 in such a manner that the line of convergence is the line $\sigma = 0$. That all the roots of all partial-sums of the series (9.1) lie in the half-plane $\sigma < 0$, follows from the theorem of Eneström-Kakeya. If the coefficients are chosen positive and increasing, then all the roots of all partial-sums lie in the half-plane $\sigma > 0$. For the sake of completeness we mention that the function

$$_1(s) = \prod_{\nu=1}^{\infty} \left[1 - \left(e^{-s-l_\nu}\right)^{2^\nu}\right],$$

where the sequence $l_\nu \to 0$ and contains an infinite number of both positive and negative terms, has an expansion of the form (8.1) (even (9.1)) with the property, that every point of the line $\sigma = 0$ is a condensation-point of zeros from the left half-plane and at the same time a condensation-point of zeros from the right half-plane.

10. Of course a direct approach to the investigation of the roots of partial-sums or arithmetical means in the half-plane $\sigma > 1$ seems to be very difficult; the stress of this paper is laid upon the connection between these questions and Riemann's hypothesis. In any case the results raise the question how the roots of the function given by a Dirichlet-series can influence the roots of its partial-sums or suitable means. In this direction no results are known so far which hold for the means of finite index. If the function is given by a Taylor-series the situation changes. If e. g.

$$f(s) = \sum_{\nu=0}^{\infty} a_\nu s^\nu$$

is an integral function of order 1, whose roots lie in the half-plane $\sigma < 0$, then all roots of all "Jensen-means" of $f(s)$

$$J_n(f) = a_0 + a_1 s + a_2\left(1 - \frac{1}{n}\right)s^2 + a_3\left(1 - \frac{1}{n}\right)\left(1 - \frac{2}{n}\right)s^3 + \cdots +$$
$$+ a_n\left(1 - \frac{1}{n}\right)\left(1 - \frac{2}{n}\right)\cdots\left(1 - \frac{n-1}{n}\right)s^n$$

lie in the half-plane $\sigma \leqq 0$. The proof is very easy and runs on known lines.

I wish to thank Mrs. Helen K. Nickerson for linguistic assistance in the preparation of the manuscript.

11. Now we pass on to the proof of theorems I—IV. Obviously it is sufficient to prove theorem IV. We base the proof on an important theorem of H. Bohr[1], combining it with a classical reasoning of Landau[2]. First we recall that given a sequence

$$\lambda_1 < \lambda_2 < \cdots < \lambda_n < \cdots \to \infty, \tag{11.1}$$

we call the sequence B of linearly independent numbers

$$\beta_1, \beta_2, \ldots \tag{11.2}$$

a basis of (11.1), if

$$\lambda_n = r_{n_1}\beta_1 + r_{n_2}\beta_2 + \cdots + r_{n_{q_n}}\beta_{q_n}$$

with rational r_{n_ν}'s. If $f(s)$ is defined in the half-plane $\sigma > A$ by the absolutely convergent series

$$f(s) = \sum_{\nu=1}^{\infty} a_\nu e^{-\lambda_\nu s}, \tag{11.3}$$

1) H. Bohr, Zur Theorie der allgemeinen Dirichletschen Reihen. Math. Ann. B 79, 1919, p. 136—156. See in particular Satz. 4.

2) E. Landau: Über einen Satz von Tschebischeff. Math. Ann. 61, 1905, p. 527—550. The whole method of **14.** is due to him. He proved moreover that the integral (14.2) converges for $\sigma > \gamma$, but we do not use this fact.

Bohr calls every function

$$g(s) = \sum_{\nu=1}^{\infty} b_\nu e^{-\lambda_\nu s}$$

equivalent to $f(s)$, if for suitable $\varphi_1, \varphi_2, \cdots$ and $n = 1, 2, \cdots$

$$b_n = a_n e^{-i(r_{n_1}\varphi_1 + r_{n_2}\varphi_2 + \cdots + r_{nq_n}\varphi_{q_n})} \tag{11.4}$$

Obviously $g(s)$ is also absolutely convergent for $\sigma > A$. Now the above-mentioned theorem of Bohr asserts that the sets of values assumed by $f(s)$ and $g(s)$ resp. in the half-plane $\sigma > A$ are identical.

We may apply this theorem to $f(s) = U_n(s)$ with

$$B \equiv (\log 2, \log 3, \cdots, \log p, \cdots), \quad \varphi_\nu = \pi \quad (\nu = 1, 2, \cdots). \tag{11.5}$$

If $\nu = p_1^{\alpha_1} p_2^{\alpha_2} \cdots p_r^{\alpha_r}$, then

$$\log \nu = \sum_{j=1}^{r} \alpha_j \log p_j;$$

hence

$$b_\nu = \exp\left(-i\pi \sum_{j=1}^{r} \alpha_j\right) = \lambda(\nu),$$

where, as usual, $\lambda(\nu)$ denotes Liouville's number theoretic function. Hence the set of values assumed by $U_n(s)$ in the half-plane

$$\sigma > 1 + Kn^{\vartheta - 1} \tag{11.6}$$

is identical with that assumed here by the polynomial

$$W_n(s) = \sum_{\nu \leq n} \frac{\lambda(\nu)}{\nu^s}. \tag{11.7}$$

If $U_n(s)$ does not assume the value c_n in the half-plane (11.6) the same holds for $W_n(s)$. But then the function

$$W_n(s) - c_n, \tag{11.8}$$

which is real on the real axis and is positive at infinity, if $n > n_1$, where

$$K_1 n_1^{\vartheta-1} < 1, \tag{11.9}$$

is necessarily positive on the whole positive axis in the half-plane (11.6) and in particular non negative for $s = 1 + Kn^{\vartheta-1}$. This gives that if

$$n > K_7 = \max(n_0, n_1),$$

$$\sum_{\nu \leqq n}' \frac{\lambda(\nu)}{\nu^{1+Kn^{\vartheta-1}}} \geqq c_n. \tag{11.10}$$

12. Using the restrictions on c_n, we may write (11.10) for $n > K_7$ in the form

$$\sum_{\nu \leqq n} \lambda(\nu)\, \nu^{-1-Kn^{\vartheta-1}} \geqq -K_1 n^{\vartheta-1}. \tag{12.1}$$

Since

$$\left| e^{-Kn^{\vartheta-1}\log\nu} - 1 + Kn^{\vartheta-1}\log\nu \right| < \frac{1}{2}\cdot\frac{K^2}{n^{2(1-\vartheta)}}\log^2\nu,$$

the error made in replacing the left hand member of (12.1) by

$$\sum_{\nu \leqq n}' \frac{\lambda(\nu)}{\nu} - Kn^{\vartheta-1} \sum_{\nu \leqq n}' \frac{\lambda(\nu)\log\nu}{\nu}$$

is in absolute value

$$\leqq \frac{K^2}{2\,n^{2(1-\vartheta)}} \sum_{\nu \leqq n}' \frac{\log^2\nu}{\nu} < \frac{K^2}{6\,n^{2(1-\vartheta)}}\log^3 n < \frac{K^2}{6} n^{\vartheta-1}$$

if $n > n_2 = K_8(\vartheta)$. Hence for $n > \max(K_7, K_8(\vartheta))$

$$\sum_{\nu \leqq n}' \frac{\lambda(\nu)}{\nu} > Kn^{\vartheta-1} \sum_{\nu \leqq n}' \frac{\lambda(\nu)\log\nu}{\nu} - \left(K_1 + \frac{K^2}{6}\right) n^{\vartheta-1}. \tag{12.2}$$

Now since for $\sigma > 0$

$$\sum_{\nu=1}^{\infty}' \frac{\lambda(\nu)\log\nu}{\nu}\cdot\frac{1}{\nu^s} = \frac{\zeta(2s+2)\,\zeta'(s+1) - 2\,\zeta(s+1)\,\zeta'(2s+2)}{\zeta(s+1)^2},$$

the usual method of complex integration shows that

$$\sum_{\nu} \frac{\lambda(\nu) \log \nu}{\nu}$$

is convergent and that its sum is $-\frac{\pi^2}{6}$. Therefore for $n > K_9$ we have $\sum_{\nu \leqq n} \frac{\lambda(\nu) \log \nu}{\nu} \geqq -2$ and for $n > \max(K_7, K_8(\vartheta), K_9) = K_{10}(\vartheta)$

$$L(n) \equiv \sum_{\nu \leqq n} \frac{\lambda(\nu)}{\nu} > -\left(2 + K_1 + \frac{K^2}{6}\right) n^{\vartheta - 1} = -K_{11} n^{\vartheta - 1}. \quad (12.3)$$

13. The inequality (12.3) can be written in the form

$$L(n) + K_{11} n^{\vartheta - 1} > 0 \quad (13.1)$$

for integral $n > K_{10}(\vartheta)$. Now we consider for a continuously varying $x \geqq K_{10}(\vartheta)$ the function

$$L(x) + 2 K_{11} x^{\vartheta - 1},$$

and we assert that this is positive for all $x \geqq K_{10}(\vartheta)$. We consider the x-values belonging to the interval $m \leqq x < m + 1$, m an integer and $\geqq K_{10}(\vartheta)$. For $x = m$ our assertion is evident. To estimate this function for other x-values we remark that $L(x)$ being a step-function is constant for $m \leqq x < m + 1$ and hence

$$L(x) + 2 K_{11} x^{\vartheta - 1} = L(m) + K_{11} m^{\vartheta - 1} + K_{11}\left(2 x^{\vartheta - 1} - m^{\vartheta - 1}\right) \geqq$$
$$\geqq \left(L(m) + K_{11} m^{\vartheta - 1}\right) + K_{11}\left(2 (m+1)^{\vartheta - 1} - m^{\vartheta - 1}\right) \geqq$$
$$\geqq K_{11} (m+1)^{\vartheta - 1}\left(2 - \left(1 + \frac{1}{m}\right)^{1 - \vartheta}\right) \geqq 0.$$

14. Now for $\sigma > 0$ we have

$$\sum_{\nu = 1}^{\infty} \frac{\lambda(\nu)}{\nu} \cdot \frac{1}{\nu^s} = \frac{\zeta(2s + 2)}{\zeta(s + 1)} = s \int_1^{\infty} \frac{L(x)}{x^{s+1}} dx,$$

$$s \int_0^{\infty} \frac{x^{\vartheta - 1}}{x^{s+1}} dx = \frac{s}{s + 1 - \vartheta};$$

therefore for $\sigma > 0$

$$\int_1^\infty \frac{L(x) + 2K_{11} x^{\vartheta - 1}}{x^{s+1}} dx = \frac{\zeta(2s+2)}{s\zeta(s+1)} + \frac{2K_{11}}{s+1-\vartheta};$$

or for $\sigma > 1$

$$\int_1^\infty \frac{L(x) + 2K_{11} x^{\vartheta - 1}}{x^{s}} dx = \frac{\zeta(2s)}{(s-1)\zeta(s)} + \frac{2K_{11}}{s-\vartheta}. \qquad (14.1)$$

From **13.** it follows that the numerator of the integrand is positive for all sufficiently large x's; hence we may apply the following theorem of Landau[1]: if a function $\varphi(s)$ is defined for $\sigma > 1$ as

$$\varphi(s) = \int_1^\infty \frac{A(x)}{x^s} dx, \qquad (14.2)$$

where $A(x)$ does not change sign for $x > x_0$ and $\varphi(s)$ is regular on the real axis for $s > \gamma\ (< 1)$, then $\varphi(s)$ is regular in the entire half-plane $\sigma > \gamma$.

Hence we have only to consider the singularities of the right of (14.1) on the real axis. The first term is regular for $s > \frac{1}{2}$, the second for $s > \vartheta$, hence their sum is regular for $s > \vartheta$. Then Landau's theorem applied to (14.1) shows that the function on the right is regular in the half-plane $\sigma > \vartheta$. But then $\zeta(s)$ cannot vanish in this half-plane and theorem IV is proved.

The basis of the proof is the observation that for given arbitrarily small positive ε and η we can find $\tau_1 = \tau_1(\varepsilon, \eta)$ such that for $\sigma > 1 + \eta$ we have

$$\left| \zeta(s + i\tau_1) - \frac{\zeta(2s)}{\zeta(s)} \right| \leqq \varepsilon.$$

15. The proof of theorem V runs on the same line but instead of Landau's theorem we use the following theorem of Pólya.[2] Considering functions of type (14.2) let

1) See above p. 13, note 2.

2) G. Pólya: Über das Vorzeichen des Restgliedes im Primzahl-Satz. Gött. Nachr. 1930, p. 19—27. A special case of his theorem is given here in a slightly altered form.

$$1 = x_0 < x_1 < x_2 < \cdots < x_n < \cdots \qquad (15.1)$$

be a sequence[1] which does not cluster to any finite positive value and with the property

$$(-1)^\nu A(x) \geqq 0 \qquad \text{for} \qquad x_\nu \leqq x < x_{\nu+1}. \qquad (15.2)$$

The values x_ν are called sign-changing values. If $B(x)$ is defined by

$$B(\omega) = \sum_{\substack{\nu \\ x_\nu \leqq \omega}} 1 \qquad (15.3)$$

he assumes that $A(x)$ has "not too many" sign-changing values, or more exactly

$$\overline{\lim_{\omega \to \infty}} \frac{B(\omega)}{\log \omega} = 0. \qquad (15.4)$$

Then Pólya's theorem asserts that if Θ is the exact regularity-abscissa of $\varphi(s)$ and $\varphi(s)$ is meromorphic in a half-plane $\sigma > \Theta - b$ $(b > 0)$, then the statement of Landau's theorem holds i. e. the point $s = \Theta$ is a singular point of $\varphi(s)$. Applying this theorem to the function on the right in (14.1) we see that the condition of meromorphism is fulfilled. If we can deduce from (3.1) that the number of the sign-changing places of $L(x) + 2K_{11}x^{\vartheta-1}$ satisfies (15.4) the proof of theorem V will be completed.

We consider first the integral x-values. If n is a value sufficiently large such that $U_n(s) \neq 0$ for $\sigma \geqq 1 + \frac{K}{\sqrt{n}}$ — or briefly if n is a "good" value—the reasoning of **11.** and **12.** gives that $L(n) + K_{11}n^{\vartheta-1} > 0$. Then the reasoning of **13.** shows that for good n's

$$L(x) + 2K_{11}x^{\vartheta-1} > 0 \qquad n \leqq x < n+1. \qquad (15.5)$$

If n is not a good value—or let us rather call it a "bad" value —then $L(n) + K_{11}n^{\vartheta-1}$ may be positive; and in this case (15.5) is true again. Finally if n is a bad value for which $L(n) + K_{11}n^{\vartheta-1}$ is $\leqq 0$, then since both of the functions[2]

$$L(x) + K_{11}x^{\vartheta-1}, \qquad L(x) + 2K_{11}x^{\vartheta-1}$$

1) This can consist of a finite number of terms or even of the single term x_0.
2) Using the fact that $L(x)$ is constant there, being a step-function.

are monotonously decreasing for $n \leqq x < n+1$, the second of them is either positive throughout $n \leqq x < n+1$ or negative throughout or finally positive for $x = n$ and decreasingly negative for $x \to (n+1)$ from the left. Hence the number of sign-changes $\leqq \omega$ for $L(x) + 2K_{11}x^{\vartheta-1}$ cannot surpass three times the number of bad n's $\leqq \omega$; but then using (3.1) we see that Pólya's further condition (15.4) is fulfilled, indeed.

We can easily show that if Riemann's hypothesis (1.5) is **not** true, then the partial-sums $U_n(s)$ vanish infinitely often in the half-plane $\sigma > 1$ or more generally for every positive ε and for an infinity of n's $U_n(s)$ vanishes in the half-plane

$$\sigma > 1 + \frac{1}{n^{1-\Theta+\varepsilon}},$$

where $\sup \sigma_\varrho = \Theta > \frac{1}{2}$. For if we have an ε_1 with $\Theta - \varepsilon_1 > \frac{1}{2}$ such that $U_n(s) \neq 0$ in the half-plane

$$\sigma > 1 + \frac{1}{n^{1-\Theta+\varepsilon_1}}$$

for all sufficiently large n, then from theorem III we could conclude that $\zeta(s) \neq 0$ in the half-plane $\sigma > \Theta - \varepsilon_1$, a contradiction. This reasoning fails completely if $\Theta = \frac{1}{2}$; the identity (14.1) valid for $\sigma > 1$

$$\int_1^\infty \frac{L(x) + 2K_{11}x^{-\frac{1}{2}}}{x^s}\,dx = \frac{(2s-1)\zeta(2s)}{2(s-1)\zeta(s)} \cdot \frac{1}{s-\frac{1}{2}} + \frac{2K_{11}}{s-\frac{1}{2}},$$

whose right-side is continuable over the whole plane and behaves asymptotically[1] if $s \to \frac{1}{2} + 0$ as

$$\frac{1}{s-\frac{1}{2}}\left(2K_{11} - \frac{1}{\zeta\left(\frac{1}{2}\right)}\right) = \left(2K_{11} + \frac{1}{\left|\zeta\left(\frac{1}{2}\right)\right|}\right)\frac{1}{s-\frac{1}{2}},$$

shows that the point $s = \frac{1}{2}$ is **certainly** a singular point.

1) We use the fact—which plays here a decisive role—that $\zeta\left(\frac{1}{2}\right)$ is **negative.**

16. Now we turn to the proof of theorem VI. The second part is only mentioned for the sake of completeness; for in the half-plane

$$\sigma \geqq 1 + 2\frac{\log\log n}{\log n}$$

$$|U_n(s)| = |\zeta(s) - r_n(s)| \geqq |\zeta(s)| - |r_n(s)| \geqq \prod \frac{1}{1+\frac{1}{p^\sigma}} -$$

$$-\sum_{\nu > n} \nu^{-\sigma} = \frac{\zeta(2\sigma)}{\zeta(\sigma)} - \frac{n^{1-\sigma}}{\sigma - 1} > K_{12}\cdot(\sigma-1) - \frac{1}{(\sigma-1)\log^2 n} >$$

$$> \frac{K_{12}}{(\sigma-1)\log^2 n}\left(4(\log\log n)^2 - \frac{1}{K_{12}}\right) > 0$$

for $n > n_0$.

For the proof of (4.1) we use an inequality of Weyl[1)] according to which for $\sigma > 0, r$ an integer, $t > 3$, $N \leqq N' < 2N$ we have

$$\left|\sum_{\nu=N}^{N'} \frac{1}{(\nu+1)^s}\right| < 2^{17} \log^{2^{1-r}} t \cdot$$

$$\cdot \left\{ N^{1-2^{1-r}-\sigma} t^{\frac{1}{(r+1)2^{r-1}}} + N^{1-\sigma} t^{-\frac{1}{(r+1)2^{r-1}}} \log^{\frac{r-1}{2^{r-1}}} N \right\}.$$

For $\sigma \geqq 1$ we have

$$\left. \begin{aligned} &\left|\sum_{\nu=N}^{N'} \frac{1}{(\nu+1)^s}\right| < 2^{17} \log^{2^{1-r}} t \cdot \\ &\cdot \left\{ N^{-2^{1-r}} t^{\frac{1}{(r+1)2^{r-1}}} + t^{-\frac{1}{(r+1)2^{r-1}}} \log^{\frac{r-1}{2^{r-1}}} N \right\}. \end{aligned} \right\} \quad (16.1)$$

To estimate $r_n(s)$ for $\sigma > 1$ and $n > t^2$ we start from

$$\left|(\nu+1)^{1-s} - \nu^{1-s} - \frac{1-s}{\nu^s}\right| < \frac{K_3\, t^2}{\nu^2}, \qquad t \geqq 4\,;$$

1) See Landau: Vorl. über Zahlenth. II. Theorem 389.

summing over ν

$$|r_n(s)| = \left|\sum_{\nu > n} \nu^{-s}\right| \leqq \frac{K_4}{t}. \tag{16.2}$$

Obviously the same estimation holds for $\sigma \geqq 1$ as we have seen in (4.5).

For $\sqrt{t} \leqq n \leqq t^2$ we have

$$|r_n(s)| < \left|\sum_{n < \nu \leqq t^2} \nu^{-s}\right| + \frac{K_4}{t} = |S_1| + \frac{K_4}{t}. \tag{16.3}$$

To estimate $|S_1|$ we split it into $O(\log t)$ sums of the form (16.1); applying (16.1) to each of them with

$$r = 2, \qquad \sqrt{t} \leqq N \leqq t^2,$$

we obtain that for $t > K_{13}$ each such sum is in absolute value

$$< K_{14}\sqrt{\log t}\left\{t^{-\frac{1}{12}} + t^{-\frac{1}{6}}\sqrt{\log t}\right\} < \frac{K_{15}}{\log^3 t},$$

i. e. $|S_1| < \frac{K_{16}}{\log^2 t}$, and from (16.3) for $t > K_{13}$ we have

$$|r_n(s)| < \frac{K_{17}}{\log^2 t}. \tag{16.4}$$

As appears from (16.2) and (16.4) this estimation is valid for all $n \geqq \sqrt{t}$.

Now we suppose only

$$n > t^{\frac{1}{\log\log t}}. \tag{16.5}$$

Then using (16.4) we have

$$\left.\begin{aligned} |r_n(s)| &\leqq \left|\sum_{n<\nu\leqq\sqrt{t}} \nu^{-s}\right| + \left|\sum_{\nu>\sqrt{t}} \nu^{-s}\right| < \frac{K_{17}}{\log^2 t} + \left|\sum_{n<\nu\leqq\sqrt{t}} \nu^{-s}\right| = \\ &= \frac{K_{17}}{\log^2 t} + |S_2|. \end{aligned}\right\} \tag{16.6}$$

Because of (16.5) we can split S_2 into $O(\log\log t)$ sums of the form

$$S_2^{(k)} = \sum_{t^{\frac{2}{k+2}} < \nu \leq t^{\frac{2}{k+1}}} \nu^{-s}, \qquad 3 \leq k \leq 2 \log\log t. \qquad (16.7)$$

$S_2^{(k)}$ can be split into $O(\log t)$ sums of the form (16.1); applying (16.1) to them with

$$r = k, \qquad t^{\frac{2}{k+2}} \leq N \leq t^{\frac{2}{k+1}}$$

we obtain that they are in absolute value

$$< K_{18} \log^{2^{1-k}} t \left\{ t^{-\frac{2}{k+2}\cdot\frac{1}{2^{k-1}}+\frac{1}{(k+1)2^{k-1}}} + t^{-\frac{1}{(k+1)2^{k-2}}} \log^{\frac{k-1}{2^{k-1}}} t \right\} < K_{19} \log^{-4} t,$$

and hence

$$|S_2| < \frac{K_{19}}{\log^2 t}.$$

Thus for $t > K_{20}$, $n > t^{\frac{1}{\log\log t}}$, $\sigma \geq 1$

$$|r_n(s)| < \frac{K_{20}}{\log^2 t}. \qquad (16.8)$$

The inequality $n > t^{\frac{1}{\log\log t}}$ can obviously be written in the form

$$|t| \leq e^{K_{21} \log n \log\log n}.$$

Finally using Gronwall's inequality (4.6) we obtain for $\sigma \geq 1$, $t > K_{22}$

$$|U_n(s)| \geq |\zeta(s)| - |r_n(s)| > \frac{1}{K_5 \log t} - \frac{K_{20}}{\log^2 t} > 0. \qquad Q.\,e.\,d.$$

17. Next we sketch the proof of theorem VII. A reasoning similar to that of **11.** shows that from our hypothesis it follows that for $n > K_{23}$

$$\sum_{\nu=1}^{n}(n-\nu+1)\frac{\lambda(\nu)}{\nu^{1+Kn^{-\frac{1}{2}}}}\geqq 0,$$

and by an argument similar to that of **12.** we obtain for these n's

$$C(n)\equiv\sum_{\nu\leqq n}(n-\nu+1)\frac{\lambda(\nu)}{\nu}\geqq -K_{24}\sqrt{n}$$

and

$$C(x)+2K_{24}\sqrt{x}\geqq 0 \qquad \text{for } x>K_{25}. \tag{17.1}$$

Next we have to find the analogue of formula (14.1). Generally if for $\sigma>0$

$$f(s)=\sum_{\nu=1}^{\infty}\frac{d_\nu}{\nu^s}, \qquad \sum_{\nu\leqq n}d_\nu=D_n,$$

$$\sum_{n\leqq m}D_n=\sum_{\nu\leqq m}(m-\nu+1)\,d_\nu=S_m,$$

then

$$\left.\begin{aligned}
f(s)&=\sum_{\nu=1}^{\infty}D_\nu\left(\nu^{-s}-(\nu+1)^{-s}\right)=\\
&=\sum_{\nu=1}^{\infty}S_\nu\left(\nu^{-s}-2(\nu+1)^{-s}+(\nu+2)^{-s}\right)=\\
&=\sum_{\nu=1}^{\infty}S_\nu\left\{s\int_{\nu}^{\nu+1}x^{-s-1}\,dx-s\int_{\nu+1}^{\nu+2}x^{-s-1}\,dx\right\}=\\
&=s\sum_{\nu=1}^{\infty}S_\nu\int_{\nu}^{\nu+1}\left(x^{-s-1}-(x+1)^{-s-1}\right)dx=\\
&=s\int_{1}^{\infty}S(x)\left(x^{-s-1}-(x+1)^{-s-1}\right)dx=\\
&=s(s+1)\int_{1}^{\infty}S(x)\left(\int_{0}^{1}(x+y)^{-s-2}\,dy\right)dx.
\end{aligned}\right\} \tag{17.2}$$

Hence for $\sigma > 0$

$$\frac{\zeta(2s+2)}{\zeta(s+1)} = s(s+1)\int_1^\infty C(x)\left(\int_0^1 (x+y)^{-s-2}\,dy\right)dx. \tag{17.3}$$

Let

$$J \equiv s(s+1)\int_1^\infty \sqrt{x}\left(\int_0^1 (x+y)^{-s-2}\,dy\right)dx =$$

$$= s(s+1)\int_1^\infty x^{-s-\frac{3}{2}}\left(\int_0^1 \left(1+\frac{y}{x}\right)^{-s-2} dy\right)dx.$$

But then

$$\frac{J-\dfrac{s(s+1)}{s+\frac{1}{2}}}{s(s+1)(s+2)} = \frac{J-s(s+1)\displaystyle\int_1^\infty x^{-s-\frac{3}{2}}\,dx}{s(s+1)(s+2)} =$$

$$= -\int_1^\infty x^{-s-\frac{3}{2}}\int_0^1 dy\left(\int_1^{1+\frac{y}{x}} \zeta^{-s-3}\,d\zeta\right)dx,$$

and since for $\sigma > -\frac{3}{2}$ the last expression is in absolute value

$$< \frac{1}{2}\int_1^\infty x^{-\sigma-\frac{5}{2}}\,dx$$

it represents a function $\vartheta(s)$ regular and bounded in every half-plane $\sigma \geqq -\frac{3}{2}+\varepsilon$. Hence for $\sigma > 0$

$$\left.\begin{aligned} &s(s+1)\int_1^\infty \sqrt{x}\left(\int_0^1 (x+y)^{-s-2}\,dy\right)dx = \\ &= \frac{s(s+1)}{s+\frac{1}{2}} + s(s+1)(s+2)\,\vartheta(s) \end{aligned}\right\} \tag{17.4}$$

and from (17.3) and (17.4)

$$s(s+1)\int_1^{\infty}\left(C(x)+2K_{24}\sqrt{x}\right)\left(\int_0^1 (x+y)^{-s-2}\,dy\right)dx =$$
$$= \frac{\zeta(2s+2)}{\zeta(s+1)} + 2K_{24}\frac{s(s+1)}{s+\frac{1}{2}} + 2K_{24}s(s+1)(s+2)\vartheta(s),$$

or for $\sigma > 1$

$$\left.\begin{aligned}&\int_1^{\infty}\left(C(x)+2K_{24}\sqrt{x}\right)\left(\int_0^1 (x+y)^{-s-1}\,dy\right)dx =\\ &= \frac{\zeta(2s)}{(s-1)\zeta(s)}\cdot\frac{1}{s} + \frac{2K_{24}}{s-\frac{1}{2}} + 2K_{24}(s+1)\vartheta(s-1).\end{aligned}\right\} \tag{17.5}$$

This is the required analogue of (17.1). Now Landau's theorem cannot be applied directly; but we can apply his method. Suppose we have proved the following Lemma. If

$$\varphi_1(s) = \int_1^{\infty} E(x)\left(\int_0^1 (x+y)^{-s-1}\,dy\right)dx \tag{17.6}$$

is convergent in the half-plane $\sigma > 1$ with $E(x)$ non negative for all sufficiently large x and is regular on the real axis for $s > \gamma\ (< 1)$, then $\varphi_1(s)$ is also regular in the whole half-plane $\sigma > \gamma$.

Then (17.1) and the representation (17.5) give that the requirements of the lemma are satisfied with $\gamma = \frac{1}{2}$; hence $(s-1)\zeta(s) \neq 0$ in the half-plane $\sigma > \frac{1}{2}$ and theorem VII is proved.

Now we prove the lemma following Landau's paradigma; we prove more, viz. that the representation (17.6) is convergent for $\sigma > \gamma$. Suppose the representation were convergent on the real axis only for $s > \delta$ where $\gamma < \delta \leqq 1$. Then $\varphi_1(s)$ is obviously absolutely convergent in the half-plane $\sigma > \delta$ and regular here. Hence the Taylor-expansion around $s_1 = 2$ is convergent at least in the circle $|s - s_1| < 2 - \delta$; since according to our hypothesis $\varphi_1(s)$ is also regular for $s = \delta$ the radius of the circle of convergence is g r e a t e r than $2 - \delta$ and hence there is a $\delta_1 < \delta$ such that the Taylor-series is convergent at $s = \delta_1$. But this Taylor-series is

$$\varphi_1(\delta_1) = \sum_{\nu=0}^{\infty} \frac{(-\delta_1+2)^\nu}{\nu!} \int_1^\infty E(x) \left(\int_0^1 (x+y)^{-3} \log^\nu (x+y)\, dy \right) dx;$$

since the integrand and all terms are positive we can interchange the summation and integration and hence

$$\varphi_1(\delta_1) = \int_1^\infty E(x) \left(\int_0^1 (x+y)^{-\delta_1-1}\, dy \right) dx,$$

i. e. the integral is convergent for $s = \delta_1 < \delta$, a contradiction. Hence the lemma is proved and the proof of theorem VII is completed.

18. Now we sketch the proof of theorem VIII. The arguments of **11.** and **12.** show that for $n > K_{25}$

$$\sum_{\nu \leqq n} (-1)^{\nu+1} \lambda(\nu) \nu^{-1-Kn^{-\frac{1}{2}}} \geqq 0;$$

for these n's

$$L_1(n) \equiv \sum_{\nu \leqq n} (-1)^{\nu+1} \frac{\lambda(\nu)}{\nu} > -K_{26}\, n^{-\frac{1}{2}}$$

and for $x > K_{27}$

$$L_1(x) + 2K_{26}\, x^{-\frac{1}{2}} > 0. \tag{18.1}$$

Next we must find the generating functions of $\sum (-1)^{\nu+1} \frac{\lambda(\nu)}{\nu^s}$.

We assert that for $\sigma > 1$

$$g(s) = \sum_{\nu=1}^{\infty} (-1)^{\nu+1} \frac{\lambda(\nu)}{\nu^s} = \left(1 + \frac{2}{2^s}\right) \frac{\zeta(2s)}{\zeta(s)}. \tag{18.2}$$

For

$$g(s) = 2 \sum_{\nu \text{ odd}} \frac{\lambda(\nu)}{\nu^s} - \sum_{\nu=1}^{\infty} \frac{\lambda(\nu)}{\nu^s} = 2 \prod_{p>2} \frac{1}{1+\frac{1}{p^s}} - \frac{\zeta(2s)}{\zeta(s)} =$$

$$= \left\{2\left(1+\frac{1}{2^s}\right) - 1\right\} \frac{\zeta(2s)}{\zeta(s)} = \left(1+\frac{2}{2^s}\right) \frac{\zeta(2s)}{\zeta(s)}.$$

Hence for $\sigma > 0$

$$s\int_1^\infty \frac{L_1(x)}{x^{s+1}}\,dx = \left(1+\frac{2}{2^{s+1}}\right)\frac{\zeta(2s+2)}{\zeta(s+1)}$$

$$\int_1^\infty \frac{L_1(x)+2K_{26}x^{-\frac{1}{2}}}{x^{s+1}}\,dx = \left(1+\frac{2}{2^{s+1}}\right)\frac{\zeta(2s+2)}{s\zeta(s+1)}+\frac{2K_{26}}{s+\frac{1}{2}},$$

or for $\sigma > 1$

$$\int_1^\infty \frac{L_1(x)+2K_{26}x^{-\frac{1}{2}}}{x^{s}}\,dx = \left(1+\frac{2}{2^{s}}\right)\frac{\zeta(2s)}{(s-1)\zeta(s)}+\frac{2K_{26}}{s-\frac{1}{2}}.$$

Now the remainder of the proof is similar to that of theorem IV.

19. Finally we show that for an infinity of n's the partial-sums[1)]

$$V_{2n}(s) = \sum_{\nu \leqq 2n}(-1)^{\nu+1}\nu^{-s}$$

vanish in the half-plane $\sigma > 1$. For this purpose we consider the values $V_{2n}(w_k)$, k fixed and $w_k = 1+\frac{2k\pi i}{\log 2} \equiv 1+iv_k$.

20. First we show that

$$\left|V_{2n}(w_k)+\frac{1}{4}(n+1)^{-1+iv_k}\right| < \frac{(2+|v_k|)^2}{n^2}. \qquad (20.1)$$

Starting from the identity

$$\left(1-\frac{2}{2^s}\right)U_n(s) = V_{2n}(s)-2^{1-s}\left(\frac{1}{(n+1)^s}+\cdots+\frac{1}{(2n)^s}\right)$$

we obtain, on setting $s = w_k$,

$$V_{2n}(w_k) = \frac{1}{(n+1)^{w_k}}+\frac{1}{(n+2)^{w_k}}+\cdots+\frac{1}{(2n)^{w_k}}. \qquad (20.2)$$

1) The restriction to even indices is only for the sake of convenience.

Now we have

$$\left(1+\frac{1}{\nu}\right)^{iv_k}-1-\binom{iv_k}{1}\frac{1}{\nu}-\binom{iv_k}{2}\frac{1}{\nu^2}=3\binom{iv_k}{3}\int_0^{1/\nu}\left(\frac{1}{\nu}-y\right)^2(1+y)^{iv_k-3}\,dy;$$

hence

$$\left|(\nu+1)^{iv_k}-\nu^{ia_k}-\binom{iv_k}{1}\nu^{-1+iv_k}-\binom{iv_k}{2}\nu^{-2+iv_k}\right|<|v_k|\frac{(2+|v_k|)^2}{6}\cdot\frac{1}{\nu^3}.$$

Putting $\nu=(n+1),(n+2),\cdots 2n$ and summing we obtain

$$\left.\begin{aligned}&\left|(2n+1)^{iv_k}-(n+1)^{iv_k}-\binom{iv_k}{1}\sum_{\nu=n+1}^{2n}\nu^{-1+iv_k}-\binom{iv_k}{2}\sum_{\nu=n+1}^{2n}\nu^{-2+iv_k}\right|<\\&\qquad<|v_k|\frac{(2+|v_k|)^2}{12}\cdot\frac{1}{n^2}.\end{aligned}\right\}\quad(20.3)$$

The first two bracketed terms can be written in the form

$$\left.\begin{aligned}&\left((2n+1)^{iv_k}-(2n+2)^{iv_k}\right)+\left((2n+2)^{iv_k}-(n+1)^{iv_k}\right)=\\&=-(2n+1)^{iv_k}\left(\left(1+\frac{1}{2n+1}\right)^{iv_k}-1\right)+(n+1)^{iv_k}\left(2^{iv_k}-1\right)=\\&=-(2n+1)^{iv_k}\left(\left(1+\frac{1}{2n+1}\right)^{iv_k}-1\right);\end{aligned}\right\}\quad(20.4)$$

hence from (20.2), (20.3) and (20.4)

$$\left.\begin{aligned}&\left|-(2n+1)^{iv_k}\left(\left(1+\frac{1}{2n+1}\right)^{iv_k}-1\right)-iv_k\,V_{2n}(w_k)-\binom{iv_k}{2}\sum_{\nu=n+1}^{2n}\nu^{-2+iv_k}\right|<\\&\qquad<|v_k|\frac{(2+|v_k|)^2}{12}\cdot\frac{1}{n^2}.\end{aligned}\right\}\quad(20.5)$$

Now

$$\left(1+\frac{1}{2n+1}\right)^{iv_k}-1-\frac{iv_k}{2n+1}=\int_0^{\frac{1}{2n+1}}\left(\frac{1}{2n+1}-y\right)(1+y)^{iv_k-2}\,dy\,(iv_k)(iv_k-1)$$

$$\left|\left(1+\frac{1}{2\,n+1}\right)^{iv_k}-1-\frac{iv_k}{2\,n+1}\right|<\frac{1}{2}\frac{|v_k|\,(1+|v_k|)}{(2\,n+1)^2};$$

hence from (20.5)

$$\left|-iv_k\,(2\,n+1)^{-1+iv_k}-iv_k\,V_{2n}(w_k)-\binom{iv_k}{2}\sum_{\nu=n+1}^{2n}{}' \nu^{-2+iv_k}\right|<$$

$$<|v_k|\frac{(2+|v_k|)^2}{12}\cdot\frac{1}{n^2}+\frac{1}{8}\frac{|v_k|\,(1+|v_k|)}{n^2}<\frac{|v_k|\,(2+|v_k|)^2}{4}\cdot\frac{1}{n^2}$$

or

$$\left|V_{2n}(w_k)+(2\,n+1)^{-1+iv_k}-\frac{1-iv_k}{2}\sum_{\nu=n+1}^{2n}{}' \nu^{-2+iv_k}\right|<\frac{(2+|v_k|)^2}{4}\cdot\frac{1}{n^2}. \qquad (20.6)$$

Further

$$\left(1+\frac{1}{\nu}\right)^{-1+iv_k}-1-(-1+iv_k)\frac{1}{\nu}=$$

$$=\int_0^{1/\nu}\left(\frac{1}{\nu}-y\right)(1+y)^{-3+iv_k}\,dy\,(-1+iv_k)(-2+iv_k)$$

$$\left|(\nu+1)^{-1+iv_k}-\nu^{-1+iv_k}+(1-iv_k)\,\nu^{-2+iv_k}\right|<\frac{1}{\nu^3}\cdot\frac{1}{2}(2+|v_k|)^2;$$

a summation over $\nu=(n+1),\cdots,2\,n$ gives

$$\left|\frac{(2\,n+1)^{-1+iv_k}}{2}+\frac{(n+1)^{-1+iv_k}}{2}-\frac{1-iv_k}{2}\sum_{\nu=n+1}^{2n}{}' \nu^{-2+iv_k}\right|<\frac{(2+|v_k|)^2}{4}\cdot\frac{1}{n^2}.$$

Putting this into (20.6) we obtain

$$\left|V_{2n}(w_k)+\left(\frac{3}{2}(2\,n+1)^{-1+iv_k}-\frac{1}{2}(n+1)^{-1+iv_k}\right)\right|<\frac{1}{2}\frac{(2+|v_k|)^2}{n^2}. \qquad (20.7)$$

If we use the transformation (20.4) again the sum of the two bracketed terms is

$$\left.\begin{aligned} &\frac{3}{2}\left((2n+1)^{-1+iv_k}-(2n+2)^{-1+iv_k}\right)+\\ &+\left(\frac{3}{2}\cdot 2^{-1+iv_k}-\frac{1}{2}\right)(n+1)^{-1+iv_k}=\frac{1}{4}(n+1)^{-1+iv_k}-\\ &-\frac{3}{2}(2n+1)^{-1+iv_k}\left(\left(1+\frac{1}{2n+1}\right)^{-1+iv_k}-1\right). \end{aligned}\right\} \quad (20.8)$$

Since the second term of the right-hand member of (20.8) is in absolute value

$$<\frac{3}{2}\cdot\frac{1}{2n+1}\left(1+|v_k|\right)\frac{1}{2n+1}<\frac{3}{8}\frac{(2+|v_k|)^2}{n^2}$$

we obtain from (20.7) and (20.8)

$$\left|V_{2n}(w_k)+\frac{1}{4}(n+1)^{-1+iv_k}\right|<\frac{(2+|v_k|)^2}{n^2}, \quad (20.9)$$

i. e. (20.1) is proved.

Now we consider $V'_{2n}(w_k)$. Obviously

$$V'_{2n}(w_k)=\left(\left(1-\frac{2}{2^s}\right)\zeta(s)\right)'_{s=w_k}+\sum_{\nu>2n}{}'_{\nu}\frac{(-1)^{\nu+1}\log\nu}{\nu^{w_k}}. \quad (21.1)$$

The first term is obviously

$$\log 2\cdot\zeta(w_k). \quad (21.2)$$

To estimate the second term of (21.1) we observe that

$$\left|\frac{\log(2l+1)}{(2l+1)^{w_k}}-\frac{\log(2l+2)}{(2l+2)^{w_k}}\right|\leq\frac{1}{2l+1}\log\frac{2l+2}{1l+1}+$$

$$+\log(2l+2)\left|(2l+1)^{-w_k}-(2l+2)^{-w_k}\right|<\frac{(2+|v_k|)\log(2l+2)}{(2l+1)^2};$$

hence for $n>10$ the second term of (21.1) is in absolute value

$$<(2+|v_k|)\sum_{\nu>n}{}'\frac{\log(2\nu+2)}{(2\nu+1)^2}<(2+|v_k|)\frac{\log n}{n}.$$

From this (21.2) and (21.1) we obtain

$$\left| V'_{2n}(w_k) - \log 2 \cdot \zeta(w_k) \right| < \left(2 + |w_k|\right) \frac{\log n}{n}. \qquad (21.3)$$

Now we consider the expression

$$F_{2n}(s) = V_{2n}(w_k) + V'_{2n}(w_k)(s - w_k), \qquad (21.4)$$

which is linear and has its zero at

$$w'_{kn} = w_k - \frac{V_{2n}(w_k)}{V'_{2n}(w_k)};$$

using (20.1) and (21.3) we get

$$w'_{kn} = w_k + \frac{1}{4 \log 2 \cdot \zeta(w_k)} (n+1)^{-1 + i v_k} + \Theta_3 K_{27}(k) \frac{\log n}{n^2}. \qquad (21.5)$$

$$|\Theta_3| \leqq 1.$$

22. We show that for fixed k and $n \to \infty$

$$w^*_{kn} = w_k + \frac{1}{4 \log 2 \cdot \zeta(w_k)} (n+1)^{-1 + i v_k} \qquad (22.1)$$

is the asymptotical expression of that root of $V_{2n}(s)$ which $\to w_k$ if $n \to \infty$. To show this we consider the circle

$$|s - w'_{kn}| = K_{27}(k) \frac{\log^2 n}{n^2}; \qquad (22.2)$$

(21.5) and (22.1) show that w^*_{kn} lies in this circle. We have

$$V_{2n}(s) = F_{2n}(s) + \sum_{j=2}^{\infty} \frac{1}{j!} V^{(j)}_{2n}(w_k)(s - w_k)^j. \qquad (22.3)$$

For $F_{2n}(s)$ we have identically

$$F_{2n}(s) = V'_{2n}(w_k)(s - w'_{kn}),$$

so that on the whole circumference of the circle (22.2), if n is sufficiently large,

$$|F_{2n}(s)| > \frac{1}{2}|\zeta(w_k)|\frac{K_{27}(k)\log^2 n}{n^2} = K_{28}(k)\frac{\log^2 n}{n^2}. \quad (22.4)$$

Since on the circle $|s - w_k| \leqq \frac{1}{2}$

$$|V_{2n}(s)| \leqq K_{29}(k)$$

independently of n we have from Cauchy's estimation of coefficients

$$\left|\frac{V_{2n}^{(j)}(w_k)}{j!}\right| < K_{29}(k)\,2^j$$

and on the circle (22.2)

$$|s - w_k| \leqq |s - w'_{kn}| + |w'_{kn} - w_k| \leqq K_{27}(k)\frac{\log^2 n}{n^2} + $$
$$+ \frac{K_{30}(k)}{n} + K_{27}(k)\frac{\log n}{n^2} < K_{31}(k)\frac{1}{n};$$

hence on the circle (22.2) the second term of (22.3) is in absolute value

$$< \sum_{j=2}^{\infty} K_{29}(k)\left(\frac{2K_{31}(k)}{n}\right)^j < K_{32}(k)\frac{1}{n^2}. \quad (22.5)$$

From (22.4) and (22.3) we obtain for $n > K_{33}(k)$ on the circumference of (22.2)

$$|F_{2n}(s)| > K_{28}(k)\frac{\log^2 n}{n^2} > \frac{K_{32}(k)}{n^2} > \left|\sum_{j=2}^{\infty}\frac{1}{j!}V_{2n}^{(j)}(w_k)(s - w_k)^j\right|;$$

hence it follows from Rouché's theorem that the circle (22.2) contains as many zeros of $V_{2n}(s)$ as of $F_{2n}(s)$, i. e. exactly one. But this circle is contained in the circle

$$|s - w^*_{kn}| < |s - w'_{kn}| + |w'_{kn} - w^*_{kn}| < 2K_{27}\frac{\log^2 n}{n^2}. \quad (22.6)$$

23. To prove finally that for an infinity of n's the polynomials $V_{2n}(s)$ vanish in the half-plane $\sigma > 1$ we have only to observe that for fixed k

$$v_k \log(n+1) - v_k \log n \to 0.$$

Then for an infinity of n's we have

$$\Re w^*_{kn} > 1 + \frac{1}{5 \log 2 \,|\, \zeta\, w_k) \,|} \cdot \frac{1}{n+1};$$

hence from (22.6) the real part of the corresponding root of $V_{2k}(s)$ is

$$> 1 + \frac{1}{5 \log 2 \,|\, \zeta(w_k) \,|} \cdot \frac{1}{n+1} - \frac{2\, K_{27}(k) \log^2 n}{n^2} > 1,$$

if n is sufficiently large.

We remark finally that for fixed k and $n \to \infty$ these roots in the half-plane $\sigma > 1$ lie in half-planes of the form

$$1 + \frac{K_{33}(k)}{n},$$

i. e. their location does not refute the hypothesis of theorem VIII. It would be interesting to study these roots if k is not fixed.

It is perhaps of some interest to note that for fixed k the behaviour of the corresponding roots of the Riesz-means is different. Denoting by w''_{kn} that root of the nth Riesz-mean of the series (6.1) which for $n \to \infty$ tends to w_k, we have

$$\left| w''_{kn} - w_k + \frac{1}{\log n} \right| < \frac{K_{34}(k)}{\log^2 n}.$$

Hence these roots converge to w_k from the left in a particularly simple way. Thus there is some chance that the behaviour of the roots of the Riesz-means is more regular.

Added in proof.

24. An easy modification of the proofs gives also a more general theorem from which I mention only two special cases.

Theorem X. If for a modulus k there is a character $\chi(n)$

such that the partial-sums of the corresponding L-function of Dirichlet

$$L(s, \chi) = \sum_{n=1}^{\infty} \frac{\chi(n)}{n^s} \tag{24.1}$$

do not wanish in the half-plane $\sigma > 1$, then Riemann's hypothesis (1.5) is true.

I cannot prove that this property of the partial-sums of (24.1) implies the non-vanishing of $L(s, \chi)$ itself.

Of course theorem X admits all refinements similarly as theorems II–V refine theorem I.

The interest of theorem X compared to theorem VIII lies obviously in the fact that the function $L(s, \chi)$ has no roots on the line $\sigma = 1$, in contrast to $\left(1 - \frac{2}{2^s}\right)\zeta(s)$.

Theorem XI. If for real sequence $\beta_1, \beta_2, \ldots$ the partial-sums of the series

$$f_\beta(s) = \prod_{\nu=1}^{\infty} \frac{1}{1 - \frac{e^{i\beta_\nu}}{p_\nu^s}} = \sum_{n=1}^{\infty} \frac{d_n}{n^s} \tag{24.2}$$

do not vanish in the half-plane $\sigma > 1$, then Riemann's hypothesis (1.5) is true.

Prof. Jessen[1)] proved that for "almost all" β-sequences the functions $f_\beta(s)$ do not vanish in the half-plane $\sigma > \frac{1}{2}$. To obtain an explicit $f_\beta(s)$ which has this property, we may choose according to a remark of Prof. A. Selberg

$$e^{i\beta_\nu} = (-1)^\nu,$$

where $p_1 = 2$, $p_2 = 3$, ... denote the increasingly ordered sequence of primes.

Theorem XI admits the same refinements as theorem X.

25. We proved implicitly that if the partial-sums

1) B. Jessen: Some analytical problems relating to probability. Journ. of Math. and Physics. Mass. Inst. Techn. vol. XIV (1935), p. 24—27.

$$L(n) = \sum_{\nu \leq n} \frac{\lambda(\nu)}{\nu} \tag{25.1}$$

are of one sign for all sufficiently large n's or are for these n's

$$> - K_{11}\, n^{\vartheta - 1}, \tag{25.2}$$

then the hypothesis (1.6) is true. Pólya[1] remarked, that if

$$L_1(n) = \sum_{\nu \leq n} \lambda(\nu) \tag{25.3}$$

is non-positive for all sufficiently large n's, then Riemann's hypothesis (1.5) is true; with the same reasoning he could prove that from the inequality

$$L_1(n) < K_{35}\, n^{\vartheta}, \tag{25.4}$$

valid for all sufficiently large n's, the conjecture (1.6) follows. It seems to me that the condition (25.2) is somewhat less deep than (25.4), i. e. one can deduce (25.2) from (25.4). If we replace, however, (25.2) and (25.4) by twosided inequalities, the corresponding statement follows by partial summation.

Pólya showed by computation the validity of (25.3) for $2 \leq n \leq 1500$; this has been extended by H. Gupta[2] up to 20,000. The young danish mathematicians Erik Eilertsen, Poul Kristensen, Aage Petersen, Niels Ove Roy Poulsen, and Aage Winther calculated the values of $L(n)$ for $n \leq 1000$. They found all of them to be positive; for $L(1000)$ they found the value

$$L(1000) = 0{,}028970560\,.$$

It is remarkable that in this range the minimal value is attained at $n = 293$ and that

$$L(293) = 0{,}005102273,$$

a much smaller value than $L(1000)$.

1) G. Pólya: Verschiedene Bemerkungen zur Zahlentheorie. Jahresb. der deutsch. Math. Ver. 28 (1919), p. 31—40.

2) H. Gupta: On a table of values of $L_1(n)$. Proc. Indian Acad. Sci. Sect. A. vol. 12 (1940), p. 407—409.

26. I conclude with two remarks. As Paul Erdös remarked, he can prove that to any given closed set H in the domain $|z| \geqq 1$ which contains the circumference of the unit-circle one can give a power-series, convergent in $|z| < 1$, such that the roots of its partial-sums cluster in $|z| \geqq 1$ to the points of H and only to those.

As Prof. Sherwood Sherman remarked, my statement on p. 13 about the Jensen-means is true only if we suppose in addition of the roots of the function $f(s)$, that the sum of their reciprocal values is convergent.

Indleveret til Selskabet den 31. Oktober 1947.
Færdig fra Trykkeriet den 9. September 1948.

12.10 A. Selberg (1949)

An Elementary Proof of the Prime Number Theorem

In this paper Atle Selberg presents an elementary proof of the prime number theorem. Together with the proof of Erdős (see Section 12.11) this is the first elementary proof of the prime number theorem. Selberg proves that

$$\sum_{p\leq x}\log^2 p+\sum_{pq\leq x}\log p\log q=2x\log x+O(x),$$

where p and q are primes. He then uses Erdős 's statement, that for any $\lambda>1$ the number of primes in $(x,\lambda x)$ is at least $Kx/\log x$ for $x>x_0$ (here K and x_0 are constants dependent on λ), to derive the prime number theorem without appealing to the theory of functions of a complex variable.

ANNALS OF MATHEMATICS
Vol. 50, No. 2, April, 1949

AN ELEMENTARY PROOF OF THE PRIME-NUMBER THEOREM

ATLE SELBERG

(Received October 14, 1948)

1. Introduction

In this paper will be given a new proof of the prime-number theorem, which is elementary in the sense that it uses practically no analysis, except the simplest properties of the logarithm.

We shall prove the prime-number theorem in the form

$$\lim_{x\to\infty} \frac{\vartheta(x)}{x} = 1 \tag{1.1}$$

where for $x > 0$, $\vartheta(x)$ is defined as usual by

$$\vartheta(x) = \sum_{p \le x} \log p, \tag{1.2}$$

p denoting the primes.

The basic new thing in the proof is a certain assymptotic formula (2.8), which may be written

$$\vartheta(x) \log x + \sum_{p \le x} \log p\, \vartheta\left(\frac{x}{p}\right) = 2x \log x + O(x). \tag{1.3}$$

From this formula there are several ways to deduce the prime-number theorem. The way I present §§2–4 of this paper, is chosen because it seems at the present to be the most direct and most elementary way.[1] But for completeness it has to be mentioned that this was not my first proof. The original proof was in fact rather different, and made use of the following result by P. Erdös, that for an arbitrary, positive fixed number δ, there exist a $K(\delta) > 0$ and an $x_0 = x_0(\delta)$ such that for $x > x_0$, there are more than

$$K(\delta)\, x/\log x$$

primes in the interval from x to $x + \delta x$.

My first proof then ran as follows: Introducing the notations

$$\varliminf \frac{\vartheta(x)}{x} = a, \qquad \varlimsup \frac{\vartheta(x)}{x} = A,$$

one can easily deduce from (1.3), using the well-known result

$$\sum_{p \le x} \frac{\log p}{x} = \log x + O(1), \tag{1.4}$$

[1] Because it avoids the concept of lower and upper limit. It is in fact easy to modify the proof in a few places so as to avoid the concept of limit at all, of course (1.1) would then have to be stated differently.

that

$$a + A = 2. \tag{1.5}$$

Next, taking a large x, with

$$\vartheta(x) = ax + o(x),$$

one can deduce from (1.3) in the modified form

$$(\vartheta(x) - ax) \log x + \sum_{p \leq x} \log p \left(\vartheta\left(\frac{x}{p}\right) - A \frac{x}{p} \right) = O(x), \tag{1.6}$$

that, for a fixed positive number δ, one has

$$\vartheta\left(\frac{x}{p}\right) > (A - \delta) \frac{x}{p}, \tag{1.7}$$

except for an exceptional set of primes $\leqq x$ with

$$\sum \frac{\log p}{p} = o(\log x).$$

Also one easily deduces that there exists an x' in the range $\sqrt{x} < x' < x$, wit

$$\vartheta(x') = Ax' + o(x').$$

Again from (1.6) with a and A interchanged, and x' instead of x, one deduces tha

$$\vartheta\left(\frac{x'}{p}\right) < (a + \delta) \frac{x'}{p}, \tag{1.8}$$

except for an exceptional set of primes $\leqq x'$ with

$$\sum \frac{\log p}{p} = o(\log x).$$

From Erdös' result it is then possible to show that one can chose primes p and p' not belonging to any of the exceptional sets, with

$$\frac{x}{p} < \frac{x'}{p'} < (1 + \delta) \frac{x}{p}.$$

Then we get from (1.7) and (1.8) that

$$(A - \delta) \frac{x}{p} < \vartheta\left(\frac{x}{p}\right) \leqq \vartheta\left(\frac{x'}{p'}\right) < (a + \delta) \frac{x'}{p'} < (a + \delta)(1 + \delta) \frac{x}{p},$$

so that

$$A - \delta < (a + \delta)(1 + \delta).$$

or making δ tend to zero

$$A \leqq a.$$

Hence since also $A \geqq a$ and $a + A = 2$ we have $a = A = 1$, which proves our theorem.

Erdös' result was obtained without knowledge of my work, except that it is based on my formula (2.8); and after I had the other parts of the above proof. His proof contains ideas related to those in the above proof, at which related ideas he had arrived independently.

The method can be applied also to more general problems. For instance one can prove some theorems proved by analytical means by Beurling, but the results are not quite as sharp as Beurlings.[2] Also one can prove the prime-number theorem for arithmetic progressions, one has then to use in addition ideas and results from my previous paper on Dirichlets theorem.[3]

Of known results we use frequently besides (1.4) also its consequence

$$\vartheta(x) = O(x). \tag{1.9}$$

Throughout the paper p, q and r denote prime numbers. $\mu(n)$ denotes Möbius' number-theoretic function, $\tau(n)$ denotes the number of divisors of n. The letter c will be used to denote absolute constants, and K to denote absolute positive constants. Some of the more trivial estimations are not carried out but left to the reader.

2. Proof of the basic formulas

We write, when x is a positive number and d a positive integer,

$$\lambda_d = \lambda_{d,x} = \mu(d) \log^2 \frac{x}{d}, \tag{2.1}$$

and if n is a positive integer,

$$\theta_n = \theta_{n,x} = \sum_{d/n} \lambda_d . \tag{2.2}$$

Then we have

$$\theta_n = \begin{cases} \log^2 x, \text{ for } n = 1, \\ \log p \log x^2/p, \text{ for } n = p^\alpha, \alpha \geqq 1, \\ 2 \log p \log q, \text{ for } n = p^\alpha q^\beta, \alpha \geqq 1, \beta \geqq 1, \\ 0, \text{ for all other } n. \end{cases} \tag{2.3}$$

The first three of these statements follow readily from (2.2) and (2.1), the fourth is easily proved by induction. Clearly it is enough to consider n square-free, then if $n = p_1 p_2 \cdots p_k$,

$$\theta_{n,x} = \theta_{n/p_k,x} - \theta_{n/p_k,x/p_k} .$$

From this the remaining part of (2.3) follows.

[2] A. Beurling: *Analyse de la loi asymptotique de la distribution des nombres premiers généralisés*, Acta Math., vol. 68, pp. 255–291 (1937).

[3] These Annals this issue, pp. 297–304.

Now consider the expression

$$(2.4)\quad \begin{aligned}\sum_{n\leqq x}\theta_n &= \sum_{n\leqq x}\sum_{d/n}\lambda_d = \sum_{d\leqq x}\lambda_d\left[\frac{x}{d}\right] = x\sum_{d\leqq x}\frac{\lambda_d}{d} + O\,(\sum_{d\leqq x}|\lambda_d|)\\ &= x\sum_{d\leqq x}\frac{\mu(d)}{d}\log^2\frac{x}{d} + O\left(\sum_{d\leqq x}\log^2\frac{x}{d}\right) = x\sum_{d\leqq x}\frac{\mu(d)}{d}\log^2\frac{x}{d} + O(x).\end{aligned}$$

This on the other hand is equal to, by (2.3),

$$(2.5)\quad \begin{aligned}\sum_{n\leqq x}\theta_n = \log^2 x &+ \sum_{p^\alpha\leqq x}\log p\log\frac{x^2}{p}\\ &+ 2\sum_{\substack{p^\alpha q^\beta\leqq x\\ p<q}}\log p\log q = \sum_{p\leqq x}\log^2 p\\ &+ \sum_{pq\leqq x}\log p\log q + O\left(\sum_{p\leqq x}\log p\log\frac{x}{p}\right)\\ &+ O\,(\sum_{\substack{p^\alpha\leqq x\\ \alpha>1}}\log^2 x) + O\,(\sum_{\substack{p^\alpha q^\beta\leqq x\\ \alpha>1}}\log p\log q)\\ &+ \log^2 x = \sum_{p\leqq x}\log^2 p + \sum_{pq\leqq x}\log p\log q + O(x).\end{aligned}$$

The remainder term being obtained by use of (1.4) and (1.9). Hence from (2.4) and (2.5),

$$(2.6)\quad \sum_{p\leqq x}\log^2 p + \sum_{pq\leqq x}\log p\log q = x\sum_{d\leqq x}\frac{\mu(d)}{d}\log^2\frac{x}{d} + O(x).$$

It remains now to estimate the sum on the right-hand-side. To this purpose we need the formulas

$$(2.7)\quad \sum_{\nu\leqq z}\frac{1}{\nu} = \log z + c_1 + O(z^{-\frac{1}{2}}),$$

and

$$(2.7')\quad \sum_{\nu\leqq z}\frac{\tau(\nu)}{\nu} = \tfrac{1}{2}\log^2 z + c_2\log z + c_3 + O(z^{-\frac{1}{2}})$$

where the c's are absolute constants, (2.7) is well known, and (2.7′) may be easily derived by partial summation from the well-known result

$$\sum_{\nu\leqq z}\tau(\nu) = z\log z + c_4 z + O(\sqrt{z}).$$

From (2.7) and (2.7′) we get

$$\log^2 z = 2\sum_{\nu\leqq z}\frac{\tau(\nu)}{\nu} + c_5\sum_{\nu\leqq z}\frac{1}{\nu} + c_6 + O(z^{-\frac{1}{2}}).$$

By taking here $z = x/d$, we get

$$\sum_{d\leq x}\frac{\mu(d)}{d}\log^2\frac{x}{d} = 2\sum_{d\leq x}\frac{\mu(d)}{d}\sum_{\nu\leq x/d}\frac{\tau(\nu)}{\nu} + c_5\sum_{d\leq x}\frac{\mu(d)}{d}\sum_{\nu\leq x/d}\frac{1}{\nu}$$

$$+ c_6\sum_{d\leq x}\frac{\mu(d)}{d} + O(x^{-\frac{1}{2}}\sum_{d\leq x}d^{-\frac{1}{2}}) = 2\sum_{d\nu\leq x}\frac{\mu(d)\tau(\nu)}{d\nu}$$

$$+ c_5\sum_{d\nu\leq x}\frac{\mu(d)}{d\nu} + c_6\sum_{d\leq x}\frac{\mu(d)}{d} + O(1)$$

$$= 2\sum_{n\leq x}\frac{1}{n}\sum_{d/n}\mu(d)\tau\left(\frac{n}{d}\right) + c_5\sum_{n\leq x}\frac{1}{n}\sum_{d/n}\mu(d)$$

$$+ O(1) = 2\sum_{n\leq x}\frac{1}{n} + c_5 + O(1) = 2\log x + O(1).$$

We used here that $\sum_{d/n}\mu(d)\tau(n/d) = 1$, and the well-known $\sum_{d\leq x}(\mu(d))/d = O(1)$. Now (2.6) yields

$$\sum_{p\leq x}\log^2 p + \sum_{pq\leq x}\log p\log q = 2x\log x + O(x). \tag{2.8}$$

This formula may also be written in the form given in the introduction

$$\vartheta(x)\log x + \sum_{p\leq x}\log p\,\vartheta\left(\frac{x}{p}\right) = 2x\log x + O(x), \tag{2.9}$$

by noticing that

$$\sum_{p\leq x}\log^2 p = \vartheta(x)\log x + O(x).$$

By partial summation we get from (2.8)

$$\sum_{p\leq x}\log p + \sum_{pq\leq x}\frac{\log p\log q}{\log pq} = 2x + O\left(\frac{x}{\log x}\right). \tag{2.10}$$

This gives

$$\sum_{pq\leq x}\log p\log q = \sum_{p\leq x}\log p\sum_{q\leq x/p}\log q = 2x\sum_{p\leq x}\frac{\log p}{p}$$

$$-\sum_{p\leq x}\log p\sum_{qr\leq x/p}\frac{\log q\log r}{\log qr} + O\left(x\sum_{p\leq x}\frac{\log p}{p\left(1+\log\frac{x}{p}\right)}\right)$$

$$= 2x\log x - \sum_{qr\leq x}\frac{\log q\log r}{\log qr}\vartheta\left(\frac{x}{qr}\right) + O(x\log\log x).$$

Inserting this for the second term in (2.8) we get

$$\vartheta(x)\log x = \sum_{pq\leq x}\frac{\log p\log q}{\log pq}\vartheta\left(\frac{x}{pq}\right) + O(x\log\log x). \tag{2.11}$$

Writing now

$$\vartheta(x) = x + R(x)$$

, (2.9) easily gives

$$R(x) \log x = -\sum_{p \leq x} \log p \, R\left(\frac{x}{p}\right) + O(x), \tag{2.12}$$

and (2.11) yields in the same manner

$$R(x) \log x = \sum_{pq \leq x} \frac{\log p \log q}{\log pq} R\left(\frac{x}{pq}\right) + O(x \log \log x), \tag{2.13}$$

since

$$\sum_{pq \leq x} \frac{\log p \log q}{pq \log pq} = \log x + O(\log \log x),$$

which follows by partial summation from

$$\sum_{pq \leq x} \frac{\log p \log q}{pq} = \tfrac{1}{2} \log^2 x + O(\log x),$$

which again follows easily from (1.4).

The (2.12) and (2.13) yield

$$2 \mid R(x) \mid \log x \leq \sum_{p \leq x} \log p \left| R\left(\frac{x}{p}\right) \right| + \sum_{pq \leq x} \frac{\log p \log q}{\log pq} \left| R\left(\frac{x}{pq}\right) \right| + O(x \log \log x).$$

From this, by partial summation,

$$2 \mid R(x) \mid \log x \leq \sum_{n \leq x} \left\{ \sum_{p \leq n} \log p + \sum_{pq \leq n} \frac{\log p \log q}{\log pq} \right\} \cdot \left\{ \left| R\left(\frac{x}{n}\right) \right| - \left| R\left(\frac{x}{n+1}\right) \right| \right\} + O(x \log \log x),$$

or by (2.10)

$$\begin{aligned} 2 \mid R(x) \mid \log x &\leq 2 \sum_{n \leq x} n \left\{ \left| R\left(\frac{x}{n}\right) \right| - \left| R\left(\frac{x}{n+1}\right) \right| \right\} \\ &\quad + O\left(\sum_{n \leq x} \frac{n}{1 + \log n} \left| R\left(\frac{x}{n}\right) - R\left(\frac{x}{n+1}\right) \right| \right) + O(x \log \log x) \\ &= 2 \sum_{n \leq x} \left| R\left(\frac{x}{n}\right) \right| + O\left(\sum_{n \leq x} \frac{n}{1 + \log n} \left\{ \vartheta\left(\frac{x}{n}\right) - \vartheta\left(\frac{x}{n+1}\right) \right\} \right) \\ &\quad + O\left(x \sum_{n \leq x} \frac{1}{n(1 + \log n)} \right) + O(x \log \log x) = 2 \sum_{n \leq x} \left| R\left(\frac{x}{n}\right) \right| \\ &\quad + O\left(\sum_{n \leq x} \frac{1}{1 + \log n} \vartheta\left(\frac{x}{n}\right) \right) + O(x \log \log x) \\ &= 2 \sum_{n \leq x} \left| R\left(\frac{x}{n}\right) \right| + O(x \log \log x), \end{aligned}$$

or

$$(2.14) \qquad |R(x)| \leqq \frac{1}{\log x} \sum_{n \leqq x} \left| R\left(\frac{x}{n}\right) \right| + O\left(x \frac{\log \log x}{\log x}\right),$$

which is the result we will use in the following.[4]

3. Some properties of $R(x)$

From (1.4) we get by partial summation that

$$\sum_{n \leqq x} \frac{\vartheta(n)}{n^2} = \log x + O(1),$$

or

$$\sum_{n \leqq x} \frac{R(n)}{n^2} = O(1).$$

This means there exists an absolute positive constant K_1, so that for all $x > 4$ and $x' > x$,

$$(3.1) \qquad \left| \sum_{x \leqq n \leqq x'} \frac{R(n)}{n^2} \right| < K_1 .$$

Accordingly we have, if $R(n)$ does not change its sign between x and x', that there is a y in the interval $x \leqq y \leqq x'$, so that

$$(3.2) \qquad \left| \frac{R(y)}{y} \right| < \frac{K_2}{\log \frac{x'}{x}}, \qquad K_2 \geqq 1.$$

This is easily seen to hold true if $R(n)$ changes the sign also.[5]

Thus for an arbitrary fixed positive $\delta < 1$ and $x > 4$, there will exist a y in the interval $x \leqq y \leqq e^{K_2/\delta} x$, with

$$(3.3) \qquad |R(y)| < \delta y.$$

From (2.10) we see that for $y < y'$,

$$0 \leqq \sum_{y < p \leqq y'} \log p \leqq 2(y' - y) + O\left(\frac{y'}{\log y'}\right),$$

from which follows that

$$|R(y') - R(y)| \leqq y' - y + O\left(\frac{y'}{\log y'}\right).$$

[4] Apparently we have here lost something in the order of the remainder-term compared to (2.8). Actually we could instead of (2.14) have used the inequality

$$|R(x)| \leqq \frac{2}{\log^2 x} \sum_{n \leqq x} \frac{\log n}{n} \left| R\left(\frac{x}{n}\right) \right| + O\left(\frac{x}{\log x}\right),$$

which can be proved in a similar way.

[5] Because there will then be a $|R(y)| < \log y$.

Hence, if $y/2 \leqq y' \leqq 2y$, $y > 4$,

$$| R(y') - R(y) | \leqq | y' - y | + O\left(\frac{y'}{\log y'}\right),$$

or

$$| R(y') | \leqq | R(y) | + | y' - y) + O\left(\frac{y'}{\log y'}\right).$$

Now consider an interval $(x, e^{K_2/\delta} x)$, according to (3.3) there exists a in this interval with

$$| R(y) | < \delta y.$$

Thus for any y' in the interval $y/2 \leqq y' \leqq 2 y$, we have

$$| R(y') | \leqq \delta y + | y' - y | + \frac{K_3 y'}{\log x},$$

or

$$\left|\frac{R(y')}{y'}\right| < 2\delta + \left|1 - \frac{y'}{y}\right| + \frac{K_3}{\log x}.$$

Hence if $x > e^{K_3/\delta}$ and $e^{-(\delta/2)} \leqq y'/y \leqq e^{\delta/2}$, we get

$$\left|\frac{R(y')}{y'}\right| < 2\delta + (e^{\delta/2} - 1) + \delta < 4\delta.$$

Thus for $x > e^{K_3/\delta}$ the interval $(x, e^{K_2/\delta}x)$ will always contain a sub-interv $(y_1, e^{\delta/2}y_1)$, such that $| R(z) | < 4\delta z$ if z belongs to this sub-interval.

4. Proof of the prime-number theorem

We are now going to prove the

THEOREM.

$$\lim_{x \to \infty} \frac{\vartheta(x)}{x} = 1.$$

Obviously this is equivalent to

$$\lim_{x \to \infty} \frac{R(x)}{x} = 0. \tag{4.1}$$

We know that for $x > 1$,

$$| R(x) | < K_4 x. \tag{4.2}$$

Now assume that for some positive number $\alpha < 8$,

$$| R(x) | < \alpha x, \tag{4.3}$$

holds for all $x > x_0$. Taking $\delta = \alpha/8$, we have according to the precedir section (since we may assume that $x_0 > e^{K_3/\delta}$), that all intervals of the type $(x, e^{K_2/\delta}x)$ with $x > x_0$, contain an interval $(y, e^{\delta/2}y)$ such that

$$| R(z) | < \alpha z/2, \tag{4.4}$$

for $y \leqq z \leqq e^{\delta/2}y$.

The inequality (2.14) then gives, using (4.2),

$$| R(x) | \leqq \frac{1}{\log x} \sum_{n \leq x} \left| R\left(\frac{x}{n}\right) \right| + O\left(\frac{x}{\sqrt{\log x}}\right)$$

$$< K_4 \frac{x}{\log x} \sum_{(x/x_0) < n \leqq x} \frac{1}{n} + \frac{x}{\log x} \sum_{n \leqq (x/x_0)} \frac{1}{n} \left| \frac{n}{x} R\left(\frac{x}{n}\right) \right| + O\left(\frac{x}{\sqrt{\log x}}\right),$$

writing now $\rho = e^{K_2/\delta}$, we get further, using (4.3) and (4.4),

$$| R(x) | < \frac{\alpha x}{\log x} \sum_{n \leqq (x/x_0)} \frac{1}{n} - \frac{\alpha x}{2 \log x} \sum_{1 \leqq \nu \leqq (\log (x/x_0)/\log \rho)}$$

$$\sum_{\substack{y_\nu \leqq n \leqq y_\nu e^{(\delta/2)} \\ \rho^{\nu-1} < y_\nu \leqq \rho^\nu e^{-(\delta/2)}}} \frac{1}{n} + O\left(\frac{x}{\sqrt{\log x}}\right) = \alpha x - \frac{\alpha x}{2 \log x} \sum_{1 \leqq \nu \leqq (\log (x/x_0)/\log \rho)} \frac{\delta}{2}$$

$$+ O\left(\frac{x}{\sqrt{\log x}}\right) = \alpha x - \frac{\alpha \delta}{4 \log \rho} x + O\left(\frac{x}{\sqrt{\log x}}\right)$$

$$= \alpha\left(1 - \frac{\alpha^2}{256K_2}\right) x + O\left(\frac{x}{\sqrt{\log x}}\right) < \alpha\left(1 - \frac{\alpha^2}{300K_2}\right) x,$$

for $x > x_1$. Since the iteration-process

$$\alpha_{n+1} = \alpha_n \left(1 - \frac{\alpha_n^2}{300K_2}\right),$$

obviously converges to zero if we start for instance with $\alpha_1 = 4$ (one sees easily that then $\alpha_n < K_5/\sqrt{n}$), this proves (4.1) and thus our theorem.

FINAL REMARK. As one sees we have actually never used the full force of (2.8) in the proof, we could just as well have used it with the remainder term $o(x \log x)$ instead of $O(x)$. It is not necessary to use the full force of (1.4) either, if we have here the remainder-term $o(\log x)$ but in addition knowing that $\vartheta(x) > Kx$ for $x > 1$ and some positive constant K, we can still prove the theorem. However, we have then to make some change in the arguments of §3.

THE INSTITUTE FOR ADVANCED STUDY
AND
SYRACUSE UNIVERSITY

12.11 P. Erdős (1949)

On a New Method in Elementary Number Theory Which Leads to an Elementary Proof of the Prime Number Theorem

Pál Erdős was one of the most notable mathematicians of the twentiest century. He was born in Hungary in 1913. Due to Erdős's Jewish heritage and the anti–semetic nature of Hungarian politics at the time, his early years were chaotic. After completing his doctoral studies at Pázmány Péter University in Budapest in 1934, Erdős moved to Manchester. At the start of the Second World War, he moved to the United States and took up a fellowship at Princeton. Erdős's tenure at Princeton was short-lived. Throughout his life, he frequently moved from university to university. Erdős spent his career posing and solving difficult mathematical problems in several disciplines. He believed strongly in collaboration; few of his publications bear his name alone. His many accomplishments and eccentricities are compiled in several biographies [120].

In this paper, Erdős presents an elementary proof of the prime number theorem. In conjunction with Selberg's paper (see Section 12.10), this paper constitutes the first elementary proof of the prime number theorem. The critical step is due to Selberg, who showed that

$$\sum_{p\leq x} \log^2 p + \sum_{pq\leq x} \log p \log q = 2x \log x + O(x).$$

Erdős's contribution is the proof that for any $\lambda > 0$ the number of primes in $(x, \lambda x)$ is at least $Kx/\log x$ for $x > x_0$, where K and x_0 are constants dependent on λ. From this he gives an account of Selberg's proof of the prime number theorem and a simplification deduced by both himself and Selberg.

$$\frac{2K}{\pi} = \frac{2^{1/3}(kk')^{-1/6}}{\sqrt{2\pi p}} \left\{ \frac{\pi \; \Gamma\left(\frac{\alpha}{p}\right)}{\pi \; \Gamma\left(\frac{\beta}{p}\right)} \right\}^{w/4} \tag{8}$$

where (p is a prime) and $w = 6$ if $p = 3$, $w = 2$ if $p > 3$; α runs through the $\frac{p-1}{2}$ quadratic residues of p that lie between 0 and p, while β runs through the remaining $\frac{p-1}{2}$ numbers between 0 and p.

Specializing again to the case $p = 7$ we obtain in the usual notation for hypergeometric series:

$$F(^1/_4, {}^1/_4, 1; {}^1/_{64}) = \sqrt{\frac{2}{7\pi}} \left\{ \frac{\Gamma(^1/_7)\Gamma(^2/_7)\Gamma(^4/_7)}{\Gamma(^3/_7)\Gamma(^5/_7)\Gamma(^6/_7)} \right\}^{1/2}$$

5. Let $G_d(s)$ denote the analytical continuation of the function defined for $\sigma > {}^3/_2$ by the series

$$\sum{}'(x^2 + y^2 + dz^2)^{-s}$$

From a formula similar to (4) it is deduced that

THEOREM: *There exists a real number θ_d such that*

$$G_d(\theta_d) = 0 \qquad [d > d_0]$$

where $\theta_d \to 0$ as $d \to \infty$, but $\theta_d \neq 0$.

ON A NEW METHOD IN ELEMENTARY NUMBER THEORY WHICH LEADS TO AN ELEMENTARY PROOF OF THE PRIME NUMBER THEOREM

BY P. ERDÖS

DEPARTMENT OF MATHEMATICS, SYRACUSE UNIVERSITY

Communicated by P. A. Smith, April 16, 1949

1. *Introduction.*—In the course of several important researches in elementary number theory A. Selberg[1] proved some months ago the following asymptotic formula:

$$\sum_{p \leqq x} (\log p)^2 + \sum_{pq \leqq x} \log p \log q = 2x \log x + O(x), \tag{1}$$

where p and q run over the primes. This is of course an immediate consequence of the prime number theorem. The point is that Selberg's ingenious proof of (1) is completely elementary. Thus (1) can be used as a starting point for elementary proofs of various theorems in analytical number theory, which previously seemed inaccessible by elementary methods.

Using (1) I proved that $p_{n+1}/p_n \to 1$ as $n \to \infty$. In fact, I proved the following slightly stronger result: To every c there exists a positive $\delta(c)$, so that for x sufficiently large we have

$$\pi[x(1 + c)] - \pi(x) > \delta(c)x/\log x \qquad (2)$$

where $\pi(x)$ is the number at primes not exceeding x.

I communicated this proof of (2) to Selberg, who, two days later, using (1), (2) and the ideas of the proof of (2), deduced the prime number theorem $\lim_{x \to \infty} \frac{\pi(x)\log x}{x} = 1$ or, equivalently[2]

$$\lim_{x \to \infty} \frac{\vartheta(x)}{x} = 1, \text{ where } \vartheta(x) = \sum_{p \leq x} \log p. \qquad (3)$$

In a few more days, Selberg simplified my proof of (2), and later we jointly simplified the proof of the prime number theorem. The new proof no longer required (2), but used the same ideas as in the proof of (2) and (3). I was also able to prove the prime number theorem for arithmetic progressions. My proof of the latter was helped by discussions with Selberg and it utilizes ideas of Selberg's previous elementary proof of Dirichlet's theorem,[3] according to which every arithmetic progression whose first term and difference are relatively prime contains infinitely many primes. This proof will be given in a separate paper.

Selberg has now a more direct proof of (3), which is not yet published. It is possible, therefore, that the present method may prove to be only of historical interest.

I now proceed to give the proofs as they occurred in chronological order. (It should be remarked that we never utilize the full strength of (1), indeed an error term $o(x \log x)$ is all that is used in the following proofs.)

We introduce the following notation:

$$A = \limsup_{x \to \infty} \frac{\vartheta(x)}{x}, \quad a = \liminf_{x \to \infty} \frac{\vartheta(x)}{x}.$$

First, we state a few elementary facts about primes which will be used subsequently. Of these, I, II and IV are well known in elementary prime number theory, while III is shown to be a simple consequence of (1).

I. $a > 0$.

II. $\sum_{p \leq x} \frac{\log p}{p} = [1 + o(1)]\log x.$

III. Let $x_2 > x_1$. Then

$$\vartheta(x_2) - \vartheta(x_1) < 2(x_2 - x_1) + o(x_2).$$

Thus, in particular, if $x_1 = 0$, we obtain $A \leqq 2$.

Put in (1) $x = x_2$ and $x = x_1$ and subtract. Then we obtain

$$\sum_{x_1 < p \leqq x_2} (\log p)^2 \leqq 2x_2 \log x_2 - 2x_1 \log x_1 + o(x_2 \log x_2) \leqq 2(x_2 - x_1)\log x_2 + o(x_2 \log x_2). \quad (4)$$

We distinguish two cases: (A) $x_1 \geqq x_2/(\log x_2)^2$. Then clearly $\log x_1 = (1 + o(1))\log x_2$ and III follows from (4) on dividing both sides by $\log x_2$. (B) $x_1 < x_2/(\log x_2)^2$. Then we have by (A)

$$\vartheta(x_2) - \vartheta(x_1) < \vartheta(x_2) - \vartheta(x_2/(\log x_2)^2) + \frac{x_2}{(\log x_2)^2}\log x_2 < 2\left(x_2 - \frac{x_2}{(\log x_2)^2}\right) + o(x_2) = 2(x_2 - x_1) + o(x_2), \text{ q. e. d.}$$

IV. $A \leqq 1.5$. This is a consequence of the known result $\vartheta(x) \leqq 1.5x$.

2. *Proof of* (2).—It is equivalent to prove that to every positive c there exists a positive $\delta(c)$ such that $\vartheta[(1 + c)x] - \vartheta(x) > \delta(c)x$ for x sufficiently large.

Suppose this not true, then there exist positive constants c' and corresponding arbitrarily large x so that

$$\vartheta[x(1 + c')] - \vartheta(x) = o(x). \quad (5)$$

Put $C = \sup c'$. It easily follows from I and the finiteness of A that $C < \infty$.

First we show that C satisfies (5), in other words, that there are arbitrarily large values of x for which

$$\vartheta[x(1 + C)] - \vartheta(x) = o(x). \quad (6)$$

Choose $c' > C - {}^1/_2\epsilon$ and let $x \to \infty$ through values satisfying (5). Then by III we have

$$\vartheta[x(1 + C)] - \vartheta(x) = \vartheta[x(1 + C)] - \vartheta[x(1 + c')] + \vartheta[x(1 + c')] - \vartheta(x) \leqq 2(C - c')x + o(x) < \epsilon x + o(x),$$

which (since ϵ can be chosen arbitrarily small) proves (6).

Now we shall show that (6) leads to a contradiction. From (1) we obtain by subtraction

$$\sum_{x < p \leqq x(1+C)} (\log p)^2 + \sum_{x < pq \leqq x(1+C)} \log p \log q = 2Cx \log x + o(x \log x).$$

From (6) we have for suitable x since $\sum_{x < p \leqq x(1+C)} (\log p)^2 = o(x \log x)$

$$\sum_{p \leq x(1+X)} \log p\left(\vartheta\left[\frac{x}{p}(1 + C)\right] - \vartheta\left(\frac{x}{p}\right)\right) = 2Cx \log x + o(x \log x) \quad (7)$$

Now we deduce the following fundamental lemma.

LEMMA 1. *Let $x \to \infty$ through values satisfying (6), then for all primes $p \leq x(1 + C)$, except possibly for a set of primes for which*

$$\sum \frac{\log p}{p} = o(\log x) \quad (8)$$

we have

$$\vartheta\left[\frac{x}{p}(1 + C)\right] - \vartheta\left(\frac{x}{p}\right) = 2C\frac{x}{p} + o\left(\frac{x}{p}\right). \quad (9)$$

Suppose the lemma is not true. Then there exist two positive constants b_1 and b_2 so that for arbitrarily large x (satisfying (6)) we have for a set of primes satisfying $\sum_{p \leq x(1+C)} \frac{\log r}{r} \sim b_1 \log x$

$$\vartheta\left[\frac{x}{r}(1 + C)\right] - \vartheta\left(\frac{x}{p}\right) < (2C - b_2)\frac{x}{r}. \quad (10)$$

But then from II, III and (10), since (9) holds at best for a set of primes satisfying $\sum \frac{\log r}{r} \sim (1 - b_1) \log x$ we have

$$\sum_{p \leq x(1+C)} \log p\left(\vartheta\left[\frac{x}{r}(1 + C)\right] - \vartheta(x)\right) \leq b_1(2C - b_2)x \log x +$$
$$2C(1 - b_1)x \log x + o(x \log x) = (2C - b_1b_2)x \log x + o(x \log x)$$

But this contradicts (7), hence the lemma is established.

The primes satisfying (9) we shall call good primes, the other primes we shall call bad primes (of course the goodness and badness of a prime depends on x).

We shall prove the existence of a sequence of good primes $p_1 < p_2 < \ldots p_k$ satisfying the following conditions:

$$10p_1 < p_k < 100p_1,\ (1 + C)(1 + t)^2 p_i > p_{i+1} >$$
$$(1 + t)p_i,\ i = 1, 2, \ldots, k - 1 \quad (11)$$

where t is a small but fixed number (small compared to C). Since $(1 + t)^k < 100$ it is clear that $k < k_o$ with constant $k_o = k_o(t)$.

Suppose we already established the existence of a sequence satisfying (11). Then we prove (2) as follows: Consider the two intervals

$$\left[\frac{x}{p_{i+1}}, \frac{x}{p_{i+1}}(1 + C)\right], \left[\frac{x}{p_i}, \frac{x}{p_i}(1 + C)\right] \tag{12}$$

If they overlap, then by (11)

$$\frac{x}{p_{i+1}}(1 + t) < \frac{x}{p_i} < \frac{x}{p_{i+1}}(1 + C)$$

Clearly

$$\vartheta\left(\frac{x}{p_i}\right) - \vartheta\left(\frac{x}{p_{i+1}}\right) = 2\left(\frac{x}{p_i} - \frac{x}{p_{i+1}}\right) + o\left(\frac{x}{p_i}\right) \tag{13}$$

since otherwise

$$\vartheta\left(\frac{x}{p_i}\right) - \vartheta\left(\frac{x}{p_{i+1}}\right) < (2 - c_1)\left(\frac{x}{p_i} - \frac{x}{p_{i+1}}\right)$$

with $c_1 > 0$ and we would have from (9)

$$\vartheta\left[\frac{x}{p_{i+1}}(1 + C)\right] - \vartheta\left(\frac{x}{p_i}\right) > (2 + c_2)\left[\frac{x}{p_{i+1}}(1 + C) - \frac{x}{p_i}\right]$$

which contradicts III. Adding (13) and (9) with $p = p_i$ we obtain

$$\vartheta\left[\frac{x}{p_i}(1 + C)\right] - \vartheta\left(\frac{x}{p_{i+1}}\right) = 2\left[\frac{x}{p_i}(1 + C)\right] - \frac{x}{p_{i+1}} + o\left(\frac{x}{p_i}\right) \tag{14}$$

If the intervals (12) do not overlap we obtain by a simple calculation (using (9) and the fact that t is small)

$$\vartheta\left[\frac{x}{p_i}(1 + C)\right] - \vartheta\left(\frac{x}{p_{i+1}}\right) > 1.9\left[\frac{x(1 + C)}{p_i} - \frac{x}{p_{i+1}}\right] \tag{15}$$

Adding all the equations (14) and (15) (for $i = 1, 2, \ldots, k$) we clearly obtain

$$\vartheta\left[\frac{x}{p_1}(1 + C)\right] - \vartheta\left(\frac{x}{p_k}\right) > 1.9\left[\frac{x}{p_1}(1 + C) - \frac{x}{p_k}\right] \tag{16}$$

Since $p_k > 10p_1$ we obtain from (16)

$$\vartheta\left[\frac{x}{p_1}(1 + C)\right] > 1.6\frac{x}{p_1}(1 + C). \tag{17}$$

But (17) contradicts IV.

Thus to complete the proof of (2) it will suffice to show the existence of a sequence of good primes satisfying (11).

Consider the intervals

$$I_r = (B^{2r}, B^{2r+1}),\ r = 0, 1, \ldots, \left[\frac{\log x}{2 \log B}\right] - 1,$$

where B is a fixed, sufficiently large number. Clearly all the intervals I_r lie in the interval $(0, x)$. First we show that with the exception of $o(\log x)$ r's the interval I_r contains good primes. From I and IV it easily follows that for sufficiently large B we have (since $\vartheta(Bx) - \vartheta(x) > cx$)

$$\sum_{p \text{ in } I_r} \frac{\log p}{p} > c_1 (c_1 > 0 \text{ independent of } r)$$

Thus if there were $c_2 \log x$ with $c_2 > 0$ of the I_r's without good primes, we would have

$$\sum_{p \text{ bad}} \frac{\log p}{p} > c_1 c_2 \log x$$

which contradicts (8).

Let now $p_1^{(r)}$ be the smallest good prime in I_3 (if it exists), and suppose that a sequence $p_1^{(r)}, p_2^{(r)}, \ldots, p_i^{(r)}$ satisfying (11) exists, but no $p_{i+1}^{(r)}$ satisfying (11) can be found. Thus, all the primes in

$$J_i^{(r)} = [p_i^{(r)}(1 + t), p_i^{(r)}(1 + t)^2(1 + C)]$$

are bad. We have, by the definition of C,

$$\sum_{p \text{ in } J^{(r)}} \log p > \eta p_i^{(r)}(1 + t)^2(1 + C),\ (\eta \text{ absolute constant}).$$

Thus

$$\sum_{p \text{ in } J^{(r)}} \frac{\log p}{p} > \eta \tag{18}$$

Clearly for $B > 100$ we have $p_i^{(r)}(1 + t)^2(1 + C) < B^{2r+2}$. Thus the intervals $J_{i_1}^{r_1}, J_{i_2}^{r_2}, \ldots$ do not overlap. Hence from (18), since the number of r's with $p_1^{(r)}$ existing is $> \dfrac{\log x}{4 \log B}$,

$$\sum_{p \text{ bad}} \frac{\log p}{p} > \frac{\eta \log x}{4 \log B}$$

which contradicts (8) and establishes (2).

3. *Selberg's deduction of the prime number theorem from* (2).—Assume $a < A$. First we prove the following lemmas.

LEMMA 2. $a + A = 2$.

Choose $x \to \infty$ so that $\vartheta(x) = Ax + o(x)$. Then a simple computation (as in the proof of III) shows that

$$\sum_{p \leqq x} (\log p)^2 = Ax \log x + o(x \log x).$$

Thus from (1)

$$\sum_{p \leqq x} (\log p)\vartheta\left(\frac{x}{r}\right) = (2 - A)x \log x + o(x \log x). \qquad (19)$$

By the definition of a and by II we obtain by a simple computation

$$\sum_{p \leqq x} (\log p)\vartheta\left(\frac{x}{r}\right) \geqq ax \sum_{p \leqq x} \frac{\log p}{p} + o(x \log x) = ax \log x + o(x \log x)$$

Thus from (19), $2 \geqq a + A$. We obtain $a + A \leqq 2$ similarly, by choosing x so that $\vartheta(x) = ax + o(x)$. Thus lemma 2 is proved.

LEMMA 3. *Let $x \to \infty$ so that $\vartheta(x) = Ax + o(x)$. Then for any prime $p_i \leqq x$ except possible for a set of primes satisfying*

$$\sum \frac{\log p}{p} = o(\log x) \qquad (20)$$

we have

$$\vartheta\left(\frac{x}{p_i}\right) = a\frac{x}{p_i} + o\left(\frac{x}{p_i}\right) \qquad (21)$$

Suppose the lemma is false. Then as in the proof of lemma 1 there exist two positive constants b_1 and b_2 so that for arbitrarily large x, satisfying $\vartheta(x) = Ax + o(x)$, and for a set of primes satisfying $\sum \frac{\log p}{p} > b_1 \log x$, *we have*

$$\vartheta\left(\frac{x}{p}\right) > (a + b_2)\frac{x}{p} \qquad (22)$$

But then we have from (22), lemma 2, (19) and II (as in the proof of lemma 1)

$$ax \log x + o(x \log x) = \sum_{p \leqq x} (\log p)\vartheta\left(\frac{x}{p}\right) > b_1(a + b_2)x \log x + (1 - b_1)ax$$
$$\log x + o(x \log x) = ax \log x + b_1b_2x \log x + o(x \log x),$$

an evident contradiction. This proves lemma 3.

LEMMA 4. *Let p_1 be the smallest prime satisfying (21). Then $p_1 < x^\epsilon$, and for all primes $p_j < x/p_1$ except possible for a set of primes satisfying*

$$\sum \frac{\log p}{p} = o(\log x) \qquad (23)$$

we have

$$\vartheta\left(\frac{x}{p_1 p_j}\right) = A\frac{x}{p_1 p_j} + o\left(\frac{x}{p_1 p_j}\right) \tag{24}$$

$p_1 < x^{\epsilon}$ follows immediately from (20) and II. The second part of lemma 4 follows by applying the argument of lemma 3 to x/p_1 instead of x and interchanging A and a.

Now the deduction of the prime number theorem. Let p_i be any prime satisfying (21). Assume $\frac{x}{p_1 p_j} < \frac{x}{p_i}$. Then (since $\vartheta(x)$ is non-decreasing) from (21) and (24)

$$a\frac{x}{p_i} + o\left(\frac{x}{p_i}\right) \leq A\frac{x}{p_1 p_j} + o\left(\frac{x}{p_1 p_j}\right)$$

or p_j cannot lie in the interval

$$I_i = \left[\frac{p_i}{p_1}, \frac{p_i}{p_1}\left(\frac{A}{a} - \delta\right)\right],$$

where $\delta > 0$ is an arbitrary fixed number. Hence all primes in I_i must be "bad," i.e., they do not satisfy (24). But it immediately follows from (2) that

$$\sum_{p \text{ in } I_i} \frac{\log p}{p} > \eta$$

To obtain a contradiction to (23) it suffices to construct $c \log x$ disjoint intervals I_i. This can be accomplished in the same way as in the end of the proof of (2) (where the disjoint intervals $J_i^{(r)}$ were constructed). This completes the first elementary proof of the prime number theorem.

4. *Sketch of Selberg's simplification of the proof of* (2).—If we can find two good primes satisfying

$$(1 + c)p_1 > p_2 > (1 + t)p_1, \quad c > \frac{C}{1 + t} \tag{25}$$

then (2) follows easily. The intervals $\left[\frac{x}{p_1}, \frac{x}{p_1}(1 + c)\right]$, $\left[\frac{x}{p_2}, \frac{x}{p_2}(1 + c)\right]$, overlap. Thus (13), with $i = 1$, holds. But then exactly as in lemma 1 there exists a prime p so that

$$\vartheta\left[\frac{x}{p_1 p}(1 + c)\right] - \vartheta\left(\frac{x}{p_2 p}\right) = o\left(\frac{x}{p_2 p}\right).$$

But this is impossible (by the definition of C) since

$$\frac{x}{p_1 p}(1 + c)/\frac{x}{p_2 p} = \frac{p_2(1 + c)}{p_1} > 1 + C.$$

Thus we only have to show that good primes satisfying (24) exist, and this can be accomplished by using III (a contradiction with $\sum_{p \text{ good}} \frac{\log p}{p} = [1 + o(1)]\log x$ can be established similarly as in the previous proof).

5. *The joint simplified proof of the prime number theorem.*—

LEMMA 5. *Let $x_2 > x_1$ and $x_1 \to \infty$. Assume that $\vartheta(x_1) = Ax_1 + o(x_1)$ and $\vartheta(x_2) = ax_2 + o(x_2)$, or $\vartheta(x_1) = ax_1 + o(x_1)$ and $\vartheta(x_2) = Ax_2 + o(x_2)$. Then*

$$x_2/x_1 \leqq A/a + o(1).$$

Since $\vartheta(x)$ is non-decreasing we have in the first case

$$ax_2 + o(x_2) \leqq Ax_1 + o(x_1) \text{ or } x_2/x_1 \leqq A/a + o(1)$$

In the second case we have by III $\vartheta(x_2) - \vartheta(x_1) \leqq 2(x_2 - x_1) + o(x_2)$

$$ax_1 + 2(x_2 - x_1) \leqq Ax_2 + o(x_2) \text{ or } (2 - A)x_2 \leqq (2 - a)x_1 + o(x_2).$$

Hence by lemma 2, $ax_2 \leqq Ax_1 + o(x_2)$. Thus again $x_2/x_1 \leqq A/a + o(1)$. q.e.d.

Put $1 + D = \frac{A}{a} + \delta$ where δ is sufficiently small, and will be determined later. Next we prove the following result.

LEMMA 6.

$$\sum_{y \leqq p \leqq (1+D)y} \frac{\log p}{p} > \eta (\eta \text{ independent of } y).$$

First we show that

$$\sum_{y \leqq p \leqq (1+D)y} \log p > \eta(1 + D)y. \qquad (26)$$

If (26) is false then for a suitable sequence of y's we have $\vartheta[(1 + D)y] - \vartheta(y) = o(y)$. But then for these y's

$$\frac{\vartheta[(1 + D)y]}{(1 + D)y} = \frac{\vartheta(y) + o(y)}{(1 + D)y} \leqq \frac{Ay + o(y)}{(1 + D)y} < a - c_1,$$

which contradicts the definition of a. Thus (26) holds and lemma 6 follows immediately.

Choose now x so that $\vartheta(x) = Ax + o(x)$. Then by lemmas 3 and 4 we obtain (p_1, p_i and p_j having the same meaning as in lemmas 3 and 4)

$$\vartheta\left(\frac{x}{p_1p_j}\right) = A\frac{x}{p_1p_j} + o\left(\frac{x}{p_1p_j}\right), \quad \vartheta\left(\frac{x}{p_i}\right) = a\frac{x}{p_i} + o\left(\frac{x}{p_i}\right)$$

From lemma 5 we obtain that for any fixed ϵ and sufficiently large x (satisfying $\vartheta(x) = Ax + o(x)$)

$$\text{either } \frac{x}{p_i} > \left(\frac{A}{a} - \epsilon\right)\frac{x}{p_1 p_j} \text{ or } \frac{x}{p_i} < \left(\frac{a}{A} + \epsilon\right)\frac{x}{p_1 p_j}$$

Hence p_j cannot lie in the interval

$$I_i = \left[\left(\frac{a}{A} + \right)\frac{p_i}{p_1}\left(\frac{A}{a} - \epsilon\right)\frac{p_i}{p_1}\right].$$

Now if δ is small enough then $1 + D \leqq \left(\frac{A}{a} - \epsilon\right)/\left(\frac{a}{A} + \epsilon\right)$. Hence by lemma 6

$$\sum_{p \text{ in } I_i} \frac{\log p}{p} > \eta$$

But by what has been said before all the primes in I_i are bad (i.e., they do not satisfy (24)). Thus to arrive at a contradiction with (23) it will suffice as in the proof of (2) to construct $c \log x$ disjoint intervals I_i. This can be accomplished as in the proof of (2), which completes the proof of the prime number theorem.

6. Perhaps this last step can be carried out slightly more easily as follows: Put

$$S = \sum \frac{\log p_i}{p_i} \sum_{p \text{ in } I_i} \frac{\log p}{p} \tag{27}$$

where p_i runs through the primes satisfying (21). As stated before all the primes in I_i are bad (i.e., they do not satisfy (24)). Thus we have from (27)

$$S > \eta \sum \frac{\log p_i}{p_i} > \frac{\eta}{2} \log x \tag{28}$$

since by II and (20) $\Sigma \frac{\log p_i}{p_i} > {}^1/_2 \log x$ for large x.

On the other hand by interchanging the order of summation we obtain

$$S = \sum \frac{\log p}{p} \sum_{p \text{ in } J_p} \frac{\log p_i}{p_i}$$

where p runs through all the primes of all the intervals I_i (each p is, of course, counted only once) and p_i runs through the primes satisfying (21) of the interval

$$J_p = \left[pp_1\left(\frac{A}{a} - \epsilon\right)^{-1}, \quad pp_1\left(\frac{a}{A} + \epsilon\right)^{-1}\right].$$

We evidently have from $A < \infty$

$$\sum_{p_i \text{ in } J_p} \frac{\log p_i}{p_i} < c$$

Hence

$$\sum \frac{\log p}{p} > S/c$$

or from (27)

$$\sum \frac{\log p}{p} > \frac{\eta}{2c} \log x,$$

which contradicts (22) and completes the proof.

[1] Selberg's proof of (1) is not yet published.

[2] See, for example, Landau, E., *Handbuch der Lehre von der Verteilung der Primzahlen*, § 19, or Ingham, A. E., *The Distribution of Prime Numbers*, p. 13.

[3] An analogous result is used in Selberg's proof of Dirichlet's theorem.

[4] See, for example, Landau, E., op. cit., §§18 and 26, or Ingham, A. E., op. cit., pp. 14, 15 and 22.

ON THE STRUCTURE OF LOCALLY COMPACT GROUPS

BY ANDREW M. GLEASON

SOCIETY OF FELLOWS, HARVARD UNIVERSITY

Communicated by Hassler Whitney, May 7, 1949

1. *Introduction*—Locally compact groups have attracted a great deal of study in the years since the introduction of invariant integration by Haar.[1] It has been shown that their structure is closely related to that of Lie groups in certain important cases (compact,[2] abelian[3] and solvable[4] groups), and it is widely conjectured that similar results are valid in general. We shall state here certain theorems which strengthen this conjecture and reduce its verification to the study of simple groups.

2. *The Extension Theorem for Lie Groups.*—THEOREM 1. *Let G be a topological group. Suppose that G has a closed normal subgroup N such that both N and G/N are Lie groups. Then G is itself a Lie group.*

In case N is abelian, Kuranishi[5] has proved this theorem under the additional hypothesis that there is a local cross-section for the cosets of N; that is, a closed set having exactly one point in common with each coset of N near the identity. The author has shown[6] that such a cross-section set always exists for abelian Lie groups. Hence our theorem is true for the special case that N is abelian.

12.12 S. Skewes (1955)

On the Difference $\pi(x) - \mathrm{Li}(x)$ *(II)*

Stanley Skewes completed his Ph.D. at Cambridge in 1938 under the supervision of J.E. Littlewood. His dissertation was titled *On the Difference* $\pi(x) - \mathrm{Li}(x)$.

This paper is the second in a series taken from Skewes's dissertation. Skewes considered the bound on $x \geq 2$ such that the inequality

$$\pi(x) - \mathrm{Li}(x) > 0$$

holds.

In a previous paper, Skewes proved that $x < 10^{10^{10^{34}}}$ (see Section 4.4). In this paper, Skewes determined a better bound under a Riemann hypothesis type assumption which he denoted hypothesis H.

Hypothesis H. *Every complex zero* $s = \sigma + it$ *of the Riemann zeta function satisfies*

$$\sigma - \frac{1}{2} \leq X^{-3} \log^{-2}(X)$$

provided that $|t| < X^3$.

Skewes showed under hypothesis H, that $x < e^{e^{e^{7.703}}}$. He also showed that under the negation of hypothesis H, $x < 10^{10^{10^{10^3}}}$.

ON THE DIFFERENCE $\pi(x) - \operatorname{li} x$ (II)

By S. SKEWES

[Received 31 December 1953.—Read 21 January 1954]

INTRODUCTION

1. LET $\pi(x)$ denote, as usual, the number of primes less than or equal to x which we suppose always to be not less than 2, and let

$$\operatorname{li} x = \lim_{\epsilon \to 0} \left(\int_0^{1-\epsilon} + \int_{1+\epsilon}^{x} \right) \frac{du}{\log u}.$$

The difference $d(x) = \pi(x) - \operatorname{li} x$ is negative for all values of x up to 10^7, and for all the special values of x for which $\pi(x)$ has been calculated (e.g. $d(x) = -1757$ for $x = 10^9$). Littlewood (**1**) proved in 1914, however, that $d(x)$ changes sign infinitely often, and in particular there exists an X such that $d(x) > 0$ for some $x < X$. This last result is our present subject. Littlewood's method depends on an 'explicit formula', as does all subsequent work, including the present paper.

If θ is the upper bound of the real parts of the zeros $\rho = \beta + i\gamma$ of the Riemann zeta-function $\zeta(s)$, the 'Riemann hypothesis' [(RH) for short] is that $\theta = \frac{1}{2}$; if this is false, then $\frac{1}{2} < \theta \leqslant 1$. In this latter case it had long been known that, for each positive ϵ, $d(x)/x^{\theta - \epsilon}$ oscillates, as x tends to infinity, over a range including ± 1. In proving the mere existence of an X it is therefore permissible to assume (RH), and Littlewood naturally did this.

Littlewood's theorem is a 'pure existence theorem', and does not provide, even when (RH) is assumed, an explicit numerical X.

When we face the problem of a numerical X, free of hypotheses, the argument falls naturally into three stages.

(i) A new method is found which assumes (RH) and provides a numerical $X = X_1$. I gave such a method in 1933 (**3**). In the meantime Ingham (**4**) has developed an alternative method (which he applies to the more general problem of the infinity of changes of sign of $d(x)$). This, adapted to our more special case (one change of sign) and with some further modifications, gives a much better X_1 than my original paper did; the argument is given in full in Part I. One of the advantages of the new method is that we can largely eliminate the ρ's beyond a given point; we operate in fact with the 269 ρ's with $0 < \gamma < 500$, whose position is approximately known.

(ii) (This is easy.) The whole argument in (i) is based on the explicit formula for $\psi_0(x) = \frac{1}{2}\{\psi(x+0)+\psi(x-0)\}$ (in the usual notation of the subject) (**2**). This is

$$\psi_0(x)-x = -\sum_\rho \frac{x^\rho}{\rho}-\frac{\zeta'(0)}{\zeta(0)}-\tfrac{1}{2}\log\left(1-\frac{1}{x^2}\right) \quad \text{for } x > 1.$$

In the course of the proof the terms of the series $\sum x^\rho/\rho$ with $|\gamma| \geqslant G = X_1^3$ can (roughly) be rejected as negligible, (RH) *or not*. It is enough, in other words, to suppose, instead of (RH), only that $\beta = \frac{1}{2}$ for those γ's satisfying $|\gamma| < G$. This hypothesis can in turn be weakened; for $x < X_1$, the $|x^{\beta+i\gamma}|$ concerned differ from $|x^{\frac{1}{2}+i\gamma}|$ by something negligible, provided the β's concerned satisfy

$$b = \beta-\tfrac{1}{2} \leqslant B = X_1^{-3}\log^{-2}X_1.$$

With minor adjustments, then, the proof in (i) can be made to provide an X_1 [actual value $\exp\exp\exp(7{\cdot}703)$] subject only to the double modification of (RH) explained above. This modification, which we will call (H), is, to repeat,

(H) *Every zero* $\rho = \beta+i\gamma$ *for which* $|\gamma| < G = X_1^3$ *is such that*

$$b = \beta-\tfrac{1}{2} \leqslant B = X_1^{-3}\log^{-2}X_1.$$

(iii) Since (H) leads to an X_1, it remains only to show that (NH), the negation of (H), leads to an X_2, i.e. that $d(x) > 0$ for some $x < X_2$. Now (NH) asserts the existence of a $\rho = \rho_0 = \beta_0+i\gamma_0$ with

$$0 < \gamma_0 < G = X_1^3, \qquad b_0 = \beta_0-\tfrac{1}{2} > B,$$

where
$$B = X_1^{-3}\log^{-2}X_1;$$

that is, it provides a more or less given ρ to the right of $\sigma = \frac{1}{2}$. In particular, it asserts that $\theta > \frac{1}{2}+B$, in which case an X_2 certainly *exists* in virtue of the old theorem about $d(x) > x^{\theta-\epsilon}$. It is natural to expect further that the proof of that theorem could use the existence of the special ρ_0 to provide a numerical X_2. But this turns out not to be so; the proof in question is another 'pure existence' one. Some further idea is called for, and I am in fact indebted to Professor Littlewood for the sketch of a method for the simpler problem of finding an X such that, for a given $h > 0$ and for some $x < X$, $\psi(x)-x > h\sqrt{x}$.

There is now a last unexpected point. In the past it has always been possible to work with the function $\psi(x)$ and its simpler explicit formula, with only a last minute switch, on established lines, to $\pi(x)$. But with (NH) this is no longer possible, and it is necessary to work, in finding X_2, with $\Pi_0(x) = \frac{1}{2}\{\Pi(x+0)+\Pi(x-0)\}$, where

$$\Pi(x) = \sum_{p^m \leqslant x} \frac{1}{m} = \sum_{m=1}^{M} \frac{1}{m}\pi(x^{1/m}), \qquad M = [\log x/\log 2].$$

The explicit formula for $\Pi_0(x)$ is, for $x > 1$,

$$\Pi_0(x) = \operatorname{li} x - \sum_{\rho} \operatorname{li} x^{\rho} + \int_x^{\infty} \frac{du}{(u^2-1)u\log u} - \log 2.$$

In the actual working out of the paper stages (i) and (ii) are telescoped, and (RH) never appears. In Part I we assume (H) from the first, and arrive at an $X_1 = \exp\exp\exp(7\cdot703)$. Part II then assumes (NH) and arrives at an X_2 differing negligibly† in expression from e^{X_1}: a (just) permissible X_2 is

$$10^{10^{10^{10^3}}}.$$

I wish in conclusion to express my humble thanks to Professor Littlewood, but for whose patient profanity this paper could never have become fit for publication.

PART I

2. We begin by collecting, in Lemma 1, some results about the zeros ρ. The fundamental theorem underlying all its results is as follows (**5**).

Let $N(T)$ be the number of roots $\rho = \beta + i\gamma$ of the ζ-function satisfying $0 < \beta < 1$, $0 < \gamma \leqslant T$. Then

$$N(T) = \frac{T}{2\pi}\log\frac{T}{2\pi e} + R(T), \tag{1}$$

where $\quad |R(T)| < (0\cdot137)\log T + (0\cdot443)\log\log T + 4\cdot350.$

We make use also of the known values of $\gamma_1, \gamma_2, \ldots, \gamma_{29}$, that is, all the γ's satisfying $0 < \gamma \leqslant 100$. We have now

LEMMA 1. *For all $T \geqslant \gamma_1 = 14\cdot13\ldots$,*

(i) $\displaystyle\sum_{0<\gamma<T} \frac{1}{\gamma} < \frac{1}{4\pi}\log^2 T,$

(ii) $\displaystyle\sum_{\gamma \geqslant T} \frac{1}{\gamma^2} < \frac{1}{2\pi}\frac{\log T}{T},$

(iii) $\displaystyle\sum_{\gamma>0} \frac{1}{\gamma^2} < 0\cdot0233.$

For $|h| < \frac{1}{2}T$,

(iv) $|N(T+h) - N(T)| < \dfrac{1}{2\pi}(|h| + 1\cdot77)\log T + 8\cdot7.$

We suppress throughout the details of purely numerical calculations.

† X_1^{100} differs negligibly (in its top index) from X_1. For this and similar reasons some of our approximations can be very crude; only in those bearing on a top index is refinement called for.

We obtain (i) from (1) and the formula†

$$\sum_{0<\gamma_n<T}\frac{1}{\gamma_n}=\sum_{n=1}^{29}\frac{1}{\gamma_n}+\int_{\gamma_{30}}^{T}\frac{N^*(x)}{x^2}\,dx+\frac{N^*(T)}{T},$$

where $N^*(T)=N(T)-29$. Since $\sum_{n=1}^{29}\frac{1}{\gamma_n}<0{\cdot}5925$, this leads by straightforward calculation to

$$\sum_{0<\gamma<T}\frac{1}{\gamma}=\frac{1}{4\pi}\log^2 T-\frac{\log 2\pi}{2\pi}\log T+R_1(T),$$

where $|R_1(T)|<0{\cdot}312$. This leads at once to (i).

We obtain (ii) similarly, from (1) and the formula†

$$\sum_{\gamma\geqslant T}\frac{1}{\gamma^2}=\int_T^{\infty}\frac{2N(x)}{x^3}\,dx-\frac{N(T)}{T^2}.$$

Of the remaining results (iii) is known, and (iv) follows at once from (1).

3. Lemma 2. *Let $\psi_0(x)$ be defined, as usual, by*

$$\psi_0(x)=\tfrac{1}{2}\{\psi(x+0)+\psi(x-0)\},$$

where $\psi(x)=\sum_{n\leqslant x}\Lambda(n)$. For $x>1$, $\psi_0(x)$ is known to possess the explicit formula‡

$$\psi_0(x)-x=-\sum_\rho\frac{x^\rho}{\rho}-\frac{\zeta'(0)}{\zeta(0)}-\tfrac{1}{2}\log\left(1-\frac{1}{x^2}\right), \tag{2}$$

where $\sum_\rho\frac{x^\rho}{\rho}$ is defined as the limit of $\sum_{|\gamma|\leqslant T}\frac{x^\rho}{\rho}$ as $T\to\infty$. If

$$\sum_\rho\frac{x^\rho}{\rho}=\sum_{|\gamma|\leqslant T}\frac{x^\rho}{\rho}+R(x,T),$$

then

(i) $|R(x,T)|<1000\dfrac{x^3}{x-1}\dfrac{\log^2 T}{T}+3\log x\quad(x\geqslant e,\ T\geqslant 9)$;

(ii) $|R(x,T)|<(0{\cdot}0001)x^{\frac{1}{2}}\quad(x\geqslant\exp(10^4),\ T\geqslant x^2)$;

(iii) $\left|\sum_\rho\dfrac{x^\rho}{\rho}\right|<3x\log x\quad(x\geqslant e)$.

The proof of (i) proceeds by straightforward calculation on the lines of the proof of (2); (ii) follows from (i); and (iii) follows from (2) and the definitions of $\psi_0(x)$ and $\psi(x)$, since

$$\sum_{n\leqslant x}\Lambda(n)\leqslant x\log x\quad\text{and}\quad|\psi_0(x)-\psi(x)|\leqslant\tfrac{1}{2}\log x.$$

† (**2**), 18, Theorem A. ‡ (**2**), 77, Theorem 29.

4. We shall for the present assume the following hypothesis, which we call (H), about the zeros $\rho = \beta+i\gamma$.

(H) *Let* $X_1 = \exp\exp\exp(7\cdot 703)$, $G = X_1^3$, $B = X_1^{-3}\log^{-2}X_1$. *Then for every zero* ρ *such that* $|\gamma| < G$, β *satisfies*

$$b = \beta-\tfrac{1}{2} \leqslant B.$$

For reference we shall prefix (H) to those results which depend on the hypothesis (H).

(H) Lemma 3. *Let* $\psi_1(x)$ *be defined, as usual, by*

$$\psi_1(x) = \int\limits_1^x \psi(u)\,du = \sum_{n\leqslant x}(x-n)\Lambda(n).$$

Then, on hypothesis (H),

(i) $|\psi_1(x)-\frac{1}{2}x^2| < 8x \quad (2 < x \leqslant e^8)$;

(ii) $|\psi_1(x)-\frac{1}{2}x^2| < \frac{1}{4}x^{\frac{3}{2}} \quad (e^8 < x < X_1)$.

For $x \geqslant 1$ we have the formula†

$$\psi_1(x)-\tfrac{1}{2}x^2 = -\sum_\rho \frac{x^{\rho+1}}{\rho(\rho+1)} - x\frac{\zeta'(0)}{\zeta(0)} + \frac{\zeta'(-1)}{\zeta(-1)} - \sum_{r=1}^{\infty}\frac{x^{1-2r}}{2r(2r-1)}. \tag{3}$$

From Lemma 1 (ii) and (iii), and assuming (H), we have, for $x < X_1$,

$$\begin{aligned} x^{-\frac{3}{2}}\left|\sum_\rho \frac{x^{\rho+1}}{\rho(\rho+1)}\right| &\leqslant \sum_{|\gamma|<G}\left|\frac{x^{\rho-\frac{1}{2}}}{\rho(\rho+1)}\right| + \sum_{|\gamma|\geqslant G}\left|\frac{x^{\rho-\frac{1}{2}}}{\rho(\rho+1)}\right| \\ &\leqslant X_1^B \sum_{|\gamma|<G}\frac{1}{\gamma^2} + X_1^{\frac{1}{2}}\frac{1}{\pi}\frac{\log G}{G} \\ &\leqslant \tfrac{1}{10}. \end{aligned}$$

Substituting in (3) and noting that

$$\zeta'(0)/\zeta(0) = \log 2\pi \quad \text{and} \quad |\zeta'(-1)/\zeta(-1)| < 1,$$

we obtain both (i) and (ii).

5. Lemma 4.

(i) $\operatorname{li} u < (1\cdot 0004)u/\log u \quad (u \geqslant \exp(4.10^3))$;

(ii) $\operatorname{li} u < 2u/\log u \quad (u \geqslant 2)$.

The value of li 2 is 1·04.... For $u > u_0 \geqslant e^2$, say,

$$\frac{1}{\log u} < \frac{d}{du}\left(\frac{u}{\log u}\right)\Big/\left(1-\frac{1}{\log u_0}\right).$$

† (2), 73, Theorem 28.

Hence
$$\mathrm{li}\,u = \mathrm{li}\,u_0 + \int_{u_0}^{u} \frac{dv}{\log v}$$
$$< u_0 + \frac{1}{1-1/\log u_0}\left[\frac{v}{\log v}\right]_{u_0}^{u},$$
and the result (i) follows by taking $u_0 = \exp(3.10^3)$. By taking $u_0 = e^2$, we find that (ii) is valid for $u \geqslant 8$, say, and, for $2 \leqslant u < 8$, (ii) is trivial.

6. We define $\Pi(x)$, as usual, by
$$\Pi(x) = \sum_{m=1}^{M} \frac{1}{m}\pi(x^{1/m}), \qquad M = [\log x/\log 2].$$

LEMMA 5. *For $x > \exp(4.10^3)$, either $\pi(\xi) > \mathrm{li}\,\xi$ for some ξ of $2 \leqslant \xi \leqslant x^{\frac{1}{2}}$, or else*
$$0 < \Pi(x)-\pi(x) < (1{\cdot}0005)x^{\frac{1}{2}}/\log x.$$

Supposing the former alternative to be false, we apply Lemma 4 (i) to the first term on the right-hand side of
$$\Pi(x)-\pi(x) = \tfrac{1}{2}\pi(x^{\frac{1}{2}}) + \sum_{m=3}^{M} \frac{1}{m}\pi(x^{1/m}),$$
and Lemma 4 (ii) to the remainder. Then
$$\Pi(x)-\pi(x) \leqslant \tfrac{1}{2}\mathrm{li}\,x^{\frac{1}{2}} + 2\sum_{m=3}^{M} x^{1/m}/\log x$$
$$< (1{\cdot}0004)x^{\frac{1}{2}}/\log x + (0{\cdot}0001)x^{\frac{1}{2}}/\log x,$$
and the desired result follows.

7. (H) LEMMA 6. *Let $P(x)$ be defined by*
$$P(x) = (\Pi(x)-\mathrm{li}\,x)-(\psi(x)-x)/\log x.$$
Then, on hypothesis (H),
$$|P(x)| < (0{\cdot}0005)x^{\frac{1}{2}}/\log x \quad (\exp(10^4) \leqslant x < X_1).$$

We have [(**2**), 64]
$$P(x) = \int_{2}^{x} \frac{\psi(u)-u}{u\log^2 u}\,du + \frac{2}{\log 2} - \mathrm{li}\,2,$$
and therefore, after integrating by parts,
$$|P(x)| \leqslant \left|\frac{\psi_1(x)-\frac{1}{2}x^2}{x\log^2 x} - \frac{\psi_1(2)-2}{2\log^2 2} + \frac{2}{\log 2} - \mathrm{li}\,2\right| + |J|, \tag{4}$$
where
$$J = \int_{2}^{x} \{\psi_1(u)-\tfrac{1}{2}u^2\}\,d\left\{\frac{1}{u\log^2 u}\right\}.$$

Now $$|J| \leqslant \int_{2}^{x} |\psi_1(u)-\tfrac{1}{2}u^2| \frac{4}{u^2\log^2 u}\,du = \int_{2}^{e^8}+\int_{e^8}^{x}.$$

Now apply Lemma 3. We have, on (H),

$$|J| < \int_{2}^{e^8} \frac{32}{u\log^2 u}\,du + \int_{e^8}^{x} \left(\frac{u^{\frac{1}{4}}}{\log^2 u}\right)\frac{1}{u^{\frac{3}{4}}}\,du.$$

Since $u^{\frac{1}{4}}/\log^2 u$ increases with u for $u > e^8$, it follows that, for

$$\exp(10^4) \leqslant x < X_1,$$

$$|J| < \frac{32}{\log 2}+\frac{x^{\frac{1}{4}}}{\log^2 x}4x^{\frac{1}{4}} < 48+0{\cdot}0004x^{\frac{1}{2}}/\log x.$$

Substituting this inequality in (4) and applying Lemma 3 (ii) to

$$(\psi_1(x)-\tfrac{1}{2}x^2)/x\log^2 x,$$

we obtain the required inequality.

(H) Lemma 7. *Assume hypothesis* (H). *Then for any given x satisfying* $\exp(10^4) \leqslant x < X_1$, *either*

(i) $\pi(\xi)-\operatorname{li}\xi > 0$ *for some ξ of* $2 \leqslant \xi \leqslant x^{\frac{1}{2}}$,

or else

(ii) '$\psi_0(x)-x > (1{\cdot}001)x^{\frac{1}{2}}$' *implies* '$\pi(x)-\operatorname{li}x > 0$'.

(i) is the first alternative of Lemma 5, and (ii) follows from the second one and Lemma 6, since

$$\begin{aligned}(\pi(x)-\operatorname{li}x)\log x &= \{\psi(x)-\psi_0(x)\}+\{\psi_0(x)-x\}-\\ &\quad -\{\psi_0(x)-x-(\Pi(x)-\operatorname{li}x)\log x\}-\{(\Pi(x)-\pi(x))\log x\}\\ &> 0+(1{\cdot}001)x^{\frac{1}{2}}-(0{\cdot}0005)x^{\frac{1}{2}}-(1{\cdot}0005)x^{\frac{1}{2}} = 0.\end{aligned}$$

8. (H) Lemma 8. *On hypothesis* (H),

$$\sum_{|\gamma|<G}\left|\frac{x^{\rho-\frac{1}{2}}}{\rho}-\frac{x^{i\gamma}}{i\gamma}\right| < 0{\cdot}0234 \quad (\exp(10^4) \leqslant x < X_1).$$

For brevity write the series on the left as $S(x)$, and let, as usual,

$$\rho-\tfrac{1}{2} = \beta-\tfrac{1}{2}+i\gamma = b+i\gamma.$$

Then

$$S(x) = \sum_{|\gamma|<G}\left|\frac{x^{b+i\gamma}}{\beta+i\gamma}-\frac{x^{i\gamma}}{i\gamma}\right| \leqslant \left(\sum_{|\gamma|<G}\frac{1}{\gamma^2}\right)\max_{|\gamma|<G}\{|\gamma(x^b-1)|+\beta\}.$$

Applying (H) and Lemma 1 (iii), we have therefore

$$\begin{aligned}S(x) &< (0{\cdot}0466)[G(2B\log X_1)+\tfrac{1}{2}+B]\\ &< 0{\cdot}0234.\end{aligned}$$

9. We now develop a modification of Ingham's argument. Consider the formula (see (**4**), 204 (6))

$$\int_a^b \chi(x)(\psi_0(x)-x)\,dx$$
$$= -\sum_\rho \frac{1}{\rho}\int_a^b \chi(x)x^\rho\,dx + \int_a^b \chi(x)\{\tfrac{1}{2}\log(1-x^{-2})^{-1}-\zeta'(0)/\zeta(0)\}\,dx, \qquad (5)$$

where $1 < a < b < \infty$, and $\chi(x)$ is any function integrable in the sense of Lebesgue. Let

$$K(y) = \left(\frac{\sin\frac{1}{2}y}{\frac{1}{2}y}\right)^2,$$

so that, for real α,

$$\frac{1}{2\pi}\int_{-\infty}^{\infty} K(y)e^{i\alpha y}\,dy = \begin{cases} 1-|\alpha| & (|\alpha| \leqslant 1), \\ 0 & (|\alpha| > 1). \end{cases} \qquad (6)$$

Let $T = 500$ and ω be any number satisfying $\omega > 2.10^4$. In (5) substitute

$$x = e^u, \qquad \chi(e^u) = e^{-\frac{1}{2}u}TK\{T(u-\omega)\}, \qquad a = e^{\frac{1}{2}\omega}, \qquad b = e^{\frac{3}{2}\omega}.$$

Then, writing for brevity

$$F(u) = \{\psi_0(e^u)-e^u\}e^{-\frac{1}{2}u}, \qquad (7)$$

we have

$$\int_{\frac{1}{2}\omega}^{\frac{3}{2}\omega} TK\{T(u-\omega)\}F(u)\,du = -\sum_\rho \frac{1}{\rho}\int_{\frac{1}{2}\omega}^{\frac{3}{2}\omega} TK\{T(u-\omega)\}e^{(\rho-\frac{1}{2})u}\,du + R, \qquad (8)$$

where, if we define $r(u)$ by

$$r(u) = e^{-\frac{1}{2}u}\{\tfrac{1}{2}\log(1-e^{-2u})^{-1}-\zeta'(0)/\zeta(0)\},$$

R is given by

$$R = \int_{\frac{1}{2}\omega}^{\frac{3}{2}\omega} TK\{T(u-\omega)\}r(u)\,du.$$

Since $\frac{1}{2}\omega \leqslant u \leqslant \frac{3}{2}\omega$ and $\omega > 2.10^4$, we have

$$|r(u)| < 2e^{-\frac{1}{4}\omega} < 0{\cdot}00001;$$

hence [in virtue of (6), with $\alpha = 0$]

$$|R| < (0{\cdot}00001)\int_{\frac{1}{2}\omega}^{\frac{3}{2}\omega} TK\{T(u-\omega)\}\,du$$
$$= (0{\cdot}00001)\int_{-\frac{1}{2}T\omega}^{\frac{1}{2}T\omega} K(y)\,dy < (0{\cdot}00001)2\pi. \qquad (9)$$

Substituting $u = \omega + y/T$ in (8), we have from (8) and (9)

$$\int_{-\frac{1}{2}T\omega}^{\frac{1}{2}T\omega} K(y)F(\omega+y/T)\,dy = -\sum_{\rho}\frac{1}{\rho}\int_{-\frac{1}{2}T\omega}^{\frac{1}{2}T\omega} K(y)e^{(\rho-\frac{1}{2})(\omega+y/T)}\,dy + R, \quad (10)$$

where $|R| < (0{\cdot}00002)\pi$.

For the infinite series on the right-hand side of (10) we shall substitute the finite series

$$\sum_{|\gamma|<G}\frac{e^{i\gamma\omega}}{i\gamma}\int_{-\infty}^{\infty} K(y)e^{i\gamma y/T}\,dy,$$

where G is the number defined in hypothesis (H), § 4. The total error introduced will be the sum of three errors e_1, e_2, e_3, where e_1 comes from discarding those terms for which $|\gamma| \geqslant G$, e_2 from replacing $(e^{(\rho-\frac{1}{2})(\omega+y/T)})/\rho$ by $(e^{i\gamma(\omega+y/T)})/i\gamma$, and e_3 from extending the limits of integration from $\pm\frac{1}{2}T\omega$ to $\pm\infty$. We shall deal with these errors in separate lemmas.

10. Lemma 9. *For* $2.10^4 \leqslant \omega \leqslant \frac{1}{3}\log G$, *the error*

$$e_1 = \sum_{|\gamma|\geqslant G}\frac{1}{\rho}\int_{-\frac{1}{2}T\omega}^{\frac{1}{2}T\omega} K(y)e^{(\rho-\frac{1}{2})(\omega+y/T)}\,dy$$

satisfies $$|e_1| < (0{\cdot}0002)\pi.$$

Since $-\frac{1}{2}T\omega \leqslant y \leqslant \frac{1}{2}T\omega$, we have $\frac{1}{2}\omega \leqslant \omega+y/T \leqslant \frac{3}{2}\omega$. Let

$$M = \sup\left|\sum_{|\gamma|\geqslant G}\frac{1}{\rho}e^{(\rho-\frac{1}{2})m}\right| \quad \text{for } \tfrac{1}{2}\omega \leqslant m \leqslant \tfrac{3}{2}\omega.$$

Then $$|e_1| = \left|\int_{-\frac{1}{2}T\omega}^{\frac{1}{2}T\omega} K(y)\left(\sum_{|\gamma|\geqslant G}\frac{1}{\rho}e^{(\rho-\frac{1}{2})(\omega+y/T)}\right)dy\right|$$

$$\leqslant M\int_{-\frac{1}{2}T\omega}^{\frac{1}{2}T\omega} K(y)\,dy \leqslant 2\pi M,$$

and this is less than $(0{\cdot}0001)2\pi$ by Lemma 2 (ii) applied to $e^{-\frac{1}{2}m}R(e^m, G)$, since $(\frac{1}{2}\omega, \frac{3}{2}\omega)$ is included in the appropriate range.

(H) Lemma 10. *On the hypothesis* (H) *and for* ω *subject to*

$$2.10^4 \leqslant \omega \leqslant \tfrac{2}{9}\log G,$$

the error

$$e_2 = \sum_{|\gamma|<G}\frac{1}{\rho}\int_{-\frac{1}{2}T\omega}^{\frac{1}{2}T\omega} K(y)e^{(\rho-\frac{1}{2})(\omega+y/T)}\,dy - \sum_{|\gamma|<G}\frac{1}{i\gamma}\int_{-\frac{1}{2}T\omega}^{\frac{1}{2}T\omega} K(y)e^{i\gamma(\omega+y/T)}\,dy$$

satisfies $$|e_2| < (0{\cdot}0468)\pi.$$

As in Lemma 9, we have $\frac{1}{2}\omega \leqslant \omega+y/T \leqslant \frac{3}{2}\omega$. Suppose here that m is that value of $\omega+y/T$ for which the value of

$$\sum_{|\gamma|<G}\left|\frac{1}{\rho}e^{(\rho-\frac{1}{2})(\omega+y/T)}-\frac{1}{i\gamma}e^{i\gamma(\omega+y/T)}\right|$$

is greatest. On (H) and for $2.10^4 \leqslant \omega \leqslant \frac{2}{9}\log G$, the error e_2 satisfies

$$\begin{aligned}|e_2| &\leqslant \int\limits_{-\frac{1}{2}T\omega}^{\frac{1}{2}T\omega} K(y)\sum_{|\gamma|<G}\left|\frac{1}{\rho}e^{(\rho-\frac{1}{2})(\omega+y/T)}-\frac{1}{i\gamma}e^{i\gamma(\omega+y/T)}\right|\,dy\\ &\leqslant \sum_{|\gamma|<G}\left|\frac{1}{\rho}e^{(\rho-\frac{1}{2})m}-\frac{1}{i\gamma}e^{i\gamma m}\right|\int\limits_{-\frac{1}{2}T\omega}^{\frac{1}{2}T\omega}K(y)\,dy\\ &< (0{\cdot}0234)2\pi\end{aligned}$$

by Lemma 8, since m again lies in the relevant range.

LEMMA 11. *For $\omega \geqslant 2.10^4$ and $T = 500$, the error*

$$e_3 = \sum_{|\gamma|<G}\frac{e^{i\gamma\omega}}{i\gamma}\left(\int\limits_{-\infty}^{-\frac{1}{2}T\omega}+\int\limits_{\frac{1}{2}T\omega}^{\infty}\right)K(y)e^{i\gamma y/T}\,dy$$

satisfies $$|e_3| < (0{\cdot}00002)\pi.$$

Since $K(y)$ is an even function of y and the γ's are symmetrically distributed, we have

$$\begin{aligned}|e_3| &\leqslant 4\sum_{0<\gamma<G}\frac{1}{\gamma}\left|\int\limits_{\frac{1}{2}T\omega}^{\infty}K(y)e^{i\gamma y/T}\,dy\right|\\ &= 4\sum_{0<\gamma<T}+4\sum_{T\leqslant\gamma<G}.\end{aligned}$$

Now we have the two inequalities†

$$\left|\int\limits_{\frac{1}{2}T\omega}^{\infty}K(y)e^{i\gamma y/T}\,dy\right|\begin{cases}\leqslant \displaystyle\int\limits_{\frac{1}{2}T\omega}^{\infty}4y^{-2}\,dy = \frac{8}{T\omega},\\ = \left|\displaystyle\int\limits_{\frac{1}{2}T\omega}^{\infty}\frac{T}{i\gamma}\{e^{i\gamma y/T}-e^{\frac{1}{2}i\gamma\omega}\}K'(y)\,dy\right| < \dfrac{2T}{|\gamma|}\dfrac{4}{\frac{1}{2}T\omega}.\end{cases}$$

Using the former inequality in $\sum\limits_{0<\gamma<T}$ and the latter in $\sum\limits_{T\leqslant\gamma<G}$, we have, from Lemma 1 (i) and (ii),

$$|e_3| \leqslant \frac{32}{T\omega}\sum_{0<\gamma<T}\frac{1}{\gamma}+\frac{64}{\omega}\sum_{T\leqslant\gamma<G}\frac{1}{\gamma^2} < \frac{32}{T\omega}\frac{1}{4\pi}\log^2 T+\frac{64}{\omega}\frac{1}{2\pi}\frac{\log T}{T}.$$

Since $\omega > 2.10^4$ and $T = 500$ the required result follows.

† We have $K'(y) = 2\sin y/y^2-8\sin^2\frac{1}{2}y/y^3$ and $|K'(y)| < 4/y^2$ in the range concerned.

11. From Lemmas 9, 10, and 11 we may now replace (10), subject to the condition
$$2.10^4 \leqslant \omega < \tfrac{2}{9}\log G, \tag{11}$$
by
$$\int_{-\frac{1}{2}T\omega}^{\frac{1}{2}T\omega} K(y)F(\omega+y/T)\,dy = -\sum_{|\gamma|<G} \frac{e^{i\gamma\omega}}{i\gamma}\int_{-\infty}^{\infty} K(y)e^{i\gamma y/T}\,dy + E, \tag{12}$$
where
$$|E| = |R-e_1-e_2+e_3| < (0\cdot0471)\pi.$$
Applying (6) to the series on the right-hand side of (12), we have, still subject to (11),
$$\int_{-\frac{1}{2}T\omega}^{\frac{1}{2}T\omega} K(y)F(\omega+y/T)\,dy = -2\sum_{0<\gamma<T} 2\pi\frac{\sin\gamma\omega}{\gamma}\left(1-\frac{\gamma}{T}\right)+E$$
$$> -2\sum_{0<\gamma<T} 2\pi\frac{\sin\gamma\omega}{\gamma}\left(1-\frac{\gamma}{T}\right)-(0\cdot0471)\pi. \tag{13}$$
Now let $F_M = F_M(\omega)$ be the upper bound of $F(\omega+y/T)$ for the range $-\frac{1}{2}T\omega \leqslant y \leqslant \frac{1}{2}T\omega$. Since $K(y) \geqslant 0$, (13) gives
$$F_M J = F_M\frac{1}{2\pi}\int_{-\frac{1}{2}T\omega}^{\frac{1}{2}T\omega} K(y)\,dy \geqslant \frac{1}{2\pi}\int_{-\frac{1}{2}T\omega}^{\frac{1}{2}T\omega} K(y)F(\omega+y/T)\,dy$$
$$> -2\sum_{0<\gamma<T} \frac{\sin\gamma\omega}{\gamma}\left(1-\frac{\gamma}{T}\right)-0\cdot0236. \tag{14}$$
Now by the definition (7) of F we have
$$F_M = \textit{upper bound of } (\psi_0(x)-x)x^{-\frac{1}{2}} \quad \textit{for } e^{\frac{1}{2}\omega} \leqslant x \leqslant e^{\frac{3}{2}\omega}. \tag{15}$$
We are therefore in a position to establish the following lemma.

(H) LEMMA 12. *On the hypothesis* (H) *a sufficient condition that*
$$\pi(x)-\operatorname{li} x > 0,$$
for some x satisfying $2 \leqslant x < X_1$,† *is that, for some ω subject to the condition* (11),
$$-\sum_{0<\gamma<500} \frac{\sin\gamma\omega}{\gamma}\left(1-\frac{\gamma}{500}\right) > 0\cdot5123. \tag{16}$$
When (16) is true we have, by (14)‡ (and the fact that $T = 500$), $F_M J > 1\cdot001$, and *a fortiori* $F_M > 1\cdot001$ since $J < \frac{1}{2\pi}\int_{-\infty}^{\infty} = 1$. Lemma 12 then follows from Lemma 7.

† We recall that X_1 is the number concerned in (H), § 4, namely $\exp\exp\exp(7\cdot703)$.
‡ Which is valid subject to (11).

12. Our problem is now to find a suitable ω. It must be chosen so that the sines in (16) are predominantly negative, and such a choice is made as follows.

The number N of terms in the series on the left-hand side of (16) is equal to the number of γ's satisfying $0<\gamma<500$; this is known to be

$$N = 269. \tag{17}$$

Let ω_0 and q be the numbers

$$\omega_0 = 2.10^4+1, \qquad q = 3600. \tag{18}$$

By Dirichlet's theorem there is a number ω' satisfying

$$\omega_0 \leqslant \omega' \leqslant \omega_0 q^N \tag{19}$$

such that

$$\left|\frac{\gamma_n \omega'}{2\pi} - r_n\right| < \frac{1}{q} \quad (n = 1, 2,..., N), \tag{20}$$

where r_n is an integer. Now let

$$\omega = \omega' - k, \tag{21}$$

where

$$k = \tfrac{3}{400}. \tag{22}$$

Then, from (20) and (21),

$$\sin\gamma_n \omega = -\sin(k\gamma_n - \phi_n),$$

where $|\phi_n| < 2\pi/q = 0°\,6'$. Now from (17), (18), (19), and (22),

$$2.10^4 < \omega < \omega_0 q^N = (2.10^4+1)3600^{269} = \exp\exp(7{\cdot}7021...) < \tfrac{2}{9}\log G.$$

The condition (11) is therefore satisfied. Hence, by Lemma 12, we shall have $\pi(x)-\mathrm{li}\,x > 0$ for some x satisfying

$$2 \leqslant x < X_1$$

provided that

$$S = \sum p(\gamma_n) = \sum_{n=1}^{269} \frac{\sin(k\gamma_n - \phi_n)}{\gamma_n}\left(1 - \frac{\gamma_n}{500}\right) > 0{\cdot}5123. \tag{23}$$

13. The inequality (23) is actually true. The right-hand side is what determines the top index of X_1 and it is here that we try to refine. It will suffice to sketch the numerical considerations involved.

The angles $k\gamma_n - \phi_n$ range from 6° to 215°, and the first 213 sines are positive. In the case of the remaining negative terms, for which

$$180° < k\gamma_n - \phi_n < 215°,$$

the γ's satisfy

$$420 < \gamma_n < 500.$$

Hence, in addition to the fact that $1/\gamma_n$ is small, either the absolute value of the sine or the factor $(1-\gamma_n/500)$ is small, and these negative terms prove to be of little importance. For the rest, sufficient is known about the values of the γ's† to enable us to obtain a lower bound to S by straightforward calculation.

In this the first 29 terms are calculated separately, the remainder are grouped in intervals (of the values of the γ's) of 10. For example, the first group contains the 4 terms for which $100 < \gamma \leqslant 110$, and the last group contains the 7 terms for which $490 < \gamma \leqslant 500$. We obtain a lower bound to the contribution of each term, or group of terms, by making use of the fact that the function $p(\gamma_n)$, defined in (23), satisfies (whatever the particular values of ϕ_n, ϕ_{n+1}) $p(\gamma_n) < p(\gamma_{n+1})$ for $\gamma_n < 457$ (approximately), and thereafter satisfies $p(\gamma_n) > p(\gamma_{n+1})$. We may replace each of the first 29 γ's by the upper bound to the interval in which it is known to lie, and for those groups for which $\gamma \leqslant 450$ we replace each of the γ's in the group by the upper bound of the interval in which it lies. The same replacement applies for the subgroup 450–457. For the subgroup 457–460 and the remaining groups for which $\gamma > 460$, since $p(\gamma_n)$ is now increasing, we replace the γ's in each group by the lower bound of the interval concerned. For example, each of the 4 γ's between 100 and 110 is replaced by 110, while each of the 7 γ's between 470 and 480 is replaced by 470. ϕ_n is replaced by $+6'$ or $-6'$ according as $\gamma_n k < 90°$ or $\gamma_n k > 90°$.

We find that $$S > 0{\cdot}5131 > 0{\cdot}5123.$$

$\pi(x) - \mathrm{li}\, x$ is therefore positive for some x satisfying

$$2 \leqslant x < X_1 = \exp\exp\exp(7{\cdot}703).$$

Part II

14. Before we can begin developing the consequences of (NH), the negation of (H), we need a number of preliminary results about the function

$$\Pi_0(x) - \mathrm{li}\, x = \tfrac{1}{2}\{\Pi(x+0) + \Pi(x-0)\} - \mathrm{li}\, x,$$

where $\Pi(x)$ is defined as in § 6. For $x > 1$ we have ((**2**), 81–82)

$$\Pi_0(x) - \mathrm{li}\, x = -\sum_\rho \mathrm{li}\, x^\rho + \int_x^\infty \frac{du}{(u^2-1)u\log u} - \log 2, \tag{24}$$

the series being boundedly convergent in any finite interval $1 < a \leqslant x \leqslant b$. The li function for a complex argument is defined by

$$\mathrm{li}\, x^\rho = \mathrm{li}\, e^{\rho \log x}, \tag{25}$$

and, for $w = u + vi$ where $v \neq 0$,

$$\mathrm{li}\, e^w = \int_{-\infty+vi}^{u+vi} \frac{e^z}{z}\, dz.$$

† (**6**), (**7**), (**8**), (**9**). In addition I have used some calculations performed by Dr. Comrie, kindly lent to me by Professor Titchmarsh.

We define the function $L(t)$ for $t>0$ by the series

$$L(t) = -e^{-\frac{1}{2}t}\sum_{\rho} \mathrm{li}\, e^{\rho t}. \tag{26}$$

From (25) and (26) we have, for $t>0$,

$$-L(t) = e^{-\frac{1}{2}t}\sum_{\rho}\int_{-\infty+i\gamma t}^{(\beta+i\gamma)t}\frac{e^z}{z}\,dz = \sum_{\rho} e^{(\rho-\frac{1}{2})t}\int_0^\infty \frac{e^{-v}}{\rho t - v}\,dv$$

$$= \sum e^{(\rho-\frac{1}{2})t}\left[\frac{1}{\rho t}+\int_0^\infty \frac{e^{-v}}{(\rho t-v)^2}\,dv\right]$$

$$= \sum \frac{e^{(\rho-\frac{1}{2})t}}{\rho t} + \sum e^{(\rho-\frac{1}{2})t}\int_0^\infty \frac{e^{-v}\,dv}{(\rho t - v)^2},$$

both series being boundedly convergent in any interval of type

$$0<a'\leqslant t\leqslant b'$$

since the first is. So

$$-L(t) = \sum_{\rho} u_1(\rho,t) + \sum_{\rho} u_2(\rho,t), \tag{27}$$

$$u_1 = \frac{e^{(\rho-\frac{1}{2})t}}{\rho t}, \qquad u_2 = e^{(\rho-\frac{1}{2})t}\int_0^\infty \frac{e^{-v}\,dv}{(\rho t-v)^2}. \tag{28}$$

15. Lemma 13. $\quad |L(t)| \leqslant 4e^{\frac{1}{2}t} \quad (t\geqslant 1).$

By Lemma 2 (iii), if $\tau = e^t \geqslant e$,

$$\left|\sum u_1(\rho,t)\right| = \left|\tau^{-\frac{1}{2}}(\log\tau)^{-1}\sum \tau^{\rho}/\rho\right| < 3\tau^{\frac{1}{2}}.$$

In $u_2(\rho,t)$ we have $|\rho t - v|^2 \geqslant |i\gamma t|^2 = \gamma^2 t^2$,

$$|u_2| \leqslant \tau^{\frac{1}{2}} t^{-2}\gamma^{-2}, \qquad \left|\sum u_2\right| \leqslant e^{\frac{1}{2}t}t^{-2}\,2\sum_{\gamma>0}\gamma^{-2} < e^{\frac{1}{2}t},$$

since $\sum \gamma^{-2} < 0{\cdot}05$. The result follows.

16. Lemma 14. *A sufficient condition that* $\pi(x)-\mathrm{li}\,x>0$ *for some* x *of* $2\leqslant x\leqslant X$ *is that, for some* y *satisfying* $10^4\leqslant y\leqslant \log X$,

$$L(y)\geqslant 1. \tag{29}$$

Suppose the condition of the lemma is satisfied for a certain y, and let $x=e^y$. Then, by Lemma 5, either $\pi(\xi)-\mathrm{li}\,\xi>0$ for some ξ of $2\leqslant\xi\leqslant x^{\frac{1}{2}}$, or else

$$\Pi_0(x)-\pi(x)\leqslant \Pi(x)-\pi(x) < (1{\cdot}0005)x^{\frac{1}{2}}/\log x < 2x^{\frac{1}{2}}/\log x.$$

In the first alternative, we have what we want at once, and we have only to consider the second. Now, from (24) and (26), the integral in (24) being positive,

$$\Pi_0(x)-\mathrm{li}\,x > x^{\frac{1}{2}}L(y)-\log 2 \geqslant x^{\frac{1}{2}}-\log 2,$$

and so, from the second alternative,

$$\pi(x)-\operatorname{li} x > x^{\frac{1}{2}}-\log 2-2x^{\frac{1}{2}}/\log x > 0,$$

as desired.

17. Let X_1 and G be, as in § 4,

$$X_1 = \exp\exp\exp(7{\cdot}703), \qquad G = X_1^3.$$

Since we cannot use O's in connexion with numerical bounds, we shall use ϑ's (ϑ', ϑ_1, etc.) to denote numbers, possibly complex, satisfying $|\vartheta| \leqslant 1$. They will in general not be the same from one occurrence to the next, but where more than one occurs in the same expression we distinguish them.

Let $y \geqslant G$, and let λ be any real number satisfying†

$$|\lambda| \leqslant G \; (\leqslant y).$$

Consider‡ the function $F(y,\lambda)$ defined by

$$F(y,\lambda) = \int_{\frac{1}{2}y}^{\infty} [-L(t)]te^{E}\,dt, \tag{30}$$

$$E = E(t,y,\lambda) = -\lambda it-\tfrac{1}{2}(t-y)^2/y. \tag{31}$$

We have the following result.

LEMMA 15. *Write* $b = \beta-\frac{1}{2}$, $r = \rho-\frac{1}{2}-i\lambda = b+i(\gamma-\lambda)$. *Then, subject to* $y \geqslant G \geqslant |\lambda|$,

$$F(y,\lambda) = \sum_{\rho} U(\rho), \tag{32}$$

where

$$U(\rho) = (2\pi y)^{\frac{1}{2}}e^{(r+\frac{1}{2}r^2)y}\left(\frac{1}{\rho}+\frac{400\vartheta}{\gamma^2 y}\right)+\vartheta'\frac{e^{-\frac{1}{10}y}}{\gamma^2}. \tag{33}$$

The proof of this is rather long, and we break it up into two subsidiary lemmas, A and B, and a short final deduction from them. We have among other things to show that the series (27) for L can be integrated term by term in (30): this involves a limit-process $T \to \infty$ for fixed y, λ. The parts of Lemmas A, B dealing with this use O's, which are accordingly uniform in the ρ (or γ), but not in the 'fixed' y, λ; the K's similarly are independent of ρ, γ, but not of y, λ.

LEMMA A. *For* $u_1(\rho,t)$, *defined by* (28), *we have*

$$\int_{T}^{\infty} u_1\,te^{E}\,dt = O\left(\frac{e^{-KT^2}}{\gamma^2}\right), \tag{34}$$

$$\int_{\frac{1}{2}y}^{\infty} u_1\,te^{E}\,dt = \frac{(2\pi y)^{\frac{1}{2}}}{\rho}e^{(r+\frac{1}{2}r^2)y}+\frac{2\vartheta e^{-\frac{1}{8}y}}{\gamma^2}. \tag{35}$$

† These conditions hold throughout the rest of the paper. Note that G is so large that any inequalities like $100y^{10}e^{-y/8} < e^{-y/10}$ that occur in the run of our argument will be true when they are 'true for large y'.

‡ The introduction of $F(y,\lambda)$ is the idea given me by Professor J. E. Littlewood.

LEMMA B. *For $u_2(\rho,t)$, defined by (28), we have*

$$\int_T^\infty u_2\, te^E\, dt = O\left(\frac{e^{-KT^2}}{\gamma^2}\right), \tag{36}$$

$$\int_{\frac{1}{4}y}^\infty u_2\, te^E\, dt = \frac{324\vartheta}{y}(2\pi y)^{\frac{1}{2}}\frac{e^{(r+\frac{1}{2}r^2)y}}{\gamma^2}+\frac{9\vartheta'}{\gamma^2}e^{-\frac{1}{8}y}. \tag{37}$$

18. In Lemmas A, B we may, by symmetry (since λ can take either sign), suppose without loss of generality that $\gamma > 0$.

Proof of Lemma A. We have

$$u_1\, te^E = \frac{1}{\rho}e^{\gamma it}f(t), \qquad f(t) = e^{(\beta-\frac{1}{2}-i\lambda)t-\frac{1}{2}(t-y)^2/y}. \tag{38}$$

For $t \geqslant T$,

$$\int_T^t u_1\, te^E\, dt = \frac{1}{\rho}\left[\frac{e^{\gamma it}}{\gamma i}f(t)\right]_T^t - \frac{1}{\rho}\int_T^t \frac{e^{\gamma it}}{\gamma i}f'(t)\, dt.$$

As $T\to\infty$ we have, uniformly in $t \geqslant T$, $f(t), f'(t) = O(e^{-KT^2})$. It follows that $\int_T^\infty u_1\, te^E\, dt$ exists, and that it is $O(\gamma^{-2}e^{-KT^2})$; and this is (34).

Next,
$$\int_{-\infty}^\infty u_1\, te^E\, dt = \frac{1}{\rho}\int_{-\infty}^\infty e^{rt-\frac{1}{2}(t-y)^2/y}\, dt = \frac{(2\pi y)^{\frac{1}{2}}}{\rho}e^{(r+\frac{1}{2}r^2)y}. \tag{39}$$

Again,

$$\int_{-\infty}^{\frac{1}{4}y} u_1\, te^E\, dt = \frac{1}{\rho}\left[\frac{e^{\gamma it}}{\gamma i}f(t)\right]_{-\infty}^{\frac{1}{4}y} - \frac{1}{\rho}\int_{-\infty}^{\frac{1}{4}y}\frac{e^{\gamma it}}{\gamma i}f'(t)\, dt = J_1+J_2, \text{ say.} \tag{40}$$

We have $f(-\infty) = 0$ and $|f(\frac{1}{4}y)| \leqslant e^{|b|\frac{1}{4}y-\frac{9}{32}y} < e^{-\frac{1}{8}y}$, so that

$$J_1 = \vartheta\gamma^{-2}e^{-\frac{1}{8}y}.$$

Also, for $-\infty < t \leqslant \frac{1}{4}y$,

$$|f'(t)| = |(\beta-\tfrac{1}{2}-i\lambda)-(t-y)/y|\,e^{bt-\frac{1}{2}(t-y)^2/y}$$
$$\leqslant (\tfrac{1}{2}+|\lambda|+|t-y|)e^{bt-\frac{1}{2}(t-y)^2/y}.$$

Writing $u = |t-y| = y-t$, and observing that $u \geqslant \frac{3}{4}y$ and

$$\tfrac{1}{2}+|\lambda| < 2y \leqslant \tfrac{8}{3}u,$$

we have
$$|f'| \leqslant 4ue^{b(y-u)-\frac{1}{2}u^2/y} \leqslant 12(by+u)e^{by}e^{-bu-\frac{1}{2}u^2/y},$$

$$|J_2| \leqslant 12\gamma^{-2}e^{by}\int_{\frac{3}{4}y}^\infty (by+u)e^{-bu-\frac{1}{2}u^2/y}\, du = 12\gamma^{-2}ye^{(\frac{1}{4}b-\frac{9}{32})y}$$
$$\leqslant 12\gamma^{-2}ye^{-\frac{5}{32}y} < \gamma^{-2}e^{-\frac{1}{8}y}.$$

So $J_1+J_2 = 2\vartheta\gamma^{-2}e^{-\frac{1}{4}y}$, which, combined with (39) and (40), gives (35) and completes the proof of Lemma A.

19. *Proof of Lemma B.* Let $t \geqslant T$ and $T \to \infty$. We have $|\rho t-v|^2 \geqslant \gamma^2t^2$, and so, from (28),

$$\left|\int_T^t u_2 te^E\,dt\right| \leqslant \int_0^\infty e^{-v}\,dv\,.\,\gamma^{-2}\int_T^t t^{-2}te^{bt-\frac{1}{2}(t-y)^2/y}\,dt$$
$$= O(\gamma^{-2}e^{-KT^2}),$$

and $\int_T^\infty$ exists and satisfies (36).

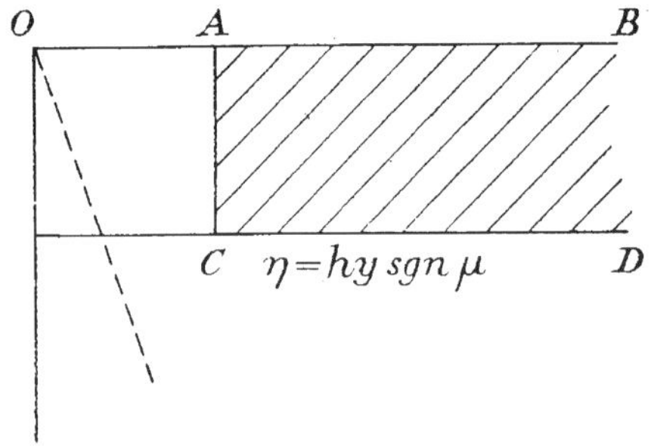

FIG. 1.

Next we have
$$\int_{\frac{1}{4}y}^\infty u_2 te^E\,dt = \int_0^\infty e^{-v}H(\rho,v)\,dv,$$

$$H = H(\rho,v) = \int_{\frac{1}{4}y}^\infty e^{\rho t-\frac{1}{2}(t-y)^2/y}\frac{t\,dt}{(\rho t-v)^2}. \tag{41}$$

We prove (37) of Lemma B by showing that for each v of $(0,\infty)$ H is of the form of the right-hand side of (37) (noting that $\int_0^\infty e^{-v}\,dv = 1$).

We deform the t-contour $\frac{1}{4}y$ to ∞, or AB, in a manner independent of v, as follows. Let $\mu = \gamma-\lambda$, $h = \min(2, |\mu|)$. With $t = \xi+i\eta$ we take a line $\eta = hy\operatorname{sgn}\mu$ $(= \pm hy)$, and replace the original path AB by ACD of the figure (drawn for the worst case, namely $\operatorname{sgn}\mu = -1$). First, the pole $t = v/\rho$ is outside the shaded area, so that† $H = \int_{ACD}$. For the pole is on a line (dotted in the figure) whose slope (tangent) is $-\gamma/\beta$; this is downward‡ and steeper absolutely than $\gamma_1 > 14$, steeper, therefore, than OC in the unfavourable case (of the figure) when C is below A.

† The integrand is uniformly $O(e^{-K\xi^2})$ as $t \to \infty$ in the shaded area.
‡ Recall that in this proof we have $\gamma > 0$.

Taking the integral for H along ACD, then, we have

$$|\rho t-v| \geqslant |\operatorname{im}(\rho t-v)| = |\gamma\xi+\beta\eta|$$
$$\geqslant \gamma\xi-1.2y \geqslant \tfrac{1}{3}\gamma\xi$$

(since $\gamma > 14$, $\xi \geqslant \frac{1}{4}y$). So

$$|H| \leqslant 9\gamma^{-2} \int_{ACD} |e^{E_1}\xi^{-2}t\,dt|, \tag{42}$$

where $$E_1 = rt-\tfrac{1}{2}(t-y)^2/y, \qquad r = b+i\mu, \tag{43}$$

and, as alternative forms,

$$E_1 = -\tfrac{1}{2}y+(r+1)t-\tfrac{1}{2}t^2/y = (r+\tfrac{1}{2}r^2)y-\tfrac{1}{2}[t-(r+1)y]^2/y, \tag{44}$$
$$\operatorname{re} E_1 = -\tfrac{1}{2}y+(b+1)\xi-\mu\eta-\tfrac{1}{2}\xi^2/y+\tfrac{1}{2}\eta^2/y. \tag{45}$$

On AC we have $\xi = \frac{1}{4}y$, $\eta = \sigma hy\operatorname{sgn}\mu$, $0 \leqslant \sigma \leqslant 1$,

$$\operatorname{re} E_1 = [-\tfrac{1}{2}+\tfrac{1}{4}(b+1)]y-\tfrac{1}{32}y-\sigma hy(|\mu|-\tfrac{1}{2}\sigma h),$$

and since the last term is non-positive and $b+1 < \frac{3}{2}$, $\operatorname{re} E_1 < -\frac{1}{7}y$, and

$$\int_{AC} |e^{E_1}\xi^{-2}t\,dt| \leqslant e^{-\frac{1}{7}y}(\tfrac{1}{4}y)^{-2}OC.AC < \tfrac{1}{2}e^{-\frac{1}{8}y}. \tag{46}$$

For CD we have two cases.

Case (i). $|\mu| \leqslant 2$. Here $\eta = \mu y$ $(= \operatorname{im} r.y)$,

$$E_1 = (r+\tfrac{1}{2}r^2)y-\tfrac{1}{2}[\xi-(b+1)y]^2/y,$$

$$\int_{CD} |e^{E_1}\xi^{-2}t\,dt| \leqslant |e^{(r+\frac{1}{2}r^2)y}| \int_{\frac{1}{4}y}^{\infty} e^{-\frac{1}{2}[\xi-(b+1)y]^2/y}\{\xi^{-2}(2y+\xi)\}\,d\xi.$$

The curly bracket is greatest for $\xi = \frac{1}{4}y$, and it is then $36y^{-1}$. Taking this outside, and then the integral from $-\infty$ to ∞, we find that $\int_{CD}$ is at most $(2\pi y)^{\frac{1}{2}}|e^{(r+\frac{1}{2}r^2)y}|36y^{-1}$.

Combining this with (42) and (46) we have, in case (i),

$$|H| \leqslant 9\gamma^{-2}(e^{-\frac{1}{8}y}+(2\pi y)^{\frac{1}{2}}|e^{(r+\frac{1}{2}r^2)y}|36y^{-1}). \tag{47}$$

Case (ii). $|\mu| > 2$. Here CD has $\eta = 2y\operatorname{sgn}\mu$. We have from (45)

$$\operatorname{re} E_1 = -\tfrac{1}{2}y+(b+1)\xi-2|\mu|y-\tfrac{1}{2}\xi^2/y+2y$$
$$\leqslant -\tfrac{5}{2}y+\tfrac{3}{2}\xi-\tfrac{1}{2}\xi^2/y, \quad \text{since } |\mu| > 2,$$
$$= -\tfrac{1}{4}y-\tfrac{1}{4}\xi^2/y-\tfrac{1}{4}(\xi-3y)^2/y \leqslant -\tfrac{1}{4}y-\tfrac{1}{4}\xi^2/y.$$

As before, $|\xi^{-2}t| \leqslant 36y^{-1}$, and so

$$\int_{CD} |e^{E_1}\xi^{-2}t\,dt| \leqslant 36y^{-1} \int_{-\infty}^{\infty} e^{-\frac{1}{4}y-\frac{1}{4}\xi^2/y}\,d\xi \leqslant \tfrac{1}{2}e^{-\frac{1}{8}y}.$$

From this and (46), $|H| \leqslant 9\gamma^{-2}e^{-\frac{1}{8}y}$, and (47) is true also in case (ii). *A fortiori* H is of the form of the right-hand side of (37), and, as we observed above, this proves (37). This completes the proof of Lemma B.

20. We now have Lemmas A and B (in which γ now is not restricted to be positive), and can take up Lemma 15. By Lemma 13 and (30) we have

$$F = \lim_{T\to\infty} \int_{\frac{1}{2}y}^{T} [-L(t)]te^{E}\,dt,$$

since $\operatorname{re}E = -\frac{1}{2}(t-y)^2/y < -Kt^2$ as $t \to \infty$. In $\int_{\frac{1}{2}y}^{T}$ we may substitute $-L(t) = \sum u_1 + \sum u_2$ from (27) and integrate term by term, since the two series are boundedly convergent. If we then replace T by ∞ in each term, the error is

$$\sum \int_{T}^{\infty} u_1\,te^{E}\,dt + \sum \int_{T}^{\infty} u_2\,te^{E}\,dt = O(e^{-KT^2}\sum \gamma^{-2}),$$

by (34) from Lemma A and (36) from Lemma B, and this tends to 0 as $T \to \infty$. Hence

$$F(y,\lambda) = \sum_{\rho} U(\rho),$$

where

$$U(\rho) = \int_{\frac{1}{2}y}^{\infty} u_1\,te^{E}\,dt + \int_{\frac{1}{2}y}^{\infty} u_2\,te^{E}\,dt,$$

and when we substitute from (35) and (37) (and make a couple of small adjustments) we arrive at Lemma 15.

21. LEMMA 16.† *For $y \geqslant G \geqslant \lambda \geqslant 0$ we have*

$$F(y,\lambda) = \sum_{|\gamma-\lambda|\leqslant 2} \frac{1}{\rho}(2\pi y)^{\frac{1}{2}}e^{(r+\frac{1}{2}r^2)y}\left(1+\frac{30\vartheta}{y}\right)+\vartheta'.$$

Since $400|\rho|/(\gamma^2 y) < 30/y < 1$, Lemma 15 shows that F is equal to something of the form of the $\sum$ in the lemma, *plus*

$$2\vartheta_1 \sum_{|\gamma-\lambda|>2} \frac{(2\pi y)^{\frac{1}{2}}|e^{r+\frac{1}{2}r^2)y}|}{|\gamma|} + \vartheta_2\,e^{-\frac{1}{10}y}\sum \frac{1}{\gamma^2}. \tag{48}$$

When $|\gamma-\lambda| > 2$ we have

$$\operatorname{re}(r+\tfrac{1}{2}r^2) = b+\tfrac{1}{2}b^2-\tfrac{1}{2}(\gamma-\lambda)^2 < -\tfrac{1}{4}(\gamma-\lambda)^2-\tfrac{3}{8},$$

† (i) From now on λ is non-negative (we normalized in the *proof* above to $\gamma > 0$ and λ of both signs). (ii) The ϑ, of course, varies with the term it occurs in.

and also $|\gamma/(\gamma-\lambda)| \leqslant 1+\lambda < 2y$. The first term in (48) is therefore

$$\vartheta \sum_{\rho} \frac{4y}{\gamma^2}(2\pi y)^{\frac{1}{2}}\{|\gamma-\lambda| e^{-\frac{1}{4}(\gamma-\lambda)^2}\} e^{-\frac{1}{8}y} = \tfrac{1}{2}\vartheta, \tag{49}$$

since the curly bracket is less than (say) 10. Lemma 16 follows.

22. We are now in a position to develop the consequences of (NH), the negation of the hypothesis (H). To assume (NH) is to assume that a zero $\beta_0+i\gamma_0$ exists (with γ_0 positive, by the symmetry) satisfying

$$\text{(NH)} \quad \begin{cases} b_0 = \beta_0-\frac{1}{2} > X_1^{-3}\log^{-2}X_1 = B, \\ 0 < \gamma_0 < X_1^3 = G. \end{cases}$$

We begin by supposing that (for an undetermined Y) the relation '$L(y) \geqslant 1$ for some y' occurring in Lemma 14 is *not* satisfied for the range $G \leqslant y \leqslant 4Y$; that is, we suppose that†

$$L(\eta) < 1 \quad \text{for } G \leqslant \eta \leqslant 4Y. \tag{50}$$

By arguing from the pair of hypotheses (NH) and (50) we find ourselves able to produce a Y_0 (actually G^{10}) such that, if the Y of (50) is Y_0, there is a contradiction. Then (NH) implies (i) that (50) is false for $Y = Y_0$; so (ii) that for *some* y of the range $G \leqslant y \leqslant 4Y_0$ we must have $L(y) \geqslant 1$, when Lemma 14 (with $4Y_0$ for $\log X$) gives $\pi(x)-\operatorname{li} x > 0$ for some x of

$$2 \leqslant x < X = \exp(4Y_0) \ [= \exp(4G^{10})].$$

This, then, is what results from (NH), and since the X is greater than the X_1 derived from (H), it is our final number.

23. Lemma 17. *If* [*in accordance with* (50)] $L(\eta) < 1$ *for* $G \leqslant \eta \leqslant 4Y$, *then for* $4G \leqslant y \leqslant Y$, $0 \leqslant \lambda \leqslant G$, *we have*

$$|F(y,0)| \leqslant 1, \tag{51}$$

$$|F(y,\lambda)| < 6Y^{\frac{1}{2}}+4. \tag{52}$$

When $\lambda = 0$, the condition $|\gamma-\lambda| \leqslant 2$ is vacuous, and (51) is a case of Lemma 16.

Next, since $L(t)$ is real for $t > 0$, we have, for λ of $0 \leqslant \lambda \leqslant G$, by (30) and (31),

$$-F(y,\lambda) = \int_{\frac{1}{4}y}^{\infty} t\, L(t)(\cos\lambda t - i\sin\lambda t)e^{-\frac{1}{2}(t-y)^2/y}\, dt$$

$$= \mathscr{R} - i\mathscr{I}, \text{ say.} \tag{53}$$

† This means 'for *all* η of the range', and similar interpretations are intended wherever we do not explicitly have 'some'. This being the usual interpretation, we may seem to be labouring the obvious, but the distinctions of 'all' and 'some' are very vital, and complicated by ranges (those in Lemma 17) that 'look' alike, but are not quite so.

Consider the four expressions

$$-F(y,0)\,\frac{\pm\mathscr{R}}{\pm\mathscr{I}} = \left(\int_{\frac{1}{2}y}^{4Y} + \int_{4Y}^{\infty}\right) tL(t)\left\{1\,\frac{\pm\cos\lambda t}{\pm\sin\lambda t}\right\} e^{-\frac{1}{2}(t-y)^2/y}\,dt$$

$$= J_1+J_2, \text{ say.} \tag{54}$$

In J_2 we substitute $|L(t)| \leqslant 4e^{\frac{1}{2}t}$ from Lemma 13, and, remembering that $4G \leqslant y \leqslant Y$, we obtain

$$|J_2| \leqslant \int_{4y}^{\infty} t\,.\,4e^{\frac{1}{2}t}\,.\,2e^{-\frac{1}{2}(t-y)^2/y}\,dt$$

$$= 8e^{-\frac{1}{2}y} \int_{4y}^{\infty} te^{-\frac{1}{2}(t-4y)-\frac{1}{2}(t-2y)^2/y}\,dt < 1. \tag{55}$$

In J_1 we have $G \leqslant t \leqslant 4Y$, and so $L(t) < 1$ by the hypothesis (50); hence, *the curly bracket in* (54) *being in all four cases non-negative*, we have, *algebraically*,

$$J_1 \leqslant \int_{\frac{1}{2}y}^{4Y} 2t\,e^{-\frac{1}{2}(t-y)^2/y}\,dt \leqslant \int_{-\infty}^{\infty} (2y+2|t-y|)e^{-\frac{1}{2}(t-y)^2/y}\,dt$$

$$= 2y(2\pi y)^{\frac{1}{2}}+8y < 6y^{\frac{3}{2}} \leqslant 6Y^{\frac{3}{2}}. \tag{56}$$

Since $|F(y,\lambda)| \leqslant |\mathscr{R}|+|\mathscr{I}|$, from (53), *and since* $|\mathscr{R}|+|\mathscr{I}|$ *is, for each* y, *one* (*varying with* y) *of the four combinations* $\pm\mathscr{R}\pm\mathscr{I}$, we have, from (54) to (56),

$$|F(y,\lambda)| \leqslant 6Y^{\frac{3}{2}}+1+2|F(y,0)| < 6Y^{\frac{3}{2}}+4,$$

the desired result.

24. We now combine Lemmas 16 and 17, and take $Y = G^{10}$ (Y has this meaning from now on). The upshot is that, subject to (NH), and to the further 'hypothesis'

$$(\mathrm{H}_1) \qquad L(\eta) < 1 \quad (G \leqslant \eta \leqslant 4Y),$$

we have, for λ, y satisfying

$$0 \leqslant \lambda \leqslant G, \tag{57}$$

$$4G \leqslant y \leqslant Y, \tag{58}$$

and for *some* set of ϑ's,

$$\left|\sum_{|\gamma-\lambda|\leqslant 2} \frac{1}{\rho} e^{(r+\frac{1}{2}r^2)y}\left(1+\frac{30\vartheta}{y}\right)\right| < (2\pi y)^{-\frac{1}{2}}(6Y^{\frac{3}{2}}+4+1) < \tfrac{1}{2}Y^{\frac{3}{2}}, \tag{59}$$

where $r = b+i(\gamma-\lambda)$.†

We now take $\lambda = \gamma_0$, where γ_0 is the number in (NH), § 22. [λ duly

† We go on to derive a contradiction from this state of things, as a result of which one of (NH) and (H_1) must be false.

satisfies (57).] So from (59) with $\lambda = \gamma_0$ and so $r = b+i(\gamma-\gamma_0)$,

$$\mathrm{re} \sum_{|\gamma-\gamma_0|\leqslant 2} \frac{i}{\rho}\left(1+\frac{30\vartheta}{y}\right)\exp[(b+\tfrac{1}{2}b^2)y-\tfrac{1}{2}(\gamma-\gamma_0)^2y+i(\gamma-\gamma_0)(1+b)y] < \tfrac{1}{2}Y^{\frac{3}{2}}. \tag{60}$$

We need to know an upper bound for the number N of terms in the sum; Lemma 1 (iv) with $h = 4$, $T = \gamma_0-2$, gives

$$N < \frac{6}{2\pi}\log\gamma_0+8{\cdot}7 < \log G. \tag{61}$$

We proceed to choose, without violating (58), a y $(= y_0)$ for which the real parts of the terms in the sum in (60) are all positive. In the first place, since $\gamma > 14$, the argument of any factor i/ρ lies between $\pm 5°$, and that of any $1+30\vartheta/y$ between $\pm 1°$. Now by Dirichlet's theorem there exists a y_0 satisfying

$$Y^{\frac{1}{2}} \leqslant y_0 \leqslant Y^{\frac{1}{2}}5^N, \tag{62}$$

and such that, for each of the N γ's satisfying $|\gamma-\gamma_0| \leqslant 2$,

$$|(\gamma-\gamma_0)(1+b)y_0-2\pi k| < \tfrac{1}{5}\,.\,2\pi,$$

where k is an integer. Further, with $Y = G^{10}$ and N satisfying (61), $y = y_0$ [satisfying (62)] duly satisfies (58). With $y = y_0$ the arguments of all the terms in the sum in (60) lie between $\pm 80°$; hence the real parts of all the terms are positive, and the sum of them is at least as great as any one term. Choosing the one term to be that with $\gamma = \gamma_0$, we have

$$\frac{\gamma_0}{\gamma_0^2+\beta_0^2}e^{(b_0+\frac{1}{2}b_0^2)y_0}\left\{1-\frac{30}{y_0}\right\} < \tfrac{1}{2}Y^{\frac{3}{2}},$$

$$e^{b_0y_0} < \tfrac{1}{2}(\gamma_0+1/\gamma_0)\{1-30/y_0\}^{-1}Y^{\frac{3}{2}} < \gamma_0 Y^{\frac{3}{2}} \leqslant GY^{\frac{3}{2}},$$

$$e^{By_0} < Y^{\frac{3}{2}}G = G^{16}.$$

With $B = X_1^{-3}\log^{-2}X_1$, $X_1^3 = G$, this contradicts $y_0 \geqslant Y^{\frac{1}{2}} = G^5$ of (62).† So either (NH) is false [and (H) true] or (H_1) is false. In the first case $\pi(x)-\mathrm{li}\,x > 0$ for an $x < X_1$; in the second it happens for an x of

$$2 \leqslant x < X_2 = \exp(4Y) = \exp(4G^{10}) = \exp(4X_1^{30}).$$

Since $X_2 > X_1 = \exp\exp\exp(7{\cdot}703)$, we conclude finally that $\pi(x)-\mathrm{li}\,x > 0$ for some $x < X$, where

$$X = \exp\exp\exp\exp(7{\cdot}705) < 10^{10^{10^{10^3}}}.$$

† There is a great deal to spare at this point: see the footnote on p. 50.

REFERENCES

1. J. E. Littlewood, 'Sur la distribution des nombres premiers', *Comptes Rendus*, 158 (1914), 263–6.
2. A. E. Ingham, *The Distribution of Prime Numbers*, Cambridge Mathematical Tracts, No. 30 (1932).
3. S. Skewes, 'On the difference $\pi(x) - \text{li}\, x$ (I)', *J. London Math. Soc.* 8 (1933), 277–83.
4. A. E. Ingham, 'A note on the distribution of primes', *Acta Arith.* 1 (1936), 201–11.
5. R. J. Backlund, 'Über die Nullstellen der Riemannschen Zetafunktion', *Acta Math.* 41 (1918), 345–75.
6. Jahnke-Emde, *Funktionentafeln* (Teubner).
7. J. I. Hutchinson, 'On the roots of the Riemann zeta function', *Trans. American Math. Soc.* 27 (1925), 49–60.
8. E. C. Titchmarsh, 'The zeros of the zeta-function', *Proc. Royal Soc.* A, 151 (1935), 234–55.
9. —— 'The zeros of the zeta-function', ibid. 157 (1936), 261–3.

University of Cape Town.

12.13 C. B. Haselgrove (1958)

A Disproof of a Conjecture of Pólya

In this paper, Haselgrove disproves the following conjecture of Pólya, which is intricately connected to the Riemann hypothesis.

Conjecture 12.1 (Pólya's Conjecture). *The function* $L(x) := \sum_{n \leq x} \lambda(n)$ *satisfies* $L(x) \leq 0$ *for* $x \geq 2$*, where* $\lambda(n)$ *is Liouville's function.*

This conjecture had been verified for all $x \leq 800000$. The conjecture implies the Riemann hypothesis, so in his disproof Haselgrove assumes the hypothesis to be valid. The disproof is based on the result of Ingham; that is, if the Riemann hypothesis is true, then

$$A_T^*(u) \leq \limsup A_T^*(u) \leq \limsup A(u).$$

Here the functions $A(u)$ and $A_T^*(u)$ are defined as

$$\begin{aligned} A(u) &:= e^{-\frac{u}{2}} L(e^u), \\ A_T^*(u) &:= \alpha_0 + 2\Re \left(\sum \left(1 - \frac{\gamma_n}{T} \right) \alpha_n e^{iu\gamma_n} \right), \end{aligned}$$

where the sum ranges over the values of n for which $0 < \gamma_n < T$ and γ_n is the imaginary part of the nth nontrivial zero, ρ_n, of the Riemann zeta function. Furthermore,

$$\alpha_0 := \frac{1}{\zeta\left(\frac{1}{2}\right)} \quad \text{and} \quad \alpha_n := \frac{\zeta(2\rho_n)}{\rho_n \zeta'(\rho_n)}.$$

Note that, if there exist T and u that give $A_T^*(u) > 0$, then the conjecture of Pólya is false. Haselgrove finds that $T = 1000$ and $u = 831.847$ give $A_T^*(u) = 0.00495$, disproving the conjecture. This is based on the calculation of the first 1500 nontrivial zeros of $\zeta(s)$. [96]

A DISPROOF OF A CONJECTURE OF PÓLYA

C. B. HASELGROVE

Let $\lambda(n)$ be Liouville's function defined by

$$\lambda(n) = (-1)^{\nu},$$

where ν is the number of prime factors of n, repeated factors being counted according to their multiplicity. Alternatively, $\lambda(n)$ may be defined by the relation

$$\zeta(2s)/\zeta(s) = \sum_{n=1}^{\infty} \lambda(n)\, n^{-s},$$

where $\zeta(s)$ is the zeta function of Riemann.

Pólya [5] conjectured that the sum

$$L(x) = \sum_{n \leqslant x} \lambda(n)$$

is negative or zero for all $x \geqslant 2$. The author has verified that this conjecture is true for $x \leqslant 250{,}000$ (Royal Society Depository for Unpublished Mathematical Tables, No. 65)†.

A number of results have been deduced on the assumption of the truth of Pólya's conjecture, in particular that all the complex zeros of $\zeta(s)$ lie on the line $s = \frac{1}{2}+it$, with t real (the Riemann hypothesis), and that all these zeros are simple. These and other results are described by Ingham [2] in a paper in which he proves a further consequence, namely that the imaginary parts of some of the zeros of the zeta function above the real axis are linearly dependent (with rational integral multipliers).

In the course of his paper Ingham, assuming the Riemann hypothesis and the simplicity of the zeros, defines two functions

$$A(u) = e^{-\frac{1}{2}u}\, L(e^u)$$

and

$$A_T^*(u) = \alpha_0 + 2\Re \sum_{0<\gamma_n<T} \left(1-\frac{\gamma_n}{T}\right) \alpha_n\, e^{i\gamma_n u},$$

where $n = 1, 2, \ldots,$ $T > 0$, $\alpha_0 = 1/\zeta(\frac{1}{2})$, γ_n runs through the imaginary parts of the zeros $\rho_n = \frac{1}{2}+i\gamma_n$ of $\zeta(s)$, and $\alpha_n = \zeta(2\rho_n)/\rho_n\, \zeta'(\rho_n)$. He then proves that

$$\underline{\lim}\, A(u) \leqslant \underline{\lim}\, A_T^*(u) \leqslant \overline{\lim}\, A_T^*(u) \leqslant \overline{\lim}\, A(u),$$

where the limits are taken as $u \to \infty$ with T fixed.

This suggests a method for investigating the truth of Pólya's conjecture. It follows from Dirichlet's theorem on Diophantine approximation, or from the fact that the function $A_T^*(u)$ is almost periodic, that $\underline{\lim}\, A_T^*(u) \leqslant A_T^*(u) \leqslant \overline{\lim}\, A_T^*(u)$. Thus if we can find T, u such that

† D. H. Lehmer has verified the conjecture for $x \leqslant 600{,}000$ (private communication).

[MATHEMATIKA 5 (1958), 141–145]

$A_T^*(u) > 0$, it will follow that $\overline{\lim} A_T^*(u) > 0$ and hence that $L(e^u) > 0$ for some u, *i.e.* that Pólya's conjecture is false. Such values of T and u have in fact been found. Since the failure of the Riemann hypothesis would in any case imply the falsehood of Pólya's conjecture, it is immaterial that his argument rests on the assumption of the Riemann hypothesis.

Now that electronic computers are available it is possible to calculate $\zeta(s)$ over large ranges with considerable accuracy. Methods of calculating have been described by Lehmer [3], and by Haselgrove and Miller [1], who give tables of $\zeta(s)$ computed on the EDSAC I at the University Mathematical Laboratory, Cambridge, and the Mark I computer at Manchester University. In the course of computing the tables the first 1500 numbers γ_n were evaluated with an error of at most 3×10^{-8}. These values were used to find $\zeta(2\rho_n)$ and $\zeta'(\rho_n)$ and hence $\zeta(2\rho_n)/\rho_n \zeta'(\rho_n)$. Table I gives $\left|\dfrac{\zeta(2\rho_n)}{\rho_n \zeta'(\rho_n)}\right|$ and $\dfrac{1}{\pi}\,\mathrm{ph}\left(\dfrac{\zeta(2\rho_n)}{\rho_n \zeta'(\rho_n)}\right)$ for the first 50 zeros above the real axis. These quantities should be accurate to within 1 in the last decimal given. Here $\mathrm{ph}(z)$ denotes the argument (or phase) of the complex number z.

In order to find a value of u such that $A_T^*(u) > 0$ we observe that the largest contributions come from the first, second and seventh zeros. We therefore select for trial such positive or negative values of u as make the contributions of these three zeros positive and as large as possible. It is desirable to find a value of u which is not too large so that errors in γ_n do not cause large errors in $\gamma_n u$. (On the other hand smaller values of u require larger values of T which increases the amount of computation necessary.) I am indebted to Mr. J. Leech for finding the set of values

$$u = 28{\cdot}148 + l \times 139{\cdot}5794 + m \times 33{\cdot}7836 + n \times 17{\cdot}3363,$$

where l, m and n are integers satisfying $|l| \leqslant 50$, $|m| \leqslant 3$, $|n| \leqslant 1$. It is found that for $u = 831{\cdot}847$ (corresponding nearly to $l = 6$, $m = -1$ and $n = 0$) and $T = 1000$ the sum $A_T^*(u)$ is positive. This value of u is not claimed to be the smallest for which this occurs. For $T = 1000$ the sum involves 649 zeros. In Table II $A_T^*(u)$ is given for $T = 1000$ and $u = 831{\cdot}8\ (0{\cdot}001)\ 831{\cdot}859$.

This result would lead us to suspect that $L(e^u)$ becomes positive in the neighbourhood of $u = 831{\cdot}847$, although there is no proof that this is where the change of sign occurs. Some idea of the similarity of behaviour of $A_T^*(u)$ and $e^{-\frac{1}{2}u} L(e^u)$ is given by Fig. 1, for the range $10{\cdot}70 \leqslant u \leqslant 10{\cdot}85$ with $T = 200$. It is well known that $L(x)$ attains the value of -2 at $x = e^u = 48{,}512$.

The numbers γ_n and $|\zeta'(\rho_n)|$ were calculated and checked on the EDSAC I. The numbers $|\alpha_n|$ and $(1/\pi)\,\mathrm{ph}\,\alpha_n$ were calculated on both the EDSAC I and the Mark I; slight discrepancies were attributable to rounding errors in the EDSAC I results [these errors were known to occur only in

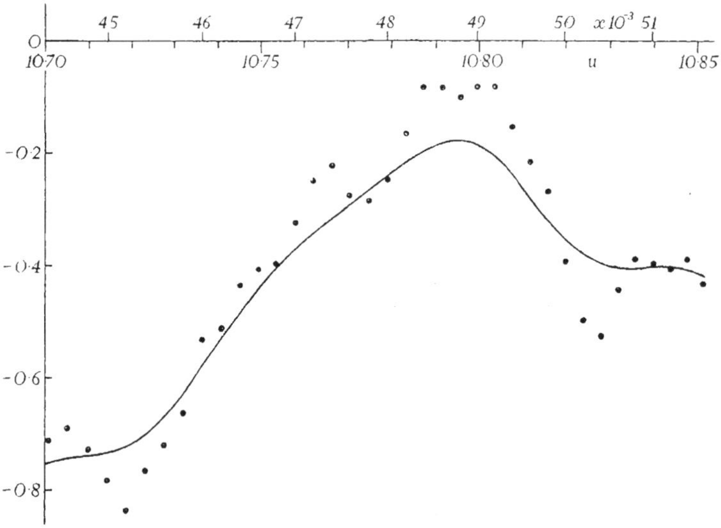

Fig. 1. The continuous line represents $A_T(u)$ with $T = 200$ and the points are values of $e^{-\frac{1}{2}u}L(e^u) = x^{-\frac{1}{2}}L(x)$ for $x = 44{,}400\ (200)\ 51{,}600$.

the calculation of $\zeta(2\rho_n)$]. The function $A_T^*(u)$ was also calculated on both machines for $u = 831{\cdot}845\ (0{\cdot}001)\ 831{\cdot}859$ with

$$T = 10{,}000\pi/32 = 981{\cdot}487\ldots.$$

The agreement of the results was within the limits set by the same rounding errors. We are thus led to the conclusion that Pólya's conjecture is false.

The methods described above may be applied to several similar problems. In particular we mention Mertens' hypothesis (Mertens [**4**], Ingham [**2**]) that

$$|M(x)| < x^{\frac{1}{2}},$$

where $M(x) = \sum_{n \leqslant x} \mu(n)$ and $\mu(n)$ is the function of Möbius, and Turán's suggestion [**6**] that

$$\sum_{n \leqslant x} \frac{\lambda(n)}{n} \geqslant 0 \quad (x \geqslant 1).$$

The functions $A_T^*(u)$ for these two problems would be

$$A_T^*(u) = 2\Re \sum_{0<\gamma_n<T} \left(1 - \frac{\gamma_n}{T}\right) \frac{1}{\rho_n \zeta'(\rho_n)} e^{i\gamma_n u}$$

and $$A_T^*(u) = -\frac{1}{\zeta(\frac{1}{2})} + 2\Re \sum_{0<\gamma_n<T} \left(1 - \frac{\gamma_n}{T}\right) \frac{\zeta(2\rho_n)}{(\rho_n - 1)\zeta'(\rho_n)} e^{i\gamma_n u}$$

respectively. In order to disprove Mertens' hypothesis it would be sufficient to find T, u such that $|A_T^*(u)| > 1$. In the case of Turán's sum it would be sufficient to find T, u for which the corresponding $A_T^*(u) < 0$.

TABLE I

$\alpha_0 = -0{\cdot}68476524$

n	γ_n	$\lvert\alpha_n\rvert$	$\frac{1}{\pi}$ ph α_n
1	14·13472513	0·17371523	−0·6474507
2	21·02203961	0·03476036	−0·3551160
3	25·01085756	0·01556708	−0·4134004
4	30·42487610	0·01297375	−0·0833945
5	32·93506159	0·01785754	−0·7548282
6	37·58617814	0·01288878	−0·7272660
7	40·91871901	0·03150963	−0·6028458
8	43·32707326	0·01230750	−0·8272898
9	48·00515087	0·00720805	0·0184893
10	49·77383246	0·02021980	−0·5793702
11	52·97032147	0·00535633	−0·4192135
12	56·44624768	0·00755698	−0·2768936
13	59·34704400	0·01237921	−0·4192967
14	60·83177851	0·00599826	−0·9329956
15	65·11254403	0·00477803	−0·2803717
16	67·07981051	0·00581876	−0·4098320
17	69·54640170	0·00478356	−0·7596008
18	72·06715766	0·00974345	−0·4975657
19	75·70469068	0·00548818	0·0765724
20	77·14484005	0·02312381	−0·5730905
21	79·33737500	0·00252000	−0·6244279
22	82·91038084	0·00387834	−0·3756371
23	84·73549297	0·00235891	−0·3687321
24	87·42527461	0·00384655	−0·4377069
25	88·80911120	0·00525947	−0·8231806
26	92·49189925	0·00138235	−0·4700115
27	94·65134403	0·00848399	−0·2819908
28	95·87063423	0·00570993	−1·0702284
29	98·83119420	0·00198488	−0·3237511
30	101·31785098	0·00179165	−0·4550882
31	103·72553802	0·00634339	−0·4263908
32	105·44662306	0·00704923	−0·7109605
33	107·16861114	0·00167392	−0·7086325
34	111·02953568	0·00541237	0·1826984
35	111·87465895	0·00526590	−0·7576648
36	114·32022105	0·00474366	−0·6894735
37	116·22668017	0·00179214	−0·3943627
38	118·79078298	0·00155757	−0·5521109
39	121·37012475	0·00260999	−0·2066128
40	122·94682964	0·01392294	−0·6669428
41	124·25681830	0·00197417	−0·8638090
42	127·51668401	0·00197639	−0·5442688
43	129·57870397	0·00195627	−0·3872602
44	131·08768873	0·00863392	−0·4815026
45	133·49773700	0·00180997	−0·3578010
46	134·75650989	0·00311216	−0·8288625
47	138·11604194	0·00104793	−0·2091791
48	139·73620909	0·00426514	−0·0393277
49	141·12370727	0·00307670	−1·0484089
50	143·11184585	0·00167694	−0·7713534

TABLE II

$T = 1000$

u	$A_T^*(u)$	u	$A_T^*(u)$	u	$A_T^*(u)$
831·800	−0·43329	831·820	−0·30119	831·840	−0·13583
·801	−0·42140	·821	−0·27534	·841	−0·12063
·802	−0·41040	·822	−0·25347	·842	−0·09590
·803	−0·40181	·823	−0·23640	·843	−0·06610
·804	−0·39640	·824	−0·22333	·844	−0·03705
·805	−0·39382	·825	−0·21269	·845	−0·01395
·806	−0·39287	·826	−0·20305	·846	0·00014
·807	−0·39220	·827	−0·19370	·847	0·00495
·808	−0·39097	·828	−0·18445	·848	0·00265
·809	−0·38918	·829	−0·17512	·849	−0·00328
831·810	−0·38762	831·830	−0·16518	831·850	−0·00950
·811	−0·38723	·831	−0·15397	·851	−0·01404
·812	−0·38853	·832	−0·14152	·852	−0·01693
·813	−0·39107	·833	−0·12920	·853	−0·01981
·814	−0·39325	·834	−0·11960	·854	−0·02493
·815	−0·39265	·835	−0·11547	·855	−0·03390
·816	−0·38674	·836	−0·11807	·856	−0·04698
·817	−0·37380	·837	−0·12600	·857	−1·56321
·818	−0·35378	·838	−0·13514	·858	−0·08124
·819	−0·32850	·839	−0·13999	·859	−0·10024

Note added September, 1958. Since submitting this paper the author has demonstrated the failure of Turán's inequality. The corresponding function $A_T^*(u)$ was shown to be negative, for $T = 1000$, at $u = 853{\cdot}853$ and $u = 996{\cdot}980$. He has not been able to disprove Mertens' conjecture, but it may well be disproved when much faster machines now being planned (about 1000 times faster than the EDSAC I and the Mark I) are available.

References.

1. C. B. Haselgrove and J. C. P. Miller, *Tables of the Riemann zeta function* (Royal Society Mathematical Tables, Vol. 6) (in the press).
2. A. E. Ingham, " On two conjectures in the theory of numbers ", *American J. of Math.*, 64 (1942), 313–319.
3. D. H. Lehmer, " Extended computation of the Riemann zeta function ", *Mathematika*, 3 (1956), 102–108.
4. F. Mertens, " Uber eine zahlentheoretische Funktion ", *Sitzungsberichte Akad. Wien.*, 106, Abt. 2a (1897), 761–830.
5. G. Pólya, " Verschiedene Bemerkungen zur Zahlentheorie ", *Jahresbericht der deutschen Math.-Vereinigung*, 28 (1919), 31–40.
6. P. Turán, " On some approximative Dirichlet-polynomials in the theory of the zeta-function of Riemann ", *Danske Vid. Selsk. Mat.-Fys. Medd.*, 24, No. 17 (1948).

The University,
Manchester 13.

(*Received 29th August*, 1958.)

12.14 H. Montgomery (1973)

The Pair Correlation of Zeros of the Zeta function

In a certain standard terminology the Conjecture may be formulated as the assertion that $1-((\sin \pi u)/\pi u)^2$ *is the pair correlation function of the zeros of the zeta function. F. J. Dyson has drawn my attention to the fact that the eigenvalues of a random complex Hermitian or unitary matrix of large order have precisely the same pair correlation function. This means that the Conjecture fits well with the view that there is a linear operator (not yet discovered) whose eigenvalues characterize the zeros of the zeta function. The eigenvalues of a random real symmetric matrix of large order have a different pair correlation, and the eigenvalues of a random symplectic matrix of large order have yet another pair correlation. In fact the "form factors"* $F_r(\alpha)$, $F_s(\alpha)$ *of these latter pair correlations are nonlinear for* $0 < \alpha < 1$, *so our Theorem enables us to distinguish the behavior of the zeros of* $\zeta(s)$ *from the eigenvalues of such matrices. Hence, if there is a linear operator whose eigenvalues characterize the zeros of the zeta function, we might expect that it is complex Hermitian or unitary.*

H. Montgomery

In this paper, Montgomery assumes the Riemann hypothesis and considers the differences between the imaginary parts of the nontrivial zeros. To this end, Montgomery defines

$$F(\alpha) = F(\alpha, T) := \left(\left(\frac{T}{2\pi}\right)\log T\right)^{-1} \sum_{\substack{0<\gamma\leq T \\ 0<\gamma'\leq T}} T^{i\alpha(\gamma-\gamma')} w(\gamma-\gamma'),$$

where α and $T \geq 2$ are real, $w(u) := \frac{4}{4+u^2}$, and γ and γ' are the imaginary parts of nontrivial zeros of the Riemann zeta function. Montgomery studies $F(\alpha)$ and makes some conjectures on the differences $\gamma - \gamma'$, which suggest a connection with the eigenvalues of a random complex Hermitian or unitary matrix of large order.

THE PAIR CORRELATION OF ZEROS OF THE ZETA FUNCTION

H. L. MONTGOMERY

1. **Statement of results.** We assume the Riemann Hypothesis (RH) throughout this paper; $\varrho=\frac{1}{2}+i\gamma$ denotes a nontrivial zero of the Riemann zeta function. Our object is to investigate the distribution of the differences $\gamma-\gamma'$ between the zeros. It would thus be desirable to know the Fourier transform of the distribution function of the numbers $\gamma-\gamma'$; with this in mind we take

$$F(\alpha)=F(\alpha,T)=\left(\frac{T}{2\pi}\log T\right)^{-1}\sum_{0<\gamma\leq T;\,0<\gamma'\leq T} T^{i\alpha(\gamma-\gamma')}\,w(\gamma-\gamma'), \tag{1}$$

where α and $T\geqq 2$ are real. Here $w(u)$ is a suitable weighting function, $w(u)=4/(4+u^2)$, so $w(0)=1$. Our results concerning $F(\alpha)$ are stated in the following

THEOREM. (*Assume RH.*) *For real* α, $T\geqq 2$, *let* $F(\alpha)$ *be defined by* (1). *Then* $F(\alpha)$ *is real, and* $F(\alpha)=F(-\alpha)$. *If* $T>T_0(\varepsilon)$ *then* $F(\alpha)\geqq-\varepsilon$ *for all* α. *For fixed* α *satisfying* $0\leqq\alpha<1$ *we have*

$$F(\alpha)=(1+o(1))\,T^{-2\alpha}\log T+\alpha+o(1) \tag{2}$$

as T tends to infinity; this holds uniformly for $0\leqq\alpha\leqq 1-\varepsilon$.

The first term on the right-hand side of the above behaves in the limit as a Dirac δ-function; it reflects the fact that if $\alpha=0$ then all the terms in (1) are positive. With more effort we could show that (2) holds uniformly throughout $0\leqq\alpha\leqq 1$.

To investigate sums involving $\gamma-\gamma'$ we have only to convolve $F(\alpha)$ with an

AMS 1970 *subject classifications*. Primary 10H05.

appropriate kernel $\hat{r}(\alpha)$; from (1) alone it is immediate that

$$\sum_{0<\gamma\leqq T;\,0<\gamma'\leqq T} r\left((\gamma-\gamma')\frac{\log T}{2\pi}\right) w(\gamma-\gamma') = \left(\frac{T}{2\pi}\log T\right)\int_{-\infty}^{+\infty} F(\alpha)\,\hat{r}(\alpha)\,d\alpha. \tag{3}$$

Here $\hat{r}$ is the Fourier transform of r,

$$\hat{r}(\alpha) = \int_{-\infty}^{+\infty} r(u)\,e(-\alpha u)\,du \qquad (e(\theta)=e^{2\pi i\theta}). \tag{4}$$

Our theorem gives us little information about $F(\alpha)$ for $\alpha \geqq 1$, so for the most part we restrict our attention to kernels $\hat{r}$ which vanish outside $[-1+\delta, 1-\delta]$. Particular choices of $\hat{r}(\alpha)$ give us

COROLLARY 1. *(Assume RH.) If $0<\alpha<1$ is fixed then*

$$\sum_{0<\gamma\leqq T;\,0<\gamma'\leqq T} \left(\frac{\sin\alpha(\gamma-\gamma')\log T}{\alpha(\gamma-\gamma')\log T}\right) w(\gamma-\gamma') \sim \left(\frac{1}{2\alpha}+\frac{\alpha}{2}\right)\frac{T}{2\pi}\log T, \tag{5}$$

and

$$\sum_{0<\gamma\leqq T;\,0<\gamma'\leqq T} \left(\frac{\sin(\alpha/2)(\gamma-\gamma')\log T}{(\alpha/2)(\gamma-\gamma')\log T}\right)^2 w(\gamma-\gamma') \sim \left(\frac{1}{\alpha}+\frac{\alpha}{3}\right)\frac{T}{2\pi}\log T. \tag{6}$$

In the latter assertion one can delete the factor $w(\gamma-\gamma')$ if one wishes. We use (6) to derive

COROLLARY 2. *(Assume RH.) As T tends to infinity*

$$\sum_{0<\gamma\leqq T;\,\varrho\ \text{simple}} 1 \geqq \left(\tfrac{2}{3}+o(1)\right)\frac{T}{2\pi}\log T. \tag{7}$$

The number of zeros of $\zeta(s)$ with $0<\gamma\leqq T$ is $\sim(T/2\pi)\log T$, so the above asserts that at least $\frac{2}{3}$ of the zeros are simple. It is known (see [**6**]) that the first 3,500,000 zeros are simple and lie on the critical line $\sigma=\frac{1}{2}$. Although one expects that all the zeros of $\zeta(s)$ are simple, the only other result in this direction is due to A. Selberg [**7**]. His result holds unconditionally; it states that a positive density of the zeros of $\zeta(s)$ are of odd order and lie on the critical line.

Let $0<\gamma_1\leqq\gamma_2\leqq\ldots$ denote the imaginary parts of the zeros of $\zeta(s)$ in the upper half-plane. The average of $\gamma_{n+1}-\gamma_n$ is $2\pi/\log\gamma_n$; our Theorem enables us to show that $\gamma_{n+1}-\gamma_n$ is not always near its average.

COROLLARY 3. (*Assume RH.*) *We can compute a constant* λ *so that*

$$\liminf_n (\gamma_{n+1}-\gamma_n)(\log\gamma_n/2\pi) \leqq \lambda < 1. \tag{8}$$

A complicated argument would permit one to show that in fact $\gamma_{n+1}-\gamma_n \leqq 2\pi\lambda/\log\gamma_n$ for a positive density of n. This, with the fact that the average value is $2\pi/\log\gamma_n$, enables one to assert that

$$\limsup_n (\gamma_{n+1}-\gamma_n)(\log\gamma_n/2\pi) \geqq \lambda' > 1. \tag{9}$$

We note that if $\zeta(s)$ has infinitely many multiple zeros then we may take $\lambda=0$ in (8). Our proof allows us to take $\lambda=0.68$. It would be of interest to have $\lambda<\frac{1}{4}$, as P. J. Weinberger and I have established the following: Let $d>0$ be square-free, and put $K=\boldsymbol{Q}((-d)^{1/2})$. Let $h(-d)$ be the class number of K, and let $\zeta_K(s)=\zeta(s)\cdot L(s,\chi)$ be the Dedekind zeta function of K. For each positive A, ε there is an effectively computable constant $d_0=d_0(A,\varepsilon)$ such that if $h(-d)\leqq A$, $d>d_0$, then all zeros of $\zeta_K(s)$ which are in the rectangle $0<\sigma<1$, $0\leqq t\leqq d^{1/2-\varepsilon}$ lie on the line $\sigma=\frac{1}{2}$; if $\frac{1}{2}+i\gamma_n$, $\frac{1}{2}+i\gamma_{n+1}$ are consecutive zeros of $\zeta_K(s)$ in this range then

$$(1-\varepsilon)\frac{2\pi}{\log d(\gamma_n+2)^2} \leqq \gamma_{n+1}-\gamma_n \leqq (1+\varepsilon)\frac{2\pi}{\log d(\gamma_n+2)^2}. \tag{10}$$

One may inquire about the behaviour of $F(\alpha)$ for $\alpha\geqq 1$. Our first observation is that (2) cannot hold uniformly for $0\leqq\alpha\leqq C$ if C is large. For if it did then (6) would hold for $0<\alpha\leqq C$. Write (6) as $G(\alpha)\sim H(\alpha)$. On one hand $|\sin 2x|\leqq 2|\sin x|$, so $G(2\alpha)\leqq G(\alpha)$ for all α. On the other hand $H(2\alpha)>\frac{3}{2}H(\alpha)$ for $\alpha\geqq 2$. This suggests that $F(\alpha)$ makes some change in its behaviour for $\alpha\geqq 1$. Further considerations of the above sort lead one to believe that certain averages of $F(\alpha)$ over large α are close to 1. At the end of § 3 we describe two heuristic arguments which suggest that

$$F(\alpha)=1+o(1) \tag{11}$$

for $\alpha\geqq 1$, uniformly in bounded intervals. This, with the Theorem, completely determines F, so an appropriate use of (3) leads immediately to a

CONJECTURE. *For fixed* $\alpha<\beta$,

$$(12)\qquad \sum_{\substack{0<\gamma\leqq T\\ 0<\gamma'\leqq T\\ 2\pi\alpha/\log T\leqq\gamma-\gamma'\leqq 2\pi\beta/\log T}} 1 \sim \left(\int_\alpha^\beta 1-\left(\frac{\sin\pi u}{\pi u}\right)^2 du+\delta(\alpha,\beta)\right)\frac{T}{2\pi}\log T$$

as T tends to infinity. Here $\delta(\alpha, \beta)=1$ *if* $0\in[\alpha, \beta]$, $\delta(\alpha, \beta)=0$ *otherwise.*

The Dirac δ-function occurs naturally in the above, for if $0\in[\alpha, \beta]$ then the sum includes terms $\gamma=\gamma'$.

The assertions (11) and (12) are essentially equivalent. From either it immediately follows that almost all zeros are simple. From (11) it is easy to see how Corollary 1 ought to be extended: If (11) is true then for $\alpha\geqq 1$,

$$(13)\qquad \sum_{0<\gamma\leqq T;\,0<\gamma'\leqq T}\left(\frac{\sin\alpha(\gamma-\gamma')\log T}{\alpha(\gamma-\gamma')\log T}\right)w(\gamma-\gamma')\sim\frac{T}{2\pi}\log T,$$

and

$$(14)\qquad \sum_{0<\gamma\leqq T;\,0<\gamma'\leqq T}\left(\frac{\sin(\alpha/2)(\gamma-\gamma')\log T}{(\alpha/2)(\gamma-\gamma')\log T}\right)^2 w(\gamma-\gamma')\sim\left(1+\frac{1}{3\alpha^2}\right)\frac{T}{2\pi}\log T.$$

In a certain standard terminology the Conjecture may be formulated as the assertion that $1-((\sin\pi u)/\pi u)^2$ is the pair correlation function of the zeros of the zeta function. F. J. Dyson has drawn my attention to the fact that the eigenvalues of a random complex Hermitian or unitary matrix of large order have precisely the same pair correlation function (see [**3**, equations (6.13), (9.61)]). This means that the Conjecture fits well with the view that there is a linear operator (not yet discovered) whose eigenvalues characterize the zeros of the zeta function. The eigenvalues of a random real symmetric matrix of large order have a different pair correlation, and the eigenvalues of a random symplectic matrix of large order have yet another pair correlation. In fact the "form factors" $F_r(\alpha)$, $F_s(\alpha)$ of these latter pair correlations are nonlinear for $0<\alpha<1$, so our Theorem enables us to distinguish the behaviour of the zeros of $\zeta(s)$ from the eigenvalues of such matrices. Hence, if there is a linear operator whose eigenvalues characterize the zeros of the zeta function, we might expect that it is complex Hermitian or unitary.

One might extend the present work to investigate the k-tuple correlation of the zeros of the zeta function. If the analogy with random complex Hermitian matrices appears to continue, then one might conjecture that the k-tuple correlation function $\hat{F}(u_1, u_2, \ldots, u_k)$ is given by

$$(15)\qquad \hat{F}(u_1, u_2, \ldots, u_k)=\det A,$$

where $A=[a_{ij}]$ is the $k\times k$ matrix with entries $a_{ii}=1$, $a_{ij}=(\sin\pi(u_i-u_j))/\pi(u_i-u_j)$ for $i\neq j$. Here the normalization is the same as in the Conjecture, which is the case $k=2$ of the above.

If one continues to draw on the analogy with random complex Hermitian matrices then one may formulate a conjecture concerning the distribution of the numbers $\gamma_{n+1}-\gamma_n$. The precise conjecture involves a complicated (but calculable) spheroidal function. Thus, or otherwise, one may conjecture that

$$\liminf_n(\gamma_{n+1}-\gamma_n)\log\gamma_n=0, \tag{16}$$

and

$$\limsup_n(\gamma_{n+1}-\gamma_n)\log\gamma_n=+\infty; \tag{17}$$

so Corollary 3 is probably far from the truth.

It would be interesting to see how numerical evidence compares with the above conjectures. The first several thousand zeros have been computed, so it would not be difficult to assemble relevant statistics. However, data on the failures of "Gram's law" indicate that the asymptotic behaviour is approached very slowly. Thus the numerical evidence may not be particularly illuminating.

2. **An explicit formula.** In proving our Theorem we require the following formula, which relates zeros of $\zeta(s)$ to prime numbers.

LEMMA. *If* $1<\sigma<2$ *and* $x\geqq 1$ *then*

$$\begin{aligned}(2\sigma-1)\sum_\gamma\frac{x^{i\gamma}}{(\sigma-\frac{1}{2})^2+(t-\gamma)^2}&=-x^{-1/2}\left(\sum_{n\leqq x}\Lambda(n)\left(\frac{x}{n}\right)^{1-\sigma+it}+\sum_{n>x}\Lambda(n)\left(\frac{x}{n}\right)^{\sigma+it}\right)\\&\quad+x^{1/2-\sigma+it}(\log\tau+O_\sigma(1))+O_\sigma(x^{1/2}\tau^{-1}),\end{aligned}\tag{18}$$

where $\tau=|t|+2$. *The implicit constants depend only on* σ.

PROOF. It is well known (see [**2**, p. 353]) that if $x>1$, $x\neq p^n$, then

$$\sum_{n\leqq x}\Lambda(n)\,n^{-s}=-\frac{\zeta'}{\zeta}(s)+\frac{x^{1-s}}{1-s}-\sum_\varrho\frac{x^{\varrho-s}}{\varrho-s}+\sum_{n=1}^{\infty}\frac{x^{-2n-s}}{2n+s}$$

provided $s\neq 1$, $s\neq\varrho$, $s\neq-2n$. This does not depend on RH, but if we assume RH then the above may be expressed as

$$\sum_{\varrho} \frac{x^{i\gamma - it}}{\sigma - \frac{1}{2} + it - i\gamma} = x^{\sigma - 1/2} \left(\frac{\zeta'}{\zeta}(s) + \sum_{n \leqq x} \Lambda(n)\, n^{-s} - \frac{x^{1-s}}{1-s} - \sum_{n=1}^{\infty} \frac{x^{-2n-s}}{2n+s} \right). \tag{19}$$

If we replace s by $1-\sigma+it$ in the above then we have

$$\begin{aligned} \sum_{\varrho} \frac{x^{i\gamma - it}}{\frac{1}{2} - \sigma + it - i\gamma} = x^{1/2-\sigma} \Bigg(& \frac{\zeta'}{\zeta}(1-\sigma+it) + \sum_{n \leqq x} \Lambda(n)\, n^{\sigma-1-it} \\ & - \frac{x^{\sigma-it}}{\sigma-it} - \sum_{n=1}^{\infty} \frac{x^{-2n-1+\sigma-it}}{2n+1-\sigma+it} \Bigg). \end{aligned} \tag{20}$$

We subtract respective sides of (20) from (19), and use the relation

$$\frac{\zeta'}{\zeta}(s) = -\sum_{n=1}^{\infty} \Lambda(n)\, n^{-s}, \tag{21}$$

which holds for $\sigma > 1$. We find that

$$\begin{aligned} (2\sigma-1) \sum_{\varrho} \frac{x^{i\gamma}}{(\sigma-\frac{1}{2})^2 + (t-\gamma)^2} = & -x^{-1/2} \left(\sum_{n \leqq x} \Lambda(n) \left(\frac{x}{n}\right)^{1-\sigma+it} + \sum_{n > x} \Lambda(n) \left(\frac{x}{n}\right)^{\sigma+it} \right) \\ & - \frac{\zeta'}{\zeta}(1-\sigma+it)\, x^{1/2-\sigma+it} + \frac{x^{1/2}(2\sigma-1)}{(\sigma-1+it)(\sigma-it)} \\ & - x^{-1/2} \sum_{n=1}^{\infty} \frac{(2\sigma-1)\, x^{-2n}}{(\sigma-1-it-2n)(\sigma+it+2n)}. \end{aligned} \tag{22}$$

Both sides of the above are continuous for all $x \geqq 1$, so we no longer exclude the values $x=1$, $x=p^n$. If $1<\sigma<2$, then from the logarithmic derivative of the functional equation of the zeta function (see [**1**, pp. 75, 82–83]) we have

$$\frac{\zeta'}{\zeta}(1-\sigma+it) = -\frac{\zeta'}{\zeta}(\sigma-it) - \log\tau + O_\sigma(1);$$

from (21) we see that this is $= -\log\tau + O_\sigma(1)$. Hence the right-hand side of (22) is

$$\begin{aligned} = -x^{-1/2} & \left(\sum_{n \leqq x} \Lambda(n) \left(\frac{x}{n}\right)^{1-\sigma+it} + \sum_{n>x} \Lambda(n) \left(\frac{x}{n}\right)^{\sigma+it} \right) \\ & + x^{1/2-\sigma+it}(\log\tau + O_\sigma(1)) + O_\sigma(x^{1/2}\tau^{-2}) + O_\sigma(x^{-2}\tau^{-1}), \end{aligned}$$

which gives the result.

3. **Proof of the Theorem.** The first assertion of the Theorem follows from the observation that we may interchange γ and γ' in (1). To prove the remaining assertions, take $\sigma=\frac{3}{2}$ in the Lemma, and write (18) briefly as $L(x,t)=R(x,t)$. We evaluate the integrals $\int_0^T |L(x,t)|^2\,dt$, $\int_0^T |R(x,t)|^2\,dt$.

We treat the left-hand side first. We have

$$\int_0^T |L(x,t)|^2\,dt = 4\sum_{\gamma,\gamma'} x^{i(\gamma-\gamma')}\int_0^T \frac{dt}{(1+(t-\gamma)^2)(1+(t-\gamma')^2)}. \tag{23}$$

We wish to exclude those numbers $\gamma\notin[0,T]$. It suffices to show that

$$\sum_{\gamma,\gamma';\,\gamma\notin[0,T]}\int_0^T \frac{dt}{(1+(t-\gamma)^2)(1+(t-\gamma')^2)} \ll \log^3 T, \tag{24}$$

for then (23) is

$$= 4\sum_{0<\gamma\leqq T;\,0<\gamma'\leqq T} x^{i(\gamma-\gamma')}\int_0^T \frac{dt}{(1+(t-\gamma)^2)(1+(t-\gamma')^2)} + O(\log^3 T). \tag{25}$$

To prove (24) we use the fact (Theorem 9.2 of [**8**]) that if $T\geqq 2$ then there are $\ll \log T$ zeros for which $T\leqq\gamma\leqq T+1$. From this it is immediate that if $0\leqq t\leqq T$ then

$$\sum_{\gamma;\,\gamma\notin[0,T]} \frac{1}{1+(t-\gamma)^2} \ll \left(\frac{1}{t+1}+\frac{1}{T-t+1}\right)\log T,$$

and

$$\sum_{\gamma'} \frac{1}{1+(t-\gamma')^2} \ll \log T.$$

On the left-hand side of (24) we take the sums inside and use the above estimates. The integration is then trivial, and we obtain (24).

Arguing similarly we may also show that

$$\sum_{0<\gamma\leqq T;\,0<\gamma'\leqq T}\int_T^\infty \frac{dt}{(1+(t-\gamma)^2)(1+(t-\gamma')^2)} \ll \log^2 T\int_T^\infty \frac{dt}{(t-T+1)^2} \ll \log^2 T.$$

The estimation of $\sum_{0<\gamma\leqq T;\,0<\gamma'\leqq T}\int_{-\infty}^{0}\ldots$ is the same, so we see that (25) is

$$=4\sum_{0<\gamma\leqq T;\,0<\gamma'\leqq T} x^{i(\gamma-\gamma')}\int_{-\infty}^{+\infty}\frac{dt}{(1+(t-\gamma)^2)\,(1+(t-\gamma')^2)}+O(\log^3 T).$$

From the calculus of residues we deduce that the definite integral is $=(\pi/2)\,w(\gamma-\gamma')$, so the above is

$$=2\pi\sum_{0<\gamma\leqq T;\,0<\gamma'\leqq T} x^{i(\gamma-\gamma')}w(\gamma-\gamma')+O(\log^3 T).$$

If we put $x=T^\alpha$ then we have

$$\int_0^T |L(T^\alpha,t)|^2\,dt = F(\alpha)\,T\log T+O(\log^3 T). \tag{26}$$

Here the left-hand side is clearly nonnegative, so we have the second assertion of the Theorem.

To complete the proof of the Theorem we prove (2); to this end we evaluate $\int_0^T |R(x,t)|^2\,dt$. In the first place

$$\int_0^T |x^{-1+it}\log\tau|^2\,dt = \frac{T}{x^2}(\log^2 T+O(\log T)) \tag{27}$$

for all $x\geqq 1$, $T\geqq 2$. To compute the mean square of the Dirichlet series on the right-hand side of (18) we use the following quantitative form (see [**5**]) of Parseval's identity for Dirichlet series:

$$\int_0^T \left|\sum_n a_n n^{-it}\right|^2 dt = \sum_n |a_n|^2(T+O(n)). \tag{28}$$

We could instead use the weaker relation

$$\int_0^T \left|\sum_{n\leqq N} a_n n^{-it}\right|^2 dt = (T+O(N))\sum_{n\leqq N}|a_n|^2;$$

this is Theorem 1.6 of [4]. However, the latter is restricted to Dirichlet polynomials, so we simplify our treatment by arguing from (28). We have

$$\frac{1}{x}\int_0^T \left|\sum_{n\leqq x} \Lambda(n)\left(\frac{x}{n}\right)^{-1/2+it} + \sum_{n>x} \Lambda(n)\left(\frac{x}{n}\right)^{3/2+it}\right|^2 dt$$

$$=\frac{1}{x}\sum_{n\leqq x} \Lambda(n)^2\left(\frac{x}{n}\right)^{-1}(T+O(n))+\frac{1}{x}\sum_{n>x} \Lambda(n)^2\left(\frac{x}{n}\right)^{3}(T+O(n)).$$

By the prime number theorem with error term this is

$$=T(\log x+O(1))+O(x\log x). \tag{29}$$

As for the error terms in (18), we see that

$$\int_0^T |x^{-1+it}|^2\, dt = \frac{T}{x^2}, \tag{30}$$

and

$$\int_0^T x\tau^{-2}\, dt \ll x. \tag{31}$$

We now combine our estimates (27), (29), (30), (31); we employ the following consequence of the Cauchy-Schwarz inequality: If $M_k=\int_0^T |A_k(t)|^2\, dt$ and $M_1\geqq M_2 \geqq M_3 \geqq M_4$, then

$$\int_0^T \left|\sum_{k=1}^{4} A_k(t)\right|^2 dt = M_1+O((M_1M_2)^{1/2}).$$

We consider three cases.

Case 1. $1\leqq x\leqq(\log T)^{3/4}$. Then our M_1 term is given by (27). Our other terms are uniformly $o(M_1)$, so our expression is $=(1+o(1))\,(T/x^2)\log^2 T$.

Case 2. $(\log T)^{3/4}<x\leqq(\log T)^{3/2}$. In this case all our estimates are uniformly $o(T\log T)$.

Case 3. $(\log T)^{3/2}<x\leqq T/\log T$. Then our M_1 term is given by (29). All our

other terms are uniformly $o(M_1)$, so our expression is $=(1+o(1))\, T\log x$.

If we put $x=T^\alpha$ then we may express our result by saying that

$$\int_0^T |R(T^\alpha, t)|^2\, dt = ((1+o(1))\, T^{-2\alpha}\log T + \alpha + o(1))\, T\log T,$$

uniformly for $0\leqq\alpha\leqq 1-\varepsilon$. This and (26) give (2), so the proof is complete.

If $\alpha>1$ in the above then $x>T$, so the second error term in (29) is no longer smaller than the main term. The error term (31) also gives problems; a little consideration reveals that what we require is to know the size of

$$\int_0^T \left|\frac{1}{x}\sum_{n\leqq x} \Lambda(n)\, n^{1/2-it} + x\sum_{n>x}\Lambda(n)\, n^{-3/2-it} - \frac{2x^{1/2-it}}{(\frac{1}{2}+it)(\frac{3}{2}-it)}\right|^2 dt. \tag{32}$$

If we multiply out the integrand and integrate terms individually, we find that there are too many nondiagonal terms to be ignored. We may, however, collect terms so that the above is expressed in terms of sums of the sort $\sum_{n\leqq y}\Lambda(n)\,\Lambda(n+h)$. There are various indications that this sum is approximately $c(h)\, y$, where $c(h)$ is a certain arithmetic constant. If we replace these sums by their conjectured approximations $c(h)\, y$, then our new expression is $\sim T\log T$. Moreover, there is a reasonable hypothesis as to the size of the differences

$$\sum_{n\leqq y}\Lambda(n)\,\Lambda(n+h) - c(h)\, y \tag{33}$$

which if true would allow us to carry out our program for $1\leqq\alpha<2$. If the differences (33) are not only reasonably small but also behave independently for different h then (32) is $\sim T\log T$ for all $\alpha\geqq 1$.

Another indication of the behaviour of the expression (32) can be obtained by considering its "q-analogue." The expression

$$\sum_{q\leqq Q}\frac{1}{\varphi(q)}\sum_{\chi\neq\chi_0}\left|\frac{1}{x}\sum_{n\leqq x}\Lambda(n)\,\chi(n)\, n^{1/2} + \sum_{n>x}\Lambda(n)\,\chi(n)\, n^{-3/2}\right|^2 \tag{34}$$

may be shown to be $\sim Q\log x$ for $Q\geqq x$, in analogy with (29). If $x(\log x)^{-A}\leqq Q\leqq x$ then we may use an established technique [**4**, Chapter 17] to show that (34) is $\sim Q\log Q$. If GRH is true then this latter asymptotic relationship holds for $x^{3/4+\varepsilon}<Q\leqq x$. This corresponds to $1\leqq\alpha<\frac{4}{3}$. One does not expect a change in the behaviour for larger α, but a more delicate error-term analysis is needed if the result is to be extended.

4. **The corollaries.** To prove Corollary 1 we use our Theorem in conjunction with (3). To obtain (5) we take $r(u)=(\sin 2\pi\alpha u)/2\pi\alpha u$. The Theorem makes it a simple task to compute

$$\int_{-\infty}^{+\infty} F(\beta)\,\hat{r}(\beta)\,d\beta=\frac{1}{2\alpha}\int_{-\alpha}^{\alpha} F(\beta)\,d\beta .$$

To obtain (6) we take $r(u)=((\sin\pi\alpha u)/\pi\alpha u)^2$. Again from the Theorem it is easy to compute

$$\int_{-\infty}^{+\infty} F(\beta)\,\hat{r}(\beta)\,d\beta=\frac{1}{\alpha^2}\int_{-\alpha}^{+\alpha}(\alpha-\beta)\,F(\beta)\,d\beta .$$

We now prove Corollary 2. Let m_ϱ be the multiplicity of the zero ϱ. In a sum over $0<\gamma\leqq T$, our convention concerning multiple zeros is that zeros are counted according to their multiplicities. This is accomplished by allowing γ to take on the same value m_ϱ times. In particular,

$$\sum_{0<\gamma\leqq T} m_\varrho=\sum_{\substack{0<\gamma\leqq T\\ 0<\gamma'\leqq T\\ \gamma=\gamma'}} 1$$

for on both sides a given zero ϱ is counted with weight m_ϱ^2. But

$$\sum_{\substack{0<\gamma\leqq T\\ 0<\gamma'\leqq T\\ \gamma=\gamma'}} 1\leqq\sum_{0<\gamma\leqq T;\,0<\gamma'\leqq T}\left(\frac{\sin(\alpha/2)(\gamma-\gamma')\log T}{(\alpha/2)(\gamma-\gamma')\log T}\right)^2 w(\gamma-\gamma'),$$

and if we take $\alpha=1-\delta$ then from (6) the above is

$$\leqq(\tfrac{4}{3}+\varepsilon)\,(T/2\pi)\log T .$$

Hence we have demonstrated that

$$\sum_{0<\gamma\leqq T} m_\varrho\leqq(\tfrac{4}{3}+o(1))\,(T/2\pi)\log T .$$

Now

$$\sum_{0<\gamma\leqq T;\,\varrho\ \mathrm{simple}} 1 \geqq \sum_{0<\gamma\leqq T} (2-m_\varrho) \geqq (2-\tfrac{4}{3}+o(1))\,\frac{T}{2\pi}\log T,$$

so we have Corollary 2. The kernel $\hat{r}(u)$ which we have used does not appear to be optimal for our purpose, so presumably one can improve slightly on the constant $\frac{2}{3}$.

We now turn to the first assertion of Corollary 3. We take $r(u)=\max(1-(|u|/\lambda),0)$ in (3), and choose λ later. Now $\hat{r}(\alpha)$ is nonnegative, and $\int_0^\infty \hat{r}(\alpha)\,d\alpha<\infty$, so our Theorem permits us to calculate a lower bound for the right-hand side of (3). We see that

$$\int_{-\infty}^{+\infty} F(\alpha)\,\hat{r}(\alpha)\,d\alpha \geqq (1+o(1))\left(\lambda+2\lambda\int_0^1 \alpha\left(\frac{\sin\pi\lambda\alpha}{\pi\lambda\alpha}\right)d\alpha\right)\frac{T}{2\pi}\log T.$$

We may assume that all but finitely many zeros are simple, so the terms $\gamma=\gamma'$ in (3) contribute an amount $\sim(T/2\pi)\log T$. Hence

$$\sum_{\substack{0<\gamma\leqq T\\ 0<\gamma'\leqq T\\ 0<\gamma-\gamma'<2\pi\lambda/\log T}} 1 \geqq (\tfrac{1}{2}+o(1))\,C(\lambda)\,\frac{T}{2\pi}\log T$$

where

$$C(\lambda)=\lambda+(1/\pi^2\lambda)\,\mathrm{Cin}(2\pi\lambda)-1.$$

Here $\mathrm{Cin}(x)$ is the "cosine integral,"

$$\mathrm{Cin}\,x=\int_0^x \frac{1-\cos u}{u}\,du.$$

Note that the integrand is nonnegative, so that $\mathrm{Cin}(x)>0$ for $x>0$. To obtain (8) we show that $C(\lambda)>0$ for some $\lambda<1$. This is easy, because $C(1)=(1/\pi^2)\cdot\mathrm{Cin}(2\pi)>0$, and $C(\lambda)$ is continuous. In fact a little calculation reveals that $C(0.68)>0$. We have not determined the optimal kernel $\hat{r}(\alpha)$, so one should be able to improve on the constant 0.68.

References

1. H. Davenport, *Multiplicative number theory*, Lectures in Advanced Math., no. 1, Markham, Chicago, Ill., 1967. MR **36** #117.

2. Edmund Landau, *Handbuch der Lehre von der Verteilung der Primzahlen*, Teubner, Berlin, 1909.

3. M. L. Mehta, *Random matrices and the statistical theory of energy levels*, Academic Press, New York, 1967. MR **36** #3554.

4. Hugh L. Montgomery, *Topics in multiplicative number theory*, Lecture Notes in Math., vol. 227, Springer-Verlag, Berlin and New York, 1971.

5. Hugh L. Montgomery and R. C. Vaughan, *Hilbert's inequality* (to appear).

6. J. B. Rosser, J. M. Yohe and L. Schoenfeld, *Rigorous computation and the zeros of the Riemann zeta-function* (*with discussion*), Information Processing 68 (Proc. IFIP Congress, Edinburgh, 1968), vol. 1, Math., Software, North-Holland, Amsterdam, 1969, pp. 70–76. MR **41** #2892.

7. Atle Selberg, *On the zeros of Riemann's zeta-function*, Skr. Norske Vid. Akad. Oslo I (1942), no. 10, 59 pp. MR **6**, 58.

8. E. C. Titchmarsh, *The theory of the Riemann zeta-function*, Clarendon Press, Oxford, 1951. MR **13**, 741.

Trinity College
Cambridge, England

12.15 D. J. Newman (1980)

Simple Analytic Proof of the Prime Number Theorem

This paper endeavors to give a simple, but not elementary, proof of the prime number theorem. Newman offers two proofs of the prime number theorem. In the first he proves the following result, due to Ingham, in a novel way.

Theorem 12.2. *Suppose* $|a_n| \leq 1$ *and form the series* $\sum a_n n^{-z}$ *which clearly converges to an analytic function* $F(z)$ *for* $\Re(z) > 1$. *If, in fact,* $F(z)$ *is analytic throughout* $\Re(z) \geq 1$, *then* $\sum a_n n^{-z}$ *converges throughout* $\Re(z) \geq 1$.

The novelty of the proof lies in the cleverly chosen contour integral

$$\int_{\Gamma} f(z) N^z \left(\frac{1}{z} + \frac{z}{R^2} \right) dz,$$

where Γ is a specific finite contour. From this result the convergence of $\sum \mu(n)/n$ follows directly (here $\mu(n)$ is the Möbius function). This is equivalent to the prime number theorem, as shown by Landau. In his second proof of the prime number theorem, Newman applies Theorem 12.2 to show that

$$\sum_{p<n} \frac{\log p}{p} - \log n$$

converges to a limit. This result is also equivalent to the prime number theorem. [34]

SIMPLE ANALYTIC PROOF OF THE PRIME NUMBER THEOREM

D. J. NEWMAN

Department of Mathematics, Temple University, Philadelphia, PA 19122

The magnificent prime number theorem has received much attention and many proofs throughout the past century. If we ignore the (beautiful) elementary proofs of Erdős [1] and Selberg [6] and focus on the analytical ones, we find that they all have some drawback. The original proofs [7] of Hadamard and de la Vallée Poussin were based, to be sure, on the nonvanishing of $\zeta(z)$ in Re $z \geqslant 1$, but they also required annoying estimates of $\zeta(z)$ at ∞, the reason being that formulas for coefficients of Dirichlet series involve integrals over *infinite* contours (unlike the situation for power series) and so effective evaluation requires estimates at ∞.

The more modern proofs, due to Wiener [2] and Ikehara [8] (see also Heins's book [3]) do get around the necessity of estimating at ∞ and are indeed based only on the appropriate nonvanishing of $\zeta(z)$, but they are tied to certain results on Fourier transforms.

We propose to return to contour integral methods so as to avoid Fourier analysis, and also to use finite contours so as to avoid estimates at ∞. Of course certain errors are introduced thereby, but the point is that these can be effectively estimated away by elementary arguments.

So let us begin with the well-known fact [7] about the ζ-function:

$$(z-1)\zeta(z) \text{ is analytic and zero free throughout Re } z \geqslant 1. \tag{1}$$

This will be assumed throughout and will allow us to give our proof of the prime number theorem.

In fact we give two proofs. The first one is the shorter and simpler of the two, but we pay a price in that we obtain one of Landau's equivalent forms of the theorem rather than the standard form, $\pi(N) \sim N/\log N$. Our second proof is a more direct assault on $\pi(N)$ but is somewhat more intricate than the first. Here we find some of Tchebychev's elementary ideas very useful.

Basically our novelty consists in using a modified contour integral,

$$\int_{\Gamma} f(z) N^z \left(\frac{1}{z} + \frac{z}{R^2}\right) dz,$$

rather than the classical one, $\int_C f(z) N^n z^{-1} dz$. The method is rather flexible, and we could use it to directly obtain $\pi(N)$ by choosing $f(z) = \log \zeta(z)$. We prefer, however, to derive both proofs from the following convergence theorem. Actually, this theorem dates back to Ingham [9], but his proof is à la Fourier analysis and is much more complicated than the contour integral method we now give.

THEOREM. *Suppose* $|a_n| \leqslant 1$ *and form the series* $\sum a_n n^{-z}$ *which clearly converges to an analytic function* $F(z)$ *for* Re $z > 1$. *If, in fact,* $F(z)$ *is analytic throughout* Re $z \geqslant 1$, *then* $\sum a_n n^{-z}$ *converges throughout* Re $z \geqslant 1$.

Proof of the convergence theorem.. Fix a w in Re $w \geqslant 1$. Thus $F(z+w)$ is analytic in Re $z \geqslant 0$. We choose an $R \geqslant 1$ and determine $\delta = \delta(R) > 0$, $\delta \leqslant \frac{1}{2}$ and an $M = M(R)$ so that

$$F(z+w) \text{ is analytic and bounded by } M \text{ in } -\delta \leqslant \text{Re}\, z, |z| \leqslant R. \tag{2}$$

Now form the counterclockwise contour Γ, bounded by the arc $|z| = R$, Re $z > -\delta$, and the

D. J. Newman received his doctorate from Harvard in 1958. He has worked mainly in Analysis, with special emphasis on Approximation Theory. Currently a Professor at Temple University, he has previously been at Yeshiva, M.I.T., and Brown.

segment Re $z=-\delta, |z|\leqslant R$. Also denote by A and B, respectively, the parts of Γ in the right and left half-planes.

By the residue theorem we have

$$2\pi i F(w)=\int_\Gamma F(z+w)N^z\left(\frac{1}{z}+\frac{z}{R^2}\right)dz. \tag{3}$$

Now on A, $F(z+w)$ is equal to its series, and we split this into its partial sum $S_N(z+w)$ and remainder $r_N(z+w)$. Again by the residue theorem we have

$$\int_A S_N(z+w)N^z\left(\frac{1}{z}+\frac{z}{R^2}\right)dz=2\pi i S_N(w)-\int_{-A} S_N(z+w)N^z\left(\frac{1}{z}+\frac{z}{R^2}\right)dz,$$

with $-A$ denoting as usual the reflection of A through the origin. Thus, changing z into $-z$, this can be written as

$$\int_A S_N(z+w)N^z\left(\frac{1}{z}+\frac{z}{R^2}\right)dz=2\pi i S_N(w)-\int_A S_N(w-z)N^{-z}\left(\frac{1}{z}+\frac{z}{R^2}\right)dz. \tag{4}$$

Combining (3) and (4) gives

$$\begin{aligned}2\pi i(F(w)-S_N(w))=&\int_A\left(r_N(z+w)N^z-\frac{S_N(w-z)}{N^z}\right)\left(\frac{1}{z}+\frac{z}{R^2}\right)dz\\&+\int_B F(z+w)N^z\left(\frac{1}{z}+\frac{z}{R^2}\right)dz,\end{aligned} \tag{5}$$

and to estimate these integrals we record the following (here as usual we write Re $z=x$, and we use the notation $\alpha\ll\beta$ to mean simply that $|\alpha|\leqslant|\beta|$):

$$\frac{1}{z}+\frac{z}{R^2}=\frac{2x}{R^2} \text{ along } |z|=R \text{ (in particular on } A), \tag{6}$$

$$\frac{1}{z}+\frac{z}{R^2}\ll\frac{1}{\delta}\left(1+\frac{|z|^2}{R^2}\right)\leqslant\frac{2}{\delta} \text{ on the line Re } z=-\delta, |z|\leqslant R, \tag{7}$$

$$r_N(z+w)\ll\sum_{n=N+1}^{\infty}\frac{1}{n^{x+1}}\leqslant\int_N^\infty\frac{dn}{n^{x+1}}=\frac{1}{xN^x}, \tag{8}$$

$$S_N(w-z)\ll\sum_{n=1}^{N}n^{x+1}\leqslant N^{x-1}+\int_0^N n^{x-1}dn=N^x\left(\frac{1}{N}+\frac{1}{x}\right). \tag{9}$$

By (6), (8), (9) we have, on A,

$$\left(r_N(z+w)N^z-\frac{S_N(w-z)}{N^z}\right)\left(\frac{1}{z}+\frac{z}{R^2}\right)\ll\left(\frac{1}{x}+\frac{1}{x}+\frac{1}{N}\right)\frac{2x}{R^2}\leqslant\frac{4}{R^2}+\frac{2}{RN},$$

and so by the "maximum times length" estimate (M-L formula) for integrals we obtain

$$\int_A\left(r_N(z+w)N^z-\frac{S_N(w-z)}{N^z}\right)\left(\frac{1}{z}+\frac{z}{R^2}\right)dz\ll\frac{4\pi}{R}+\frac{2\pi}{N}. \tag{10}$$

Next by (2), (6), and (7) we obtain

$$\begin{aligned}\int_B F(z+w)N^z\left(\frac{1}{z}+\frac{z}{R^2}\right)dz&\ll\int_{-R}^{R}M\cdot N^{-\delta}\cdot\frac{2}{\delta}dy+2M\int_{-\delta}^{0}N^x\frac{2|x|}{R^2}\frac{3}{2}dx\\&\leqslant\frac{4MR}{\delta N^\delta}+\frac{6M}{R^2\log^2 N}.\end{aligned} \tag{11}$$

Inserting the estimates (10) and (11) into (5) gives

$$F(w)-S_N(w) \ll \frac{2}{R}+\frac{1}{N}+\frac{MR}{\delta N^{\delta}}+\frac{M}{R^2 \log^2 N}$$

and if we fix $R=3/\epsilon$ we note that this right-hand side is $<\epsilon$ for all large N. We have verified the very definition of convergence!

First Proof of the Prime Number Theorem. Landau has pointed out that the convergence of $\Sigma\,\mu(n)/n$ is equivalent to the prime number theorem. Since $\Sigma\,\mu(n)/n^z=1/\zeta(z)$ for $\operatorname{Re} z>1$, however, (1) ensures that the hypotheses of our theorem hold, and Landau's form of the prime number theorem follows immediately.

Second Proof of the Prime Number Theorem. In this section we begin with Tchebychev's observation [5] that

$$\sum_{p<n} \frac{\log p}{p}-\log n \quad \text{is bounded,} \tag{12}$$

which he derives in a direct elementary way from the prime factorization of $n!$.

The point is that the prime number theorem is easily derived from

$$\sum_{p<n} \frac{\log p}{p}-\log n \quad \text{converges to a limit,} \tag{13}$$

by a simple summation by parts, which we leave to the reader. Nevertheless the transition from (12) to (13) is not a simple one and we turn to this now.

So form, for $\operatorname{Re} z>1$, the function

$$f(z)=\sum_{n=1}^{\infty} \frac{1}{n^z}\left(\sum_{p<n} \frac{\log p}{p}\right)=\sum_{p} \frac{\log p}{p}\left(\sum_{n>p} \frac{1}{n^z}\right).$$

Now

$$\sum_{n>p} \frac{1}{n^z}=\frac{1}{(z-1)p^{z-1}}+z\int_p^{\infty} \frac{1-\{t\}}{t^{z+1}}\,dt=\frac{p}{(z-1)}\left(\frac{1}{p^z-1}+A_p(z)\right)$$

where $A_p(z)$ is analytic for $\operatorname{Re} z>0$ and is bounded by

$$\frac{1}{p^x(p^x-1)}+\frac{|z(z-1)|}{xp^{x+1}}.$$

Hence

$$f(z)=\frac{1}{z-1}\left(\sum_{p} \frac{\log p}{p^z-1}+A(z)\right),$$

where $A(z)$ is analytic for $\operatorname{Re} z>\frac{1}{2}$ by the Weierstrass M-test.

By Euler's factorization formula, however, we recognize that

$$\sum_{p} \frac{\log p}{p^z-1}=\frac{-d}{dz} \log \zeta(z); \tag{14}$$

and so we deduce, by (1), that $f(z)$ is analytic in $\operatorname{Re} z \leqslant 1$ except for a double pole with principal part $1/(z-1)^2+c/(z-1)$, at $z=1$. Thus if we set

$$F(z)=f(z)+\zeta'(z)-c\zeta(z)=\sum \frac{a_n}{n^z}, \qquad \text{where } a_n=\sum_{p<n} (\log p)/p-\log n-c,$$

we deduce that $F(z)$ is analytic in $\operatorname{Re} z \geqslant 1$.

From (12) and our convergence theorem, then, we conclude that

$$\sum \frac{a_n}{n} \text{ converges,}$$

and from this and the fact, from (14), that $a_n + \log n$ is nondecreasing we proceed to prove $a_n \to 0$.

By applying the Cauchy criterion we find that, for N large, we have both

$$\sum_{N}^{N(1+\epsilon)} \frac{a_n}{n} \leqslant \epsilon^2, \tag{15}$$

and

$$\sum_{N(1-\epsilon)}^{N} \frac{a_n}{n} \geqslant -\epsilon^2. \tag{16}$$

In the range N to $N(1+\epsilon)$ we have, by (14), that $a_n \geqslant a_N + \log(N/n) \geqslant a_N - \epsilon$ and so $\Sigma_N^{N(1+\epsilon)} a_n/n \geqslant (a_N - \epsilon)\Sigma_N^{N(1+\epsilon)} 1/n$ and (15) yields

$$a_N \leqslant \epsilon + \frac{\epsilon^2}{\sum_{N}^{N(1+\epsilon)} \frac{1}{n}} \leqslant \epsilon + \frac{\epsilon^2}{N\epsilon/N(1+\epsilon)} = 2\epsilon + \epsilon^2. \tag{17}$$

Similarly in $[N(1-\epsilon), N]$ we have $a_n \leqslant a_N + \log(N/n) \leqslant a_N + \epsilon/(1-\epsilon)$ so that

$$\sum_{N(1-\epsilon)}^{N} \frac{a_n}{n} \leqslant \left(a_N + \frac{\epsilon}{1-\epsilon}\right) \sum_{N(1-\epsilon)}^{N} \frac{1}{n}$$

and (16) gives

$$a_N \geqslant \frac{-\epsilon}{1-\epsilon} - \frac{\epsilon^2}{\sum_{N(1-\epsilon)}^{N} \frac{1}{n}} \geqslant -\frac{\epsilon}{1-\epsilon} - \frac{\epsilon^2}{N\epsilon/N} = \frac{\epsilon^2 - 2\epsilon}{1-\epsilon}. \tag{18}$$

Taken together (17) and (18) establish that $a_N \to 0$ and so (13) is proved.

The research for this paper was supported in part by NSF Grant No. MCSF8-02171.

References

1. P. Erdős, On a new method in elementary number theory, Proc. Nat. Acad. Sci. U.S.A., 35 (1949) 374–384.
2. G. H. Hardy, Divergent Series, Clarendon Press, Oxford, 1949, pp. 283, et seq.
3. Maurice Heins, Complex Function Theory, Academic Press, 1968, pp. 243–249.
4. E. Landau, Über einige neuere Grenzwertsätze, Rend. Circ. Mat. Palermo, 34 (1912) 121–131.
5. W. J. LeVeque, Topics in Number Theory, vol. 1, Addison-Wesley, 1956, p. 108.
6. A. Selberg, An elementary proof of the prime number theorem, Ann. Math. (2), vol. 50, pp. 305–313.
7. E. C. Titchmarsh, The Theory of the Riemann Zeta-Function, Clarendon Press, Oxford, pp. 38 et seq.
8. D. V. Widder, The Laplace Transform, Princeton University Press, 1941, pp. 233 et seq.
9. A. E. Ingham, On Wiener's method in Tauberian theorems, Proc. London Math. Soc. (2) 38 (1935) 458–480.

MISCELLANEA

43. If you ask mathematicians what they do, you always get the same answer; they think. They are trying to solve difficult and novel problems. (They never think about ordinary problems—they just write down the answers.)

—M. Evgrafov, *Literaturnaya Gazeta*, no. 49 (1979) 12.

12.16 J. Korevaar (1982)

On Newman's Quick Way to the Prime Number Theorem

In this paper, Korevaar presents Newman's simple proof of the prime number theorem (see Section 12.15). Korevaar's paper is expository in nature, and he presents Newman's method in great detail. He starts with a brief historical note on the prime number theorem and presents some elementary properties of the Riemann zeta function that Newman takes as given. Korevaar proceeds to prove the prime number theorem by applying Newman's method to Laplace integrals, as opposed to Dirichlet series, and replacing the Ikehara–Wiener Tauberian theorem by a "poor man's" version. Korevaar's presentation is clear and detailed, and is an excellent starting point for studying proofs of the prime number theorem.

108

On Newman's Quick Way to the Prime Number Theorem

J. Korevaar

1. Introduction and Overview

There are several interesting functions in number theory whose tables look quite irregular, but which exhibit surprising asymptotic regularity as $x \to \infty$. A notable example is the function $\pi(x)$ which counts the number of primes p not exceeding x.

1.1. The Famous Prime Number Theorem

$$\pi(x) = \sum_{p \leqslant x} 1 \sim \frac{x}{\log x} \quad \text{as} \quad x \to \infty, \tag{1.1}$$

was surmised already by Legendre and Gauss. However, it took a hundred years before the first proofs appeared, one by Hadamard and one by de la Vallée Poussin (1896). Their and all but one of the subsequent proofs make heavy use of the Riemann zeta function. (The one exception is the long so-called elementary proof by Selberg [11] and Erdös [4].)

For Re $s > 1$ the zeta function is given by the Dirichlet series

$$\zeta(s) = \sum_{1}^{\infty} \frac{1}{n^s}. \tag{1.2a}$$

By the unique representation of positive integers n as products of prime powers, the series may be converted to the Euler product (cf. [5])

$$\zeta(s) = \left(1 + \frac{1}{p_1^s} + \frac{1}{p_1^{2s}} + \ldots\right)\left(1 + \frac{1}{p_2^s} + \frac{1}{p_2^{2s}} + \ldots\right)\ldots$$
$$= \prod_p \frac{1}{1 - p^{-s}}. \tag{1.2b}$$

The above function element is analytic for Re $s > 1$ and can be continued across the line Re $s = 1$ (Fig. 1). More precisely, the difference

$$\zeta(s) - \frac{1}{s-1}$$

can be continued analytically to the half-plane Re $s > 0$ (cf. § B.1 in the box on p. 111) and in fact to all of $\mathbb{C}$. The essential property of $\zeta(s)$ in the proofs of the prime number theorem is its non-vanishing on the line Re $s = 1$

D. J. Newman

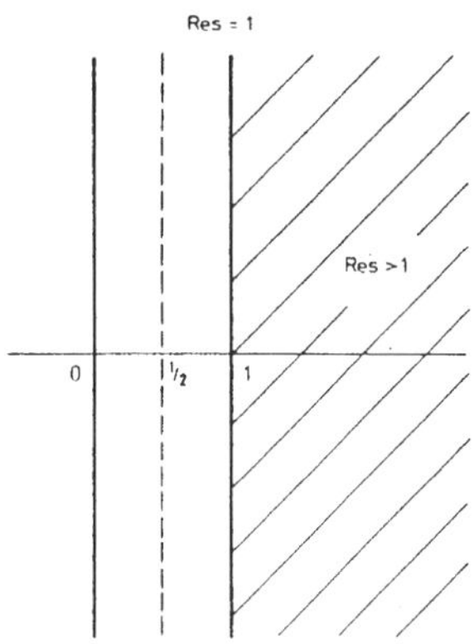

Figure 1

(cf. § B.2). [The zeta function has many zeros in the strip $0 < \text{Re } s < 1$. Riemann's conjecture (1859) that they all lie on the central line Re $s = \frac{1}{2}$ remains unproven to this day.]

For about fifty years now, the standard proofs of the prime number theorem have involved some form of Wiener's Tauberian theory for Fourier integrals, usually the Ikehara-Wiener theorem of § 1.2 (see Wiener [14] and cf. various books, for example Doetsch [2], Chandrasekharan [1], Heins [6]). Thus the proof of the prime number theorem has remained quite difficult until the recent breakthrough by D. J. Newman [10].

In 1980, he succeeded in replacing the Wiener theory in the proof by an ingenious application of complex integration theory, involving nothing more difficult than Cauchy's integral formula, together with suitable estimates. We present Newman's method in § 2 (applying it to Laplace integrals instead of Dirichlet series). In this method, the Ikehara-Wiener Tauberian theorem is replaced by a poor man's version which also readily leads to the prime number theorem.

Excellent accounts of the history of the prime number theorem and the zeta function may be found in the books of Landau [9], Ingham [7], Titchmarsh [13] and Edwards [3].

1.2. A Gem from Ingham's Work

Newman's method leads directly to the following pretty theorem which is already contained in work of Ingham [8]. However, Ingham used Wiener's method to prove his (more general) results.

Auxiliary Tauberian theorem. *Let $F(t)$ be bounded on $(0, \infty)$ and integrable over every finite subinterval, so that the Laplace transform*

$$G(z) = \int_0^\infty F(t)e^{-zt}dt \tag{1.3}$$

is well-defined and analytic throughout the open half-plane Re $z > 0$. *Suppose that $G(z)$ can be continued analytically to a neighborhood of every point on the imaginary axis. Then*

$$\int_0^\infty F(t)dt \text{ exists} \tag{1.4}$$

as an improper integral [and is equal to $G(0)$].

Under the given hypothesis, the Laplace integral (1.3) will converge everywhere on the imaginary axis. For the conclusion (1.4), it is actually sufficient that $G(z)$ have a continuous extension to the closed half-plane Re $z \geqslant 0$ which is smooth at $z = 0$: see § 2.

At first glance, the above theorem looks quite different from the

Ikehara-Wiener theorem [14]: Let $f(x)$ be nonnegative and nondecreasing on $[1, \infty)$ and such that the Mellin transform

$$g_0(s) = \int_1^\infty x^{-s}df(x) = -f(1) + s\int_1^\infty f(x)x^{-s-1}dx$$

exists for Re $s > 1$. Suppose that for some constant c, the function

$$g_0(s) - \frac{c}{s-1}$$

has a continuous extension to the closed half-plane Re $s \geqslant 1$. Then

$$f(x)/x \to c \quad \text{as} \quad x \to \infty.$$

This is an extremely useful theorem, but what could we do with the auxiliary theorem in the same direction? We will show that the latter has a corollary which is just as good for the application that we want to make.

1.3. A Poor Man's Ikehara-Wiener Theorem

We will establish the following

Corollary to the auxiliary theorem. *Let $f(x)$ be nonnegative, nondecreasing and $O(x)$ on $[1, \infty)$, so that its Mellin transform*

$$g(s) = s\int_1^\infty f(x)x^{-s-1}dx \tag{1.5}$$

is well-defined and analytic throughout the half-plane Re $s > 1$. *Suppose that for some constant c, the function*

$$g(s) - \frac{c}{s-1} \tag{1.6}$$

can be continued analytically to a neighborhood of every point on the line Re $s = 1$. *Then*

$$f(x)/x \to c \quad \text{as} \quad x \to \infty. \tag{1.7}$$

110

Derivation from the auxiliary theorem. Let $f(x)$ and $g(s)$ satisfy the hypotheses of the corollary. We set $x = e^t$ and define

$$e^{-t}f(e^t) - c = F(t),$$

so that $F(t)$ is bounded on $(0, \infty)$. Its Laplace transform will be

$$G(z) = \int_0^\infty \{e^{-t}f(e^t) - c\}\, e^{-zt}dt$$

$$= \int_1^\infty f(x)x^{-z-2}dx - \frac{c}{z} = \frac{1}{z+1}\left\{g(z+1) - \frac{c}{z} - c\right\}.$$

Thus by the hypothesis of the corollary, $G(z)$ can be continued analytically to a neighborhood of every point on the imaginary axis. We may now apply the auxiliary theorem from § 1.2.

What does its conclusion tell us? Setting $t = \log x$ we find that the improper integrals

$$\int_0^\infty \{e^{-t}f(e^t) - c\}\, dt = \int_1^\infty \frac{f(x) - cx}{x^2}\, dx \tag{1.8}$$

exist. Using the fact that $f(x)$ is an increasing function, one readily derives that $f(x) \sim cx$ in the sense of (1.7).

Indeed, suppose for a moment that $\limsup f(x)/x > c$ $(\geqslant 0)$. Then there would be a positive constant δ such that for certain arbitrarily large numbers y

$$f(y) > (c + 2\delta)y.$$

It would follow that

$$f(x) > (c + 2\delta)y > (c + \delta)x \quad \text{for} \quad y < x < \rho y$$

where $\rho = (c + 2\delta)/(c + \delta)$. But then

$$\int_y^{\rho y} \frac{f(x) - cx}{x^2}\, dx > \int_y^{\rho y} \frac{\delta}{x}\, dx = \delta \log \rho$$

for those same numbers y, contradicting the existence of (1.8).

One similarly disposes of the contingency $\liminf f(x)/x < c$ (in this case c would have to be positive and one would consider intervals $\theta y < x < y$ with $\theta < 1$ where $f(x) < (c - \delta)x$). Thus $f(x)/x \to c$.

1.4. Corollary ⇒ Prime Number Theorem

This step is routine to number theorists. One takes $f(x) = \psi(x)$, where $\psi(x)$ is that well-known function from prime number theory,

$$\psi(x) = \sum_{p^m \leqslant x} \log p \tag{1.9}$$

(the summation is over all prime powers not exceeding x). It is a simple fact (first noticed by Chebyshev) that $\pi(x) = O(x/\log x)$ or equivalently, $\psi(x) = O(x)$ (cf. § B.4 in the box for more details). Thus $f(x)$ is as the corollary wants it.

What about its Mellin transform $g(s)$? A standard calculation based on the Euler product in (1.2) shows that

$$g(s) = -\frac{\zeta'(s)}{\zeta(s)}, \quad \text{Re}\, s > 1$$

(cf. § B.3). Since $\zeta(s)$ behaves like $1/(s-1)$ around $s = 1$, the same is true for $g(s)$. The analyticity of $\zeta(s)$ at the points of the line Re $s = 1$ (different from $s = 1$) and its non-vanishing there imply that $g(s)$ can be continued analytically to a neighborhood of every one of those points (cf. §§ B.1, B.2). Thus

$$g(s) - \frac{1}{s-1}$$

has an analytic continuation to a neighborhood of the closed half-plane Re $s \geqslant 1$.

The conclusion of the corollary now tells us that

$$\psi(x)/x \to 1 \quad \text{as} \quad x \to \infty,$$

and this is equivalent to the prime number theorem (1.1) (cf. § B.4).

2. Newman's Beautiful Method

2.1. Proof of the Auxiliary Tauberian Theorem

Let $F(t)$ be bounded on $(0, \infty)$ and such that its Laplace transform $G(z)$ can be continued to a function (still called $G(z)$) which is analytic in a neighborhood of the closed half-plane Re $z \geqslant 0$. We may and will assume that

$$|F(t)| \leqslant 1, \quad t > 0.$$

For $0 < \lambda < \infty$ we write

$$G_\lambda(z) = \int_0^\lambda F(t)e^{-zt}dt. \tag{2.1}$$

Observe that $G_\lambda(z)$ is analytic for all z. We will show that

$$G_\lambda(0) = \int_0^\lambda F(t)dt \to G(0) \quad \text{as} \quad \lambda \to \infty.$$

111

Some details left out in 1.4

We begin with the necessary facts about the zeta function.

B.1. Analytic continuation of $\zeta(s)$. Simple transformations show that for Re $s > 2$

$$\zeta(s) = \sum_1^\infty \frac{n}{n^s} - \sum_1^\infty \frac{n-1}{n^s} = \sum_1^\infty \frac{n}{n^s} - \sum_1^\infty \frac{n}{(n+1)^s} = \sum_1^\infty n \left\{ \frac{1}{n^s} - \frac{1}{(n+1)^s} \right\} = \sum_1^\infty ns \int_n^{n+1} x^{-s-1}dx = s \sum_1^\infty \int_n^{n+1} [x] x^{-s-1}dx =$$

$$= s \int_1^\infty [x] x^{-s-1}dx, \tag{B.1}$$

where $[x]$ denotes the largest integer $\leqslant x$. Since first and final member are analytic for Re $s > 1$, the integral formula holds throughout that half-plane.

It is reasonable to compare the integral with

$$s \int_1^\infty x \cdot x^{-s-1}dx = \frac{s}{s-1} = 1 + \frac{1}{s-1}. \tag{B.2}$$

Combination of (B.1) and (B.2) gives

$$\zeta(s) - \frac{1}{s-1} = 1 + s \int_1^\infty ([x] - x) x^{-s-1}dx. \tag{B.3}$$

The new integral converges and represents an analytic function throughout the half-plane Re $s > 0$. Thus (B.3) provides an analytic continuation of the left-hand side to that half-plane.

B.2. Non-vanishing of $\zeta(s)$ *for* Re $s \geqslant 1$. The Euler product in (1.2) shows that $\zeta(s) \neq 0$ for Re $s > 1$. For Re $s = 1$ we will use Mertens's clever proof of 1898. The key fact is the inequality

$$3 + 4 \cos\theta + \cos 2\theta = 2(1 + \cos\theta)^2 \geqslant 0, \quad \theta \text{ real}. \tag{B.4}$$

Suppose that $\zeta(1 + ib)$ would be equal to 0, where b is real and $\neq 0$. Then the auxiliary analytic function

$$\varphi(s) = \zeta^3(s)\zeta^4(s + ib)\zeta(s + 2ib)$$

would have a zero for $s = 1$: the pole of $\zeta^3(s)$ could not cancel the zero of $\zeta^4(s + ib)$. It would follow that

$$\log|\varphi(s)| \to -\infty \quad \text{as} \quad s \to 1. \tag{B.5}$$

We now take s *real* and > 1. By the Euler product,

$$\log|\zeta(s + it)| = -\text{Re} \sum_p \log(1 - p^{-s-it}) = \text{Re} \sum_p \{p^{-s-it} + \tfrac{1}{2}(p^2)^{-s-it} + \tfrac{1}{3}(p^3)^{-s-it} + \ldots\} = \text{Re} \sum_1^\infty a_n n^{-s-it} \quad \text{with} \quad a_n \geqslant 0.$$

Thus

$$\log|\varphi(s)| = \text{Re} \sum_1^\infty a_n n^{-s}(3 + 4n^{-ib} + n^{-2ib}) = \sum_1^\infty a_n n^{-s}\{3 + 4\cos(b \log n) + \cos(2b \log n)\} \geqslant 0$$

because of (B.4), contradicting (B.5).

B.3. Representations for $\zeta'(s)/\zeta(s)$. Logarithmic differentiation of the Euler product in (1.2) gives

$$-\frac{\zeta'(s)}{\zeta(s)} = \sum_p \frac{p^{-s}}{1 - p^{-s}} \log p = \sum_p (p^{-s} + p^{-2s} + \ldots) \log p = \sum_1^\infty \Lambda(n) n^{-s}, \tag{B.6}$$

where $\Lambda(n)$ is the von Mangoldt function,

112

$$\Lambda(n) = \begin{cases} \log p \text{ if } n = p^m, \\ 0 \quad \text{ if } n \text{ is not a prime power.} \end{cases}$$

The corresponding partial sum function is equal to $\psi(x)$:

$$\psi(x) = \sum_{p^m \leqslant x} \log p = \sum_{n \leqslant x} \Lambda(n). \tag{B.7}$$

Proceeding as in (B.1), the series (B.6) leads to the integral representation

$$-\frac{\zeta'(s)}{\zeta(s)} = s \int_1^\infty \psi(x) x^{-s-1} dx, \quad \operatorname{Re} s > 1. \tag{B.8}$$

The integral converges and is analytic for Re $s > 1$ since by (B.7), $\psi(x) \leqslant x \log x$.

B.4. Relation between $\psi(x)$ and $\pi(x)$. By (B.7), $\psi(x)$ counts $\log p$ (for fixed p) as many times as there are powers $p^m \leqslant x$, hence

$$\psi(x) = \sum_{p \leqslant x} \left[\frac{\log x}{\log p}\right] \log p \leqslant \log x \sum_{p \leqslant x} 1 = \pi(x) \log x. \tag{B.9}$$

On the other hand, when $1 < y < x$,

$$\pi(x) = \pi(y) + \sum_{y<p\leqslant x} 1 \leqslant \pi(y) + \sum_{y<p\leqslant x} \frac{\log p}{\log y} < y + \frac{\psi(x)}{\log y}.$$

Taking $y = x/\log^2 x$ one thus finds that

$$\pi(x)\frac{\log x}{x} < \frac{1}{\log x} + \frac{\psi(x)}{x} \frac{\log x}{\log x - 2 \log\log x}. \tag{B.10}$$

Combination of (B.9) and (B.10) shows that

$$\lim \pi(x)\frac{\log x}{x} = 1 \quad \text{if and only if} \quad \lim \frac{\psi(x)}{x} = 1. \tag{B.11}$$

We finally indicate a standard proof of the estimate

$$\psi(x) = 0(x). \tag{B.12}$$

For positive integral n, the binomial coefficient $\binom{2n}{n}$ must be divisible by all primes p on $(n, 2n]$. Hence

$$\prod_{n<p\leqslant 2n} p \leqslant \binom{2n}{n} < 2^{2n},$$

so that

$$\sum_{2^{k-1}<p\leqslant 2^k} \log p \leqslant 2^k \log 2.$$

It follows that

$$\sum_{p\leqslant 2^k} \log p \leqslant (2^k + 2^{k-1} + \ldots + 1) \log 2 < 2^{k+1} \log 2$$

and hence there is a constant C such that

$$\sum_{p\leqslant x} \log p \leqslant Cx.$$

Since the prime powers higher than the first contribute at most a term $0(x^{1/2+\epsilon})$ to $\psi(x)$, inequality (B.12) follows.

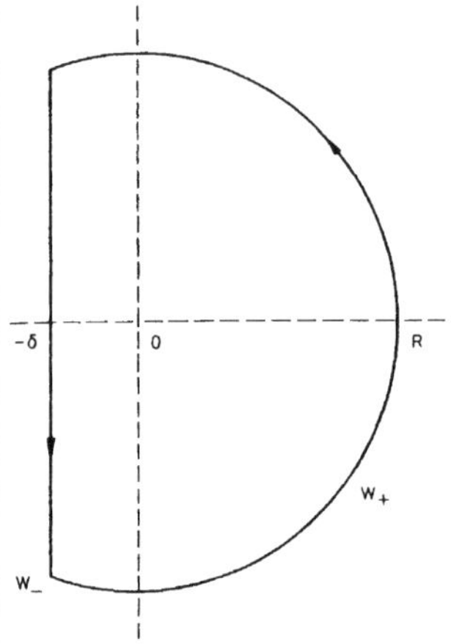

Figure 2

First idea. We try to estimate $G(0) - G_\lambda(0)$ with the aid of Cauchy's formula. Thus we look for a suitable path of integration W around 0. The simplest choice would be a circle, but we can not go too far into the left half-plane because we know nothing about $G(z)$ there. So for given $R > 0$, the positively oriented path W will consist of an arc of the circle $|z| = R$ and a segment of the vertical line $\mathrm{Re}\, z = -\delta$ (Fig. 2). Here the number $\delta = \delta(R) > 0$ is chosen so small that $G(z)$ is analytic on and inside W. We denote the part of W in $\mathrm{Re}\, z > 0$ by W_+, the part in $\mathrm{Re}\, z < 0$ by W_-. By Cauchy's formula,

$$G(0) - G_\lambda(0) = \frac{1}{2\pi i} \int_W \{G(z) - G_\lambda(z)\} \frac{1}{z}\, dz. \tag{2.2}$$

We have the following simple estimates:

for $x = \mathrm{Re}\, z > 0$,

$$|G(z) - G_\lambda(z)| = |\int_\lambda^\infty F(t) e^{-zt} dt| \leqslant \int_\lambda^\infty e^{-xt} dt = \frac{1}{x} e^{-\lambda x}; \tag{2.3}$$

for $x = \mathrm{Re}\, z < 0$,

$$|G_\lambda(z)| = |\int_0^\lambda F(t) e^{-zt} dt| \leqslant \int_0^\lambda e^{-xt} dt < \frac{1}{|x|} e^{-\lambda x}. \tag{2.4}$$

Second idea. Observe the similarity between the bounds obtained in (2.3) and (2.4)! It will be advantageous to multiply $G(z)$ and $G_\lambda(z)$ in (2.2) by $e^{\lambda z}$. This will not affect the left-hand side, but in estimating on W, the exponential $e^{-\lambda x}$ (large on W_-) will disappear from (2.3) and (2.4).

Third idea. Could we also get rid of the troublesome factor $1/|x|$ in the estimates which is bad near the imaginary axis? Yes, this can be done by adding the term z/R^2 to $1/z$ in (2.2), again without affecting the left-hand side. (For the experts: this trick is used also in Carleman's formula for the zeros of an analytic function in a half-plane, cf. [12].) The resulting modified formula is

$$G(0) - G_\lambda(0) = \frac{1}{2\pi i} \int_W \{G(z) - G_\lambda(z)\}\, e^{\lambda z} \left(\frac{1}{z} + \frac{z}{R^2}\right) dz. \tag{2.5}$$

Let us start harvesting. On the circle $|z| = R$,

$$\frac{1}{z} + \frac{z}{R^2} = \frac{2x}{R^2}. \tag{2.6}$$

Thus on W_+ the integrand $I(z)$ in (2.5) may be estimated as follows (see (2.3)):

$$|I(z)| \leqslant \frac{1}{x} e^{-\lambda x} e^{\lambda x} \frac{2x}{R^2} = \frac{2}{R^2}.$$

The corresponding integral is harmless:

$$|\frac{1}{2\pi i} \int_{W_+} I(z) dz| \leqslant \frac{1}{2\pi} \frac{2}{R^2} \pi R = \frac{1}{\mathrm{R}}! \tag{2.7}$$

Fourth idea. We now turn to the part of (2.5) due to W_-. Since $G_\lambda(z)$ is analytic for all z, we may replace the integral over W_- involving $G_\lambda(z)$ by the corresponding integral over the semi-circle

$$W_-^* : \{|z| = R\} \cap \{\mathrm{Re}\, z < 0\}$$

(Fig. 3). Cauchy's theorem and inequality (2.4) readily give

$$|\frac{1}{2\pi i} \int_{W_-} G_\lambda(z) e^{\lambda z} \left(\frac{1}{z} + \frac{z}{R^2}\right) dz| = |\frac{1}{2\pi i} \int_{W_-^*} \dots dz| < \frac{1}{\mathrm{R}}. \tag{2.8}$$

We finally tackle the remaining integral

$$\frac{1}{2\pi i} \int_{W_-} G(z) e^{\lambda z} \left(\frac{1}{z} + \frac{z}{R^2}\right) dz. \tag{2.9}$$

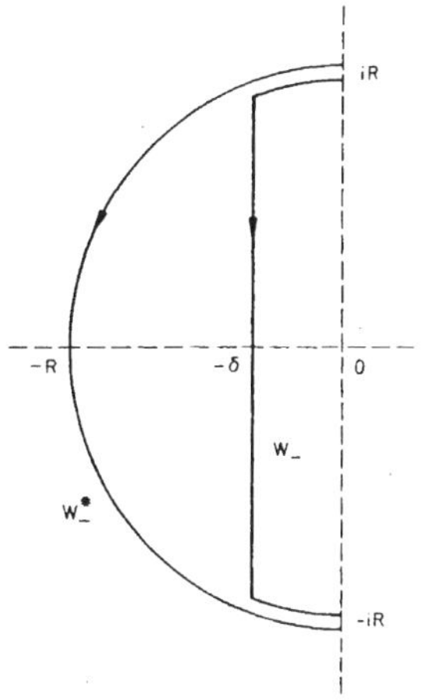

Figure 3

By the analyticity of $G(z)$ on W_- there will be a constant $B = B(R, \delta)$ such that

$$|G(z)\left(\frac{1}{z} + \frac{z}{R^2}\right)| \leqslant B \quad \text{on} \quad W_-.$$

It follows that

$$|G(z)e^{\lambda z}\left(\frac{1}{z} + \frac{z}{R^2}\right)| \leqslant Be^{\lambda x}.$$

Hence on the part of W_- where $x \leqslant -\delta_1 < 0$, the integrand in (2.9) tends to zero uniformly as $\lambda \to \infty$. On the remaining small part of W_- (we take $\delta_1 < \delta$ small), the integrand is bounded by B. Thus for fixed W, the integral in (2.9) tends to zero as $\lambda \to \infty$.

Conclusion. For given $\epsilon > 0$ one may choose $R = 1/\epsilon$. One next chooses δ so small that $G(z)$ is analytic on and inside W. One finally determines λ_0 so large that (2.9) is bounded by ϵ for all $\lambda > \lambda_0$. Then by (2.5) and (2.7)–(2.9),

$$|G(0) - G_\lambda(0)| < 3\epsilon \quad \text{for} \quad \lambda > \lambda_0.$$

In other words, $G_\lambda(0) \to G(0)$ as $\lambda \to \infty$.

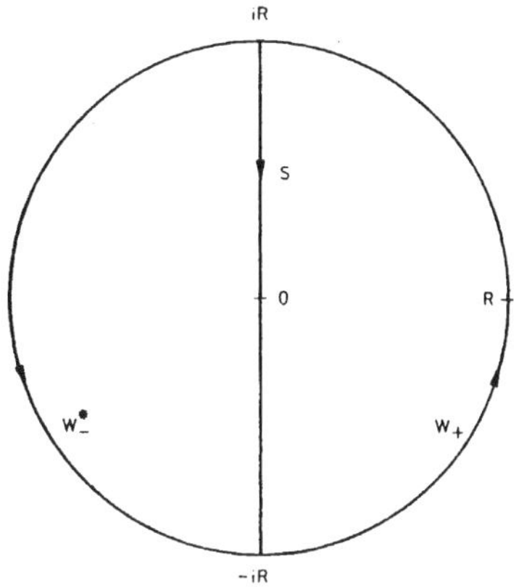

Figure 4

2.2. Relaxing the Conditions on $G(z)$

In the above proof, it is not really necessary to take $G(z)$ into the left half-plane. By modifying $F(t)$ on some finite interval one may assume that $G(0) = 0$. Then $G(z)/z$ will be analytic for $\operatorname{Re} z \geqslant 0$ and thus

$$G(0) = 0 = \frac{1}{2\pi i} \int_{W_+ \cup S} G(z)e^{\lambda z}\left(\frac{1}{z} + \frac{z}{R^2}\right)dz,$$

where S is the segment $[iR, -iR]$ of the imaginary axis (Fig. 4). For $G_\lambda(0)$ we integrate over the circle $|z| = R$. Subtracting, one obtains

$$G(0) - G_\lambda(0) = \frac{1}{2\pi i} \int_{W_+} \{G(z) - G_\lambda(z)\}e^{\lambda z}\left(\frac{1}{z} + \frac{z}{R^2}\right)dz$$
$$+ \frac{1}{2\pi i} \int_{iR}^{-iR} G(z)e^{\lambda z} \ldots dz - \frac{1}{2\pi i} \int_{W_-^*} G_\lambda(z) \ldots dz. \tag{2.10}$$

The first and third integral are just as before. To deal with the second integral one may apply integration by parts or the Riemann-Lebesgue lemma.

In order to arrive at (2.10), we have not used any analyticity of $G(z)$ at points of the imaginary axis. It would be more than enough to know that $G(z)/z$ can be extended continuously to $\operatorname{Re} z \geqslant 0$. The Riemann-Lebesgue lemma will then handle the second integral.

115

Conclusion. In the auxiliary Tauberian theorem, it is sufficient to require that $\{G(z) - G(0)\}/z$ can be extended continuously to the closed half-plane $\operatorname{Re} z \geqslant 0$.

References

1. K. Chandrasekharan: Introduction to analytic number theory. Springer, Berlin, 1968
2. G. Doetsch: Theorie und Anwendung der Laplace-Transformation. Springer, Berlin, 1937
3. H. M. Edwards: Riemann's zeta function. Academic Press, New York, 1974
4. P. Erdös: On a new method in elementary number theory. Proc. Nat. Acad. Sci. U. S. A. *35* (1949) 374–384
5. G. H. Hardy, E. M. Wright: An introduction to the theory of numbers. Oxford Univ. Press, London, 1954
6. M. Heins: Complex function theory. Academic Press, New York, 1968
7. A. E. Ingham: The distribution of prime numbers. Cambridge Univ. Press, 1932; reprinted by Hafner, New York 1971
8. A. E. Ingham: On Wiener's method in Tauberian theorems. Proc. London Math. Soc. (2) *38* (1935) 458–480
9. E. Landau: Handbuch der Lehre von der Verteilung der Primzahlen (2 volumes). Teubner, Leipzig, 1909; reprinted by Chelsea, New York, 1953
10. D. J. Newman: Simple analytic proof of the prime number theorem. American Math. Monthly *87* (1980) 693–696
11. A. Selberg: An elementary proof of the prime number theorem. Ann. of Math. (2) *50* (1949) 305–313
12. E. C. Titchmarsh: The theory of functions. Oxford Univ. Press, London, 1939
13. E. C. Titchmarsh: The theory of the Riemann zeta-function. Oxford Univ. Press, London, 1951
14. N. Wiener: Tauberian theorems. Ann. of Math. *33* (1932) 1–100

J. Korevaar
Mathematisch Instituut
Universiteit van Amsterdam
Roeterstraat 15
Amsterdam

12.17 H. Daboussi (1984)

Sur le Théorème des Nombres Premiers

In this paper, Daboussi gives another elementary proof of the prime number theorem. His proof differs from those of Selberg and Erdős (see Sections 12.10 and 12.11), in that it makes no use of the Selberg formula,

$$\sum_{p\leq x}\log^2 p+\sum_{pq\leq x}\log p\log q=2x\log x+O(x),$$

where p and q are primes. Daboussi uses the identity $\log n=\sum_{m|n}\Lambda(m)$ of Chebyshev and a sieving technique. He does not prove the classical form of the prime number theorem directly. Instead, he proves an equivalent statement, $M(x)/x\to 0$ as $x\to\infty$, where $M(x)=\sum_{n\leq x}\mu(n)$ and $\mu(n)$ is the Möbius function. [50]

C. R. Acad. Sc. Paris, t. 298, Série I, n° 8, 1984 161

THÉORIE DES NOMBRES. — Sur le Théorème des Nombres Premiers. Note de **Hédi Daboussi**, présentée par Jean-Pierre Kahane.

Remise le 19 décembre 1983.

Nous donnons une nouvelle démonstration du théorème des nombres premiers n'utilisant pas l'inégalité de Selberg.

NUMBER THEORY. — On the Prime Number Theorem.

We give a new elementary proof of the prime number theorem which does not use Selberg's inequality.

A H. Delange et P. Erdōs à l'occasion de leur 70ᵉ anniversaire.

I. 1. Soit $y \geqq 2$ et v_y, u_y deux fonctions complètement multiplicatives définies par :

$$v_y(p) = \begin{cases} 1 & \text{si } p \leqq y \\ 0 & \text{si } p > y \end{cases}, \qquad u_y(p) = \begin{cases} 1 & \text{si } p > y \\ 0 & \text{si } p \leqq y \end{cases}$$

la lettre p désignant des nombres premiers.

Λ désigne la fonction de Von Mangoldt, $\mathbf{1}$ la fonction constante égale à 1, μ la fonction de Möbius; ainsi, par exemple $\log n = (\Lambda * \mathbf{1})(n)$, où le signe $*$ désigne la convolution de Dirichlet. On notera $V_y(t) = \sum_{n \leqq t} v_y(n)\mu(n)$, $V_y^*(t) = \sum_{n \leqq t} v_y(n)$ et $M(t) = \sum_{n \leqq t} \mu(n)$. Nous montrerons que $\lim_{x \to \infty} |M(x)/x| = 0$.

I. 2. *Aperçu de la méthode.* — Nous démontrerons que pour tout $y \geqq 2$:

$$\overline{\lim_{x \to \infty}} |M(x)/x| \leqq \Big\{ \prod_{p \leqq y} (1 - (1/p)) \Big\} \int_1^\infty (|V_y(t)|/t^2)\, dt. \tag{1}$$

Soit $\alpha = \overline{\lim} |M(x)|/x$; évidemment $\alpha \leqq 1$.

Nous établirons qu'il existe $\delta > 1$ tel que pour tout β, $\alpha < \beta < 2$, on a :

$$\int_1^y (|V_y(t)|/t^2)\, dt \leqq \beta/\delta \log y + O(1), \tag{2}$$

et que l'on a :

$$\int_y^\infty (|V_y(t)|/t^2)\, dt \leqq \beta(C-1) \log y + o(\log y), \tag{3}$$

où $C = \lim_{y \to \infty} (\log y)^{-1} \prod_{p \leqq y} (1 - (1/p))^{-1}$ (Il est connu que $C = e^\gamma$, où γ est la constante d'Euler, nous n'en ferons pas usage).

(3) entraînera que $\alpha \leqq \beta(1 - C^{-1}(1 - (1/\delta))$, le facteur de β étant < 1, il en résulte, en faisant tendre β vers α, que $\alpha = 0$).

II. 1. Les séries $\Sigma v_y(n)/n$ et $\Sigma(v_y(n)/n)\mu(n)$ sont absolument convergentes avec pour sommes $\prod_{p \leqq y} (1 - (1/p))^{-1}$ et $\prod_{p \leqq y} (1 - (1/p))$.

On en déduit (en partant de $u_y = v_y \mu * \mathbf{1}$) que :

$$\lim_{x \to \infty} (1/x) \sum_{n \leqq x} u_y(n) \quad \text{existe et est égale à} \quad \prod_{p \leqq y} (1 - (1/p)).$$

D'après les définitions de v_y et u_y, il est clair que $\mu(n) = (\mu u_y * \mu v_y)(n)$ pour tout entier

n. Ainsi :

$$\mathrm{M}(x)=\sum_{n\leq x}\mu(n)\,u_y(n)\,\mathrm{V}_y(x/n).$$

Notons $d_1=1<d_2\ldots<d_q$ la suite finie des entiers sans facteur carré ayant tous leurs diviseurs premiers $\leqq y$, et remarquons que, si n vérifie $x/d_{j+1}<n\leqq x/d_j$, alors $\mathrm{V}_y(x/n)=\mathrm{V}_y(d_j)$. On obtient ainsi :

$$\mathrm{M}(x)=\sum_{j=1}^{q-1}\mathrm{V}_y(d_j)\sum_{x/d_{j+1}<n\leqq x/d_j}u_y(n)\,\mu(n)+\mathrm{V}_y(d_q)\sum_{n\leqq x/d_q}u_y(n)\,\mu(n),$$

$$\overline{\lim_{x\to\infty}}\,|\mathrm{M}(x)/x|\leqq\sum_{j=1}^{q-1}|\mathrm{V}_y(d_j)|\lim_{x\to\infty}(1/x)\sum_{x/d_{j+1}<n\leqq x/d_j}u_y(n)+|\mathrm{V}_y(d_q)|\lim_{x\to\infty}(1/x)\sum_{n\leqq x/d_q}u_y(n).$$

Ce qui fournit (1).

II. 2. On sait qu'il existe $\mathrm{M}>0$ tel que, pour tous nombres a et b positifs :

$$\left|\int_a^b(\mathrm{M}(t)/t^2)\,dt\right|\leqq\mathrm{M}.$$

Prenons $\alpha<\beta<2$ et x_β tel que pour $x\geqq x_\beta$, $|\mathrm{M}(x)|\leqq\beta x$.

Soit $\delta=\min(2,\ 1+(\alpha^2/4\,\mathrm{M}))$.

Puisque $v_y(n)=1$ si $n\leqq y$ et donc $\mathrm{V}_y(t)=\mathrm{M}(t)$ si $t\leqq y$, l'inégalité (2) s'écrit :

$$(2)'\qquad\int_1^y(|\mathrm{M}(t)|/t^2)\,dt\leqq\beta/\delta\,\log y+O(1).$$

Une telle inégalité intervient dans la méthode de Selberg ([1], [3]). L'inégalité (2)′ s'établit par la méthode utilisée en [2] pour prouver le lemme 5. 8.

III. Quelques lemmes.

III. 1. Lemme 1. — *Soit h une fonction définie sur $[y,+\infty[$, positive, décroissante et possédant une dérivée continue. On a :*

Pour tout $t\geqq y$: $$\sum_{p\leqq y}(\log p/p)\,h(pt)=\int_t^{yt}(h(v)/v)\,dv+O(h(y)).$$

Pour tout $t\geqq 1$: $$\sum_{y/t<p\leqq y}(\log p/p)\,h(pt)=\int_y^{yt}(h(v)/v)\,dv+O(h(y)).$$

Ce lemme s'obtient par intégration par parties grâce à la relation :

$$\sum_{p\leqq t}\log p/p=\log t+O(1).$$

III. 2. Lemme 2. — *Posons, pour $s>0$:*

$$k(s)=\int_0^\infty e^{-sx}\,e^{f(x)}\,dx,\qquad où\quad f(x)=\int_0^x((1-e^{-u})/u)\,du.$$

Alors la fonction k est positive, décroissante et indéfiniment dérivable. De plus :

$$(4)\qquad sk(s)-\int_s^{s+1}k(u)\,du=1\quad pour\ tout\ s>0.$$

[Il est immédiat que $\int_s^{s+1}k(u)\,du=\int_0^\infty e^{f(x)}\,e^{-sx}\,f'(x)\,dx$. En intégrant par parties on obtient (4)].

III.3. LEMME 3. — *Soit k la fonction définie au lemme* 2, *on a* :

$$\int_1^2 k(u)\,(2-u)\,du = C-1. \tag{5}$$

Nous établirons ce lemme par une méthode purement arithmétique, une méthode analogue nous fournira (3). De la relation $\log = \Lambda * \mathbf{1}$, nous déduisons $v_y \log = v_y \Lambda * v_y$, et donc :

$$\sum_{n \leqq t} v_y(n) \log n = \sum_{n \leqq t} v_y(n)\,\Lambda(n)\,V_y^*(t/n);$$

ou encore, $V_y^*(t)\log t = \sum_{n \leqq t} v_y(n)\,\Lambda(n)\,V_y^*(t/n) + \sum_{n \leqq t} v_y(n)\log(t/n)$.

Par définition de Λ et v_y, on a :

$$V_y^*(t)\log t = \sum_{\substack{p \leqq t\\ p \leqq y}} \log p\, V_y^*(t/p) + \sum_{\substack{p \leqq y\\ p^r \leqq t,\, r \geqq 2}} \log p\, V_y^*(t/p^r) + \sum_{n \leqq t} v_y(n)\log(t/n).$$

Posons pour $t > y$: $h(t) = (1/\log y)\,k(\log t/\log y)$. Alors :

$$\int_y^\infty (V_y^*(t)/t^2)\log t \,.\, h(t)\,dt = \int_y^\infty \sum_{\substack{p \leqq t\\ p \leqq y}} \log p\, V_y^*(t/p)\,(h(t)/t^2)\,dt + E_1 + E_2, \tag{6}$$

où

$$E_1 = \int_y^\infty \sum_{\substack{p \leqq y\\ p^r \leqq t,\, r \geqq 2}} \log p\, V_y^*(t/p^r)\,(h(t)/t^2)\,dt, \qquad E_2 = \int_y^\infty \sum_{n \leqq t} v_y(n)\log(t/n)\,(h(t)/t^2)\,dt.$$

La décroissance de h entraîne que :

$$E_1 \leqq h(y).\Big(\sum_{\substack{r \geqq 2\\ p}} \log p/p^r\Big).\int_1^\infty (V_y^*(u)/u^2)\,du, \qquad E_2 \leqq h(y).\Big(\sum_n v_y(n)/n\Big).\int_1^\infty (\log t/t^2)\,dt.$$

Par ailleurs $\sum v_y(n)/n = \int_1^\infty (V_y^*(u)/u^2)\,du = O(\log y)$, ce qui implique que $E_1 = O(1)$ et $E_2 = O(1)$. L'intégrale à droite de (6) s'écrit :

$$\int_y^\infty \sum_{\substack{p \leqq t\\ p \leqq y}} \log p\, V_y^*(t/p)\,(h(t)/t^2)\,dt = \sum_{p \leqq y} \log p/p \int_{y/p}^{y} (V_y^*(t)/t^2)\,h(pt)\,dt$$

$$+ \sum_{p \leqq y} \log p/p \int_y^\infty (V_y^*(t)/t^2)\,h(pt)\,dt$$

$$= \int_1^y V_y^*(t)/t^2 \sum_{y/t < p \leqq y} (\log p/p)\,h(pt)\,dt + \int_y^\infty V_y^*(t)/t^2 \sum_{p \leqq y} (\log p/p)\,h(pt)\,dt,$$

ce qui, grâce au lemme 1, donne :

$$\int_y^\infty V_y^*(t)/t^2 \left\{ \log t \,.\, h(t) - \int_t^{yt} (h(v)/v)\,dv \right\} dt = \int_1^y V_y^*(t)/t^2 \left(\int_y^{yt} (h(v)/v)\,dv \right) dt + O(1).$$

Il découle du lemme 2 et de la définition de h que :

$$\log t \,.\, h(t) - \int_t^{yt} (h(v)/v)\,dv = 1 \quad \text{pour tout } t \geqq 1.$$

En utilisant également le fait que $V_y^*(t) = [t] = t + O(1)$ pour tout $t \leqq y$, on obtient par un calcul simple que :

$$\int_y^\infty (V_y^*(t)/t^2)\,dt = \left(\int_1^2 k(u)(2-u)\,du\right)\log y + O(1).$$

Par ailleurs :

$$\int_y^\infty (V_y^*(t)/t^2)\,dt = \sum_n v_y(n)/n - \sum_{n\leq y} v_y(n)/n = (C+o(1))\log y - \log y + O(1).$$

Ces deux formes de l'intégrale fournissent l'égalité (5).

IV. Preuve de l'inégalité (3). — De la relation : $-\mu\log = \mu * \Lambda$, nous déduisons que $-v_y\mu\log = v_y\mu * v_y\Lambda$.

En raisonnant comme au paragraphe précédent, nous avons successivement :

$$|V_y(t)|\log t \leqq \sum_{n\leqq t} v_y(n)\Lambda(n)|V_y(t/n)| + \sum_{n\leqq t} v_y(n)\log(t/n),$$

et, avec la fonction h définie plus haut,

$$\int_y^\infty |V_y(t)|/t^2\left\{\log t \,.\, h(t) - \int_t^{yt}(h(v)/v)\,dv\right\}dt$$

$$\leqq \int_1^y |V_y(t)|/t^2\left(\int_y^{yt}(h(v)/v)\,dv\right)dt + O(1),$$

et finalement :

$$\int_y^\infty (|V_y(t)|/t^2)\,dt \leqq \int_1^y |M(t)|/t^2\left(\int_y^{yt}(h(v)/v)\,dv\right)dt + O(1).$$

En majorant $|M(t)|$ par βt pour $t\geqq x_\beta$ et en effectuant l'intégration à droite, on a :

$$\int_y^\infty (|V_y(t)|/t^2)\,dt \leqq \beta\left(\int_1^2 k(u)(2-u)\,du\right)\log y + O(1),$$

et donc l'inégalité (3) grâce au lemme 3.

Références bibliographiques

[1] P. Erdős, On a new method in elementary number theory which leads to an elementary proof of the prime Number Theorem, *Proc. Nat. Acad. Sc. U.S.A.*, 35, 1949, p. 374-384.

[2] N. Levinson, A motivated account of an elementary proof of the prime number theorem, *Amer. Math. Monthly*, 76, 1969, p. 225-245.

[3] A. Selberg, An elementary proof of the prime number theorem, *Ann. of Math.*, (2), 50, 1949, p. 305-313.

H. D. : *Université Paris-Sud, Département de Mathématiques*, bât. 425, 91405 *Orsay*.

12.18 A. Hildebrand (1986)

The Prime Number Theorem via the Large Sieve

In this paper, Hildebrand gives another elementary proof of the prime number theorem by proving the equivalent statement that $\frac{1}{x}\sum_{n\le x}\mu(n) = o(1)$, where $\mu(n)$ is the Möbius function. His proof, like that of Daboussi, does not use Selberg's formula. Hildebrand's proof makes use of the large sieve to show that $\frac{1}{x}\sum_{n\le x}\mu(n) - \frac{1}{x'}\sum_{n\le x'}\mu(n) \to 0$ uniformly as $x \to \infty$, for $x \le x' \le x^{1+\eta(x)}$, where $\eta(x) \to 0$. It is interesting to write that the function $M(x) := \frac{1}{x}\sum_{n\le x}\mu(n)$ in this paper is slightly different from the usual Mertens' function (see Section 5.7).

THE PRIME NUMBER THEOREM VIA THE LARGE SIEVE

ADOLF HILDEBRAND

§1. *Introduction.* In the last three decades there appeared a number of elementary proofs of the prime number theorem (PNT) in the literature (see [3] for a survey). Most of these proofs are based, at least in part, on ideas from the original proof by Erdős [5] and Selberg [12]. In particular, one of the main ingredients of the Erdős-Selberg proof, Selberg's formula

$$\sum_{p\leqslant x} \log^2 p + \sum_{pq\leqslant x} \log p \log q = 2x \log x + O(x) \tag{1}$$

(where p and q run through primes) appears, in some form, in almost all these proofs.

Several authors [**1, 2, 8, 10**] have given direct elementary proofs of the relation (μ being the Moebius function)

$$M(x) = \frac{1}{x} \sum_{n\leqslant x} \mu(n) = o(1) \qquad (x\to\infty), \tag{2}$$

which is known to be "equivalent" to the PNT. With the exception of a recent proof by Daboussi [**2**], these proofs are based on identities such as

$$M(x) \log x = -\sum_{p\leqslant x} \frac{\log p}{p} M\left(\frac{x}{p}\right) + O(1), \tag{3}$$

which are similar in structure of Selberg's formula (1) and play the same role in the proof of (2) as Selberg's formula does in the direct proof of the PNT. Daboussi's elementary proof of (2) does not use Selberg's formula (1) nor its analogue (3), and constitutes in fact the first elementary proof of the PNT, which is fundamentally different from the original Erdős-Selberg proof.

We shall give here a new elementary proof of (2), again without using Selberg type formulae. Its main feature is the application of a large sieve type inequality in order to show that $M(x)$ varies slowly over fairly large intervals. More precisely, we shall show in Lemma 4 below that $M(x)-M(x')\to 0$ as $x\to\infty$, uniformly for $x\leqslant x'\leqslant x^{1+\eta(x)}$, provided $\eta(x)\to 0$. Relation (2) is an almost immediate consequence of this result.

The central idea of the proof, the application of the large sieve, has already been exploited in [7] to obtain a new proof for Wirsing's mean value theorem. This result contains (2) as a special case, but since the proof in [7] made use of the PNT, it did not yield a new proof of the PNT. Our main task here, which will take up the largest part of the proof, is to eliminate the application of the PNT and replace it by the elementary prime number estimates from Mertens' theory, together with a sieve upper bound for primes in short intervals.

[MATHEMATIKA, **33** (1986), 23-30]

We remark that it is possible to prove, by the same method, a quantitative version of (2), namely

$$M(x) \ll (\log\log x)^{-1/2} \qquad (x \geq 3).$$

Since this result is relatively weak in comparison with the known elementary error term estimates for the PNT, and in order to keep the proof more transparent, we shall confine ourselves to the proof of the asymptotic relation (2).

§2. *Lemmas.* In the first lemma, we state the mentioned large sieve type inequality. It is due to Elliott [**4**, Lemma 4.7], who gave a simple and elementary proof for it via the Turán–Kubilius inequality.

LEMMA 1. *There exists an absolute constant $C>0$ such that for all complex numbers a_n, $1 \leq n \leq x$, the inequality*

$$\sum_{p \leq x} \frac{1}{p} \left| \frac{p}{x} \sum_{\substack{n \leq x \\ p|n}} a_n - \frac{1}{x} \sum_{n \leq x} a_n \right|^2 \leq C \frac{1}{x} \sum_{n \leq x} |a_n|^2$$

holds.

The next lemma gives an upper bound for primes in short intervals. The result is, in a stronger form, a standard consequence of Selberg's upper bound sieve, see, *e.g.*, [**6**, Theorem 3.7]. It may also be deduced from Montgomery's version of the large sieve [**9**, Corollary 3.2] or from Selberg's formula (1).

LEMMA 2. *The estimate*

$$\pi(x+y) - \pi(x) \leq (2+o(1)) \frac{y}{\log y}$$

holds, as $y \to \infty$, uniformly for $x \geq 1$.

COROLLARY. *The estimate*

$$\sum_{x<p \leq x+y} \frac{\log p}{p} \leq (2+o(1)) \log \frac{x+y}{x}$$

holds, as $y \to \infty$, uniformly for $x \geq y$.

Proof. Dividing the interval $(x, x+y]$ into subintervals of length y/k, where $k = k(y)$ tends to infinity sufficiently slowly, as $y \to \infty$, the result follows from Lemma 2.

LEMMA 3. *Let $0 < \varepsilon \leq 1$ be given. For any $x' \geq x \geq 2$, there exists a number λ, $1 \leq \lambda \leq \lambda_0$ such that the inequality*

$$\sum_{y<p \leq y(1+\varepsilon)} \frac{\log p}{p} \geq \delta \tag{4}$$

holds both for $y=\lambda x$ and for $y=\lambda x'$. Here $\delta=\delta(\varepsilon)>0$ and $\lambda_0=\lambda_0(\varepsilon)\geq 1$ are constants depending only on ε.

Remark. In the course of his proof of the PNT, Erdős [**5**] established (4) (in an equivalent form) for *all* sufficiently large y. His proof relied on Selberg's formula (1). We need however only the weaker result stated in the lemma, the proof of which is simpler and does not require Selberg's formula.

Proof. Given $0<\varepsilon\leq 1$ and $x\geq 2$, put $\varepsilon_1=\varepsilon/3$ and

$$x_i=x(1+\varepsilon_1)^i \qquad (i\geq 0).$$

Fix a number $\delta>0$ and let I be the set of indices $i\geq 0$, for which

$$\sum_{x_i<p\leq x_{i+1}}\frac{\log p}{p}\geq\delta$$

holds. Moreover, put

$$\bar{I}=\{i\geq 0:\quad i\in I \quad\text{or}\quad i+1\in I\}.$$

Since

$$x_i<x_{i+1}<x_{i+2}=(1+\varepsilon_1)^2x_i\leq(1+\varepsilon)x_i \qquad (i\geq 0),$$

we see that, for every $i\in\bar{I}$, (4) holds with $y=x_i$. We shall show that if δ is sufficiently small in terms of ε, then for some i_0, depending only on ε, $\bar{I}$ contains more than $i_0/2$ indices $i<i_0$. From this assertion it follows that any two sets $\bar{I}$ and $\bar{I}'$, defined as before with respect to given numbers $x'\geq x\geq 2$, have a common element, say i, in the interval $[0, i_0)$, and in view of the above remark we obtain the conclusion of the lemma with $\lambda=(1+\varepsilon_1)^i$ and $\lambda_0=(1+\varepsilon_1)^{i_0}$.

To prove our claim, let $i_0\geq 1$ be for the moment unspecified and denote by N and $\bar{N}$ the cardinalities of $I\cap[0, i_0)$ and $\bar{I}\cap[0, i_0)$ respectively. From Mertens' formula

$$\sum_{p\leq y}\frac{\log p}{p}=\log y+O(1) \qquad (y\geq 2)$$

and the Corollary to Lemma 2 we get

$$\begin{aligned}\log(1+\varepsilon_1)^{i_0}+O(1)&=\sum_{x<p\leq x(1+\varepsilon_1)^{i_0}}\frac{\log p}{p}=\sum_{0\leq i<i_0}\sum_{x_i<p\leq x_{i+1}}\frac{\log p}{p}\\&\leq 2N\log(1+\varepsilon_1)+o(i_0)+(i_0-N)\delta,\end{aligned}$$

whence

$$N\geq i_0\left(\tfrac{1}{2}+o(1)-\frac{\delta}{2\log(1+\varepsilon_1)}+O\left(\frac{1}{i_0\log(1+\varepsilon_1)}\right)\right). \tag{5}$$

(Here the "o"-notation refers to the limiting behaviour as $i_0\to\infty$, and is uniform with respect to $x\geq 2$). For $\delta\leq\varepsilon/10$ and sufficiently large i_0 the right-hand side of (6) becomes positive, and we now fix such an index $i_0=i_0(\varepsilon)$. Thus the interval $[0, i_0)$ contains an element of I, and replacing $x=x_0$ by x_i, $i\geq 1$, we see that every interval on the positive real axis of length i_0 contains an element of I.

Next, let $i_0' > 6i_0$ and define N' and $\bar{N}'$, as before with respect to i_0'. If now

$$\bar{N}' - N' \leqslant \frac{i_0'}{3i_0},$$

then there are at most $(i_0'/3i_0)+1$ "blocks" of consecutive indices $i \leqslant i_0'$ not belonging to I. Since by the above remark each of these blocks has length $\leqslant i_0$, it follows that

$$\bar{N}' \geqslant N' \geqslant i_0' - \tfrac{1}{3}i_0' - i_0 > i_0'/2.$$

But if

$$\bar{N}' - N' > \frac{i_0'}{3i_0},$$

then we have by (5) (with i_0' and N' in place of i_0 and N)

$$\bar{N}' \geqslant N' + \frac{i_0'}{3i_0} \geqslant i_0'\left(\tfrac{1}{2} + o(1) - \frac{\delta}{2\log(1+\varepsilon_1)} + \frac{1}{3i_0} + O\left(\frac{1}{i_0'\log(1+\varepsilon_1)}\right)\right),$$

and choosing $\delta = \delta(\varepsilon)$ sufficiently small and i_0' sufficiently large, we get again $\bar{N}' > i_1'/2$.

This proves our initial claim and hence the lemma.

§3. *Proof of the estimate* (2). The proof is based on the following lemma.

LEMMA 4. *Let $\eta(x)$ be a non-negative function tending to zero as x tends to infinity. Then the relation*

$$M(x') = M(x) + o(1)$$

holds, as $x \to \infty$, uniformly for $x \leqslant x' \leqslant x^{1+\eta(x)}$.

The estimate (2) is an almost immediate consequence of this result. In fact, with $\eta = \eta(x) = (\log x)^{-1/2}$, the lemma yields, as $x \to \infty$,

$$\begin{aligned} M(x) &= \frac{1}{\eta \log x} \int_x^{x^{1+\eta}} \frac{M(x')}{x} \, dx' + o(1) \\ &= \frac{1}{\eta \log x} \sum_{n \leqslant x^{1+\eta}} \mu(n)\left(\frac{1}{\max(n,x)} - x^{-1-\eta}\right) + o(1) \\ &= \frac{1}{\sqrt{\log x}} \sum_{x < n \leqslant x^{1+\eta}} \frac{\mu(n)}{n} + o(1), \end{aligned}$$

and in view of the well-known elementary estimate

$$\sum_{n \leqslant y} \frac{\mu(n)}{n} = O(1) \qquad (y \geqslant 1)$$

(2) follows.

Proof of Lemma 4. Fix $0<\varepsilon\leqslant 1$. We shall show that if $\eta>0$ is sufficiently small in terms of ε, then we have

$$|M(x)-M(x')|\ll\varepsilon \tag{6}$$

uniformly for all sufficiently large x and $x\leqslant x'\leqslant x^{1+\eta}$. By letting $\varepsilon\to 0$, this implies the assertion of the lemma.

Let now $0<\eta\leqslant\frac{1}{2}$ and fix $2\leqslant x\leqslant x'\leqslant x^{1+\eta}$. We apply Lemma 1 with $a_n=\mu(n)$. Noting that, for all $p\leqslant x$,

$$\frac{p}{x}\sum_{\substack{n\leqslant x\\ p\mid n}}\mu(n)=\frac{p}{x}\mu(p)\sum_{\substack{n\leqslant x/p\\ p\nmid n}}\mu(n)+O\left(\frac{1}{p}\right)=-M\left(\frac{x}{p}\right)+O\left(\frac{1}{p}\right),$$

we obtain

$$\sum_{p\leqslant x}\frac{1}{p}\left|M(x)+M\left(\frac{x}{p}\right)\right|^2\ll 1.$$

By the Cauchy-Schwarz inequality it follows that, for any set $\mathscr{P}$ of primes $\leqslant x$,

$$\begin{aligned}\left(\sum_{p\in\mathscr{P}}\frac{1}{p}\right)M(x)&=-\sum_{p\in\mathscr{P}}\frac{M(x/p)}{p}+O\left(\sum_{p\in\mathscr{P}}\frac{1}{p}\left|M(x)+M\left(\frac{x}{p}\right)\right|\right)\\&=-\sum_{p\in\mathscr{P}}\frac{M(x/p)}{p}+O\left(\left(\sum_{p\in\mathscr{P}}\frac{1}{p}\right)^{1/2}\right).\end{aligned} \tag{7}$$

An analogous identity, (7)$'$ say, holds with x' instead of x and an arbitrary set $\mathscr{P}'$ of primes $\leqslant x'$. We shall show that, with an appropriate choice of the sets $\mathscr{P}$ and $\mathscr{P}'$, the right-hand sides of (7) and (7)$'$ nearly cancel each other, and $\mathscr{P}$ and $\mathscr{P}'$ have approximately the same (large) sum of reciprocals. This will lead to the desired estimate (6).

Put

$$x_0=x^{\sqrt{\eta}},\qquad x_0'=x_0(x'/x).$$

By a repeated application of Lemma 3 we obtain a sequence $(x_j)_{j\geqslant 1}$, satisfying

$$x_j(1+\varepsilon)\leqslant x_{j+1}\leqslant\lambda_0 x_j(1+\varepsilon)\qquad(j\geqslant 0),$$

such that (4) holds for $y=x_j$ and $y=x_j'=x_j(x'/x)$ and all $j\geqslant 1$, where $\lambda_0=\lambda_0(\varepsilon)$ and $\delta=\delta(\varepsilon)$ are the constants of Lemma 3. Define j_0 by $x_{j_0}\leqslant x<x_{j_0}+1$; note that $j_0\geqslant 2$, if x is sufficiently large, as we may assume. Put

$$I_j=(x_j,x_j(1+\varepsilon)],\qquad I_j'=(x_j',x_j'(1+\varepsilon)].$$

By construction, the intervals I_j, $1\leqslant j<j_0$, and I_j', $1\leqslant j<j_0$, are pairwise disjoint and contained in $(x_0,x]$ and $(x_0',x']$, respectively. Now choose sets of primes

$$\mathscr{P}\subset\bigcup_{1\leqslant j<j_0}I_j\quad\text{and}\quad\mathscr{P}'\subset\bigcup_{1\leqslant j<j_0}I_j'$$

such that, for $1\leqslant j<j_0$,

$$\left|\sum_{p\in\mathscr{P}\cap I_j}\frac{\log p}{p}-\delta\right|\leqslant\frac{\log x_j(z+\varepsilon)}{x_j}, \tag{8}$$

and

$$\left| \sum_{p \in \mathcal{P}' \cap I_j'} \frac{\log p}{p} - \delta \right| \leq \frac{\log x_j'(1+\varepsilon)}{x_j'}. \tag{8)'$$

Since (4) holds with $y = x_j$ and $y = x_j'$, this can be achieved by discarding from each I_j or I_j' a suitable number of primes.

Let

$$S = \sum_{p \in \mathcal{P}} \frac{1}{p}, \qquad S' = \sum_{p \in \mathcal{P}'} \frac{1}{p}.$$

From (8) and (8)′ we get

$$\begin{aligned} S &= \left(1 + O\left(\frac{1}{\log x_0}\right)\right) \sum_{1 \leq j < j_0} \frac{1}{\log x_j} \sum_{p \in \mathcal{P} \cap I_j} \frac{\log p}{p} \\ &= \left(1 + O\left(\frac{1}{\delta \log x_0}\right)\right) \sum_{1 \leq j < j_0} \frac{\delta}{\log x_j}, \end{aligned}$$

$$S' = \left(1 + O\left(\frac{1}{\delta \log x_0}\right)\right) \sum_{1 \leq j < j_0} \frac{\delta}{\log x_j'}.$$

Since

$$0 < \frac{\log x_j' - \log x_j}{\log x_j} = \frac{\log (x'/x)}{\log x_j} \leq \frac{\eta \log x}{\log x_0} = \sqrt{\eta},$$

it follows that

$$S = S'\left(1 + O(\sqrt{\eta}) + O\left(\frac{1}{\delta \log x_0}\right)\right) \tag{9}$$

and, in particular,

$$S \ll S' \ll S, \tag{9)'$$

if η is sufficiently small and x sufficiently large, as we may assume. Moreover, we have

$$\begin{aligned} S' \gg S \gg_\varepsilon & \sum_{1 \leq j < j_0} \frac{1}{\log x_j} \\ \geq & \sum_{1 \leq j < j_0} \frac{1}{\log \{x_0(\lambda_0(1+\varepsilon))^j\}} \gg_\varepsilon \log \frac{\log x}{\log x_0} = \log \frac{1}{\sqrt{\eta}}. \end{aligned} \tag{10}$$

We now subtract the identity (7)′ from (7). Using (8), (8)′, (9)′ and the trivial bound

$$|M(y+z) - M(y)| \ll \frac{z+1}{y} \qquad (y \geq 1,\, z \geq 0),$$

we obtain

$$SM(x) - S'M(x') = -\sum_{1\le j<j_0}\left\{\sum_{p\in\mathscr{P}\cap I_j}\frac{M(x/p)}{p} - \sum_{p\in\mathscr{P}'\cap I'_j}\frac{M(x'/p)}{p}\right\} + O(\sqrt{S})$$

$$= -\sum_{1\le j<j_0}\frac{M(x/x_j)}{\log x_j}\left\{\sum_{p\in\mathscr{P}\cap I_j}\frac{\log p}{p} - \sum_{p\in\mathscr{P}'\cap I'_j}\frac{\log p}{p}\right\}$$

$$+ O\left(\varepsilon S + \frac{S}{\log x_0} + \sqrt{S}\right)$$

$$= O\left(\frac{j_0}{x_0} + \varepsilon S + \frac{S}{\log x_0} + \sqrt{S}\right),$$

whence, by (9),

$$|M(x) - M(x')| \ll \varepsilon + \sqrt{\eta} + \frac{1}{\delta \log x_0} + \frac{j_0}{x_0 S} + \frac{1}{\sqrt{S}}.$$

In view of (10) and the definitions of x_0 and j_0, the right-hand side is of order $O(\varepsilon)$, provided x is sufficiently large and η sufficiently small. This proves (6) and hence the lemma.

§4. *Concluding remarks.* It is interesting to compare the identity (7), on which our proof is based, with the identity (3), which has been used in previous proofs of (2). In both instances, $M(x)$ is expressed as a certain average over values $-M(x/p)$, plus an error term. The essential difference between (3) and (7) lies in the fact that in (3) this average is extended over all primes $p \le x$, while in (7) the average is taken over primes belonging to an arbitrary set $\mathscr{P}$. The possibility of choosing this set freely was crucial for our proof. It enabled us to treat the sum on the right-hand side of (7) effectively without using the prime number theorem and thus avoid the problems arising from the "quadratic" terms in identities like (1) and (3).

We deduced (7) from the large sieve inequality given by Lemma 1. Elliott established this inequality by showing that it is the "dual" of the Turán-Kubilius inequality. This approach is probably the most elementary and at the same time one of the simplest proofs of a large sieve inequality. A slightly weaker inequality than that of Lemma 1, namely with the sum being extended over the smaller range $p \le x^{1/2}$ (which would be sufficient for our purposes), follows from the standard version of the large sieve, *viz.*

$$\sum_{p\le x^{1/2}}\frac{1}{p}\sum_{r=1}^{p}\left|\frac{p}{x}\sum_{\substack{n\le x\\ n\equiv r \bmod p}} a_n - \frac{1}{x}\sum_{n\le x} a_n\right|^2 \le C\frac{1}{x}\sum_{n\le x}|a_n|^2. \tag{11}$$

However, the proof of (11), as given for example in [**9**, Chapter 3], involves exponential sums and can hardly be called elementary. Renyi [**11**] proved (11) with the exponent $\frac{1}{3}$ instead of $\frac{1}{2}$ and only for the case $a_n = 0$ or 1. His proof is fairly simple and elementary. It can be generalized to arbitrary complex coefficients a_n and thus provides an alternative elementary access to the large sieve inequality of Lemma 1 (with the summation extended over $p \le x^{1/3}$).

As has been pointed out to the author by G. Halász, it is also possible to deduce (7) using the Turán–Kubilius inequality itself instead of its dual, the large sieve inequality of Lemma 1. The argument, which is essentially due to I. Z. Ruzsa, runs as follows: It is easily checked that, in the notation of (7),

$$M(x)\sum_{p\in\mathcal{P}}\frac{1}{p}+\sum_{p\in\mathcal{P}}\frac{1}{p}M(x/p)=-\frac{1}{x}\sum_{n\leq x}\mu(n)\left(\omega_{\mathcal{P}}(n)-\sum_{p\in\mathcal{P}}\frac{1}{p}\right)+O\left(\sum_{p\in\mathcal{P}}\frac{1}{p^2}\right)$$

where

$$\omega_{\mathcal{P}}(n)=\sum_{\substack{p\mid n\\ p\in\mathcal{P}}}1.$$

By the Cauchy-Schwarz inequality and the Turán–Kubilius inequality [4, Lemma 4.5], the last expression is, in absolute value,

$$\ll\left(\frac{1}{x}\sum_{n\leq x}\left(\omega_{\mathcal{P}}(n)-\sum_{p\in\mathcal{P}}\frac{1}{p}\right)^2\right)^{1/2}+\sum_{p\in\mathcal{P}}\frac{1}{p^2}\ll\left(\sum_{p\in\mathcal{P}}\frac{1}{p}\right)^{1/2},$$

and (7) follows.

References

1. K. Corrádi. A remark on the theory of multiplicative functions. *Acta Sci. Math.* (*Szeged*), 28 (1967), 83–92.
2. H. Daboussi. Sur le Théorème des Nombres Premiers. *C. R. Acad. Sc. Paris, Série I*, 298 (1984), 161–164.
3. H. Diamond. Elementary methods in the study of the distribution of prime numbers. *Bull. Amer. Math. Soc.*, 7 (1982), 553–589.
4. P. D. T. A. Elliott. *Probabilistic Number Theory I* (Springer, New York, 1979).
5. P. Erdős. On a new method, which leads to an elementary proof of the prime number theorem. *Proc. Nat. Acad. Sci. U.S.A.*, 35 (1949), 374–384.
6. H. Halberstam and H.-E. Richert. *Sieve Methods* (Academic Press, London, 1974).
7. A. Hildebrand. On Wirsing's mean value theorem for multiplicative functions. *Bull. London Math. Soc.*, 18 (1986), 147–152.
8. M. Kalecki. A simple elementary proof of $M(x)=\sum_{n\leq x}\mu(n)=o(x)$. *Acta Arith.*, 13 (1967), 1–7.
9. H. L. Montgomery. *Topics in Multiplicative Number Theory. Lecture Notes in Mathematics*, Vol. 227 (Springer, Berlin, 1971).
10. A. G. Postnikov and N. P. Romanov. A simplilcation of Selberg's elementary proof of the asymptotic law of distribution of primes. *Uspehi Mat. Nauk*, 10 (1955), 75–87.
11. A. Renyi. On the large sieve of Ju. V. Linnik. *Compositio Math.*, 8 (1950), 68–75.
12. A. Selberg. An elementary proof of the prime number theorem. *Ann. Math.*, 50 (1949), 305–313.

Prof. A. Hildebrand,
Department of Mathematics,
University of Illinois,
1409 West Green Street,
Urbana, Illinois 61801,
U.S.A.

10H15: *NUMBER THEORY; Multiplicative theory; Distribution of primes.*

Received on the 11*th of September*, 1985.

12.19 D. Goldston and H. Montgomery (1987)

Pair Correlation of Zeros and Primes in Short Intervals

In this paper, the authors prove two main results. First, they prove if the Riemann hypothesis is true, then

$$\int_1^X (\psi(x+\delta x) - \psi(x) - \delta x)^2 x^{-2} dx \ll \delta(\log X)\left(\log\left(\frac{2}{\delta}\right)\right)$$

is valid for $0 < \delta \leq 1$, $X \geq 2$. Here the function $\psi(x)$ is defined as

$$\psi(x) := \sum_{p^m \leq x} \log p.$$

This extends a result of Selberg from 1943. The second result builds on the work of Montgomery presented in Section 12.14. The authors show that

$$\int_1^X (\psi(x+\delta x) - \psi(x) - \delta x)^2 dx \sim \frac{1}{2}\delta X^2 \log\left(\frac{1}{\delta}\right)$$

holds uniformly for $X^{-B_2} \leq \delta \leq X^{-B_1}$, under the assumptions of the Riemann hypothesis and the hypothesis that for $0 < B_1 \leq B_2 < 1$,

$$F(X,T) = \sum_{0<\gamma,\gamma'\leq T} X^{i(\gamma-\gamma')} w(\gamma-\gamma') \sim \frac{T}{2\pi}\log T$$

holds uniformly for $X^{B_1}(\log X)^{-3} \leq T \leq X^{B_2}(\log X)^3$, where $w(u) = 4/(4+u^2)$ and γ, γ' are the imaginary parts of nontrivial zeros of $\zeta(s)$. [81]

PAIR CORRELATION OF ZEROS AND PRIMES IN SHORT INTERVALS

Daniel A. Goldston and Hugh L. Montgomery*

1. Statement of results.

In 1943, A. Selberg [15] deduced from the Riemann Hypothesis (RH) that

$$\int_1^X (\psi((1+\delta)x) - \psi(x) - \delta x)^2 x^{-2}\, dx \ll \delta(\log X)^2 \qquad (1)$$

for $X^{-1} \leqslant \delta \leqslant X^{-1/4}$, $X \geqslant 2$. Selberg was concerned with small values of δ, and the constraint $\delta \leqslant X^{-1/4}$ was imposed more for convenience than out of necessity. For larger δ we have the following result.

Theorem 1. *Assume* RH. *Then*

$$\int_1^X (\psi((1+\delta)x) - \psi(x) - \delta x)^2 x^{-2}\, dx \ll \delta(\log X)(\log 2/\delta) \qquad (2)$$

for $0 < \delta \leqslant 1$, $X \geqslant 2$.

In this estimate, the error term for the number of primes in the interval $(x, (1+\delta)x]$ is damped by the factor x^{-2}, and the length of the interval, δx, varies with x. Saffari and Vaughan [14] considered the undamped integral, and derived from RH the estimates

$$\int_1^X (\psi((1+\delta)x) - \psi(x) - \delta x)^2\, dx \ll \delta X^2 (\log 2/\delta)^2 \qquad (3)$$

for $0 < \delta \leqslant 1$, and

$$\int_1^X (\psi(x+h) - \psi(x) - h)^2\, dx \ll hX(\log 2X/h)^2 \qquad (4)$$

---

*Research supported in part by NSF Grant MCS82-01602.

for $0 < h \leqslant X$. It may be similarly shown that RH gives the estimate

$$\int_1^X (\psi(x) - x)^2\, dx \ll X^2 . \qquad (5)$$

Gallagher and Mueller [5] showed that if one assumes not only RH but also the pair correlation conjecture

$$\# \{(\gamma,\gamma') : 0 < \gamma \leqslant T,\ 0 < \gamma - \gamma' \leqslant 2\pi a/\log T\}$$
$$= \Big(\frac{1}{2\pi} \int_0^a 1 - \Big(\frac{\sin \pi u}{\pi u}\Big)^2 du + o(1)\Big)\, T \log T \qquad (6)$$

then it can be deduced that

$$\int_1^X (\psi((1 + \delta)x) - \psi(x) - \delta x)^2\, x^{-2}\, dx \sim \delta(\log 1/\delta)(\log X\sqrt{\delta}) \qquad (7)$$

for $X^{-1} \leqslant \delta \leqslant X^{-\varepsilon}$. Here γ denotes the ordinate of a non-trivial zero of the Riemann zeta function. Thus it seems likely that the estimate of Theorem 1 is best possible.

In the course of formulating the conjecture (6), Montgomery [13] also proposed a more precise estimate, namely that

$$F(X,T) \sim \frac{1}{2\pi}\, T \log T \qquad (8)$$

uniformly for $T \leqslant X \leqslant T^A$, for any fixed $A > 1$, where

$$F(X,T) = \sum_{0 < \gamma,\gamma' \leqslant T} X^{i(\gamma-\gamma')} w(\gamma-\gamma') \qquad (9)$$

and $w(u) = 4/(4 + u^2)$. We now relate this conjecture to the size of the integral in (3).

Theorem 2. *Assume* RH. *If* $0 < B_1 \leqslant B_2 \leqslant 1$, *then*

$$\int_1^X (\psi((1 + \delta)x) - \psi(x) - \delta x)^2\, dx \sim \frac{1}{2}\, \delta X^2 \log 1/\delta \qquad (10)$$

uniformly for $X^{-B_2} \leqslant \delta \leqslant X^{-B_1}$, *provided that* (8) *holds uniformly*

185

for

$$X^{B_1} (\log X)^{-3} \leqslant T \leqslant X^{B_2} (\log X)^3 . \qquad (11)$$

Conversely, if $1 \leqslant A_1 \leqslant A_2 < \infty$, *then* (8) *holds uniformly for* $T^{A_1} \leqslant X \leqslant T^{A_2}$, *provided that* (10) *holds uniformly for*

$$X^{-1/A_1} (\log X)^{-3} \leqslant \delta \leqslant X^{-1/A_2} (\log X)^3 . \qquad (12)$$

Previously Mueller [12] derived (10) from RH and a strong quantitative form of (8). Heath-Brown and Goldston [11] showed that RH and (8) for $T^a \leqslant X \leqslant T^b$, $a < 2 < b$, imply

$$p_{n+1} - p_n = o(p_n^{1/2} (\log p_n)^{1/2}) .$$

This estimate follows easily from Theorem 2 by taking $\delta = \varepsilon X^{-1/2} (\log X)^{1/2}$ in (10). In deriving (10) from (8) we also use the weaker estimate (3). In the case of very small δ, say $\delta \approx (\log X)/X$, we can do better by appealing instead to the bound

$$\int_1^X (\psi((1 + \delta)x) - \psi(x) - \delta x)^2 \, dx \ll \delta X^2 \log X + \delta^2 X^3 \qquad (13)$$

which follows from sieve estimates (see the proof of Lemma 7). In this way we could show that

$$\int_1^X (\pi(x + h) - \pi(x) - h/\log x)^2 \, dx \sim hX/\log X \qquad (14)$$

for $h \approx \log X$, given RH and (8) for $T \leqslant X \leqslant f(T)T \log T$. Here $f(T)$ tends to infinity arbitrarily slowly with T. From this it follows easily that

$$\liminf \, (p_{n+1} - p_n) / \log p_n = 0 .$$

Heath-Brown [10] derived this from a slightly stronger hypothesis.

In assessing the depth of the estimates (8) and (10), we note that (10) is a logarithm sharper than (3), and that (8) is a logarithm sharper than the trivial bound

$$|F(X,T)| \leq F(1,T) \sim \frac{1}{2\pi} T(\log T)^2 \quad . \tag{15}$$

(See Lemma 8.) As in (4), we can relate (10) to primes in intervals of constant length. In summary we have the following

Corollary. *Assume* RH. *Then the following assertions are equivalent:*

(a) *For every fixed* $A \geq 1$, (8) *holds uniformly for* $T \leq X \leq T^A$.

(b) *For every fixed* $\varepsilon > 0$, (10) *holds uniformly for* $X^{-1} \leq \delta \leq X^{-\varepsilon}$.

(c) *For every fixed* $\varepsilon > 0$,

$$\int_0^X (\psi(x+h) - \psi(x) - h)^2 \, dx \sim hX \log X/h \tag{16}$$

holds uniformly for $1 \leq h \leq X^{1-\varepsilon}$.

It is not hard to show that either (b) or (c) implies RH. Gallagher [4] has shown that a weak quantitative form of the prime k-tuple hypothesis gives (16) when $h \approx \log X$.

The path we take between (8) and (10) involves elementary arguments of Abelian and Tauberian character; these are of two sorts. First, we consider the connection between the assertion

$$\int_{-\infty}^{+\infty} e^{-2|y|} f(Y+y) \, dy = 1 + o(1) \tag{17}$$

as $Y \to +\infty$, and the more general assertion

$$\int_a^b R(y) f(Y+y) \, dy = \int_a^b R(y) \, dy + o(1) \tag{18}$$

as $Y \to +\infty$ where R is any Riemann-integrable function. (These two statements are equivalent if f is bounded and non-negative.) This interplay reflects the choice of the weighting function w(u) in the definition (9) of F(X,T). Second, and more intrinsically, we consider a question of Riemann summability (R_2), namely the

187

connection between the two assertions

$$\int_0^\infty \left(\frac{\sin \kappa u}{u}\right)^2 f(u)du = (\pi/2 + o(1))\kappa \log 1/\kappa \tag{19}$$

as $\kappa \to 0^+$, and

$$\int_0^U f(u)du = (1 + o(1))\, U \log U \tag{20}$$

as $U \to +\infty$. Because of the intricacies of the (R_2) method, neither of these assertions implies the other, although they are equivalent for non-negative functions f. The lemmas we formulate below are complicated by the fact that we specify the relation between the parameters κ and U.

2. Lemmas of summability.

Lemma 1. *If*

$$I(Y) = \int_{-\infty}^{+\infty} e^{-2|y|} f(Y + y)dy = 1 + \varepsilon(Y) ,$$

and if $f(y) \geq 0$ *for all* y, *then for any Riemann-integrable function* R(y),

$$\int_a^b R(y)\, f(Y + y)dy = \left(\int_a^b R(y)dy\right)\left(1 + \varepsilon'(y)\right) . \tag{21}$$

If R *is fixed then* $|\varepsilon'(Y)|$ *is small provided that* $|\varepsilon(y)|$ *is small uniformly for* $Y + a - 1 \leq y \leq Y + b + 1$.

In terms of Wiener's general Tauberian theorem, the truth of this lemma hinges on the fact that the Fourier transform of the kernel $k(y) = e^{-2|y|}$, namely the function

$$\hat{k}(t) = \int_{-\infty}^{+\infty} k(y)\, e(-ty)dy = \frac{1}{\pi^2 t^2 + 1} , \quad \left(e(u) = e^{2\pi i u}\right),$$

never vanishes.

Proof. Let $K_c(y) = \max(0, c - |y|)$. By comparing Fourier transforms, or by direct calculation, we see that

$$K_c(y) = \frac{1}{2} e^{-2|y|} - \frac{1}{4} e^{-2|y-c|} - \frac{1}{4} e^{-2|y+c|} + \int_{-c}^{c} (c - |z|)\, e^{-2|y-z|}\, dz .$$

Hence

$$\int_{-c}^{c} K_c(y)\, f(Y + y)\, dy = \frac{1}{2} I(Y) - \frac{1}{4} I(Y + c) - \frac{1}{4} I(Y - c) + \int_{-c}^{c} (c - |z|) I(Y + z)\, dz$$
$$= c^2 + \varepsilon_1(Y)$$

where $|\varepsilon_1|$ is small if $c > 0$ is fixed and if $|\varepsilon(y)|$ is small for $Y - c \leqslant y \leqslant Y + c$. Since

$$\frac{1}{\eta}\big(K_c(y) - K_{c-\eta}(y)\big) \leqslant \chi_{[-c,c]}(y) \leqslant \frac{1}{\eta}\big(K_{c+\eta}(y) - K_c(y)\big) ,$$

and since $f \geqslant 0$, we deduce that (21) holds in the case of the step function $R(y) = \chi_{[-c,c]}(y)$. Since the general R can be approximated above and below by step functions, we obtain (21).

Lemma 2. *Suppose that* $f(t)$ *is a continuous non-negative function defined for all* $t \geqslant 0$, *with* $f(t) \ll \log^2 (t + 2)$. *If*

$$J(T) = \int_0^T f(t)dt = \big(1 + \varepsilon(T)\big)T \log T ,$$

then

$$\int_0^\infty \Big(\frac{\sin \kappa u}{u}\Big)^2 f(u)du = \big(\pi/2 + \varepsilon'(\kappa)\big)\kappa \log 1/\kappa \qquad (22)$$

where $|\varepsilon'(\kappa)|$ *is small as* $\kappa \to 0^+$ *if* $|\varepsilon(T)|$ *is small uniformly for* $\kappa^{-1} (\log \kappa)^{-2} \leqslant T \leqslant \kappa^{-1}(\log \kappa)^2$.

Proof. We divide the range of integration in (22) into four subintervals: $0 \leqslant u \leqslant \kappa^{-1}(\log \kappa)^{-2} = U_1$, $U_1 \leqslant u \leqslant C\kappa^{-1} = U_2$, $U_2 \leqslant u \leqslant \kappa^{-1}(\log \kappa)^2 = U_3$, and $U_3 \leqslant u < \infty$.
Since $f(t) \ll \log^2(t + 2)$, we see that

189

$$\int_0^{U_1} \ll \int_0^{U_1} \kappa^2 \log^2(u+2)\, du \ll \kappa^2 U_1 \log^2 U_1 \ll \kappa ,$$

and similarly that

$$\int_{U_3}^{\infty} \ll \int_{U_3}^{\infty} u^{-2} \log^2 u\, du \ll U_3^{-1} \log^2 U_3 \ll \kappa .$$

By writing $f(u) = \log 1/\kappa + \log \kappa u + (f(u) - \log u)$, we express the integral from U_1 to U_2 as a sum of three integrals. We note that

$$\int_{U_1}^{U_2} \left(\frac{\sin \kappa u}{u}\right)^2 du = \int_0^{\infty} \left(\frac{\sin \kappa u}{u}\right)^2 du + O\left(\kappa(\log \kappa)^{-2}\right)$$

$$= \frac{\pi}{2} \kappa\left(1 + O(\log \kappa)^{-2}\right),$$

and that

$$\int_{U_1}^{U_2} \left(\frac{\sin \kappa u}{u}\right)^2 \log \kappa u\, du \ll \int_0^{\infty} \min(\kappa^2, u^{-2}) \log \kappa u\, du \ll \kappa .$$

Put $r(u) = J(u) - u \log u + u$. Then by integrating by parts we see that

$$\int_{U_1}^{U_2} \left(\frac{\sin \kappa u}{u}\right)^2 (f(u) - \log u) du \ll \kappa\left(1 + (\log \frac{1}{\kappa}) \max_{U_1 \le u \le U_2} |\varepsilon(u)|\right) \log(C+2) .$$

As for the range $U_2 \le u \le U_3$, we see that if $\varepsilon(u) \le 1$ then

$$\int_{U_2}^{U_3} \le \int_{U_2}^{U_3} f(u) u^{-2}\, du \ll U_2^{-1} \log U_2 \ll C^{-1} \kappa \log 1/\kappa .$$

We make this small by taking C large. Then the remaining error terms are small if $\varepsilon(u)$ is small.

Lemma 3. *If* K *is even,* K" *continuous,* $\int_{-\infty}^{+\infty} |K| < \infty$, $K(x) \to 0$ *as* $x \to +\infty$, $K' \to 0$ *as* $x \to +\infty$, *and if* $K''(x) \ll x^{-3}$ *as* $x \to +\infty$, *then*

$$\hat{K}(t) = \int_0^{\infty} K''(x) \left(\frac{\sin \pi t x}{\pi t}\right)^2 dx . \quad (23)$$

Proof. Integrate by parts twice.

Lemma 4. *If f is a non-negative function defined on* $[0, +\infty)$, $f(t) \ll \log^2(t+2)$, *and if*

$$I(\kappa) = \int_0^\infty \left(\frac{\sin \kappa t}{t}\right)^2 f(t)\,dt = (\pi/2 + \varepsilon(\kappa))\kappa \log 1/\kappa$$

then

$$J(T) = \int_0^T f(t)\,dt = (1 + \varepsilon')T \log T$$

where $|\varepsilon'|$ *is small if* $|\varepsilon(\kappa)| \leqslant \varepsilon$ *uniformly for*

$$T^{-1}(\log T)^{-1} \leqslant \kappa \leqslant T^{-1}(\log T)^2 .$$

Proof. Let K be a kernel with the properties specified in Lemma 3. Replace t by t/T in (23), multiply by $f(t) - \log t$, and integrate over $0 \leqslant t < \infty$. Then we find that

$$\int_0^\infty \left(f(t) - \log t\right) \hat{K}(t/T)\,dt = \pi^{-2}\, T^2 \int_0^\infty K''(x)\, R(\pi x/T)\,dx$$

where

$$R(\kappa) = I(\kappa) - \int_0^\infty \left(\frac{\sin \kappa t}{t}\right)^2 \log t\, dt$$

$$= I(\kappa) - \frac{1}{2}\pi\kappa \log 1/\kappa + O(\kappa).$$

Since

$$I(\kappa) < \int_0^\infty \min(\kappa^2, t^{-2})\log^2(t+2)\,dt < \kappa \log^2(2 + 1/\kappa)$$

for all $\kappa > 0$, on taking $x_1 = (\log T)^{-1}$ we see that

$$\int_0^{x_1} K''R < \int_0^{x_1} xT^{-1} \log^2 T/x \,dx \ll T^{-1}.$$

On taking $x_2 = \frac{1}{4}(\log T)^2$ we find that

$$\int_{x_2}^\infty K''R < \int_{x_2}^\infty x^{-3}(x/T) \log^2 T\, dx \ll T^{-1}.$$

Assuming, as we may, that $\varepsilon > (\log T)^{-1}$, we have $R(\pi x/T) \ll \varepsilon x T^{-1} \log T$ for $x_1 \leqslant x \leqslant x_2$. Hence

$$\int_{x_1}^{x_2} K''R \ll \varepsilon T^{-1}(\log T) \int_0^\infty \min(1, x^{-3})\, x\, dx \ll \varepsilon T^{-1} \log T.$$

For $\eta > 0$ take

$$K(x) = K_\eta(x) = \left(\sin 2\pi x + \sin 2\pi(1+\eta)x\right)\left(2\pi x(1 - 4\eta^2 x^2)\right)^{-1},$$

so that

$$\hat{K}(t) = \begin{cases} 1 & \text{if } |t| \leqslant 1, \\ \cos^2\left(\pi(|t| - 1)/(2\eta)\right) & \text{if } 1 \leqslant |t| \leqslant 1+\eta, \\ 0 & \text{if } |t| \geqslant 1+\eta . \end{cases}$$

Thus

$$\int_0^\infty f(t)\hat{K}_\eta(t/T)dt = (1 + O(\eta))T \log T + O_\eta(T) + O_\eta(\varepsilon T \log T) .$$

Since f is non-negative, we see that

$$\int_0^\infty f(t)\hat{K}_\eta\left((1+\eta)t/T\right)dt \leqslant J(T) \leqslant \int_0^\infty f(t)\hat{K}_\eta(t/T)dt ,$$

and we obtain the desired result by taking η small.

In this argument we have made free use of existing treatments of Riemann summability. We note especially Hardy [8, pp. 301, 316, 365] and Hardy and Rogosinski [9, Theorem III].

3. Lemmas of analytic number theory.

As is customary, we write $s = \sigma + it$, and we let $\rho = \beta + i\gamma$ be a typical non-trivial zero of the Riemann zeta function. We first note a simple result of Gallagher [3]:

Lemma 5. *Let* $S(t) = \sum_{\mu \in M} c(\mu)e(\mu t)$ *where* M *is a countable set of real numbers and* $\sum |c(\mu)| < \infty$ *. Then*

$$\int_{-T}^{T} |S(t)|^2\, dt < T^2 \int_{-\infty}^{+\infty} \Big| \sum_{\substack{\mu \in M \\ |\mu - u| \leqslant (4T)^{-1}}} c(\mu)\Big|^2\, du.$$

When a main term is desired, we use the following more elaborate estimate.

Lemma 6. *Let* S(t) *be as above. If* $\delta \geqslant T^{-1}$ *then*

$$\int_0^T |S(t)|^2\, dt = (T + O(\delta^{-1})) \sum_{\mu \in M} |c(\mu)|^2$$

$$+ O\Big(T \sum_{\substack{\mu, \nu \in M \\ 0 < |\mu - \nu| < \delta}} |c(\mu)c(\nu)|\Big) .$$

Proof. Selberg (see Vaaler[17]) has constructed functions $F_-(t)$ and $F_+(t)$ such that $F_-(T) \leqslant \chi_{[0,T]}(t) \leqslant F_+(t)$, $\hat{F}_\pm(x) = 0$ for $|x| \geqslant \delta$, and $\int_{-\infty}^{+\infty} F_\pm(t)dt = T \pm \delta^{-1}$. Hence

$$\int_0^T |S|^2 \leqslant \int_{-\infty}^{+\infty} |S|^2 F_+ = \sum_{\mu, \nu} c(\mu)\overline{c(\nu)}\, \hat{F}_+(\nu - \mu) .$$

The terms $\mu = \nu$ contribute $(T + \delta^{-1}) \sum_\mu |c(\mu)|^2$. Since

$$|\hat{F}_+| \leqslant \int |F_+| = T + \delta^{-1} \leqslant 2T ,$$

the terms $\mu \neq \nu$ contribute at most

$$2T \sum_{0 < |\mu - \nu| < \delta} |c(\mu)c(\nu)|.$$

This gives an upper bound, and a corresponding lower bound is derived similarly using F_-.

Lemma 7. *Let* $C(x) > 0$ *be a continuous function such that* $C(x) \approx C(y)$ *whenever* $x \approx y$. If $|c(p)| \leqslant C(p)$ *for all primes* p, *and if* $\delta \geqslant T^{-1}$, *then*

$$\int_0^T |\sum_p c(p)p^{it}|^2 \, dt = (T + O(\delta^{-1})) \sum_p |c(p)|^2$$

$$+ O\Big(\delta T \int_{\delta^{-1}}^{\infty} C(u)^2 u(\log u)^{-2} \, du\Big)$$

Proof. We appeal to the previous lemma. In the second error term, the primes $p \in (X,2X]$ contribute

$$T\, C(X)^2 \sum_{X<p\leqslant 2X} \sum_{p<p'\leqslant(1+2\delta)p} 1 \ll T\, C(X)^2 \sum_{1\leqslant k\leqslant 4\delta X} \pi_2(2X,k)$$

where $\pi_2(x,k)$ denotes the number of primes $p \leqslant x$ for which $p + k$ is also prime. It is well-known (see Halberstam and Richert [7, p.117]) that

$$\pi_2(x,k) \ll \big(k/\phi(k)\big)x(\log x)^{-2}$$

uniformly for $x \geqslant 2$, $k \neq 0$. Since $\sum_{k \leqslant K} k/\phi(k) \ll K$, it follows that our upper bound is

$$\ll T\, C(X)^2 \, \delta X^2 \, (\log X)^{-2} \ll \delta T \int_X^{2X} C(u)^2 \, u(\log u)^{-2} \, du \, .$$

We put $X = \delta^{-1} 2^r$ and sum over $r \geqslant 0$ to obtain the desired result.

We now present the main known properties of $F(X,T)$.

Lemma 8. *Assume* RH, *and let* $F(X,T)$ *be as in* (9). *Then* $F(X,T) \geqslant 0$, $F(X,T) = F(1/X,T)$, *and*

$$F(X,T) = T\big(X^{-2}(\log T)^2 + \log X\big)\Big(\frac{1}{2\pi} + O\big((\log T)^{-1/2} (\log\log T)^{1/2}\big)\Big) \tag{24}$$

uniformly for $1 \leqslant X \leqslant T$.

Proof. The first assertion is an immediate consequence of either of the two identities

$$F(X,T) = 2\pi \int_{-\infty}^{+\infty} e^{-4\pi|u|} \Big| \sum_{0<\gamma\leqslant T} X^{i\gamma} e(\gamma u) \Big|^2 du, \tag{25}$$

or

$$F(X,T) = \frac{2}{\pi} \int_{-\infty}^{+\infty} \Big| \sum_{0<\gamma\leqslant T} \frac{X^{i\gamma}}{1 + (t-\gamma)^2} \Big|^2 dt.$$

The observation that F is non-negative has also been made by Mueller (unpublished). The second assertion is obvious from the definition of F. The estimate (24) is substantially due to Goldston [6, Lemma B], and may be proved by substituting an appeal to Lemma 7 in the argument of Montgomery [13].

Lemma 9. *If* $0 \leqslant h \leqslant T$ *then*

$$\#\{(\gamma,\gamma') : 0 \leqslant \gamma \leqslant T, |\gamma - \gamma'| \leqslant h\} \ll (1 + h \log T)T \log T . \tag{27}$$

Proof. We argue unconditionally, although if RH is assumed then the above follows easily from Lemma 8 (see (6) of Montgomery [13]). Let $N(T) = \#\{\gamma: 0 < \gamma \leqslant T\}$. Following Selberg, Fujii [2] showed that

$$\int_0^T \Big(N(t+h) - N(t) - \frac{1}{2\pi} h \log t\Big)^2 dt \ll T \log(2 + h \log T)$$

for $0 \leqslant h \leqslant 1$. Hence

$$\int_0^T \big(N(t+h) - N(t)\big)^2 dt \ll h^2 T(\log T)^2$$

for $(\log T)^{-1} \leqslant h \leqslant 1$. This gives (27) in this case. To derive (27) when $0 \leqslant h \leqslant (\log T)^{-1}$, it suffices to consider $h = (\log T)^{-1}$. As for the range $1 \leqslant h \leqslant T$, it suffices to use the bound

$$N(T + 1) - N(T) \ll \log T \tag{28}$$

(see Titchmarch [16,p. 178]) .

Lemma 10. *For* $0 < \delta \leqslant 1$ *let*

$$a(s) = ((1 + \delta)^s - 1)/s \ . \tag{29}$$

If $|c(\gamma)| \leqslant 1$ *for all* γ *then*

$$\int_{-\infty}^{+\infty} |a(it)|^2 \left| \sum_{\gamma} \frac{c(\gamma)}{1 + (t-\gamma)^2} \right|^2 dt = \int_{-\infty}^{+\infty} \left| \sum_{|\gamma| \leqslant Z} \frac{a(1/2 + i\gamma)c(\gamma)}{1 + (t-\gamma)^2} \right|^2 dt$$

$$+ O(\delta^2(\log 2/\delta)^3) + O(Z^{-1}(\log Z)^3) \tag{30}$$

provide that $Z \geqslant 1/\delta$.

Proof. By (28), the sum that occurs in the integral on the left is $\ll \log(2 + |t|)$. Since

$$a(s) \ll \min(\delta, |s|^{-1}) \tag{31}$$

in the strip $|\sigma| \leqslant 1/\delta$, it follows by Cauchy's formula or by direct calculation that

$$a'(s) \ll \min(\delta^2 , \delta/|s|) \tag{32}$$

for $|\sigma| \leqslant (2\delta)^{-1}$. Hence in particular,

$$a(it) - a(1/2 + it) \ll \min(\delta^2 , \delta/|t|) ,$$

and consequently

$$|a(it)|^2 - |a(1/2 + it)|^2 \ll \min(\delta^3, \delta/t^2) \ .$$

Let I denote the integral on the left in (30), and J the corresponding integral with a(it) replaced by $a(1/2 + it)$. Then

$$I - J \ll \int \min(\delta^3, \delta/t^2)(\log(2 + |t|))^2 \, dt \ll \delta^2 (\log 2/\delta)^2 \ .$$

Write J in the form $J = \int |A|^2$. From (28) and (31) we see that

$$A \ll \min(\delta,\ |t|^{-1}) \log(2 + |t|) \tag{33}$$

Now let K be the integral with $a(\frac{1}{2} + it)$ replaced by $a(\frac{1}{2} + i\gamma)$, and write $K = \int |B|^2$. Then B also satisfies the estimate (33). From (31) and (32) we see that

$$a(\tfrac{1}{2} + i\gamma) - a(\tfrac{1}{2} + it) \ll |t - \gamma| \min(\delta^2,\ \delta/|t|) \quad .$$

Thus

$$A - B \ll \min(\delta^2,\ \delta/|t|)(\log(2/\delta + |t|))^2,$$

so that

$$|A|^2 - |B|^2 \ll \min(\delta^3,\ \delta/t^2)(\log (2/\delta + |t|))^3 \ ,$$

and hence

$$J - K \ll \delta^2(\log 2/\delta)^3 \quad .$$

Finally, let $L = \int |C|^2$ be the integral on the right in (30). We note that C also satisfies the estimate (33). Since

$$B - C \ll \min(Z^{-1},\ |t|^{-1}) \log(2Z + |t|),$$

we find that

$$|B|^2 - |C|^2 \ll \min(Z^{-1}(1 + |t|)^{-1},\ t^{-2})(\log(2Z + |t|))^2 \ .$$

Thus

$$K - L \ll Z^{-1} (\log 2Z)^3 \quad ,$$

and the proof is complete.

4. Proof of Theorem 1.

Although we arrange the technical details differently, the ideas are entirely the same as in Selberg's paper. If $\delta X \leqslant 1$ then there is at most one prime power in the interval $(x, (1 + \delta)x]$, so

197

that our integral is

$$\ll \delta \sum_{n \leqslant X} \Lambda(n)^2/n + \delta^2 X \ll \delta(\log X)^2 ,$$

which suffices. We now suppose that $\delta X > 1$. By the above argument we see that

$$\int_0^{1/\delta} \ldots \ll \delta(\log 2/\delta)^2.$$

Thus it suffices to consider the range $1/\delta \leqslant x \leqslant X$. Here we apply the explicit formula for $\psi(x)$ (see Davenport [1, 17]), which gives

$$\begin{aligned}\psi\big((1+\delta)x\big) - \psi(x) - \delta x = - &\sum_{|\rho| \leqslant Z} a(\rho)x^{\rho} \qquad (34)\\ &+ O\Big((\log x)\min(1, \frac{x}{Z\,\|x\|})\Big)\\ &+ O\Big((\log x)\min(1, \frac{x}{Z\,\|(1+\delta)x\|})\Big)\\ &+ O(x\,Z^{-1}(\log xZ)^2)\end{aligned}$$

where $a(s)$ is given in (29), and $\|\theta\| = \min_n \|\theta - n\|$ is the distance from θ to the nearest integer. The error terms contribute a negligible amount if we take $Z = X(\log X)^2$. Writing $\rho = \frac{1}{2} + i\gamma$, $x = e^y$, $Y = \log X$, we see that it remains to show that

$$\int_{\log 1/\delta}^{Y} \Big| \sum_{|\gamma| \leqslant Z} a(\rho)\, e^{i\gamma y}\Big|^2 dy \ll \delta Y \log 2/\delta. \qquad (35)$$

By Lemma 5 we see that this integral is

$$\ll Y^2 \int_{-\infty}^{\infty} \Big(\sum_{\substack{|\gamma| \leqslant Z\\ |\gamma - 2\pi u| \leqslant 2/Y}} a(\rho)^2\Big)\, du \ll Y \sum_{\substack{|\gamma| \leqslant Z\\ |\gamma'| \leqslant Z\\ |\gamma-\gamma'| \leqslant 4/Y}} |a(\rho)a(\rho')| .$$

By (31) and Lemma 9 this gives (35), and the proof is complete.

5. Proof of Theorem 2.

We first assume (8) as needed, and derive (10). Let

$$J(T) = J(X,T) = 4 \int_0^T \Big| \sum_\gamma \frac{X^{i\gamma}}{1 + (t-\gamma)^2} \Big|^2 \, dt \quad .$$

Montgomery [13] (see his (26), but beware of the changes in notation) used (28) to show that

$$J(X,T) = 2\pi \, F(X,T) + O\big((\log T)^3\big) \ .$$

Thus (8) is equivalent to

$$J(X,T) = \big(1 + o(1)\big) T \log T \quad . \tag{36}$$

With a(s) defined in (29), we note that

$$|a(it)|^2 = 4\Big(\frac{\sin \kappa t}{t}\Big)^2$$

where $\kappa = \frac{1}{2} \log (1 + \delta)$. Then by Lemma 2 we deduce that

$$\int_0^\infty |a(it)|^2 \Big| \sum_\gamma \frac{X^{i\gamma}}{1 + (t-\gamma)^2} \Big|^2 dt = (\pi/2 + o(1))\kappa \log 1/\kappa$$
$$= (\pi/4 + o(1))\delta \log 1/\delta \ . \tag{37}$$

The values of T for which we have used (8) lie in the range

$$\delta^{-1}(\log 1/\delta)^{-2} \leqslant T \leqslant 3\delta^{-1}(\log 1/\delta)^2 \ . \tag{38}$$

The integrand is even, so that the value is doubled if we integrate over negative values of t as well. Then by Lemma 10

$$\int_{-\infty}^{+\infty} \Big| \sum_{|\gamma| \leqslant Z} \frac{a(\rho)\, X^{i\gamma}}{1 + (t-\gamma)^2} \Big|^2 dt = \big(\pi/2 + o(1)\big)\delta \log 1/\delta$$

provided that $Z \geqslant \delta^{-1}(\log 1/\delta)^3$. Let S(t) denote the above sum over γ. Its Fourier transform is

$$\hat{S}(u) = \int_{-\infty}^{+\infty} S(t)\, e(-tu)dt = \pi \sum_{|\gamma| \leqslant Z} a(\rho)\, X^{i\gamma}\, e(-\gamma u) e^{-2\pi|u|} \ .$$

Hence by Plancherel's identity the integral above is

$$= \pi^2 \int_{-\infty}^{+\infty} \Big| \sum_{|\gamma| \leqslant Z} a(\rho) X^{i\gamma}\, e(-\gamma u) \Big|^2 e^{-4\pi|u|} \, du \ .$$

On writing $Y = \log X$, $-2\pi u = y$, we find that

$$\int_{-\infty}^{+\infty} \Big| \sum_{|\gamma| \leq Z} a(\rho)\, e^{i\gamma(Y+y)} \Big|^2 e^{-2|y|}\, dy = (1 + o(1))\delta \log 1/\delta \, . \tag{39}$$

In Lemma 1 we take

$$R(y) = \begin{cases} e^{2y} & 0 \leq y \leq \log 2, \\ 0 & \text{otherwise} \, . \end{cases}$$

On making the change of variable $x = e^{Y+y}$ we deduce that

$$\int_{X}^{2X} \Big| \sum_{|\gamma| \leq Z} a(\rho) x^{\rho} \Big|^2 dx = \big(3/2 + o(1) \big)\, \delta X^2 \log 1/\delta \, .$$

We replace X by $X2^{-k}$, sum over k, $1 \leq k \leq K$, and use the explicit formula (34) with $Z = X(\log X)^3$ to see that

$$\int_{X2^{-K}}^{X} \big(\psi((1+\delta)x) - \psi(x) - \delta x\big)^2 dx = \tfrac{1}{2} \big(1-2^{-2K} + o(1)\big)\, \delta X^2 \log 1/\delta.$$

We take $K = [\log\log X]$, and note that it suffices to have (8) in the range (11). To bound the contribution of the range $1 \leq x \leq X2^{-K}$, we appeal to (3) with X replaced by $X2^{-K}$. Thus we have (10) .

We now deduce (8) from (10). By integrating (10) by parts from X_1 to $X_2 = X_1(\log X_1)^{2/3}$, we find that

$$\int_{X_1}^{X_2} \big(\psi((1+\delta)x) - \psi(x) - \delta x\big)^2 x^{-4} dx = \big(\tfrac{1}{2} + o(1)\big)\delta(\log 1/\delta) X_1^{-2}.$$

From (3) we similarly deduce that

$$\int_{X_2}^{\infty} \big(\psi((1+\delta)x) - \psi(x) - \delta x\big)^2 x^{-4}\, dx \ll \delta\, (\log 1/\delta)^2\, X_2^{-2}$$

$$= O\big(\delta\, (\log 1/\delta)\, X_1^{-2} \big) \, .$$

We add these relations, and multiply through by X_1^2 . By making a further appeal to (10) with $X = X_1$ we deduce that

$$\int_{0}^{\infty} \min(x^2/X_1^2, X_1^2/x^2)\big(\psi((1+\delta)x) - \psi(x) - \delta x\big)^2 x^{-2}\, dx$$

$$= (1 + o(1))\delta \log 1/\delta \, .$$

We write X for X_1, put $Y = \log X$, $x = e^{Y+y}$, and appeal to the explicit formula (34) with $Z = X(\log X)^3$, and we find that we have (39). Retracing our steps, we find that we have (37). Then by Lemma 4 we obtain (36), and hence (8). The values of δ and X for which we have used (10) also satisfy (12).

6. Proof of the Corollary.

We note that Lemma 8 gives (8) when

$$X(\log X)^{-3} \leqslant T \leqslant X,$$

and that (10) is trivial when

$$X^{-1}(\log X)^{-3} \leqslant \delta \leqslant X^{-1}.$$

Thus the equivalence of (a) and (b) follows immediately from Theorem 2.

We now show that (b) implies (c). We suppress the converse argument, which is similar. The method here is that of Saffari and Vaughan [14]. Our first goal is to deduce from (b) that

$$\int_0^H \int_0^X (\psi(x+h) - \psi(x) - h)^2\, dxdh \sim \frac{1}{2} H^2 X \log X/H \tag{40}$$

uniformly for $1 \leqslant H \leqslant X^{1-\varepsilon}$. To this end it suffices to show that

$$\int_{\frac{1}{2}X}^X \int_0^H (\psi(x+h) - \psi(x) - h)^2\, dh\, dx \sim \frac{1}{4} H^2 X \log X/H \tag{41}$$

In this integral we replace h by $\delta = h/x$, and invert the order of integration. Thus the left hand side above is

$$\int_0^{H/X} \int_{\frac{1}{2}X}^X f(x,\delta x)^2\, x\, dx\, d\delta + \int_{H/X}^{2H/X} \int_{\frac{1}{2}X}^{H/\delta} f(x,\delta x)^2\, x\, dx\, d\delta$$

where $f(x,y) = \psi(x+y) - \psi(x) - y$. By integrating by parts, we see from (b) that if $A \approx B \approx X$ then

$$\int_A^B f(x,\delta x)^2\, x\, dx = \frac{1}{3}(B^3 - A^3)\, \delta \log 1/\delta + O(X^3 \delta \log 1/\delta).$$

This yields (41). Then (40) follows by replacing X by $X2^{-k}$ in (41), summing over $0 \leqslant k \leqslant K = [2 \log\log X]$, and by appealing to (4) with X replaced by $X2^{-K-1}$.

We now deduce (c) from (40). Suppose that $0 < \eta < 1$. By differencing in (40) we see that

$$\int_H^{(1+\eta)H} \int_0^X f(x,h)^2 \, dx \, dh = (\eta + \tfrac{1}{2}\eta^2 + o(1))XH^2 \log X/H \quad .$$

Let $g(x,h) = f(x,H)$. From the identity

$$f^2 - g^2 = 2f(f-g) - (f-g)^2$$

and the Cauchy-Schwartz inequality we find that

$$\iint f^2 - g^2 \ll \left(\iint f^2\right)^{1/2} \left(\iint (f-g)^2\right)^{1/2} + \iint (f-g)^2 \quad .$$

But $f(x,h) - g(x,h) = f(x+H, h-H)$, so that

$$\iint (f-g)^2 = \int_0^{\eta H} \int_H^{X+H} f(x,h)^2 \, dx \, dh$$

$$\ll \eta^2 H^2 X \log X/H$$

by (40). Hence we see that

$$\eta H \int_0^X (\psi(x+H) - \psi(x) - H)^2 \, dx = \iint g^2$$

$$= \iint f^2 + O(\eta^{3/2} XH^2 \log X/H)$$

$$= \left(\eta + O(\eta^{3/2}) + o(1)\right) XH^2 \log X/H \; .$$

We now divide both sides by ηH, and obtain the desired result by letting $\eta \to 0^+$ sufficiently slowly.

References.

1. H. Davenport, Multiplicative Number Theory, Second Edition, Springer-Verlag, 1980.

2. A. Fujii, On the zeros of Dirichlet L-functions, I, *Trans. Amer. Math. Soc.* **196** (1974), 225-235. (Corrections to this paper are noted in *Trans. Amer. Math. Soc.* **267** (1981), pp 38-39, and in [5; pp. 219-220].)

3. P.X. Gallagher, A large sieve density estimate near $\sigma = 1$, *Invent. Math.* **11** (1970), 329-339.

4. P.X. Gallagher, On the distribution of primes in short intervals, *Mathematika* **23** (1976), 4-9.

5. P.X. Gallagher and Julia H. Mueller, Primes and zeros in short intervals, *J. Reine Agnew. Math.* **303/304** (1978), 205-220.

6. Daniel A. Goldston, Large differences between consecutive prime numbers, Thesis, University of California Berkeley, 1981.

7. H.Halberstam and H.-E. Richert, Sieve Methods. Academic Press, London, 1974.

8. G.H. Hardy, Divergent Series, Oxford University Press, 1963.

9. G.H. Hardy and W.W. Rogosinski, Notes on Fourier series (I): On sine series with positive coefficients, *J. London Math. Soc.* **18** (1943), 50-57.

10. D.R. Heath-Brown, Gaps between primes, and the pair correlation of zeros of the zeta-function, *Acta Arith.* **41** (1982), 85-99.

11. D.R. Heath-Brown and D.A. Goldston, A note on the difference between consecutive primes, *Math. Ann.* **266** (1984), 317-320.

12. Julia Huang (=J.H. Mueller), Primes and zeros in short intervals, Thesis, Columbia University, 1976.

13. H.L Montgomery, The pair correlation of zeros of the zeta function, *Proc. Sympos. Pure Math.* **24** (1973), 181-193.

203

14. B. Saffari and R.C. Vaughan, On the fractional Parts of x/n and related sequences II, *Ann. Inst. Fourier* (Grenoble) **27**, no. 2, (1977), 1-30.

15. A. Selberg, On the normal density of primes in small intervals, and the difference between consecutive primes, *Arch. Math. Naturvid.* **47**, no. 6, (1943), 87-105.

16. E.C. Titchmarsh, The theory of the Riemann zeta-function, Oxford University Press, 1951.

17. J.D. Vaaler, Some extremal functions in Fourier analysis, *Bull. Amer. Math. Soc.*, **12**, No. 2, (1985), 183-216.

D. A. Goldston
San Jose State University,
San Jose, CA 95192,
U.S.A.

H. L. Montgomery
University of Michigan,
Ann Arbor, MI 48109,
U.S.A.

12.20 M. Agrawal, N. Kayal, and N. Saxena (2004)

PRIMES is in P

The majority of the papers presented so far are focused on the prime number theorem. This paper gives an approximation to the distribution of prime numbers. However, it does not give any information as to whether a specific number is prime. This question is practical, since prime numbers (or determining which numbers are prime) are essential for gathering computational evidence for some of the conjectures we have presented, and more modernly, for cryptography and Internet security. More specifically, the algorithm presented in this paper removes the extended Riemann hypothesis as a condition to performing primality testing in deterministic polynomial time. The authors present a rigorous proof, and discuss further improvements to the running time of the algorithm. Particularly striking is the simplicity of both the algorithm and its proof. [128]

Annals of Mathematics, **160** (2004), 781–793

PRIMES is in P

By Manindra Agrawal, Neeraj Kayal, and Nitin Saxena*

Abstract

We present an unconditional deterministic polynomial-time algorithm that determines whether an input number is prime or composite.

1. Introduction

Prime numbers are of fundamental importance in mathematics in general, and number theory in particular. So it is of great interest to study different properties of prime numbers. Of special interest are those properties that allow one to determine efficiently if a number is prime. Such efficient tests are also useful in practice: a number of cryptographic protocols need large prime numbers.

Let PRIMES denote the set of all prime numbers. The definition of prime numbers already gives a way of determining if a number n is in PRIMES: try dividing n by every number $m \leq \sqrt{n}$—if any m divides n then it is composite, otherwise it is prime. This test was known since the time of the ancient Greeks—it is a specialization of the *Sieve of Eratosthenes* (ca. 240 BC) that generates all primes less than n. The test, however, is inefficient: it takes $\Omega(\sqrt{n})$ steps to determine if n is prime. An efficient test should need only a polynomial (in the size of the input $= \lceil \log n \rceil$) number of steps. A property that *almost* gives an efficient test is Fermat's Little Theorem: for any prime number p, and any number a not divisible by p, $a^{p-1} = 1 \pmod{p}$. Given an a and n it can be efficiently checked if $a^{n-1} = 1 \pmod{n}$ by using repeated squaring to compute the $(n-1)^{\text{th}}$ power of a. However, it is not a correct test since many composites n also satisfy it for some a's (*all* a's in case of *Carmichael* numbers [Car]). Nevertheless, Fermat's Little Theorem became the basis for many efficient primality tests.

Since the beginning of complexity theory in the 1960s—when the notions of complexity were formalized and various complexity classes were defined—

*The last two authors were partially supported by MHRD grant MHRD-CSE-20010018.

this problem (referred to as the *primality testing* problem) has been investigated intensively. It is trivial to see that the problem is in the class co-NP: if n is not prime it has an easily verifiable short certificate, viz., a nontrivial factor of n. In 1974, Pratt observed that the problem is in the class NP too [Pra] (thus putting it in NP $\cap$ co-NP).

In 1975, Miller [Mil] used a property based on Fermat's Little Theorem to obtain a deterministic polynomial-time algorithm for primality testing assuming the *Extended Riemann Hypothesis* (ERH). Soon afterwards, his test was modified by Rabin [Rab] to yield an unconditional but randomized polynomial-time algorithm. Independently, Solovay and Strassen [SS] obtained, in 1974, a different randomized polynomial-time algorithm using the property that for a prime n, $\left(\frac{a}{n}\right) = a^{\frac{n-1}{2}} \pmod{n}$ for every a ($\left(\frac{\cdot}{\cdot}\right)$ is the Jacobi symbol). Their algorithm can also be made deterministic under ERH. Since then, a number of randomized polynomial-time algorithms have been proposed for primality testing, based on many different properties.

In 1983, Adleman, Pomerance, and Rumely achieved a major breakthrough by giving a deterministic algorithm for primality that runs in $(\log n)^{O(\log\log\log n)}$ time (all the previous deterministic algorithms required exponential time). Their algorithm was (in a sense) a generalization of Miller's idea and used higher reciprocity laws. In 1986, Goldwasser and Kilian [GK] proposed a randomized algorithm based on elliptic curves running in expected polynomial-time, on almost all inputs (*all* inputs under a widely believed hypothesis), that produces an easily verifiable short certificate for primality (until then, all randomized algorithms produced certificates for compositeness only). Based on their ideas, a similar algorithm was developed by Atkin [Atk]. Adleman and Huang [AH] modified the Goldwasser-Kilian algorithm to obtain a randomized algorithm that runs in expected polynomial-time on all inputs.

The ultimate goal of this line of research has been, of course, to obtain an unconditional deterministic polynomial-time algorithm for primality testing. Despite the impressive progress made so far, this goal has remained elusive. In this paper, we achieve this. We give a deterministic, $O^{\sim}(\log^{15/2} n)$ time algorithm for testing if a number is prime. Heuristically, our algorithm does better: under a widely believed conjecture on the density of Sophie Germain primes (primes p such that $2p+1$ is also prime), the algorithm takes only $O^{\sim}(\log^6 n)$ steps. Our algorithm is based on a generalization of Fermat's Little Theorem to polynomial rings over finite fields. Notably, the correctness proof of our algorithm requires only simple tools of algebra (except for appealing to a sieve theory result on the density of primes p with $p-1$ having a large prime factor—and even this is not needed for proving a weaker time bound of $O^{\sim}(\log^{21/2} n)$ for the algorithm). In contrast, the correctness proofs of earlier algorithms producing a certificate for primality [APR], [GK], [Atk] are much more complex.

In Section 2, we summarize the basic idea behind our algorithm. In Section 3, we fix the notation used. In Section 4, we state the algorithm and present its proof of correctness. In Section 5, we obtain bounds on the running time of the algorithm. Section 6 discusses some ways of improving the time complexity of the algorithm.

2. The idea

Our test is based on the following identity for prime numbers which is a generalization of Fermat's Little Theorem. This identity was the basis for a randomized polynomial-time algorithm in [AB]:

LEMMA 2.1. *Let $a \in \mathcal{Z}$, $n \in \mathcal{N}$, $n \geq 2$, and $(a, n) = 1$. Then n is prime if and only if*

$$(X + a)^n = X^n + a \pmod{n}. \tag{1}$$

Proof. For $0 < i < n$, the coefficient of x^i in $((X + a)^n - (X^n + a))$ is $\binom{n}{i} a^{n-i}$.

Suppose n is prime. Then $\binom{n}{i} = 0 \pmod{n}$ and hence all the coefficients are zero.

Suppose n is composite. Consider a prime q that is a factor of n and let $q^k || n$. Then q^k does not divide $\binom{n}{q}$ and is coprime to a^{n-q} and hence the coefficient of X^q is not zero $(\text{mod } n)$. Thus $((X + a)^n - (X^n + a))$ is not identically zero over Z_n. □

The above identity suggests a simple test for primality: given an input n, choose an a and test whether the congruence (1) is satisfied. However, this takes time $\Omega(n)$ because we need to evaluate n coefficients in the LHS in the worst case. A simple way to reduce the number of coefficients is to evaluate both sides of (1) modulo a polynomial of the form $X^r - 1$ for an appropriately chosen small r. In other words, test if the following equation is satisfied:

$$(X + a)^n = X^n + a \pmod{X^r - 1, n}. \tag{2}$$

From Lemma 2.1 it is immediate that all primes n satisfy the equation (2) for all values of a and r. The problem now is that some composites n may also satisfy the equation for a few values of a and r (and indeed they do). However, we can almost restore the characterization: we show that for appropriately chosen r if the equation (2) is satisfied for several a's then n must be a prime power. The number of a's and the appropriate r are both bounded by a polynomial in $\log n$ and therefore, we get a deterministic polynomial time algorithm for testing primality.

3. Notation and preliminaries

The class P is the class of sets accepted by deterministic polynomial-time Turing machines [Lee]; see [Lee] for the definitions of classes NP, co-NP, etc.

Z_n denotes the ring of numbers modulo n and F_p denotes the finite field with p elements, where p is prime. Recall that if p is prime and $h(X)$ is a polynomial of degree d and irreducible in F_p, then $F_p[X]/(h(X))$ is a finite field of order p^d. We will use the notation $f(X) = g(X) \pmod{h(X), n}$ to represent the equation $f(X) = g(X)$ in the ring $Z_n[X]/(h(X))$.

We use the symbol $O^{\sim}(t(n))$ for $O(t(n) \cdot \text{poly}(\log t(n)))$, where $t(n)$ is any function of n. For example, $O^{\sim}(\log^k n) = O(\log^k n \cdot \text{poly}(\log\log n)) = O(\log^{k+\varepsilon} n)$ for any $\varepsilon > 0$. We use log for base 2 logarithms, and ln for natural logarithms.

$\mathcal{N}$ and $\mathcal{Z}$ denote the set of natural numbers and integers respectively. Given $r \in \mathcal{N}$, $a \in \mathcal{Z}$ with $(a, r) = 1$, the *order of a modulo r* is the smallest number k such that $a^k = 1 \pmod{r}$. It is denoted as $o_r(a)$. For $r \in \mathcal{N}$, $\phi(r)$ is *Euler's totient function* giving the number of numbers less than r that are relatively prime to r. It is easy to see that $o_r(a) \mid \phi(r)$ for any a, $(a, r) = 1$.

We will need the following simple fact about the lcm of the first m numbers (see, e.g., [Nai] for a proof).

LEMMA 3.1. *Let* LCM(m) *denote the* lcm *of the first* m *numbers. For* $m \geq 7$:

$$\text{LCM}(m) \geq 2^m.$$

4. The algorithm and its correctness

Input: integer $n > 1$.

1. If ($n = a^b$ for $a \in \mathcal{N}$ and $b > 1$), output COMPOSITE.
2. Find the smallest r such that $o_r(n) > \log^2 n$.
3. If $1 < (a, n) < n$ for some $a \leq r$, output COMPOSITE.
4. If $n \leq r$, output PRIME.[1]
5. For $a = 1$ to $\lfloor \sqrt{\phi(r)} \log n \rfloor$ do
   if ($(X + a)^n \neq X^n + a \pmod{X^r - 1, n}$), output COMPOSITE;
6. Output PRIME.

Algorithm for Primality Testing

THEOREM 4.1. *The algorithm above returns* PRIME *if and only if n is prime.*

In the remainder of the section, we establish this theorem through a sequence of lemmas. The following is trivial:

LEMMA 4.2. *If n is prime, the algorithm returns* PRIME.

Proof. If n is prime then steps 1 and 3 can never return COMPOSITE. By Lemma 2.1, the `for` loop also cannot return COMPOSITE. Therefore the algorithm will identify n as PRIME either in step 4 or in step 6. □

The converse of the above lemma requires a little more work. If the algorithm returns PRIME in step 4 then n must be prime since otherwise step 3 would have found a nontrivial factor of n. So the only remaining case is when the algorithm returns PRIME in step 6. For the purpose of subsequent analysis we assume this to be the case.

The algorithm has two main steps (2 and 5): step 2 finds an appropriate r and step 5 verifies the equation (2) for a number of a's. We first bound the magnitude of the appropriate r.

LEMMA 4.3. *There exists an $r \leq \max\{3, \lceil \log^5 n \rceil\}$ such that $o_r(n) > \log^2 n$.*

Proof. This is trivially true when $n = 2$; $r = 3$ satisfies all conditions. So assume that $n > 2$. Then $\lceil \log^5 n \rceil > 10$ and Lemma 3.1 applies. Let r_1, r_2, $\dots$, r_t be all numbers such that either $o_{r_i}(n) \leq \log^2 n$ or r_i divides n. Each of these numbers must divide the product

$$n \cdot \prod_{i=1}^{\lfloor \log^2 n \rfloor} (n^i - 1) < n^{\log^4 n} \leq 2^{\log^5 n}.$$

By Lemma 3.1, the lcm of the first $\lceil \log^5 n \rceil$ numbers is at least $2^{\lceil \log^5 n \rceil}$ and therefore there must exist a number $s \leq \lceil \log^5 n \rceil$ such that $s \notin \{r_1, r_2, \dots, r_t\}$. If $(s, n) = 1$ then $o_s(n) > \log^2 n$ and we are done. If $(s, n) > 1$, then since s does not divide n and $(s, n) \in \{r_1, r_2, \dots, r_t\}$, $r = \frac{s}{(s,n)} \notin \{r_1, r_2, \dots, r_t\}$ and so $o_r(n) > \log^2 n$. □

Since $o_r(n) > 1$, there must exist a prime divisor p of n such that $o_r(p) > 1$. We have $p > r$ since otherwise either step 3 or step 4 would decide about the primality of n. Since $(n, r) = 1$ (otherwise either step 3 or step 4 will correctly identify n), $p, n \in Z_r^*$. Numbers p and r will be fixed in the remainder of this section. Also, let $\ell = \lfloor \sqrt{\phi(r)} \log n \rfloor$.

[1]Lemma 4.3 shows that $r \leq \lceil \log^5 n \rceil$, so that Step 4 is relevant only when $n \leq 5,690,034$.

Step 5 of the algorithm verifies ℓ equations. Since the algorithm does not output COMPOSITE in this step, we have:

$$(X + a)^n = X^n + a \pmod{X^r - 1, n}$$

for every a, $0 \le a \le \ell$ (the equation for $a = 0$ is trivially satisfied). This implies:

$$(X + a)^n = X^n + a \pmod{X^r - 1, p} \tag{3}$$

for $0 \le a \le \ell$. By Lemma 2.1, we have:

$$(X + a)^p = X^p + a \pmod{X^r - 1, p} \tag{4}$$

for $0 \le a \le \ell$. From equations 3 and 4 it follows that:

$$(X + a)^{\frac{n}{p}} = X^{\frac{n}{p}} + a \pmod{X^r - 1, p} \tag{5}$$

for $0 \le a \le \ell$. Thus both n and $\frac{n}{p}$ behave like prime p in the above equation. We give a name to this property:

Definition 4.4. For polynomial $f(X)$ and number $m \in \mathcal{N}$, we say that m is *introspective* for $f(X)$ if

$$[f(X)]^m = f(X^m) \pmod{X^r - 1, p}.$$

It is clear from equations (5) and (4) that both $\frac{n}{p}$ and p are introspective for $X + a$ when $0 \le a \le \ell$.

The following lemma shows that introspective numbers are closed under multiplication:

LEMMA 4.5. *If m and m' are introspective numbers for $f(X)$ then so is $m \cdot m'$.*

Proof. Since m is introspective for $f(X)$ we have:

$$[f(X)]^{m \cdot m'} = [f(X^m)]^{m'} \pmod{X^r - 1, p}.$$

Also, since m' is introspective for $f(X)$, we have (after replacing X by X^m in the introspection equation for m'):

$$\begin{aligned}[f(X^m)]^{m'} &= f(X^{m \cdot m'}) \pmod{X^{m \cdot r} - 1, p} \\ &= f(X^{m \cdot m'}) \pmod{X^r - 1, p} \text{ (since } X^r - 1 \text{ divides } X^{m \cdot r} - 1).\end{aligned}$$

Putting together the above two equations we get:

$$[f(X)]^{m \cdot m'} = f(X^{m \cdot m'}) \pmod{X^r - 1, p}.$$ □

For a number m, the set of polynomials for which m is introspective is also closed under multiplication:

LEMMA 4.6. *If m is introspective for $f(X)$ and $g(X)$ then it is also introspective for $f(X) \cdot g(X)$.*

Proof. We have:

$$\begin{aligned}[f(X) \cdot g(X)]^m &= [f(X)]^m \cdot [g(X)]^m \\ &= f(X^m) \cdot g(X^m) \pmod{X^r - 1, p}.\end{aligned}$$ □

The above two lemmas together imply that every number in the set $I = \{(\frac{n}{p})^i \cdot p^j \mid i, j \geq 0\}$ is introspective for every polynomial in the set $P = \{\prod_{a=0}^{\ell}(X+a)^{e_a} \mid e_a \geq 0\}$. We now define two groups based on these sets that will play a crucial role in the proof.

The first group is the set of all residues of numbers in I modulo r. This is a subgroup of Z_r^* since, as already observed, $(n, r) = (p, r) = 1$. Let G be this group and $|G| = t$. G is generated by n and p modulo r and since $o_r(n) > \log^2 n$, $t > \log^2 n$.

To define the second group, we need some basic facts about cyclotomic polynomials over finite fields. Let $Q_r(X)$ be the rth cyclotomic polynomial over F_p. Polynomial $Q_r(X)$ divides $X^r - 1$ and factors into irreducible factors of degree $o_r(p)$ [LN]. Let $h(X)$ be one such irreducible factor. Since $o_r(p) > 1$, the degree of $h(X)$ is greater than one. The second group is the set of all residues of polynomials in P modulo $h(X)$ and p. Let $\mathcal{G}$ be this group which is generated by elements X, $X+1$, $X+2$, …, $X+\ell$ in the field $F = F_p[X]/(h(X))$ and is a subgroup of the multiplicative group of F.

The following lemma proves a lower bound on the size of the group $\mathcal{G}$. It is a slight improvement on a bound shown by Hendrik Lenstra Jr. [Len], which, in turn, improved a bound shown in an earlier version of our paper [AKS].[2]

LEMMA 4.7 (Hendrik Lenstra Jr.). $|\mathcal{G}| \geq \binom{t+\ell}{t-1}$.

Proof. First note that since $h(X)$ is a factor of the cyclotomic polynomial $Q_r(X)$, X is a primitive r^{th} root of unity in F.

We now show that any two distinct polynomials of degree less than t in P will map to different elements in $\mathcal{G}$. Let $f(X)$ and $g(X)$ be two such polynomials in P. Suppose $f(X) = g(X)$ in the field F. Let $m \in I$. We also have $[f(X)]^m = [g(X)]^m$ in F. Since m is introspective for both f and g, and $h(X)$ divides $X^r - 1$, we get:

$$f(X^m) = g(X^m)$$

in F. This implies that X^m is a root of the polynomial $Q(Y) = f(Y) - g(Y)$ for every $m \in G$. Since $(m, r) = 1$ (G is a subgroup of Z_r^*), each such X^m is a

[2]Macaj [Mac] also proved this lemma independently.

primitive r^{th} root of unity. Hence there will be $|G| = t$ distinct roots of $Q(Y)$ in F. However, the degree of $Q(Y)$ is less than t by the choice of f and g. This is a contradiction and therefore, $f(X) \neq g(X)$ in F.

Note that $i \neq j$ in F_p for $1 \leq i \neq j \leq \ell$ since $\ell = \lfloor \sqrt{\phi(r)} \log n \rfloor < \sqrt{r} \log n < r$ and $p > r$. So the elements X, $X+1$, $X+2$, $\ldots$, $X+\ell$ are all distinct in F. Also, since the degree of h is greater than one, $X + a \neq 0$ in F for every a, $0 \leq a \leq \ell$. So there exist at least $\ell + 1$ distinct polynomials of degree one in $\mathcal{G}$. Therefore, there exist at least $\binom{t+\ell}{t-1}$ distinct polynomials of degree $< t$ in $\mathcal{G}$. □

In case n is not a power of p, the size of $\mathcal{G}$ can also be upper bounded:

LEMMA 4.8. *If n is not a power of p then $|\mathcal{G}| \leq n^{\sqrt{t}}$.*

Proof. Consider the following subset of I:

$$\hat{I} = \{(\frac{n}{p})^i \cdot p^j \mid 0 \leq i, j \leq \lfloor \sqrt{t} \rfloor\}.$$

If n is not a power of p then the set $\hat{I}$ has $(\lfloor \sqrt{t} \rfloor + 1)^2 > t$ distinct numbers. Since $|G| = t$, at least two numbers in $\hat{I}$ must be equal modulo r. Let these be m_1 and m_2 with $m_1 > m_2$. So we have:

$$X^{m_1} = X^{m_2} \pmod{X^r - 1}.$$

Let $f(X) \in P$. Then,

$$\begin{aligned}
[f(X)]^{m_1} &= f(X^{m_1}) \pmod{X^r - 1, p} \\
&= f(X^{m_2}) \pmod{X^r - 1, p} \\
&= [f(X)]^{m_2} \pmod{X^r - 1, p}.
\end{aligned}$$

This implies

$$[f(X)]^{m_1} = [f(X)]^{m_2}$$

in the field F. Therefore, $f(X) \in \mathcal{G}$ is a root of the polynomial $Q'(Y) = Y^{m_1} - Y^{m_2}$ in the field F.[3] As $f(X)$ is an arbitrary element of $\mathcal{G}$, the polynomial $Q'(Y)$ has at least $|\mathcal{G}|$ distinct roots in F. The degree of $Q'(Y)$ is $m_1 \leq (\frac{n}{p} \cdot p)^{\lfloor \sqrt{t} \rfloor} \leq n^{\sqrt{t}}$. This shows $|\mathcal{G}| \leq n^{\sqrt{t}}$. □

Armed with these estimates on the size of $\mathcal{G}$, we are now ready to prove the correctness of the algorithm:

LEMMA 4.9. *If the algorithm returns* PRIME *then n is prime.*

[3]This formulation of the argument is by Adam Kalai, Amit Sahai, and Madhu Sudan [KSS].

Proof. Suppose that the algorithm returns PRIME. Lemma 4.7 implies that for $t = |G|$ and $\ell = \lfloor\sqrt{\phi(r)}\log n\rfloor$:

$$\begin{aligned}|\mathcal{G}| &\geq \binom{t+\ell}{t-1} \\ &\geq \binom{\ell+1+\lfloor\sqrt{t}\log n\rfloor}{\lfloor\sqrt{t}\log n\rfloor} && (\text{since } t > \sqrt{t}\log n) \\ &\geq \binom{2\lfloor\sqrt{t}\log n\rfloor+1}{\lfloor\sqrt{t}\log n\rfloor} && (\text{since } \ell = \lfloor\sqrt{\phi(r)}\log n\rfloor \geq \lfloor\sqrt{t}\log n\rfloor) \\ &> 2^{\lfloor\sqrt{t}\log n\rfloor+1} && (\text{since } \lfloor\sqrt{t}\log n\rfloor > \lfloor\log^2 n\rfloor \geq 1) \\ &\geq n^{\sqrt{t}}.\end{aligned}$$

By Lemma 4.8, $|\mathcal{G}| \leq n^{\sqrt{t}}$ if n is not a power of p. Therefore, $n = p^k$ for some $k > 0$. If $k > 1$ then the algorithm will return COMPOSITE in step 1. Therefore, $n = p$. □

This completes the proof of Theorem 4.1.

5. Time complexity analysis and improvements

It is straightforward to calculate the time complexity of the algorithm. In these calculations we use the fact that addition, multiplication, and division operations between two m bits numbers can be performed in time $O^\sim(m)$ [vzGG]. Similarly, these operations on two degree d polynomials with coefficients at most m bits in size can be done in time $O^\sim(d \cdot m)$ steps [vzGG].

THEOREM 5.1. *The asymptotic time complexity of the algorithm is* $O^\sim(\log^{21/2} n)$.

Proof. The first step of the algorithm takes asymptotic time $O^\sim(\log^3 n)$ [vzGG].

In step 2, we find an r with $o_r(n) > \log^2 n$. This can be done by trying out successive values of r and testing if $n^k \neq 1 \pmod r$ for every $k \leq \log^2 n$. For a particular r, this will involve at most $O(\log^2 n)$ multiplications modulo r and so will take time $O^\sim(\log^2 n \log r)$. By Lemma 4.3 we know that only $O(\log^5 n)$ different r's need to be tried. Thus the total time complexity of step 2 is $O^\sim(\log^7 n)$.

The third step involves computing the gcd of r numbers. Each gcd computation takes time $O(\log n)$ [vzGG], and therefore, the time complexity of this step is $O(r\log n) = O(\log^6 n)$. The time complexity of step 4 is just $O(\log n)$.

In step 5, we need to verify $\lfloor\sqrt{\phi(r)}\log n\rfloor$ equations. Each equation requires $O(\log n)$ multiplications of degree r polynomials with coefficients of size $O(\log n)$. So each equation can be verified in time $O^{\sim}(r \log^2 n)$ steps. Thus the time complexity of step 5 is $O^{\sim}(r\sqrt{\phi(r)} \log^3 n) = O^{\sim}(r^{\frac{3}{2}} \log^3 n) = O^{\sim}(\log^{21/2} n)$. This time dominates all the others and is therefore the time complexity of the algorithm. □

The time complexity of the algorithm can be improved by improving the estimate for r (done in Lemma 4.3). Of course the best possible scenario would be when $r = O(\log^2 n)$ and in that case the time complexity of the algorithm would be $O^{\sim}(\log^6 n)$. In fact, there are two conjectures that support the possibility of such an r (below ln is the natural logarithm):

Artin's Conjecture. Given any number $n \in \mathcal{N}$ that is not a perfect square, the number of primes $q \leq m$ for which $o_q(n) = q - 1$ is asymptotically $A(n) \cdot \frac{m}{\ln m}$ where $A(n)$ is Artin's constant with $A(n) > 0.35$.

Sophie-Germain Prime Density Conjecture. The number of primes $q \leq m$ such that $2q + 1$ is also a prime is asymptotically $\frac{2C_2 m}{\ln^2 m}$ where C_2 is the twin prime constant (estimated to be approximately 0.66). Primes q with this property are called Sophie-Germain primes.

Artin's conjecture—if it becomes effective for $m = O(\log^2 n)$—immediately shows that there is an $r = O(\log^2 n)$ with the required properties. There has been some progress towards proving Artin's conjecture [GM], [GMM], [HB], and it is also known that this conjecture holds under the Generalized Riemann Hypothesis.

If the second conjecture holds, we can conclude that $r = O^{\sim}(\log^2 n)$:

> By the density of Sophie-Germain primes, there must exist at least $\log^2 n$ such primes between $8 \log^2 n$ and $c \log^2 n (\log \log n)^2$ for a suitable constant c. For any such prime q, either $o_q(n) \leq 2$ or $o_q(n) \geq \frac{q-1}{2}$. Any q for which $o_q(n) \leq 2$ must divide $n^2 - 1$ and so the number of such q is bounded by $O(\log n)$. This implies that there must exist a prime $r = O^{\sim}(\log^2 n)$ such that $o_r(n) > \log^2 n$. Such an r will yield an algorithm with time complexity $O^{\sim}(\log^6 n)$.

There has been progress towards proving this conjecture as well. Let $P(m)$ denote the greatest prime divisor of number m. Goldfeld [Gol] showed that primes q with $P(q-1) > q^{\frac{1}{2}+c}$, $c \approx \frac{1}{12}$, occur with positive density. Improving upon this, Fouvry has shown:

LEMMA 5.2 ([Fou]). *There exist constants $c > 0$ and n_0 such that, for all $x \geq n_0$:*

$$|\{q \mid q \text{ is prime}, q \leq x \text{ and } P(q-1) > q^{\frac{2}{3}}\}| \geq c\frac{x}{\ln x}.$$

The above lemma is now known to hold for exponents up to 0.6683 [BH]. Using the above lemma, we can improve the analysis of our algorithm:

THEOREM 5.3. *The asymptotic time complexity of the algorithm is* $O^{\sim}(\log^{15/2} n)$.

Proof. As argued above, a high density of primes q with $P(q-1) > q^{\frac{2}{3}}$ implies that step 2 of the algorithm will find an $r = O(\log^3 n)$ with $o_r(n) > \log^2 n$. This brings the complexity of the algorithm down to $O^{\sim}(\log^{15/2} n)$. □

Recently, Hendrik Lenstra and Carl Pomerance [LP1] have come up with a modified version of our algorithm whose time complexity is provably $O^{\sim}(\log^6 n)$.

6. Future work

In our algorithm, the loop in step 5 needs to run for $\lfloor \sqrt{\phi(r)} \log n \rfloor$ times to ensure that the size of the group $\mathcal{G}$ is large enough. The number of iterations of the loop could be reduced if we could show that a still smaller set of $(X+a)$'s generates a group of the required size. This seems very likely.

One can further improve the complexity to $O^{\sim}(\log^3 n)$ if the following conjecture—given in [BP] and verified for $r \leq 100$ and $n \leq 10^{10}$ in [KS]—is proved:

CONJECTURE 6.1. *If r is a prime number that does not divide n and if*

$$(X-1)^n = X^n - 1 \pmod{X^r - 1, n}, \tag{6}$$

then either n is prime or $n^2 = 1 \pmod{r}$.

If this conjecture is true, we can modify the algorithm slightly to search first for an r which does not divide $n^2 - 1$. Such an r can assuredly be found in the range $[2, 4\log n]$. This is because the product of prime numbers less than x is at least e^x (see [Apo]). Thereafter we can test whether the congruence (6) holds or not. Verifying the congruence takes time $O^{\sim}(r \log^2 n)$. This gives a time complexity of $O^{\sim}(\log^3 n)$.

Recently, Hendrik Lenstra and Carl Pomerance [LP2] have given a heuristic argument which suggests that the above conjecture is false. However, some variant of the conjecture may still be true (for example, if we force $r > \log n$).

Acknowledgments. We wish to express our gratitude to Hendrik Lenstra Jr. for allowing us to use his observation on improving the lower bound on the size of the group $\mathcal{G}$. This has made the proof completely elementary (an earlier version required the density bound of Lemma 5.2), simplified the proof *and* improved the time complexity!

We are also thankful to Adam Kalai, Amit Sahai, and Madhu Sudan for allowing us to use their proof of Lemma 4.8. This has made the proofs of both upper and lower bounds on the size of $\mathcal{G}$ similar (both are now based on the number of roots of a polynomial in a field).

We thank Somenath Biswas, Rajat Bhattacharjee, Jaikumar Radhakrishnan, and V. Vinay for many useful discussions.

We thank Erich Bach, Abhijit Das, G. Harman, Roger Heath-Brown, Pieter Moree, Richard Pinch, and Carl Pomerance for providing us with useful references.

Since our preprint appeared, a number of researchers took the trouble of pointing out various omissions in our paper. We are thankful to all of them. We have tried to incorporate their suggestions in the paper and our apologies if we missed out on some.

Finally, we thank the anonymous referee of the paper whose suggestions and observations have been very useful.

Department of Computer Science & Engineering, Indian Institute of Technology Kanpur, Kanpur, India
E-mail addresses: manindra@cse.iitk.ac.in
nitinsa@cse.iitk.ac.in
kayaln@cse.iitk.ac.in

References

[AH] L. M. Adleman and M.-D. Huang, *Primality Testing and Abelian Varieties over Finite Fields*, *Lecture Notes in Math.* **1512**, Springer-Verlag, New York, 1992.

[APR] L. M. Adleman, C. Pomerance, and R. S. Rumely, On distinguishing prime numbers from composite numbers, *Ann. of Math.* **117** (1983), 173–206.

[AB] M. Agrawal and S. Biswas, Primality and identity testing via Chinese remaindering, *Journal of the ACM* **50** (2003), 429–443.

[AKS] M. Agrawal, N. Kayal, and N. Saxena, PRIMES is in P, preprint (http://www.cse.iitk.ac.in/news/primality.ps), August 2002.

[Apo] T. M. Apostol, *Introduction to Analytic Number Theory*, Springer-Verlag, New York, 1997.

[Atk] A. O. L. Atkin, Lecture notes of a conference, Boulder (Colorado), Manuscript, August 1986.

[BH] R. C. Baker and G. Harman, The Brun-Titchmarsh Theorem on average, in *Analytic Number Theory*, *Volume* I (Allerton Park, IL, 1995), *Progr. Math.* **138**, 39–103, Birkhäuser Boston, Boston, MA, 1996.

[BP] R. Bhattacharjee and P. Pandey, Primality testing, Technical report, IIT Kanpur, 2001; available at http://www.cse.iitk.ac.in/research/btp2001/primality.html.

[Car] R. D. Carmichael, Note on a number theory function, *Bull. Amer. Math. Soc.* **16** (1910), 232–238.

[Fou] E. Fouvry, Théorème de Brun-Titchmarsh; application au théorème de Fermat, *Invent. Math.* **79** (1985), 383–407.

[GK] S. Goldwasser and J. Kilian, Almost all primes can be quickly certified, in *Proc. Annual ACM Symposium on the Theory of Computing*, 316–329, 1986.

[GM] R. Gupta and M. Ram Murty, A remark on Artin's conjecture, *Invent. Math.* **78** (1984), 127–130.

[GMM] R. Gupta, V. Kumar Murty, and M. Ram Murty, The Euclidian algorithm for S-integers, *Number Theory* (Montreal, Que., 1985), *CMS Conf. Proc.* **7** (1987), 189–202.

[Gol] M. Goldfeld, On the number of primes p for which $p + a$ has a large prime factor, *Mathematika* **16** (1969), 23–27.

[HB] D. R. Heath-Brown, Artin's conjecture for primitive roots, *Quart. J. Math. Oxford* **37** (1986), 27–38.

[KS] N. Kayal and N. Saxena, Towards a deterministic polynomial-time test, Technical report, IIT Kanpur, 2002; available at http://www.cse.iitk.ac.in/research/btp2002/primality.html.

[KSS] A. Kalai, A. Sahai, and M. Sudan, Notes on primality test and analysis of AKS, Private communication, August 2002.

[Lee] J. V. Leeuwen (ed.), *Handbook of Theoretical Computer Science, Volume A*, MIT Press, Cambridge, MA, 1990.

[Len] H. W. Lenstra, Jr., Primality testing with cyclotomic rings, unpublished (see http://cr.yp.to/papers.html#aks for an exposition of Lenstra's argument), August 2002.

[LN] R. Lidl and H. Niederreiter, *Introduction to Finite Fields and their Applications*, Cambridge Univ. Press, Cambridge, 1986.

[LP1] H. W. Lenstra, Jr. and C. Pomerance, Primality testing with gaussian periods, Private communication, March 2003.

[LP2] ———, Remarks on Agrawal's conjecture, unpublished (http://www.aimath.org/WWN/ primesinp/articles/html/50a/), March 2003.

[Mac] M. Macaj, Some remarks and questions about the AKS algorithm and related conjecture, unpublished (http://thales.doa.fmph.uniba.sk/macaj/aksremarks.pdf), December 2002.

[Mil] G. L. Miller, Riemann's hypothesis and tests for primality, *J. Comput. Sys. Sci.* **13** (1976), 300–317.

[Nai] M. Nair, On Chebyshev-type inequalities for primes, *Amer. Math. Monthly* **89** (1982), 126–129.

[Pra] V. Pratt, Every prime has a succinct certificate, *SIAM Journal on Computing* **4** (1975), 214–220.

[Rab] M. O. Rabin, Probabilistic algorithm for testing primality, *J. Number Theory* **12** (1980), 128–138.

[SS] R. Solovay and V. Strassen, A fast Monte-Carlo test for primality, *SIAM Journal on Computing* **6** (1977), 84–86.

[vzGG] J. von zur Gathen and J. Gerhard, *Modern Computer Algebra*, Cambridge Univ. Press, Cambridge, 1999.

(Received January 24, 2003)

References

1. M. Abramowitz and I. A. Stegun (eds.), *Handbook of Mathematical Functions, with Formulas, Graphs, and Mathematical Tables*, Dover Publications Inc., New York, 1966.
2. M. Agrawal, N. Kayal, and N. Saxena, *PRIMES is in P*, Ann. of Math. (2) **160** (2004), no. 2, 781–793.
3. *Workshop website network of the American Institute of Mathematics*, `http://www.aimath.org/WWN/rh/articles/html/18a/`.
4. *The Riemann Hypothesis*, `http://www.aimath.org/WWN/rh/rh.pdf`.
5. J. Alcántara-Bode, *An integral equation formulation of the Riemann Hypothesis*, Integral Equations Operator Theory **17** (1993), no. 2, 151–168.
6. F. Amoroso, *On the heights of a product of cyclotomic polynomials*, Rend. Sem. Mat. Univ. Politec. Torino **53** (1995), no. 3, 183–191, Number theory, I (Rome, 1995).
7. ———, *Algebraic numbers close to* 1 *and variants of Mahler's measure*, J. Number Theory **60** (1996), no. 1, 80–96.
8. E. Bach and J. Shallit, *Factoring with cyclotomic polynomials*, Math. Comp. **52** (1989), no. 185, 201–219.
9. ———, *Algorithmic Number Theory. Vol. 1*, Foundations of Computing Series, vol. 1, MIT Press, Cambridge, MA, 1996, Efficient algorithms.
10. R. Backlund, *Über die Nullstellen der Riemannschen Zetafunktion. (Swedish)*, Acta Math. **41** (1918), 345–375.
11. ———, *Über die Beziehung zwischen Anwachsen und Nullstellen der Zetafunktion. (Finnish)*, Öfversigt Finska Vetensk. Soc. **61** (1918–1919), no. 9.
12. R. Balasubramanian and K. Ramachandra, *Proof of some conjectures on the mean-value of Titchmarsh series. I*, Hardy-Ramanujan J. **13** (1990), 1–20.
13. M. Balazard and E. Saias, *The Nyman-Beurling equivalent form for the Riemann Hypothesis*, Expo. Math. **18** (2000), no. 2, 131–138.
14. M. Balazard, E. Saias, and M. Yor, *Notes sur la fonction* ζ *de Riemann. II*, Adv. Math. **143** (1999), no. 2, 284–287.
15. W. Barratt, R. W. Forcade, and A. D. Pollington, *On the spectral radius of a* $(0,1)$ *matrix related to Mertens' function*, Linear Algebra Appl. **107** (1988), 151–159.
16. C. Bays and R. H. Hudson, *A new bound for the smallest* x *with* $\pi(x) > li(x)$, Math. Comp. (2000), no. 69, 1285–1296.

17. M. V. Berry and J. P. Keating, *The Riemann zeros and eigenvalue asymptotics*, SIAM Rev. **41** (1999), no. 2, 236–266.
18. F. Beukers, *A note on the irrationality of $\zeta(2)$ and $\zeta(3)$*, Bull. London Math. Soc. **11** (1979), no. 3, 268–272. MR MR554391 (81j:10045).
19. E. Bombieri, *Problems of the Millennium: The Riemann Hypothesis*, `http://www.claymath.org/millennium/Riemann_Hypothesis/Official_Problem_Description.pdf`.
20. ———, *Prime territory: Exploring the infinite landscape at the base of the number system*, The Sciences (1992), 30–36.
21. E. Bombieri and J. C. Lagarias, *Complements to Li's criterion for the Riemann Hypothesis*, J. Number Theory **77** (1999), no. 2, 274–287.
22. J. Borwein and P. Borwein, *Pi and the AGM*, Canadian Mathematical Society Series of Monographs and Advanced Texts, John Wiley and Sons, New York, 1987.
23. J. Borwein, D. Bradley, and R. Crandall, *Computational strategies for the Riemann zeta function*, J. Comput. Appl. Math. **121** (2000), no. 1–2, 247–296.
24. P. Borwein, *An efficient algorithm for the Riemann zeta function*, Constructive, Experimental, and Nonlinear Analysis (Limoges, 1999), CMS Conf. Proc., vol. 27, Amer. Math. Soc., Providence, RI, 2000, pp. 29–34.
25. C. Boutin, *Purdue mathematician claims proof for Riemann Hypothesis*, `http://news.uns.purdue.edu/UNS/html4ever/2004/040608.DeBranges.Riemann.html`.
26. R. P. Brent, *On the zeros of the Riemann zeta function in the critical strip*, Math. Comp. **33** (1979), no. 148, 1361–1372.
27. R. P. Brent, J. van de Lune, H. J. J. te Riele, and D. T. Winter, *On the zeros of the Riemann zeta function in the critical strip. II*, Math. Comp. **39** (1982), no. 160, 681–688.
28. C. Calderón, *Riemann's zeta function*, Rev. Acad. Cienc. Zaragoza (2) **57** (2002), 67–87.
29. ———, *The Riemann Hypothesis*, Millennium Problems (Spanish), Monogr. Real Acad. Ci. Exact. Fís.-Quím. Nat. Zaragoza, 26, Acad. Cienc. Exact. Fís. Quím. Nat. Zaragoza, Zaragoza, 2004, pp. 1–30.
30. J. R. Chen, *On the representation of a large even integer as the sum of a prime and the product of at most two primes*, Sci. Sinica **16** (1973), 157–176.
31. S. Chowla, *The Riemann Hypothesis and Hilbert's Tenth Problem*, Mathematics and Its Applications, Vol. 4, Gordon and Breach Science Publishers, New York, 1965.
32. R. V. Churchill and J. W. Brown, *Complex Variables and Applications*, fourth ed., McGraw-Hill Book Co., New York, 1984.
33. B. Cipra, *A prime case of chaos*, What's Happening in the Mathematical Sciences, vol. 4, 1999, pp. 2–17.
34. G. L. Cohen, *MathSciNet review of MR0602825*, 2007, `http://www.ams.org/mathscinet/`.
35. J. B. Conrey, *More than two-fifths of the zeros of the Riemann zeta function are on the critical line*, J. Reine Angew. Math. **399** (1989), 1–26.
36. ———, *The Riemann Hypothesis*, Notices Amer. Math. Soc. **50** (2003), no. 3, 341–353.
37. J. B. Conrey and A. Ghosh, *On mean values of the zeta-function*, Mathematika **31** (1984), no. 1, 159–161.

38. J. B. Conrey, A. Ghosh, and S. M. Gonek, *Simple zeros of the Riemann zeta-function*, Proc. London Math. Soc. (3) **76** (1998), no. 3, 497–522.
39. J. B. Conrey and X.-J. Li, *A note on some positivity conditions related to zeta and L-functions*, Internat. Math. Res. Notices (2000), no. 18, 929–940.
40. G. Csordas, A. Odlyzko, W. Smith, and R. S. Varga, *A new Lehmer pair of zeros and a new lower bound for the de Bruijn-Newman constant Λ*, Electron. Trans. Numer. Anal. **1** (1993), 104–111 (electronic only).
41. G. Csordas, W. Smith, and R. S. Varga, *Lehmer pairs of zeros, the de Bruijn-Newman constant and the Riemann hypothesis*, Constr. Approx. **10** (1994), 107–109.
42. H. Davenport, *Multiplicative Number Theory*, third ed., Graduate Texts in Mathematics, vol. 74, Springer-Verlag, New York, 2000, Revised and with a preface by Hugh L. Montgomery.
43. L. de Branges, *Homepage of Louis de Branges*, `http://www.math.purdue.edu/~branges/`.
44. ———, *Hilbert Spaces of Entire Functions*, Prentice-Hall Inc, Englewood Cliffs, N.J., 1968.
45. ———, *The Riemann Hypothesis for Hilbert spaces of entire functions*, Bull. Amer. Math. Soc. **15** (1986), 1–17.
46. ———, *The convergence of Euler products*, J. Funct. Anal. **107** (1992), 122–210.
47. ———, *A conjecture which implies the Riemann Hypothesis*, J. Funct. Anal. **121** (1994), 117–184.
48. J. Derbyshire, *Prime Obsession*, Joseph Henry Press, Washington, 2003.
49. J. M. Deshouillers, G. Effinger, H. J. J. te Riele, and D. Zinoviev, *A complete Vinogradov 3-primes theorem under the Riemann Hypothesis*, Electronic Research Announcements of the American Mathematical Society **3** (1997), 99–104.
50. H. G. Diamond, *MathSciNet review of MR741085*, 2007, `http://www.ams.org/mathscinet/`.
51. H. M. Edwards, *Riemann's Zeta Function*, Academic Press (Reprinted by Dover Publications, 2001), 1974.
52. A. Y. Fawaz, *The explicit formula for $L_0(x)$*, Proc. London Math. Soc. **1** (1951), 86–103.
53. ———, *On an unsolved problem in the analytic theory of numbers*, Quart. J. Math., Oxford Ser. **3** (1952), 282–295.
54. J. Franel, *Les suites de Farey et le problème des nombres premiers*, Gött. Nachr. (1924), 198–206.
55. L. J. Goldstein, *A history of the prime number theorem*, Amer. Math. Monthly **80** (1973), 599–615.
56. D. Goldston and H. L. Montgomery, *Pair correlation of zeros and primes in short intervals*, Analytic Number Theory and Diophantine Problems (Stillwater, OK, 1984), Progr. Math., vol. 70, Birkhäuser Boston, Boston, MA, 1987, pp. 183–203.
57. X. Gourdon, *The 10^{13} first zeros of the Riemann zeta function, and zeros computation at very large height*, `http://numbers.computation.free.fr/Constants/Miscellaneouszetazeros1e13-1e24.pdf`.
58. J. P. Gram, *Note sur les zéros de la fonction de Riemann*, Acta Math. **27** (1903), 289–304.

59. A. Granville and K. Soundararajan, *Upper bounds for* $|L(1,\chi)|$, Q. J. Math. **3** (2002), no. 3, 265–284.
60. Richard K. Guy, *Unsolved Problems in Number Theory*, second ed., Problem Books in Mathematics, Springer-Verlag, New York, 1994, Unsolved Problems in Intuitive Mathematics, I. MR MR1299330 (96e:11002).
61. Petros Hadjicostas, *Some generalizations of Beukers' integrals*, Kyungpook Math. J. **42** (2002), no. 2, 399–416. MR MR1942200 (2003h:11098).
62. G. H. Hardy, *Sur les zéros de la fonction* $\zeta(s)$ *de Riemann*, C. R. Acad. Sci. Paris **158** (1914), 1012–1014.
63. ______, *Ramanujan: Twelve Lectures on Subjects Suggested by His Life and Work.*, Chelsea Publishing Company, New York, 1959. MR MR0106147 (21 #4881).
64. ______, *A Mathematician's Apology*, Canto, Cambridge University Press, Cambridge, 1992.
65. G. H. Hardy and J. E. Littlewood, *Contributions to the theory of the Riemann zeta-function and the theory of the distribution of the primes*, Acta Math. **41** (1918), no. 1, 119–196.
66. ______, *Some problems of "Partitio Numerorum" (III): On the expression of a number as a sum of primes*, Acta Math. **44** (1922), 1–70.
67. ______, *Some problems of "Partitio Numerorum" (V): A further contribution to the study of Goldbach's problem*, Proc. London Math. Soc. (2) **22** (1924), 46–56.
68. G. H. Hardy and E. M. Wright, *An Introduction to the Theory of Numbers*, fifth ed., The Clarendon Press Oxford University Press, New York, 1979. MR MR568909 (81i:10002).
69. J. Havil, *Gamma: Exploring Euler's Constant*, Princeton University Press, Princeton, NJ, 2003, with a foreword by Freeman Dyson. MR MR1968276 (2004k:11195).
70. D. R. Heath-Brown, *Artin's conjecture for primitive roots*, Quart. J. Math. Oxford Ser. (2) **37** (1986), no. 145, 27–38.
71. ______, *Siegel zeros and the least prime in an arithmetic progression*, Quart. J. Math. Oxford Ser. (2) **41** (1990), no. 164, 405–418.
72. ______, *Fractional moments of the Riemann zeta-function. II*, Quart. J. Math. Oxford Ser. (2) **44** (1993), no. 174, 185–197.
73. G. Hitchcock, *Six things to do with a conjecture you can't prove*, Zimaths (2004), no. 8.2, `http://www.uz.ac.zw/science/maths/zimaths/82/conject.html`.
74. C. Hooley, *On Artin's conjecture*, J. Reine Angew. Math. **225** (1967), 209–220.
75. J. Horgan, *The death of proof*, Scientific American **269** (1993), no. 4, 92–103.
76. J. I. Hutchinson, *On the roots of the Riemann zeta function*, Trans. Amer. Math. Soc. **27** (1925), no. 1, 49–60.
77. A. E. Ingham, *Mean-value theorems in the theory of the Riemann zeta-function*, Proc. Lond. Math. Soc. **27** (1926), 273–300.
78. ______, *On two conjectures in the theory of numbers*, Amer. J. Math. **64** (1942), 313–319.
79. A. Ivić, *The Riemann Zeta-Function–The Theory of the Riemann Zeta-Function with Applications*, John Wiley & Sons Inc., New York, 1985.
80. A. Ivić, *On Some Results Concerning the Riemann Hypothesis*, Analytic Number Theory (Kyoto, 1996), London Math. Soc. Lecture Note Ser., vol. 247, Cambridge Univ. Press, Cambridge, 1997, pp. 139–167.

81. ______, *MathSciNet review of MR1018376*, 2007, `http://www.ams.org/mathscinet/`.
82. H. Iwaniec and E. Kowalski, *Analytic Number Theory*, American Mathematical Society, Providence, Rhode Island, 2004.
83. H. Iwaniec and P. Sarnak, *Perspectives on the analytic theory of L-functions*, Geom. Funct. Anal. (2000), 705–741, Special Volume, Part II, GAFA 2000 (Tel Aviv, 1999).
84. N. M. Katz and P. Sarnak, *Zeroes of zeta functions and symmetry*, Bull. Amer. Math. Soc. (N.S.) **36** (1999), no. 1, 1–26.
85. J. P. Keating and N. C. Snaith, *Random matrix theory and L-functions at $s = 1/2$*, Comm. Math. Phys. **214** (2000), no. 1, 91–110.
86. E. Klarriech, *Prime time*, New Scientist (2000), no. 2264.
87. N. M. Korobov, *On zeros of the $\zeta(s)$ function*, Dokl. Akad. Nauk SSSR (N.S.) **118** (1958), 431–432.
88. J. C. Lagarias, *On a positivity property of the Riemann ξ-function*, Acta Arith. **89** (1999), no. 3, 217–234.
89. ______, *An elementary problem equivalent to the Riemann Hypothesis*, Amer. Math. Monthly (2002), no. 109, 534–543.
90. R. S. Lehman, *On Liouville's function*, Math. Comp. (1960), no. 14, 311–320.
91. ______, *On the difference $\pi(x) - li(x)$*, Acta Arith. **11** (1966), 397–410.
92. ______, *Separation of zeros of the Riemann zeta-function*, Math. Comp. **20** (1966), 523–541.
93. ______, *On the distribution of zeros of the Riemann zeta-function*, Proc. London Math. Soc. (3) **20** (1970), 303–320.
94. D. H. Lehmer, *Extended computation of the Riemann zeta-function*, Mathematika **3** (1956), 102–108.
95. ______, *On the roots of the Riemann zeta-function*, Acta Math. **95** (1956), 291–298.
96. ______, *MathSciNet review of MR0104638*, 2007, `http://www.ams.org/mathscinet/`.
97. N. Levinson, *More than one third of zeros of Riemann's zeta-function are on $\sigma = 1/2$*, Advances in Math. **13** (1974), 383–436.
98. X.-J. Li, *The positivity of a sequence of numbers and the Riemann Hypothesis*, J. Number Theory **65** (1997), no. 2, 325–333.
99. J. E. Littlewood, *Quelques conséquences de l'hypothèse que la fonction $\zeta(s)$ de Riemann n'a pas de zéros dans le demi-plan $\Re(s) > 1/2$*, Comptes Rendus de l'Acad. des Sciences (Paris) **154** (1912), 263–266.
100. ______, *On the class number of the corpus $P(\sqrt{-k})$*, Proc. London Math. Soc. **28** (1928), 358–372.
101. ______, *The Riemann Hypothesis*, The Scientist Speculates: An Anthology of Partly-Baked Ideas, Basic Books, New York, 1962.
102. J.-P. Massias, J.-L. Nicolas, and G. Robin, *Évaluation asymptotique de l'ordre maximum d'un élément du groupe symétrique*, Acta Arith. **50** (1988), no. 3, 221–242.
103. M. L. Mehta, *Random Matrices and the Statistical Theory of Energy Levels*, Academic Press, New York-London, 1967.
104. N. A. Meller, *Computations connected with the check of Riemann's Hypothesis. (Russian)*, Dokl. Akad. Nauk SSSR **123** (1958), 246–248.
105. F. Mertens, *Über eine Zahlentheoretische Funktion*, Sitzungsber. Akad. Wiss. Wien IIa **106** (1897), 761–830.

106. G. L. Miller, *Riemann's hypothesis and tests for primality*, Seventh Annual ACM Symposium on the Theory of Computing (1975), 234–239.
107. H. L. Montgomery, *The pair correlation of zeros of the zeta function*, Analytic Number Theory (Proc. Sympos. Pure Math., Vol. XXIV, St. Louis Univ., St. Louis, MO., 1972), Amer. Math. Soc., Providence, R.I., 1973, pp. 181–193.
108. ———, *Zeros of approximations to the zeta function*, Studies in Pure Mathematics, Birkhäuser, Basel, 1983, pp. 497–506.
109. M. R. Murty, *Artin's Conjecture for primitive roots*, Math. Intelligencer **10** (1988), no. 4, 59–67.
110. M. R. Murty and R. Gupta, *A remark on Artin's conjecture*, Invent. Math. **78** (1984), no. 1, 127–130.
111. Władysław Narkiewicz, *The Development of Prime Number Theory*, Springer Monographs in Mathematics, Springer-Verlag, Berlin, 2000, From Euclid to Hardy and Littlewood.
112. C. M. Newman, *Fourier transforms with only real zeros*, Proc. Amer. Math. Soc. **61** (1976), no. 2, 245–251.
113. J. J. O'Connor and E. F. Robertson, *André Weil*, `http://www-history.mcs.st-andrews.ac.uk/Biographies/Weil.html`.
114. ———, *Bernhard Riemann*, `http://.../Riemann.html`.
115. ———, *Charles de la Vallée Poussin*, `http://.../Vallee_Poussin.html`.
116. ———, *Godfrey Harold Hardy*, `http://.../Hardy.html`.
117. ———, *Jacques Hadamard*, `http://.../Hadamard.html`.
118. ———, *John Edensor Littlewood*, `http://.../Littlewood.html`.
119. ———, *Pafnuty Lvovich Chebyshev*, `http://.../Chebyshev.html`.
120. ———, *Paul Erdős*, `http://.../Erdos.html`.
121. ———, *Paul Turán*, `http://.../Turan.html`.
122. A. Odlyzko, *Homepage of Andrew Odlyzko*, `http://www.dtc.umn.edu/~odlyzko/`.
123. ———, *On the distribution of spacings between zeros of the zeta function*, Math. Comp. **48** (1987), no. 177, 273–308.
124. ———, *Supercomputers and the Riemann zeta function*, Supercomputing '89: Supercomputing Structures & Computations, Proc. 4-th Intern. Conf. on Supercomputing (L. P. Kartashev and S. I. Kartashev, eds.), Intern. Supercomputing Inst., 1989.
125. ———, *An improved bound for the de Bruijn-Newman constant*, Numer. Algorithms **25** (2000), no. 1-4, 293–303.
126. A. Odlyzko and A. Schönhage, *Fast algorithms for multiple evaluations of the Riemann zeta function*, Trans. Amer. Math. Soc. **309** (1988), no. 2, 797–809.
127. A. Odlyzko and H. J. J. te Riele, *Disproof of the Mertens Conjecture*, J. Reine Angew. Math. **357** (1985), 138–160.
128. C. Pomerance, *MathSciNet review of MR2123939*, 2007, `http://www.ams.org/mathscinet/`.
129. G. P. Pugh, *The Riemann-Siegel formula and large scale computations of the Riemann zeta function*, Master's thesis, The University of British Columbia, Vancouver, BC, 1992.
130. L. D. Pustyl′nikov, *On a property of the classical zeta function associated with the Riemann conjecture on zeros. (Russian)*, Uspekhi Mat. Nauk **54** (1999), no. 1(325), 259–260.
131. H. Rademacher, *Higher Mathematics from an Elementary Point of View*, Birkhäuser Boston, Mass., 1983.

132. R. M. Redheffer, *Eine explizit lösbare Optimierungsaufgabe*, Internat. Schriftenreihe Numer. Math. **142** (1977), 141–152.
133. B. Riemann, *On the number of prime numbers less than a given quantity*, Translated by David R. Wilkins. Available at `http://www.claymath.org/millenniumRiemann_Hypothesis/1859_manuscript/EZeta.pdf`.
134. ———, *Ueber die Anzahl der Primzahlen unter einer gegebenen Grösse*, Monatsberichte der Berliner Akademie (1859).
135. ———, *Gesammelte Werke*, Teubner, Leipzig, 1892.
136. G. Robin, *Grandes valeurs de la fonction somme des diviseurs et Hypothèse de Riemann*, J. Math. Pures Appl. (9) **63** (1984), no. 2, 187–213.
137. J. B. Rosser, J. M. Yohe, and L. Schoenfeld, *Rigorous computation and the zeros of the Riemann zeta-function (with discussion)*, Information Processing 68 (Proc. IFIP Congress, Edinburgh, 1968), Vol. 1: Mathematics, Software, North-Holland, Amsterdam, 1969, pp. 70–76.
138. K. Sabbagh, *Dr. Riemann's Zeros*, Atlantic Books, London, 2002.
139. ———, *The Riemann Hypothesis*, Farrar, Straus and Giroux, New York, 2002.
140. R. Salem, *Sur une proposition équivalente à l'hypothèse de Riemann*, C. R. Acad. Sci. Paris **236** (1953), 1127–1128.
141. P. Sarnak, *Problems of the Millennium: The Riemann Hypothesis*, (2004), `http://www.claymath.org/millennium/Riemann_Hypothesis/Sarnak_RH.pdf`.
142. A. Selberg, *On the zeros of Riemann's zeta-function*, Skr. Norske Vid. Akad. Oslo I. **1942** (1942), no. 10, 59.
143. ———, *Collected Papers. Vol. I*, Springer-Verlag, Berlin, 1989, With a foreword by K. Chandrasekharan.
144. C. L. Siegel, *Über Riemann's Nachlass zur Analytischen Zahlentheorie*, Quellen und Studien zur Geschichte der Math. Astr. Phys. **2** (1932), 275–310.
145. S. Skewes, *On the difference* $\pi(x) - li(x)$., J. London Math. Soc. **8** (1933), 277–283.
146. ———, *On the difference* $\pi(x) - li(x)$. *II.*, Proc. London Math. Soc. **5** (1955), 48–70.
147. J. Snell et al., *Chance in the primes: Part 3*, `http://www.dartmouth.edu/~chance/chance_news/recent_news/primes_part2.pdf`.
148. K. Soundararajan, *Mean-values of the Riemann zeta-function*, Mathematika **42** (1995), no. 1, 158–174.
149. A. Speiser, *Geometrisches zur Riemannschen Zetafunktion*, Math. Annalen **110** (1934), 514–521.
150. M. Tanaka, *A numerical investigation on cumulative sum of the Liouville function*, Tokyo J. Math. **3** (1980), no. 1, 187–189.
151. H. J. J. te Riele, *On the sign of the difference* $\pi(x) - li(x)$, Math. Comp. **48** (1987), no. 177, 323–328.
152. *As you were*, Time **XLV** (1945), no. 18.
153. E. C. Titchmarsh, *The zeros of the Riemann zeta-function*, Proceedings of the Royal Society of London, A - Mathematical and Physical Sciences, vol. 151, 1935, pp. 234–255.
154. ———, *Introduction to the Theory of Fourier Integrals*, third ed., Chelsea Publishing Co., New York, 1986.
155. ———, *The Theory of the Riemann Zeta-Function*, second ed., The Clarendon Press Oxford University Press, New York, 1986, Edited and with a preface by D. R. Heath-Brown.

156. P. Turán, *On some approximative Dirichlet-polynomials in the theory of the zeta-function of Riemann*, Danske Vid. Selsk. Mat.-Fys. Medd. **24** (1948), no. 17, 1–36.
157. A. M. Turing, *A method for the calculation of the zeta-function*, Proc. London Math. Soc. (2) **48** (1943), 180–197.
158. ———, *Some calculations of the Riemann zeta-function*, Proc. London Math. Soc. (3) **3** (1953), 99–117.
159. J. van de Lune and H. J. J. te Riele, *On the zeros of the Riemann zeta function in the critical strip. III*, Math. Comp. **41** (1983), no. 164, 759–767.
160. J. van de Lune, H. J. J. te Riele, and D. T. Winter, *On the zeros of the Riemann zeta function in the critical strip. IV*, Math. Comp. **46** (1986), no. 174, 667–681.
161. A. Verjovsky, *Arithmetic geometry and dynamics in the unit tangent bundle of the modular orbifold*, Dynamical Systems (Santiago, 1990), Pitman Res. Notes Math. Ser., vol. 285, Longman Sci. Tech., Harlow, 1993, pp. 263–298.
162. ———, *Discrete measures and the Riemann Hypothesis*, Kodai Math. J. **17** (1994), no. 3, 596–608, Workshop on Geometry and Topology (Hanoi, 1993).
163. I. M. Vinogradov, *Representation of an odd number as a sum of three primes*, Dokl. Akad. Nauk SSSR **15** (1937), 291–294.
164. ———, *A new estimate of the function $\zeta(1+it)$*, Izv. Akad. Nauk SSSR. Ser. Mat. **22** (1958), 161–164.
165. V. V. Volchkov, *On an equality equivalent to the Riemann hypothesis*, Ukraïn. Mat. Zh. **47** (1995), no. 3, 422–423.
166. H. von Koch, *Sur la distribution des nombres premiers*, Acta Math. **24** (1901), 159–182.
167. H. von Mangoldt, *Zur Verteilung der Nullstellen der Riemannschen Funktion $\xi(s)$*, Math. Ann. **60** (1905), 1–19.
168. S. Wedeniwski, *Zetagrid - computational verification of the Riemann hypothesis*, In *Conference in Number Theory in Honour of Professor H.C. Williams*, Banff, Alberta, Canda, May 2003. `http://www.zetagrid.net/zeta/ZetaGrid-Conference_in_honour_of_Hugh_Williams_2003.pdf`.
169. A. Weil, *Sur les fonctions algébriques à corps de constantes fini*, C. R. Acad. Sci. Paris **210** (1940), 592–594.
170. ———, *On the Riemann Hypothesis in function fields*, Proc. Nat. Acad. Sci. U. S. A. **27** (1941), 345–347.
171. ———, *Basic Number Theory*, Springer, New York, 1973.
172. E. T. Whittaker and G. N. Watson, *A course of modern analysis. an introduction to the general theory of infinite processes and of analytic functions: with an account of the principal transcendental functions*, fourth edition. Reprinted, Cambridge University Press, New York, 1990.

References

Adleman, L. M. and Huang, M.-D., *Primality Testing and Abelian Varieties over Finite Fields*, Lecture Notes in Mathematics, vol. 1512, Springer-Verlag, Berlin, 1992.

Adleman, L. M., Pomerance, C., and Rumely, R. S., *On distinguishing prime numbers from composite numbers*, Ann. of Math. (2) **117** (1983), no. 1, 173–206.

Agrawal, M. and Biswas, S., *Primality and identity testing via Chinese remaindering*, J. ACM **50** (2003), no. 4, 429–443 (electronic).

Agrawal, M., Kayal, N., and Saxena, N., *PRIMES is in P*, Ann. of Math. (2) **160** (2004), no. 2, 781–793.

Anderson, R. J. and Stark, H. M., *Oscillation theorems*, Analytic Number Theory (Philadelphia, PA., 1980), Lecture Notes in Math., vol. 899, Springer, Berlin, 1981, pp. 79–106.

Apostol, T. M., *Introduction to Analytic Number Theory*, Springer-Verlag, New York, 1976, Undergraduate Texts in Mathematics.

Théorie des Topos et Cohomologie Étale des Schémas. Tome 1: Théorie des topos, Springer-Verlag, Berlin, 1972, Séminaire de Géométrie Algébrique du Bois-Marie 1963–1964 (SGA 4), Dirigé par M. Artin, A. Grothendieck, et J. L. Verdier. Avec la collaboration de N. Bourbaki, P. Deligne et B. Saint-Donat, Lecture Notes in Mathematics, Vol. 269.

Théorie des Topos et Cohomologie Étale des Schémas. Tome 2, Springer-Verlag, Berlin, 1972, Séminaire de Géométrie Algébrique du Bois-Marie 1963–1964 (SGA 4), Dirigé par M. Artin, A. Grothendieck et J. L. Verdier. Avec la collaboration de N. Bourbaki, P. Deligne et B. Saint-Donat, Lecture Notes in Mathematics, Vol. 270.

Théorie des Topos et Cohomologie Étale des Schémas. Tome 3, Springer-Verlag, Berlin, 1973, Séminaire de Géométrie Algébrique du Bois-Marie 1963–1964 (SGA 4), Dirigé par M. Artin, A. Grothendieck et J. L. Verdier. Avec la collaboration de P. Deligne et B. Saint-Donat, Lecture Notes in Mathematics, Vol. 305.

Atkinson, F. V., *The mean-value of the Riemann zeta function*, Acta Math. **81** (1949), 353–376.

Backlund, R. J., *Über die Nullstellen der Riemannschen Zetafunktion*, Acta. Math. **41** (1918), 345–375.

Baker, R. C. and Harman, G., *The Brun-Titchmarsh theorem on average*, Analytic Number Theory, Vol. 1 (Allerton Park, IL., 1995), Progr. Math., vol. 138, Birkhäuser Boston, Boston, MA, 1996, pp. 39–103.

Balasubramanian, R., *On the frequency of Titchmarsh's phenomenon for $\zeta(s)$. IV*, Hardy-Ramanujan J. **9** (1986), 1–10.

Balasubramanian, R. and Ramachandra, K., *On the frequency of Titchmarsh's phenomenon for $\zeta(s)$. III*, Proc. Indian Acad. Sci. Sect. A **86** (1977), no. 4, 341–351.

Bombieri, E., *Problems of the Millennium: The Riemann Hypothesis*, `http://www.claymath.org/millennium/Riemann_Hypothesis/Official_Problem_Description.pdf`.

Bombieri, E., *On the large sieve*, Mathematika **12** (1965), 201–225.

Bombieri, E., *A variational approach to the explicit formula*, Comm. Pure Appl. Math. **56** (2003), no. 8, 1151–1164, Dedicated to the memory of Jürgen K. Moser.

Bombieri, E. and Hejhal, D. A., *On the zeros of Epstein zeta functions*, C. R. Acad. Sci. Paris Sér. I Math. **304** (1987), no. 9, 213–217.

Carmichael, R. D., *Note on a number theory function*, Bull. Amer. Math. Soc. **16** (1910), 232–238.

Chandrasekharan, K., *Introduction to analytic number theory*, Die Grundlehren der mathematischen Wissenschaften, Band 148, Springer-Verlag New York, Inc., New York, 1968.

Connes, A., *Trace formula in noncommutative geometry and the zeros of the Riemann zeta function*, Selecta Math. (N.S.) **5** (1999), no. 1, 29–106.

Conrey, J. B., *More than two fifths of the zeros of the Riemann zeta function are on the critical line*, J. Reine Angew. Math. **399** (1989), 1–26.

Conrey, J. B., *L-functions and random matrices*, Mathematics Unlimited—2001 and Beyond, Springer, Berlin, 2001, pp. 331–352.

Conrey, J. B., *The Riemann Hypothesis*, Notices Amer. Math. Soc. **50** (2003), no. 3, 341–353.

Conrey, J. B., Farmer, D. W., Keating, J. P., Rubinstein, M. O., and Snaith, N. C., *Integral moments of L-functions*, Proc. London Math. Soc. (3) **91** (2005), no. 1, 33–104.

Conrey, J. B. and Li, X.-J., *A note on some positivity conditions related to zeta and L-functions*, Internat. Math. Res. Notices (2000), no. 18, 929–940.

Corrádi, K. A., *A remark on the theory of multiplicative functions*, Acta Sci. Math. (Szeged) **28** (1967), 83–92.

Daboussi, H., *Sur le théorème des nombres premiers*, C. R. Acad. Sci. Paris Sér. I Math. **298** (1984), no. 8, 161–164.

Davenport, H., *Multiplicative Number Theory*, second ed., Graduate Texts in Mathematics, vol. 74, Springer-Verlag, New York, 1980, Revised by Hugh L. Montgomery.

Davenport, H. and Heilbronn, H., *On the zeros of certian Dirichlet series I*, J. London Math. Soc. **11** (1936), 181–185.

Davenport, H. and Heilbronn, H., *On the zeros of certian Dirichlet series II*, J. London Math. Soc. **11** (1936), 307–312.

Deift, P. A., *Orthogonal polynomials and random matrices: a Riemann-Hilbert approach*, Courant Lecture Notes in Mathematics, vol. 3, New York University Courant Institute of Mathematical Sciences, New York, 1999.

Deligne, P., *La conjecture de Weil. I*, Inst. Hautes Études Sci. Publ. Math. (1974), no. 43, 273–307.

Deligne, P., *La conjecture de Weil. II*, Inst. Hautes Études Sci. Publ. Math. (1980), no. 52, 137–252.

Deninger, C., *Some analogies between number theory and dynamical systems on foliated spaces*, Proceedings of the International Congress of Mathematicians, Vol. I (Berlin, 1998), no. Extra Vol. I, 1998, pp. 163–186.

Diaconu, A., Goldfeld, D., and Hoffstein, J., *Multiple Dirichlet series and moments of zeta and L-functions*, Compositio Math. **139** (2003), no. 3, 297–360.

Diamond, H. G., *Elementary methods in the study of the distribution of prime numbers*, Bull. Amer. Math. Soc. (N.S.) **7** (1982), no. 3, 553–589.

Doetsch, G., *Theorie und Anwendung der Laplace-Transformation*, Dover Publication, N. Y., 1943.

Duke, W. and Schulze-Pillot, R., *Representation of integers by positive ternary quadratic forms and equidistribution of lattice points on ellipsoids*, Invent. Math. **99** (1990), no. 1, 49–57.

Edwards, H. M., *Riemann's Zeta Function*, Academic Press, 1974, Pure and Applied Mathematics, Vol. 58.

Elliott, P. D. T. A., *Probabilistic Number Theory, I*, Grundlehren der Mathematischen Wissenschaften [Fundamental Principles of Mathematical Science], vol. 239, Springer-Verlag, New York, 1979, Mean-value theorems.

Erdős, P., *On a new method in elementary number theory which leads to an elementary proof of the prime number theorem*, Proc. Nat. Acad. Sci. U. S. A. **35** (1949), 374–384.

Fouvry, É., *Théorème de Brun-Titchmarsh: application au théorème de Fermat*, Invent. Math. **79** (1985), no. 2, 383–407.

Fujii, A., *On the zeros of Dirichlet L-functions. I*, Trans. Amer. Math. Soc. **196** (1974), 225–235.

Fujii, A., *On the distribution of the zeros of the Riemann zeta function in short intervals*, Bull. Amer. Math. Soc. **81** (1975), 139–142.

Gallagher, P. X., *A large sieve density estimate near $\sigma = 1$*, Invent. Math. **11** (1970), 329–339.

Gallagher, P. X., *On the distribution of primes in short intervals*, Mathematika **23** (1976), no. 1, 4–9.

Gallagher, P. X. and Mueller, J. H., *Primes and zeros in short intervals*, J. Reine Angew. Math. **303/304** (1978), 205–220.

Gelbart, S. S., Lapid, E. M., and Sarnak, P., *A new method for lower bounds of L-functions*, C. R. Math. Acad. Sci. Paris **339** (2004), no. 2, 91–94.

Gel′fand, I. M. and Shilov, G. E., *Generalized Functions. Vol. 2. Spaces of Fundamental and Generalized Functions*, Translated from the Russian by Morris D. Friedman, Amiel Feinstein, and Christian P. Peltzer, Academic Press, New York, 1968.

Gel′fond, A. O., *Ischislenie Konechnykh Raznostei*, Third corrected edition, Izdat. "Nauka," Moscow, 1967.

Goldfeld, M., *On the number of primes p for which $p + a$ has a large prime factor*, Mathematika **16** (1969), 23–27.

Goldwasser, S. and Kilian, J., *Almost all primes can be quickly certified*, Proc. Annual ACM Symposium on the Theory of Computing, 1986, pp. 316–329.

Gupta, R. and Murty, M. R., *A remark on Artin's conjecture*, Invent. Math. **78** (1984), no. 1, 127–130.

Gupta, R., Murty, M. R., and Murty, V. K., *The Euclidean algorithm for S-integers*, Number Theory (Montreal, PQ, 1985), CMS Conf. Proc., vol. 7, Amer. Math. Soc., Providence, RI, 1987, pp. 189–201.

Hafner, J. L. and Ivić, A., *On some mean value results for the Riemann zeta-function*, Théorie des Nombres (Quebec, PQ, 1987), de Gruyter, Berlin, 1989, pp. 348–358.

Hafner, J. L. and Ivić, A., *On the mean-square of the Riemann zeta-function on the critical line*, J. Number Theory **32** (1989), no. 2, 151–191.

Halberstam, H. and Richert, H.-E., *Sieve Methods*, Academic Press [A subsidiary of Harcourt Brace Jovanovich, Publishers], London-New York, 1974, London Mathematical Society Monographs, No. 4.

Haran, S., *Index Theory, Potential Theory, and the Riemann Hypothesis*, L-Functions and Arithmetic (Durham, 1989), London Math. Soc. Lecture Note Ser., vol. 153, Cambridge Univ. Press, Cambridge, 1991, pp. 257–270.

Hardy, G. H., *Divergent Series*, Oxford, at the Clarendon Press, 1949.

Hardy, G. H. and Rogosinski, W. W., *Notes on Fourier series. I. On sine series with positive coefficients*, J. London Math. Soc. **18** (1943), 50–57.

Hardy, G. H. and Wright, E. M., *An Introduction to the Theory of Numbers*, Oxford, at the Clarendon Press, 1954, 3rd ed.

Haselgrove, C. B. and Miller, J. C. P., *Tables of the Riemann zeta function*, Royal Society Mathematical Tables, Vol. 6, Cambridge University Press, New York, 1960.

Heath-Brown, D. R., *Gaps between primes, and the pair correlation of zeros of the zeta function*, Acta Arith. **41** (1982), no. 1, 85–99.

Heath-Brown, D. R., *Artin's conjecture for primitive roots*, Quart. J. Math. Oxford Ser. (2) **37** (1986), no. 145, 27–38.

Heath-Brown, D. R. and Goldston, D. A., *A note on the differences between consecutive primes*, Math. Ann. **266** (1984), no. 3, 317–320.

Heins, M., *Complex Function Theory*, Pure and Applied Mathematics, Vol. 28, Academic Press, New York, 1968.

Hildebrand, A., *On Wirsing's mean value theorem for multiplicative functions*, Bull. London Math. Soc. **18** (1986), no. 2, 147–152.

Hooley, C., *On Artin's conjecture*, J. Reine Angew. Math. **225** (1967), 209–220.

Hutchinson, J. I., *On the roots of the Riemann zeta function*, Trans. Amer. Math. Soc. **27** (1925), no. 1, 49–60.

Huxley, M. N., *Exponential sums and the Riemann zeta function. IV*, Proc. London Math. Soc. (3) **66** (1993), no. 1, 1–40.

Huxley, M. N., *A note on exponential sums with a difference*, Bull. London Math. Soc. **26** (1994), no. 4, 325–327.

Ingham, A. E., *On Wiener's method in Tauberian theorems*, Proc. London Math. Soc. (2) **38** (1935), 458–480.

Ingham, A. E., *A note on the distribution of primes*, Acta. Arith. **1** (1936), 201–211.

Ingham, A. E., *On two conjectures in the theory of numbers*, Amer. J. Math. **64** (1942), 313–319.

Ingham, A. E., *The Distribution of Prime Numbers*, Cambridge Tracts in Mathematics and Mathematical Physics, No. 30, Stechert-Hafner, Inc., New York, 1964.

Ivić, A., *Large values of the error term in the divisor problem*, Invent. Math. **71** (1983), no. 3, 513–520.

Ivić, A., *The Riemann Zeta-Function*, A Wiley-Interscience Publication, John Wiley & Sons Inc., New York, 1985, The theory of the Riemann zeta-function with applications.

Ivić, A., *On a problem connected with zeros of $\zeta(s)$ on the critical line*, Monatsh. Math. **104** (1987), no. 1, 17–27.

Ivić, A., *Large values of certain number-theoretic error terms*, Acta Arith. **56** (1990), no. 2, 135–159.

Ivić, A., *Lectures on Mean Values of the Riemann Zeta Function*, Tata Institute of Fundamental Research Lectures on Mathematics and Physics, vol. 82, Published for the Tata Institute of Fundamental Research, Bombay, 1991.

Ivić, A., *On a class of convolution functions connected with* $\zeta(s)$, Bull. Acad. Serbe Sci. Arts Cl. Sci. Math. Natur. (1995), no. 20, 29–50.

Ivić, A., *On the fourth moment of the Riemann zeta-function*, Publ. Inst. Math. (Beograd) (N.S.) **57(71)** (1995), 101–110.

Ivić, A., *On the distribution of zeros of a class of convolution functions*, Bull. Cl. Sci. Math. Nat. Sci. Math. (1996), no. 21, 61–71.

Ivić, A., *On the ternary additive divisor problem and the sixth moment of the zeta-function*, Sieve Methods, Exponential Sums, and their Applications in Number Theory (Cardiff, 1995), London Math. Soc. Lecture Note Ser., vol. 237, Cambridge Univ. Press, Cambridge, 1997, pp. 205–243.

Ivić, A., *On the multiplicity of zeros of the zeta-function*, Bull. Cl. Sci. Math. Nat. Sci. Math. (1999), no. 24, 119–132.

Ivić, A. and Motohashi, Y., *The mean square of the error term for the fourth power moment of the zeta-function*, Proc. London Math. Soc. (3) **69** (1994), no. 2, 309–329.

Ivić, A. and te Riele, H. J. J., *On the zeros of the error term for the mean square of* $|\zeta(\frac{1}{2}+it)|$, Math. Comp. **56** (1991), no. 193, 303–328.

Iwaniec, H., *Fourier coefficients of modular forms of half-integral weight*, Invent. Math. **87** (1987), no. 2, 385–401.

Iwaniec, H., *Introduction to the Spectral Theory of Automorphic Forms*, Biblioteca de la Revista Matemática Iberoamericana. [Library of the Revista Matemática Iberoamericana], Revista Matemática Iberoamericana, Madrid, 1995.

Iwaniec, H., *Topics in Classical Automorphic Forms*, Graduate Studies in Mathematics, vol. 17, American Mathematical Society, Providence, RI, 1997.

Iwaniec, H. and Kowalski, E., *Analytic Number Theory*, American Mathematical Society Colloquium Publications, vol. 53, American Mathematical Society, Providence, RI, 2004.

Iwaniec, H. and Sarnak, P., *Perspectives on the analytic theory of L-functions*, Geom. Funct. Anal., Special Volume, Part II (2000), 705–741, GAFA 2000 (Tel Aviv, 1999).

Jacquet, H., *Principal L-functions of the linear group*, Automorphic Forms, Representations and L-Functions (Proc. Sympos. Pure Math., Oregon State Univ., Corvallis, Ore., 1977), Part 2, Proc. Sympos. Pure Math., XXXIII, Amer. Math. Soc., Providence, R.I., 1979, pp. 63–86.

Jutila, M., *On the value distribution of the zeta function on the critical line*, Bull. London Math. Soc. **15** (1983), no. 5, 513–518.

Kalecki, M., *A simple elementary proof of* $M(x) = \sum_{n\leq x}\mu(n) = o(x)$, Acta Arith. **13** (1967/1968), 1–7.

Karatsuba, A. A., *On the zeros of the Davenport-Heilbronn function lying on the critical line (Russian)*, Izv. Akad. Nauk SSSR ser. mat. **54** (1990), no. 2, 303–315.

Karatsuba, A. A. and Voronin, S. M., *The Riemann Zeta-Function*, de Gruyter Expositions in Mathematics, vol. 5, Walter de Gruyter & Co., Berlin, 1992, Translated from the Russian by Neal Koblitz.

Katz, N. M., *An overview of Deligne's proof of the Riemann hypothesis for varieties over finite fields*, Mathematical Developments Arising from Hilbert Problems (Proc. Sympos. Pure Math., Vol. XXVIII, Northern Illinois Univ., De Kalb, Ill., 1974), Amer. Math. Soc., Providence, R.I., 1976, pp. 275–305.

Katz, N. M. and Sarnak, P., *Random Matrices, Frobenius Eigenvalues, and Monodromy*, American Mathematical Society Colloquium Publications, vol. 45, American Mathematical Society, Providence, RI, 1999.

Katz, N. M. and Sarnak, P., *Zeroes of zeta functions and symmetry*, Bull. Amer. Math. Soc. (N.S.) **36** (1999), no. 1, 1–26.

Kuznetsov, N. V., *Sums of Kloosterman sums and the eighth power moment of the Riemann zeta-function*, Number Theory and Related Topics (Bombay, 1988), Tata Inst. Fund. Res. Stud. Math., vol. 12, Tata Inst. Fund. Res., Bombay, 1989, pp. 57–117.

Landau, E., *Über einige neuere Grenzwertsätze*, Rend. Circ. Mat. Palermo **34** (1912), 121–131.

Landau, E., *Handbuch der Lehre von der Verteilung der Primzahlen. 2 Bände*, Chelsea Publishing Co., New York, 1953, 2d ed, With an appendix by Paul T. Bateman.

Langlands, R. P., *On the Functional Equations Satisfied by Eisenstein Series*, Springer-Verlag, Berlin, 1976, Lecture Notes in Mathematics, Vol. 544.

Lavrik, A. A., *Uniform approximations and zeros in short intervals of the derivatives of the Hardy Z-function*, Anal. Math. **17** (1991), no. 4, 257–279.

Lehman, R. S., *On the difference* $\pi(x) - \mathrm{li}(x)$, Acta Arith. **11** (1966), 397–410.

Lehmer, D. H., *Extended computation of the Riemann zeta-function*, Mathematika **3** (1956), 102–108.

Lehmer, D. H., *On the roots of the Riemann zeta-function*, Acta Math. **95** (1956), 291–298.

LeVeque, W. J., *Topics in Number Theory. Vols. I, II*, Dover Publications Inc., Mineola, NY, 2002, Reprint of the 1956 original [Addison-Wesley Publishing Co., Inc., Reading, Mass.; MR0080682 (18,283d)], with separate errata list for this edition by the author.

Levinson, N., *A motivated account of an elementary proof of the prime number theorem.*, Amer. Math. Monthly **76** (1969), 225–245.

Levinson, N., *More than one third of zeros of Riemann's zeta-function are on* $\sigma = 1/2$, Advances in Math. **13** (1974), 383–436.

Lidl, R. and Niederreiter, H., *Introduction to Finite Fields and Their Applications*, first ed., Cambridge University Press, Cambridge, 1994.

Lindenstrauss, E., *Invariant measures and arithmetic quantum unique ergodicity*, Ann. of Math. (2) **163** (2006), no. 1, 165–219.

Littlewood, J. E., *Sur la distribution des nombres premiers*, Comptes rendus Académie Sci. (Paris) **158** (1914), 1869–1872.

Mehta, M. L., *Random Matrices and the Statistical Theory of Energy Levels*, Academic Press, New York, 1967.

Mehta, M. L., *Random Matrices*, second ed., Academic Press Inc., Boston, MA, 1991.

Mertens, F., *Uber eine Zahlentheoretische Funktion*, Sitzungsberichte Akad. Wien. **106** (1897), 761–830.

Meyer, R., *On a representation of the idele class group related to primes and zeros of L-functions*, Duke Math. J. **127** (2005), no. 3, 519–595.

Miller, G. L., *Riemann's hypothesis and tests for primality*, J. Comput. System Sci. **13** (1976), no. 3, 300–317, Working papers presented at the ACM-SIGACT Symposium on the Theory of Computing (Albuquerque, N.M., 1975).

Molteni, G., *Upper and lower bounds at $s = 1$ for certain Dirichlet series with Euler product*, Duke Math. J. **111** (2002), no. 1, 133–158.

Montgomery, H. L., *Topics in Multiplicative Number Theory*, Springer-Verlag, Berlin, 1971, Lecture Notes in Mathematics, Vol. 227.

Montgomery, H. L., *The pair correlation of zeros of the zeta function*, Analytic Number Theory (Proc. Sympos. Pure Math., Vol. XXIV, St. Louis Univ., St. Louis, MO., 1972), Amer. Math. Soc., Providence, R.I., 1973, pp. 181–193.

Montgomery, H. L., *Distribution of the zeros of the Riemann zeta function*, Proceedings of the International Congress of Mathematicians (Vancouver, B. C., 1974), Vol. 1, Canad. Math. Congress, Montreal, Que., 1975, pp. 379–381.

Montgomery, H. L., *Extreme values of the Riemann zeta function*, Comment. Math. Helv. **52** (1977), no. 4, 511–518.

Montgomery, H. L. and Vaughan, R. C., *Hilbert's inequality*, J. London Math. Soc. (2) **8** (1974), 73–82.

Motohashi, Y., *The fourth power mean of the Riemann zeta-function*, Proceedings of the Amalfi Conference on Analytic Number Theory (Maiori, 1989) (Salerno), Univ. Salerno, 1992, pp. 325–344.

Motohashi, Y., *An explicit formula for the fourth power mean of the Riemann zeta-function*, Acta Math. **170** (1993), no. 2, 181–220.

Motohashi, Y., *A relation between the Riemann zeta-function and the hyperbolic Laplacian*, Ann. Scuola Norm. Sup. Pisa Cl. Sci. (4) **22** (1995), no. 2, 299–313.

Motohashi, Y., *The Riemann zeta-function and the non-Euclidean Laplacian [translation of Sūgaku* **45** *(1993), no. 3, 221–243; MR1256466 (95b:11081)]*, Sugaku Expositions **8** (1995), no. 1, 59–87, Sugaku Expositions.

Motohashi, Y., *Spectral Theory of the Riemann Zeta-Function*, Cambridge Tracts in Mathematics, vol. 127, Cambridge University Press, Cambridge, 1997.

Nair, M., *On Chebyshev-type inequalities for primes*, Amer. Math. Monthly **89** (1982), no. 2, 126–129.

Newman, D. J., *Simple analytic proof of the prime number theorem*, Amer. Math. Monthly **87** (1980), no. 9, 693–696.

Odlyzko, A. M., `http://www.dtc.umn.edu/~odlyzko/`.

Odlyzko, A. M., *On the distribution of spacings between zeros of the zeta function*, Math. Comp. **48** (1987), no. 177, 273–308.

Odlyzko, A. M., *Supercomputers and the Riemann zeta function*, Supercomputing 89: Supercomputing Structures and Computations, Proc. 4-th Intern. Conf. on Supercomputing (Kartashev, L. P. and Kartashev, S. I., eds.), International Supercomputing Institute, 1989, pp. 348–352.

Odlyzko, A. M., *Analytic computations in number theory*, Mathematics of Computation 1943–1993: A Half-Century of Computational Mathematics (Vancouver, BC, 1993), Proc. Sympos. Appl. Math., vol. 48, Amer. Math. Soc., Providence, RI, 1994, pp. 451–463.

Odlyzko, A. M. and te Riele, H. J. J., *Disproof of the Mertens conjecture*, J. Reine Angew. Math. **357** (1985), 138–160.

Ono, K. and Soundararajan, K., *Ramanujan's ternary quadratic form*, Invent. Math. **130** (1997), no. 3, 415–454.

Pólya, G., *Verschiedene Bemerkungen zur Zahlentheorie*, Jahresbericht der deutschen Math.-Vereinigung **28** (1919), 31–40.

Postnikov, A. G. and Romanov, N. P., *A simplification of A. Selberg's elementary proof of the asymptotic law of distribution of prime numbers*, Uspehi Mat. Nauk (N.S.) **10** (1955), no. 4(66), 75–87.

Pratt, V. R., *Every prime has a succinct certificate*, SIAM J. Comput. **4** (1975), no. 3, 214–220.

Rabin, M. O., *Probabilistic algorithm for testing primality*, J. Number Theory **12** (1980), no. 1, 128–138.

Ramachandra, K., *Progress towards a conjecture on the mean-value of Titchmarsh series*, Recent Progress in Analytic Number Theory, Vol. 1 (Durham, 1979), Academic Press, London, 1981, pp. 303–318.

Ramachandra, K., *On the Mean-Value and Omega-Theorems for the Riemann Zeta-Function*, Tata Institute of Fundamental Research Lectures on Mathematics and Physics, vol. 85, Published for the Tata Institute of Fundamental Research, Bombay, 1995.

Rényi, A., *On the large sieve of Ju V. Linnik*, Compositio Math. **8** (1950), 68–75.

Riemann, B., *Ueber die Anzahl der Primzahlen unter einer gegebenen Grösse*, November 1859.

Rosser, J. B., Yohe, J. M., and Schoenfeld, L., *Rigorous computation and the zeros of the Riemann zeta-function (with discussion)*, Information Processing 68 (Proc. IFIP Congress, Edinburgh, 1968), Vol. 1: Mathematics, Software, North-Holland, Amsterdam, 1969, pp. 70–76.

Rudnick, Z. and Sarnak, P., *Zeros of principal L-functions and random matrix theory*, Duke Math. J. **81** (1996), no. 2, 269–322, A celebration of John F. Nash, Jr.

Saffari, B. and Vaughan, R. C., *On the fractional parts of x/n and related sequences. II*, Ann. Inst. Fourier (Grenoble) **27** (1977), no. 2, v, 1–30.

Selberg, A., *On the zeros of Riemann's zeta-function*, Skr. Norske Vid. Akad. Oslo I. **1942** (1942), no. 10, 59.

Selberg, A., *On the zeros of the zeta-function of Riemann*, Norske Vid. Selsk. Forh., Trondhjem **15** (1942), no. 16, 59–62.

Selberg, A., *On the normal density of primes in small intervals, and the difference between consecutive primes*, Arch. Math. Naturvid. **47** (1943), no. 6, 87–105.

Selberg, A., *On the remainder in the formula for $N(T)$, the number of zeros of $\zeta(s)$ in the strip $0 < t < T$*, Avh. Norske Vid. Akad. Oslo. I. **1944** (1944), no. 1, 27.

Selberg, A., *Contributions to the theory of the Riemann zeta-function*, Arch. Math. Naturvid. **48** (1946), no. 5, 89–155.

Selberg, A., *An elementary proof of the prime-number theorem*, Ann. of Math. (2) **50** (1949), 305–313.

Selberg, A., *Harmonic analysis and discontinuous groups in weakly symmetric Riemannian spaces with applications to Dirichlet series*, J. Indian Math. Soc. (N.S.) **20** (1956), 47–87.

Selberg, A., *Old and new conjectures and results about a class of Dirichlet series*, Proceedings of the Amalfi Conference on Analytic Number Theory (Maiori, 1989) (Salerno), Univ. Salerno, 1992, pp. 367–385.

Serre, J.-P., *Quelques applications du théorème de densité de Chebotarev*, Inst. Hautes Études Sci. Publ. Math. (1981), no. 54, 323–401.

Severi, F., *Sulla totalità delle curve algebriche tracciate sopra una superficie algebrica*, Math. Ann. **62** (1906), no. 2, 194–225.

Siegel, C. L., *Über Riemanns Nachlass zur analytischen Zahlentheorie*, Quellen und Studien zur Geschichte der Mathematik, Astronomie und Physik **2** (1932), 45–80.

Skewes, S., *On the difference $\pi(x) - \operatorname{li} x$. I*, J. London Math. Soc. **8** (1933), 277–283.

Solovay, R. and Strassen, V., *A fast Monte-Carlo test for primality*, SIAM J. Comput. **6** (1977), no. 1, 84–85.

Spira, R., *Some zeros of the Titchmarsh counterexample*, Math. Comp. **63** (1994), no. 208, 747–748.

Tate, J., *Number theoretic background*, Automorphic Forms, Representations and *L*-Functions (Proc. Sympos. Pure Math., Oregon State Univ., Corvallis, Ore., 1977), Part 2, Proc. Sympos. Pure Math., XXXIII, Amer. Math. Soc., Providence, R.I., 1979, pp. 3–26.

Taylor, R. and Wiles, A., *Ring-theoretic properties of certain Hecke algebras*, Ann. of Math. (2) **141** (1995), no. 3, 553–572.

te Riele, H. J. J., *On the sign of the difference* $\pi(x) - \mathrm{li}(x)$, Math. Comp. **48** (1987), no. 177, 323–328.

te Riele, H. J. J. and van de Lune, J., *Computational number theory at CWI in 1970–1994*, CWI Quarterly **7** (1994), no. 4, 285–335.

Titchmarsh, E. C., *The zeros of the zeta-function*, Proc. Royal Soc. A **151** (1935), 234–255.

Titchmarsh, E. C., *The zeros of the zeta-function*, Proc. Royal Soc. A **157** (1936), 261–263.

Titchmarsh, E. C., *The Theory of Functions*, Oxford Univ. Press, London, 1939.

Titchmarsh, E. C., *The Theory of the Riemann Zeta-Function*, second ed., The Clarendon Press Oxford University Press, New York, 1986, edited and with a preface by D. R. Heath-Brown.

Tsang, K. M., *Some* Ω*-theorems for the Riemann zeta-function*, Acta Arith. **46** (1986), no. 4, 369–395.

Tsang, K. M., *The large values of the Riemann zeta-function*, Mathematika **40** (1993), no. 2, 203–214.

Turán, P., *On some approximative Dirichlet-polynomials in the theory of the zeta-function of Riemann*, Danske Vid. Selsk. Mat.-Fys. Medd. **24** (1948), no. 17, 36.

Vaaler, J. D., *Some extremal functions in Fourier analysis*, Bull. Amer. Math. Soc. (N.S.) **12** (1985), no. 2, 183–216.

van de Lune, J., te Riele, H. J. J., and Winter, D. T., *On the zeros of the Riemann zeta function in the critical strip. IV*, Math. Comp. **46** (1986), no. 174, 667–681.

van der Waerden, B. L., *Zur Algebraischen Geometrie. XIII*, Math. Ann. **115** (1938), no. 1, 359–378.

van der Waerden, B. L., *Zur algebraischen Geometrie. XIV. Schnittpunktszahlen von algebraischen Mannigfaltigkeiten*, Math. Ann. **115** (1938), no. 1, 619–642.

van Leeuwen, J. (ed.), *Handbook of Theoretical Computer Science. Vol. A*, Elsevier Science Publishers B.V., Amsterdam, 1990, Algorithms and complexity.

Vinogradov, A. I., *The density hypothesis for Dirichet L-series*, Izv. Akad. Nauk SSSR Ser. Mat. **29** (1965), 903–934.

von zur Gathen, J. and Gerhard, J., *Modern Computer Algebra*, second ed., Cambridge University Press, Cambridge, 2003.

Waldspurger, J.-L., *Sur les coefficients de Fourier des formes modulaires de poids demi-entier*, J. Math. Pures Appl. (9) **60** (1981), no. 4, 375–484.

Weil, A., *Sur les fonctions algébriques à corps de constantes fini*, C. R. Acad. Sci. Paris **210** (1940), 592–594.

Weil, A., *Sur les Courbes Algébriques et les Variétés qui s'en dÉduisent*, Actualités Sci. Ind., no. 1041 = Publ. Inst. Math. Univ. Strasbourg **7** (1945), Hermann et Cie., Paris, 1948.

Weil, A., *Sur les "formules explicites" de la théorie des nombres premiers*, Comm. Sém. Math. Univ. Lund [Medd. Lunds Univ. Mat. Sem.] **1952** (1952), no. Tome Supplementaire, 252–265.

Weil, A., *Basic Number Theory*, third ed., Springer-Verlag, New York, 1974, Die Grundlehren der Mathematischen Wissenschaften, Band 144.

Weil, A., *Scientific Works. Collected Papers. Vol. I (1926–1951)*, Springer-Verlag, New York, 1979.

Widder, D. V., *The Laplace Transform*, Princeton Mathematical Series, v. 6, Princeton University Press, Princeton, N. J., 1941.

Wiener, N., *Tauberian theorems*, Ann. of Math. (2) **33** (1932), no. 1, 1–100.

Wiles, A., *Modular elliptic curves and Fermat's last theorem*, Ann. of Math. (2) **141** (1995), no. 3, 443–551.

Index

1898526R00292

Printed in Germany
by Amazon Distribution
GmbH, Leipzig